NETWORK THEORY

NETWORK THEORY

Edited by

R. BOITE

Faculté Polytechnique de Mons, Belgium

GORDON AND BREACH SCIENCE PUBLISHERS

New York London Paris

GORDON AND BREACH SCIENCE PUBLISHERS INC.
440 Park Avenue South, New York, N.Y. 10016

Library of Congress Catalog Card number: 74-115193

Editorial Office for Great Britain:

Gordon and Breach Science Publishers Ltd.
42 William IV Street
London W.C.2

Editorial Office for France:

Gordon & Breach
7-9 rue Emile Dubois
Paris 14e

ISBN 0 677 14170 X

Printed in Great Britain

PREFACE

The second NATO Advanced Institute on Network Theory was held in Knokke, Belgium, from September 1 to September 10, 1969, three years after the first one, held in Trieste, Italy.

Sixteen well-known experts from different NATO countries contributed to the Institute, which was devoted to conventional Network Theory.

These proceedings contain most of the texts of their lectures, in addition to some of the informal evening lectures. The most important areas of research in 1969 - and, I hope, for some years to come - have been treated during the Institute and are reported in the proceedings. The beginner will find here valuable indications for choosing a well-motivated field of research in Network Theory, and the established researcher will be happy to find information on topics surrounding his own field of interest.

I do not intend to make a full review of the topics treated in the proceedings, but I would like to mention a few.

Sandberg's contribution to non-linear transistors networks is a fundamental step toward a unified theory; on the other hand new roads for the application of Distribution Theory to networks problems have been opened by Professor Zemanian. Professor Fettweis has succeeded in generalizing the Belevitch treatment of the synthesis of LC networks by the factorization of the transfer matrix. The field of distributed networks is covered by Carlin, Civalleri, Ridella and Bianco and some controversial problems are discussed by Kamp.

Unfortunately I am unable to make specific mention of all the contributions, but this in no way implies that the other topics lack interest.

I must now recognise the very important contribution of the "Scientific Committee" composed of Professors O. Beaufays, J. Neirynck, and especially V. Belevitch, to the organization of the Institute, and also the appreciated collaboration of Miss Cosyn, from the Belgian Organizing Centre.

I also wish to thank the members of the Committee of the Scientific Affairs Division of NATO, whose financial support allowed us to carry out the second NATO Advanced Institute on Network Theory.

R. BOITE

CONTRIBUTORS

K.M. ADAMS, Technische Hogeschool Delft, Afdeling Elektrotechniek, Delft, Holland

V. AMOIA, Politecnico di Milano, Istituto di Elettrotecnica ed, Elettronica, Milan, Italy

B. BIANCO, Istituto Elettrotecnica Nazionale Galileo Ferraris, Torino, Italy

M.L. BLOSTEIN, McGill University, Dept. of Electrical Engineering, Montreal, Canada

H.J. CARLIN, Cornell University, School of Electrical Engineering, New York, U.S.A.

P.P. CIVALLERI, Istituto Elettrotecnica Nazionale Galileo Ferraris, Torino, Italy

J.E. COLIN, Société anonyme des Télécommunications, Paris, France

C. DESOER, University of California, Dept. of Electrical Engineering and Computer Sciences, Berkeley, Calif., U.S.A.

A. FETTWEIS, Ruhr Universität Bochum, Lehrstuhl für Nachrichtentechnik, Bochum, Germany

Om P. GUPTA, Cornell University, School of Electrical Engineering, New York, U.S.A.

D.P. HOWSON, University of Bradford, Postgraduate School of Studies in Electrical and Electronic Engineering, Bradford, Great Britain

Y. KAMP, M.B.L.E. Research Laboratory, Avenue Van Becelaere, 2, 1170-Brussels, Belgium

D.J. MILLAR, McGill University, Dept. of Electrical Engineering, Montreal, Canada

D. MÖNCH, Institut für Elektrische Nachrichtentechnik, Aachen, Germany

R. NOUTA, Technische Hogeschool Delft, Afdeling Elektrotechniek, Delft, Holland

S. RIDELLA, Università degli studi di Genova, Istituto di Elettrotecnica, Genoa, Italy

I.W. SANDBERG, Bell Telephone Laboratories, Murray Hill, N.J., U.S.A.

W. SARAGA, The General Electric Co., Ltd., Hirst Research Centre, Wembley, Great Britain

A.F. SCHWARZ, Technische Hogeschool Delft, Afdeling der Elektrotechniek, Delft, Holland

J.K. SKWIRZYNSKI, The Marconi Company, Ltd., Great-Baddow Laboratories, Great-Baddow, Great Britain

T.E. STERN, Columbia University, Dept. of Electrical Engineering, New York, U.S.A.

J.P. THIRAN, M.B.L.E. Research Laboratory, 1170-Brussels, Belgium

A.H. ZEMANIAN, State University of New York, College of Engineering, Dept. of Applied Analysis, New York, U.S.A.

CONTENTS

ON THE REALIZABILITY THEORY OF THREE-TERMINAL REACTANCE NETWORKS

K. M. ADAMS AND R. NOUTA

Delft University of Technology, Netherlands

1. INTRODUCTION

The realization of three-terminal networks consisting of two kinds of elements - but not containing transformers - is an old problem, which has absorbed a large amount of mental effort from various workers that is out of all proportion to the meager quantity of really significant results that have been obtained. Nevertheless, recent developments in electrotechnology - in particular the arrival of integrated circuits - have shown that this and related difficult problems are likely to be of considerable practical importance, so that, even though new and useful results are not easy to find, it is still probably worth the effort to try and extend the existing body of knowledge.

There is, for example, the group of synthesis problems in which circuits of minimum sensitivity are to be realized by some n-port RC network, in which some of the ports are terminated in operational amplifiers, controlled sources or some other active devices. Most of what is known about the synthesis of n-port RC networks is fragmentary and relates to very special cases. Moreover, there is an intimate connexion between the three-terminal problem and certain special cases of the n-port problem. The n-port problem certainly deserves more attention than it has received to date.

One such special case, related in a very special way to the three-terminal network problem, is that in which each inductance of a three-terminal reactance network is replaced by a gyrator terminated at one port by a capacitance. As the gyrator is frequently realized by active devices - complete with all parasitic effects inherent in such devices - the problem belongs to the general class considered. It is known that in many filter applications such a network possesses favourable sensitivity properties [1,2].

In this paper, an attempt will be made to explain some of the known results of the realization and realizability theory of three-terminal reactance networks and to show how they can be applied to practical problems. Unfortunately, much of the existing theory involves long and complicated algebraic calculations which, although basically elementary, are not easily comprehensible to non-specialists. Consequently, this tedious material will be avoided as much as possible.

2. BASIC PARAMETERS

Reciprocal two-port networks can be conveniently characterized by Tellegen's polynomials [3], which can be introduced as follows: we define the zeros of the polynomials A, B, C, D, H as the natural (complex) frequencies of the free oscillations of the two-port network under the port termination conditions indicated in Fig. 1.

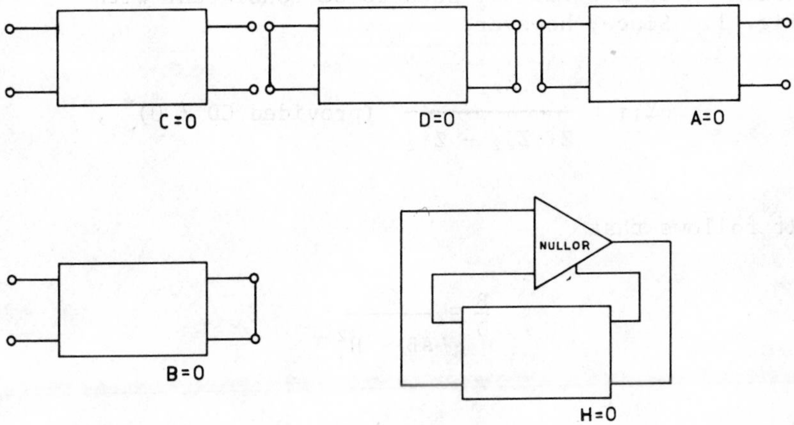

FIG. 1

The characteristic equations under these circumstances are then given by equating the polynomials to zero. If now we arbitrarily fix the leading coefficient of one of the polynomials, e.g. C, then the leading coefficients of A, B and H follow from the relations

$$Z_{11} = \frac{A}{C}, \quad Z_{12} = \frac{H}{C}, \quad Z_{22} = \frac{B}{C},$$

provided $C \neq 0$. That these relations are consistent with Fig. 1 follows readily from the fact that, e.g.,

$$Z_{11} = \frac{U_1}{I_1} \quad (I_2 = 0),$$

so that $AI_1 = CU_1$, which leads to the same characteristic equations as indicated in Fig. 1 when port 2 is open.

We can now determine the leading coefficient of D from one of the relations

$$Y_{11} = \frac{B}{D}, \quad Y_{12} = \frac{-H}{D}, \quad Y_{22} = \frac{A}{D},$$

which again are readily seen to be consistent with Fig. 1. Since, however,

$$Y_{11} \equiv \frac{Z_{22}}{Z_{11}Z_{22} - Z_{12}^2} \quad \text{(provided } CD \neq 0\text{)}^* ,$$

it follows that

$$\frac{B}{D} \equiv \frac{BC}{AB - H^2} ,$$

i.e.,

$$AB - H^2 = CD. \tag{1}$$

Alternatively, if Fig. 1 and Eq. (1) are given, then the polynomials are determined up to a constant factor common to all five polynomials if, in addition, the impedance levels at the two ports are specified. This follows from a consideration of Eq. (1) at the zeros of H and at the zeros of AB.

From Fig. 1 we can readily deduce that the denominators of the various two-port matrices are as follows:

Z-matrix : C	Y-matrix : D
H-matrix : B	G-matrix : A

chain matrices: H.

*If $CD \neq 0$, then Eq. (1) can be derived from the relation between two of the hybrid or chain matrices.

The two-port network is thus always characterizable by the five polynomials except in the highly degenerate (and pathological) case when all five polynomials are identically zero.

When dealing specifically with three-terminal networks, it is convenient to introduce two additional polynomials F and G defined by

$$A = F + H, \quad B = G + H.$$

Eq. (1) can then be written

$$GH + HF + FG = CD. \tag{2}$$

The network representation of the five polynomials F, G, H, C, D is shown in Fig. 2, where the one-port immittances are not necessarily realizable [4].

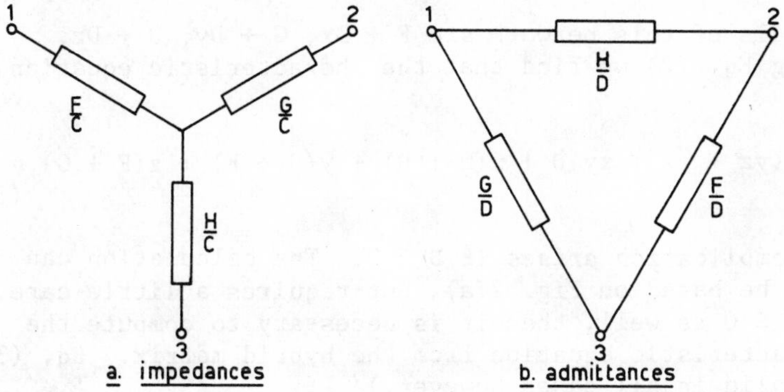

FIG. 2

If any of the polynomials is identically zero, at least one of the equivalent circuits will still be valid.

We note in passing that the zeros of the polynomial G + H are the natural frequencies for the network formed by short-circuiting terminals 1 and 2.

Finally, we note that for lossless networks either A, B, F, G, H are even and C, D are odd or vice-versa.

3. UNOBSERVABLE MODES

Consider a network formed by connecting three conductances x, y, z across the three ports of the three-terminal network, formed by the terminals 23, 31, 12. This is evidently representative of the most general resistive terminations. To find the characteristic equation of the resulting network we need to calculate the "C" polynomial of this network. From Fig. 2(b), the three transfer admittances of this network become

$\frac{F}{D} + x$, $\frac{G}{D} + y$, $\frac{H}{D} + z$, so that the "F", "G", "H" polynomials of this network are $F + Dx$, $G + Dy$, $H + Dz$. Using Eq. (2) we find that the characteristic equation is

$$C + (yz + zx + xy)D + x(G + H) + y(H + F) + z(F + G) = 0. \tag{3}$$

(A complication arises if $D \equiv 0$. The calculation can then be based on Fig. 2(a), but requires a little care. If $C \equiv 0$ as well, then it is necessary to compute the characteristic equation from the hybrid matrix. Eq. (3) is valid in all cases however.)

It is evident that a factor common to all five polynomials specifies a free oscillation that is independent of the conductances x, y, z. Such a free oscillation must be sinusoidal, since by the definitions of Fig. 1 it also occurs in a lossless network. It cannot therefore be associated with currents and voltages in the resistances, as is readily established by Tellegen's theorem. Thus the port currents and voltages are zero in this mode, i.e., the mode is unobservable at the terminals. If now the resistive terminations are removed and arbitrary voltages and currents are imposed on the terminals, various modes will be excited. Any unobservable mode can be superimposed to an arbitrary degree on any distribution of currents and voltages in the network without conflicting with terminal requirements. The reason is that in a lossless, passive, reciprocal system there is no mode interaction, even when two natural frequencies coalesce, so that there are no secular terms in the free oscillations. It follows that such an unobservable mode is also uncontrollable.

A simple example is that of a simple string vibrating transversely in some mode whereby excitation can only be applied at a node of the vibration. In such a case the mode can neither be observed nor excited at the node.

Another example which throws some light on synthesis problems is that of the well-known symmetrical parallel-T network [5] (Fig. 3) in which the three terminals are nodes 1, 2, 3. Several modes with natural frequency given by $\omega = 1$ are possible when all three ports are open or short-circuited. One of them is unobservable. The existence of these modes is in accordance with the fact that both C and D contain the factor $\lambda(\lambda^2 + 1)^2$ where λ is the complex frequency. Only the factor $\lambda^2 + 1$, however, is common also to F + G, F + H and H + F.

In various synthesis procedures, common factors are often introduced as an essential step in the procedure. It is important to realize the physical significance of such factors and not to complicate the realization unduly by introducing common factors unnecessarily. Conversely it is important not to discard common factors when they are needed in the realization.

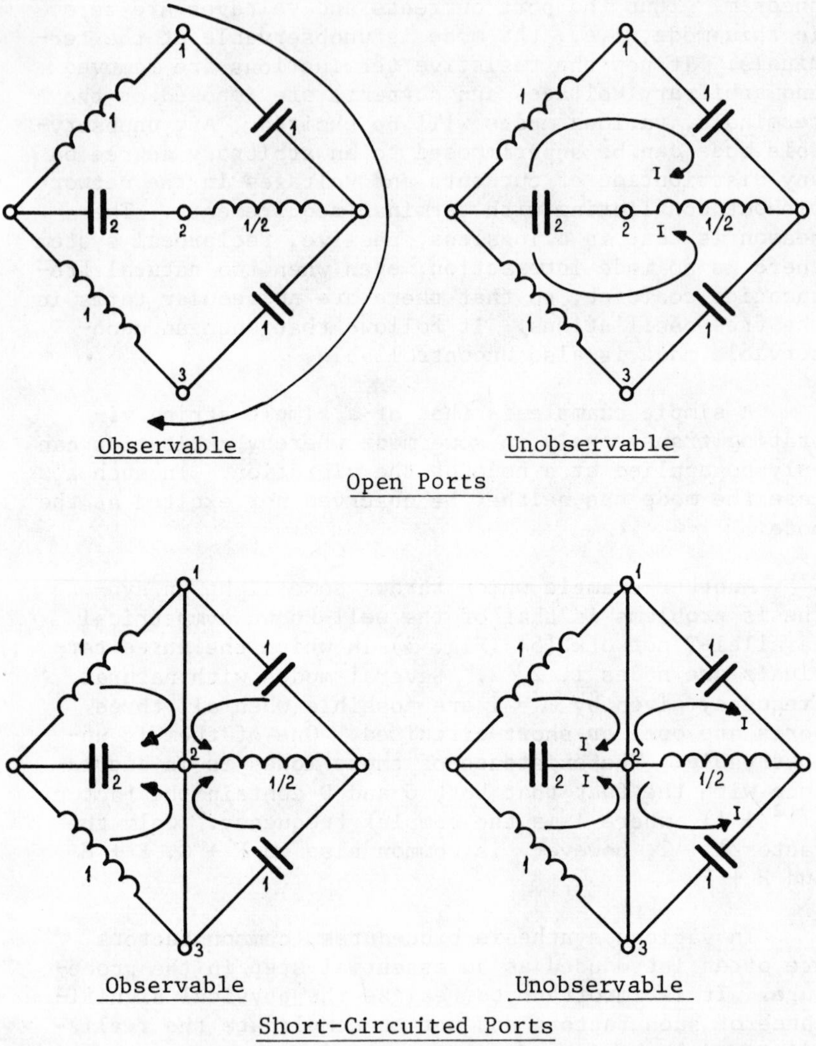
1
1
1
2
2
1/2
1
1
3
Observable
1
1
1
I
2
2
I
1/2
1
1
3
Unobservable
Open Ports
1
1
1
2
2
1/2
1
1
3
Observable
1
1
1
I
I
2
2
I
1/2
I
1
1
3
Unobservable
Short-Circuited Ports

FIG. 3

4. SYMMETRIC AND ANTIMETRIC NETWORKS

In filter applications the scattering matrix of the lossless two-port is especially useful. Calculation from one of the standard formulae shows that

$$S = \frac{1}{\Delta}\begin{bmatrix} D - R_1R_2C - R_1B + R_2A & 2\sqrt{R_1R_2}\,H \\ 2\sqrt{R_1R_2}\,H & D - R_1R_2C + R_1B - R_2A \end{bmatrix}$$

where $\Delta = D + R_1R_2C + R_1B + R_2A$ and R_1, R_2 are the normalizing resistances of ports 1 and 2 *.

However, the following alternative method is instructive. The poles of S are the natural frequencies of the network formed by terminating the ports in the normalizing resistances ($a_1 = a_2 = 0$). It is then readily found that the "C" polynomial corresponding to this case follows from the "A" and "B" polynomials, viz. $A + D/R_2$, $B + D/R_1$ and (1) or directly from (3). The zeros of S_{11} are the natural frequencies for the case $b_1 = a_2 = 0$, i.e., when port 1 is terminated in $-R_1$ and port 2 in R_2. The numerator of S_{11} thus follows from the denominator by replacing R_1 by $-R_1$. A similar result applies to S_{22}. Since $S_{11} \to -1$ as $R_1 \to \infty$, S_{11} and S_{22} can be completely determined. Finally $S_{12} (= S_{21})$ follows from the unitarity of S, the evenness-oddness properties of the polynomials, and (1).

From (1) it follows that if A, B, C, D = 0, then H = 0, so that the existence of S ($\Delta \neq 0$) is equivalent to the condition that A, B, C, D, H are not all zero, provided $R_1, R_2 \geq 0$ and the network is passive (implying that all leading coefficients of A, B, C, D have the same sign).

---

*These expressions should not be confused with the deceptively similar expression for S in terms of the elements of the chain matrix, which are still frequently denoted by A, B, C, D.

When the numerator of S_{11} is even or odd - in particular when all the zeros of S_{11} lie on the imaginary axis - then, except in one special case, either $R_2A = R_1B$ or $D = R_1R_2C$. When the terminations are equal, this results in either $A = B$ or $D = R^2C$; i.e., the two-port is either symmetric or antimetric. Since a large and important class of filter networks satisfies this criterion, this shows why symmetric and antimetric networks are so important in filter design. If the terminations are not equal, we shall call the two-port quasi-symmetric or antimetric respectively*. In the further discussion it will be convenient to work with primed quantities of the same dimension:

$$D' = D,\ A' = R_2A,\ B' = R_1B,\ C' = R_1R_2C,\ H' = \sqrt{R_1R_2}\ H.$$

The exceptional case occurs when the numerator and denominator of S_{11} have a common factor. Such a common factor describes a free oscillation under the conditions $a_2 = 0$, a_1 and b_1 arbitrary, where the a's and b's are normalized scattering variables. This free oscillation is thus independent of the relation between a_1 and b_1 and thus of the termination at port 1. It is hence unobservable at port 1 when port 2 is terminated by its normalizing resistance**.

If, under these conditions, the free oscillation is observable at port 2, the corresponding natural frequency must be in the open, left-half plane. The common factor is thus a Hurwitz polynomial. The numerator of $S_{11} = D' - C' - B' + A' = \Phi\psi$, where ϕ is either even or odd and ψ is the Hurwitz polynomial in question. Thus $\phi\psi$ is neither odd nor even and so the network is neither

*There is no need for a term quasi-antimetric, since in both cases $D = KC$ where K has the dimensions Ω^2 and the network is structurally self-dual when it is planar.

**The presence of such a common factor implies that S can be realized [16] as a cascade of a two-port with a scattering matrix of lower degree than S and an all-pass two-port.

quasi-symmetric nor antimetric.

If, however, the free oscillation is also unobservable at port 2, then all the zeros of the common factor lie on the imaginary axis and the network is either quasi-symmetric or antimetric. The presence of unobservable oscillations implies that the realization is not minimal in the sense of a minimum number of reactances. In transformerless networks such realizations are the exception and not the rule. In many well-known ladder networks used in filter technique all the unobservable free oscillations have zero natural frequencies, corresponding to the presence of capacitor cut-sets and inductor loops. These facts are frequently lost sight of in the usual synthesis procedure.

5. ORDER AND DEGREE

The order of the differential equation that describes the free oscillations of the network - terminated with finite, but otherwise arbitrary, resistances - observable at the terminals, is called the *order of the three-pole or two-port* [3]. It is thus equal to the highest power of the complex frequency present in the five polynomials F, G, H, C, D after all common factors have been divided out. (The order of the network* is the order of the differential equation describing all the free oscillations observable as well as unobservable at the terminals.) The terms McMillan degree and weak degree are also used for these concepts [17]. Here we reserve the word degree for quite a different concept that is significant only for transformerless reactance three-poles. After all common factors have been divided out, some negative coefficients may appear in F, G, H. By multiplying all five polynomials again by common factors with zeros exclusively on the imaginary axis, we can generate a set of polynomials with exclusively *non-negative*

---

*Alternative terms are: order of complexity (Bryant) and strong degree (Belevitch) [17].

coefficients. Of all the permissible common polynomial factors, we select one of the lowest possible degree (in the ordinary sense) that achieves the desired result. The highest power of the complex frequency in the resulting polynomials is called the *degree of the three-pole* [6].

This concept is especially significant when we come to consider the most widely used synthesis procedure. Nearly all the known three-poles are *series-parallel* three-poles. These are three-poles that can be constructed from transformerless two-poles by applying some sequence of the following three operations: connexion of a two-pole in series with a three-pole, connexion of a two-pole in parallel with a three-pole, and connexion of two three-poles in parallel [6]. These connexions are illustrated in Fig. 4.

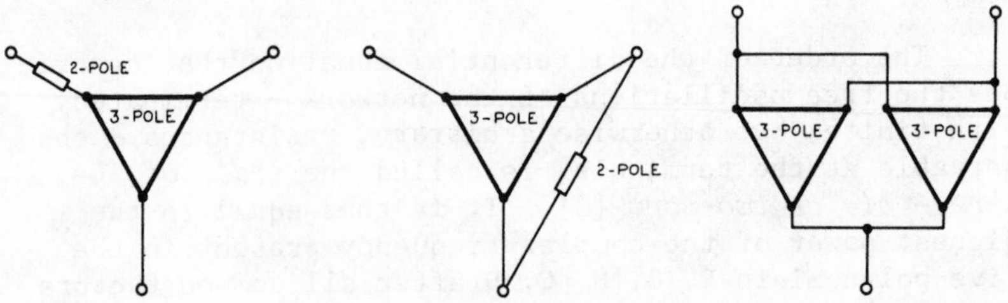

FIG. 4

The realizability conditions for reactance three-poles are not yet completely known. What we do know is that the following conditions are necessary:

I. The well-known Cauer residue conditions pertaining to all reactance two-ports.

II. The Fialkow-Gerst conditions, viz. F, G, H are positive on the open, real, positive axis. This applies only to transformerless networks.

III. In the case of series-parallel networks the condition: the order and degree of the three-pole are equal [7,8].

Condition II follows from the fact that at positive, real values of the complex frequency the three-pole behaves as a resistance network. For such a network the equivalent circuits of Fig. 2 can be calculated by eliminating internal nodes by successive applications of the star-polygon transformation. Since only additions and multiplications of positive numbers occur in these operations, F, G, H must be positive. The result also follows from physical reasoning based on the consideration that in resistive networks the potential attains its maximum and minimum values at the terminals of an energy source.

Condition III can be proved as follows: we consider what happens when, starting from a given series-parallel three-pole, we reverse the sequence of operations by which the three-pole was constructed. Removal of a reactance three-pole in series or in parallel with a component three-pole corresponds to removing poles on the imaginary axis from one of the functions F/C, G/C, H/C, F/D, G/D, H/D, as readily follows from Fig. 2. If after cancelling the common factors the polynomials of the given three-pole contain negative coefficients, then the polynomials of the component three-pole will also contain negative coefficients owing to the subtraction of pole terms with positive residues.

Next we consider the operation of partitioning the given three-pole into two component three-poles in parallel. This corresponds to writing the polynomials in the form

$$\frac{F}{D} = \frac{F_1}{D} + \frac{F_2}{D}\ ,\ \frac{G}{D} = \frac{G_1}{D} + \frac{G_2}{D}\ ,\ \frac{H}{D} = \frac{H_1}{D} + \frac{H_2}{D}\ ,$$

where the subscripts 1 and 2 refer to the two component networks. Suppose $(\lambda^2 + \alpha)^r$ is a common factor of F, G, H, C, D. If r = 1, then the residues f, g, h of F/D, G/D, H/D at the poles $\lambda = \pm j\alpha$ are all zero. The corresponding residues in the partitioned polynomials must satisfy

$$f_1 + g_1 \geq 0, \quad f_2 + g_2 \geq 0, \quad g_1 + h_1 \geq 0,$$

$$g_2 + h_2 \geq 0, \quad h_1 + f_1 \geq 0, \quad h_2 + f_2 \geq 0,$$

on account of the properties of reactance functions.

These results, together with the Cauer residue conditions

$$g_1h_1 + h_1f_1 + f_1g_1 \geq 0, \quad g_2h_2 + h_2f_2 + f_2g_2 \geq 0,$$

imply that the only possibility is $f_1 = g_1 = h_1 = f_2 = g_2 = h_2 = 0$. Thus the same factor divides the partitioned polynomials also. If $r > 1$, the same conclusion applies owing to the impossibility of multiple poles in reactance functions.

If the common factors are first cancelled and then the partitioning is carried out, then negative coefficients in the partitions will result. By the foregoing, the same result ensues if the partitioning is first performed and then the common factors are cancelled. Thus in all cases for at least one of the component three-poles the order is not equal to the degree. Further application of the reversed construction operations leads eventually to a three-pole with polynomials consisting of single terms and containing negative coefficients. No permissible common factor can cause the negative coefficients to disappear. We are then in conflict with condition II, so that such a component three-pole is unrealizable. It follows that if the three-pole is series-parallel, the order must equal the degree.

It may seem puzzling that in the discussion of partitioning above, C is not mentioned explicitly even though we insist that a common factor must divide all five polynomials. However, C enters essentially into the picture as a result of (2) and the properties of reactance functions [8].

As is generally well known, it was conjectured by

Darlington [9] that the series-parallel configuration is canonical for the whole class of three-poles. This conjecture has since been disproved in a highly significant paper by Fialkow [10] by showing how to construct a class of counter-examples for which the degree of the three-pole is higher than the order. Fialkow's result clears up a question which had been hanging fire for years, but it raises a host of new ones. At this point it is sufficient to note that we shall have to develop effective methods of non-series-parallel synthesis if we are to exploit fully the possibilities inherent in reactance three-poles. Most of the so-called methods that have been published on this general class of networks are not systematic, work only for a few carefully chosen examples, and in any case are intractable for networks containing more than six nodes, in spite of the optimistic remarks of the authors to the contrary.

Fialkow's proof is very instructive and suggestive for many new investigations. It is unfortunately too long and difficult to reproduce here, and a simple equivalent has not yet been found. However, we shall return briefly to this matter in Section 10.

6. THE SIXTH AND LOWER DEGREES

For three-poles of degree five or lower, the order is automatically equal to the degree, and conditions I, II, III of Section 5 are sufficient for the realization [8]. The series-parallel configuration is canonical and any desired set of realizable polynomials can be realized in practice. In the case of the sixth degree, most of the sub-cases are realizable without any further condition. There is, however, one subcase where the series-parallel configuration is no longer canonical [9]. It occurs when *one* of the polynomials F, G, H (e.g. F) is of the form $a_6\lambda^6 + a_4\lambda^4 + a_2\lambda^2 + a_0$ and D is of the form $\lambda(\lambda^2 + \alpha_1)(\lambda^2 + \alpha_2)$, where λ is the complex frequency and $\alpha_2 \geq \alpha_1 > 0$. (Note that the other polynomials G, H cannot contain a term λ^6, since otherwise C would contain a term λ^7 and the system would not be of the 6th degree.) The condition is: if the following inequalities hold

$$a_4 < \alpha_2 a_6 \ , \quad \alpha_1 a_2 < a_0 \ , \tag{4}$$

and at least one of the inequalities

$$a_4 < \alpha_1 a_6 \ , \quad \alpha_2 a_2 < a_0 \tag{5}$$

where $\alpha_2 \geq \alpha_1 > 0$, then a series-parallel realization is impossible *. Conversely, if one of conditions (4), (5) is violated and if I and II of Section 5 hold, then the sixth-degree polynomials are always realizable. Thus a series-parallel realization requires not only that the coefficients be negative but also that some of them must be greater than certain positive quantities. Thus in order to disprove Darlington's conjecture it would have been sufficient to construct a counter-example violating these coefficient conditions, without the order necessarily differing from the degree. Such counter-examples certainly exist but it is not yet certain whether the complete class given by (2) is realizable by non-series-parallel networks [13]. This is a matter worthy of further investigation.

7. THE SEVENTH DEGREE

Because of its importance and because it has not appeared in print before, we shall now include some discussion of the seventh-degree polynomials. We can write the polynomials in the following standard form:

$$\frac{F}{D} = f_\infty \lambda + f_0 \mu + \frac{f_1}{\lambda + \alpha_1 \mu} + \frac{f_2}{\lambda + \alpha_2 \mu}$$

$$\frac{G}{D} = \qquad g_0 \mu + \frac{g_1}{\lambda + \alpha_1 \mu} + \frac{g_2}{\lambda + \alpha_2 \mu}$$

---

*These conditions [12] are equivalent to the conditions given in [11]. The proof of the equivalence though elementary is not trivial.

$$\frac{H}{D} = \qquad h_0\mu + \frac{h_1}{\lambda + \alpha_1\mu} + \frac{h_2}{\lambda + \alpha_2\mu}$$

where λ is the complex frequency and μ its reciprocal - thus $\lambda\mu \equiv 1$. It is readily verified that any set of 7th degree polynomials satisfying (2) can be expressed in this form - if necessary after interchanging λ and μ, F, G and H or C and D - where the so-called compactness conditions hold, viz.,

$$g_1h_1 + h_1f_1 + f_1g_1 = g_2h_2 + h_2f_2 + f_2g_2 = 0.$$

[If the compactness conditions do not hold, then C will contain higher powers than the seventh. If $\alpha_1 = \alpha_2$, then F, G, H, D (but not C) have a common factor. The residues at the poles $\lambda = \pm j\alpha$ are then $f_1 + f_2$, $g_1 + g_2$, $h_1 + h_2$. Then

$$(g_1+g_2)(h_1+h_2) + (h_1+h_2)(f_1+f_2) + (f_1+f_2)(g_1+g_2)$$
$$= \frac{(g_1h_2 - g_2h_1)^2}{(g_1 + h_1)(g_2 + h_2)} > 0,$$

unless $\frac{g_1}{g_2} = \frac{h_1}{h_2}$, in which case the common factor also divides C. Thus in general non-compactness results when two or more compact poles coalesce. As a result of these relations it is convenient to regard non-compactness as a special case of compactness and not vice versa.]

Interchanging λ and μ corresponds to interchanging inductances and capacitances and interchange of F, G or H corresponds to a renumbering of the terminals. Interchange of C and D corresponds to replacing the network by its dual (when it exists). Otherwise, if carefully used it can still simplify the classification of the various sub-cases considerably.

In the following we shall consider series-parallel operations that can be employed to reduce the degree to a case of lower degree whose realization is already known. We employ the case classification of [11], in which the first figure denotes the degree. Most of the tedious algebraic details will be suppressed.

CASE 7.1. If $f_0 + f_1 + f_2 \geq 0,$

$$\text{and} \quad (\alpha_1 + \alpha_2)f_0 + \alpha_2 f_1 + \alpha_1 f_2 \geq 0, \tag{6}$$

then $f_\infty\lambda$ can be removed. The realizable case 6.121 results. The removal of this term corresponds to realizing the network as a sixth-degree three-pole with a capacitance connected between terminals 2 and 3 (Fig. 4). If D is replaced by C, the term $f_\infty\lambda$ corresponds to an inductance connected to terminal 1. (Interchange of λ and μ is of course possible with the appropriate interpretation.)

CASE 7.2. If $(\alpha_1 + \alpha_2)f_\infty + f_1 + f_2 \geq 0,$

$$\text{and} \quad \alpha_1\alpha_2 f_\infty + \alpha_2 f_1 + \alpha_1 f_2 \geq 0, \tag{7}$$

then $f_0\mu$ can be removed. Further, at least one of $h_0\mu$ and $g_0\mu$ can be removed (as is readily proved); the realizable case 6.211 then results.

CASE 7.3. At least one of the inequalities (6) and at least one of the inequalities (7) is violated. The only possibility of eventually reducing the degree to less than 6 is to partition the system into two sets of polynomials, corresponding to the operation of connecting two three-poles in parallel, according to the following scheme:

$$\frac{F}{D} = f_\infty\lambda + \frac{f_{11}}{\lambda + \alpha_1\mu} + \frac{f_{21}}{\lambda + \alpha_2\mu} + f_0\mu + \frac{f_{12}}{\lambda + \alpha_1\mu} + \frac{f_{22}}{\lambda + \alpha_2\mu}$$

$$\frac{G}{D} = g_{01}\mu + \frac{g_{11}}{\lambda + \alpha_1\mu} + \frac{g_{21}}{\lambda + \alpha_2\mu} + g_{02}\mu + \frac{g_{12}}{\lambda + \alpha_1\mu} + \frac{g_{22}}{\lambda + \alpha_2\mu}$$

$$\frac{H}{D} = h_{01}\mu + \frac{h_{11}}{\lambda + \alpha_1\mu} + \frac{h_{21}}{\lambda + \alpha_2\mu} + h_{02}\mu + \frac{h_{12}}{\lambda + \alpha_1\mu} + \frac{h_{22}}{\lambda + \alpha_2\mu}$$

Any other form of partitioning either does not result in two systems of lower degree which can be realized as series-parallel networks without further conditions, or is possible only if this scheme is possible.

The right-hand partition is case 6.121. The left-hand partition can be reduced to case 6.211 by removing either $g_{01}\mu$ or $h_{01}\mu$, since in one of the expressions G/D, H/D all the residues must be positive. As in case 6.221.2, this partitioning is possible only if either one of the inequalities (4) or both inequalities (5) is violated. The proof of this result is as involved as that of case 6.221.2 and requires showing that if the conditions (4) and (5) hold, they will continue to hold after carrying out any possible pole-removal at $\lambda = \infty$ or $\mu = \infty$ on the expressions F/C, G/C, H/C. Thus, provided D does not contain more than two binomial factors, the conditions for realization are the same as those for case 6.221.21.

CASE 7.4 In the foregoing C and D are interchanged. Then D becomes a polynomial containing three binomial factors. If it is not possible by pole-removal at $\lambda = \infty$ or $\mu = \infty$, performed on the functions F/C, G/C, H/C, to reduce the system to a system of lower degree that is series-parallel realizable, then it is necessary to consider the various partitioning possibilities of F/D, G/D and H/D. Since there are now three finite poles in these expressions, this case divides up into a large number of sub-cases, not all of which are series-parallel realizable without the imposition of extra conditions. This material is too complicated for presentation here.

It is worth noting that the reduction to lower degree would be considerably simplified if a standard procedure for realizing case 6.221.2, when conditions (4) and (5) hold, were available - that is of course some non-series-parallel synthesis procedure.

8. REALIZATION OF QUASI-SYMMETRIC AND ANTIMETRIC THREE-POLES

In this section we consider the realization of a given scattering matrix that is known to be realizable by a reactance three-pole without transformers and with equal terminating resistances. We shall investigate the realization of the *same* scattering matrix by a reactance three-pole terminated in finite resistances. This problem is frequently mentioned in the literature. The procedures that are usually given, however, either do not result in the realization of the same S_{21} but introduce some extra flat loss or require the use of a transformer or are applicable only to special network structures. When the results are satisfactory or the procedures applicable, these methods have the advantage that use can be made of existing tables to compute the element values simply [14].

We consider a given scattering matrix that is realizable as a reactance network terminated in resistances of 1Ω. Denote the polynomials of this network by A', B', C', D', H' . If the same scattering matrix is realizable with terminations R_1, R_2, then it follows from the expression for S that the polynomials of the reactance network are given by *

$$D - R_1R_2C - R_1B + R_2A = D' - C' - B' + A' ,$$

$$D - R_1R_2C + R_1B - R_2A = D' - C' + B' - A' ,$$

$$\sqrt{R_1R_2}\, H = H' ,$$

$$D + R_1R_2C + R_1B + R_2A = D' + C' + B' + A' .$$

Comparing the even and odd parts, we find

$$D = D' , \quad R_2A = A' , \quad R_1B = B' , \quad R_1R_2C = C' ,$$

in accordance with the definition for the normalized polynomials employed earlier. Then $R_2(F + H) = F' + H'$, etc. Denoting $1/R_1$ and $1/R_2$ by k_1 and k_2 respectively, we have

---

*Strictly speaking, an arbitrary common factor can be introduced on either side of these equations. We do not, however, wish to increase the complexity of the networks unnecessarily.

$$F = k_2F' + (k_2 - \sqrt{k_1k_2})H', \ G = k_1G' + (k_1 - \sqrt{k_1k_2})H'. \quad (8)$$

If now in the given network we replace the 1Ω resistors by equivalent one-ports consisting of ideal transformers with the secondary ports terminated in R_1 and R_2 respectively, the two-port formed by the original reactance two-port cascaded fore and aft by the ideal transformers is a realizable two-port that therefore satisfies the Cauer conditions, Accordingly, any transformerless equivalent two-port also satisfies the Cauer conditions and any such equivalent constitutes a solution of the realization problem. We next examine the poles at $\lambda = \infty$ and $\mu = \infty$.

We suppose $k_1 > k_2$. It is evident from (8) that G will then have non-negative coefficients. Hoever, F can have non-negative coefficients only if the highest and lowest powers of λ present in H' are also present in F'. Unfortunately, for a large class of practical filter networks, this restriction does not hold, so that realizations with equal terminations are then the only possibilities.

If the highest and lowest powers of λ present in H' are also present in F' , and if $1 - \sqrt{k_1/k_2}$ is chosen sufficiently small to ensure that the coefficients of F are non negative (cf. (8)), then the system will be realizable if the degree is 5 or lower or if the degree is 6, with the exception of case 6.221. In case 6.221 and cases 7.1, 7.2, 7.3 it will be necessary to apply conditions (4) and (5) to F, in order to determine the maximum value of $1 - \sqrt{k_1/k_2}$ that permits a series-parallel realization.

We consider next the symmetric three-poles, i.e. F' = G'. The following cases (applied to the primed polynomials) are now excluded: 4.1, 6.121.2, 6.21, 6.22 and all odd degrees. The restriction of symmetry combined with the previously mentioned coefficient conditions in F' and H' is thus severe. However, no additional restriction on the realizability conditions of the cases themselves are required.

We consider next the antimetric three-poles. The

condition $C' = D'$ does not lead to any real simplification of the coefficient relations in F', G', H', so that the foregoing conclusions for the general case apply here. However, an antimetric lossless two-port can only be of even order. This follows simply from (1) with $C = D$, i.e. $AB - H^2 = C^2$.

If the highest power of λ in C is n, then the highest power in A is n+1 and in B is n-1 (or with A and B interchanged), because of the Cauer conditions. If the lowest power in C is zero, then the lowest powers in A and B must also be zero. This, however, is in conflict with the evenness-oddness condition. Thus the lowest power in C is one, which means that n must be odd and n+1 - which is the highest power - must be even. If the order is even, then the degree is also even.

Finally, we consider quasi-symmetric three-poles, (independently of whether their scattering matrices are also realizable by symmetric networks with equal terminations or not).

We have $A = n^2B$, so that $F = n^2G + (n^2-1)H$. At any pole (assumed compact) of F/D, G/D, H/D we have

$$f = n^2g + \frac{1 - n^2}{f + g} fg,$$

where f, g, h are the corresponding residues. Then $f = \pm ng$; unless $f = g = 0$, in which case $h = 0$ or $n = 1$; i.e. this exceptional case is not significant. Without loss of generality we can take $n > 1$. (If $n < 1$, interchange F and G. There is no need for an $n < 0$.) If $f = ng$, then $f > 0$, $g > 0$, $h < 0$; if $f = - ng$, then $f > 0$, $g < 0$, $h > 0$, on account of the Cauer conditions. If the pole is not compact, then $f = \pm ng(1 + \eta)$ where $\eta > 0$ is necessary in this case also. Thus all residues of F/D are positive, so that F/D is a Foster function and F satisfies the Fialkow-Gerst condition. Next we express G in terms of elements of the scattering matrix. We find

$$G = \frac{k_2}{n^2} (A' - nH'),$$

where $n^2 = k_2/k_1$, $A = k_2A'$, $H = \sqrt{k_1k_2}\,H'$, and A' and H' follow from the numerators of S_{11} and S_{12}. The coefficient conditions on H are thus independent of n. If the coefficients of G are non-negative for n > 1, then they are also non-negative for n = 1. It can also be shown that if (4) and (5) are violated for n > 1, they are also violated for n = 1.

We thus see that the symmetric two-port is a preferred transformerless realization which is possible if any quasi-symmetric realization is possible.

9. EQUIVALENT NETWORKS OF SOME WELL-KNOWN FILTER CONFIGURATIONS

In this section we apply the realization theory of series-parallel three-poles to the study of alternative realizations of various filter characteristics.

We restrict our discussion to the elliptic-function low-pass filters of odd degree with equal termination resistances and to the band-stop filters which can be derived from those filters.

It is known that negative elements may occur if the realization is of the ladder form. On the other hand series-parallel equivalents can be obtained which do not have this drawback and which also may have fewer elements than the corresponding ladder.

Since all the zeros of the reflection factors lie on the imaginary axis it follows from Sections 3, 8 that these filter two-ports are symmetric. In addition to this the equivalent circuits which have been obtained are also structurally symmetric. Next we give some equivalent realizations (but not all) for the 5th and 7th degrees.

5th degree: Figs. 5 and 6 show the well-known ladder realizations. If we restrict ourselves further to the filters given in Saal's tables [18], we find that at least one of the Figs. 7, 8 and 9 can always be used to replace Fig. 5. They have been obtained by resynthetising the polynomials describing Fig. 5.

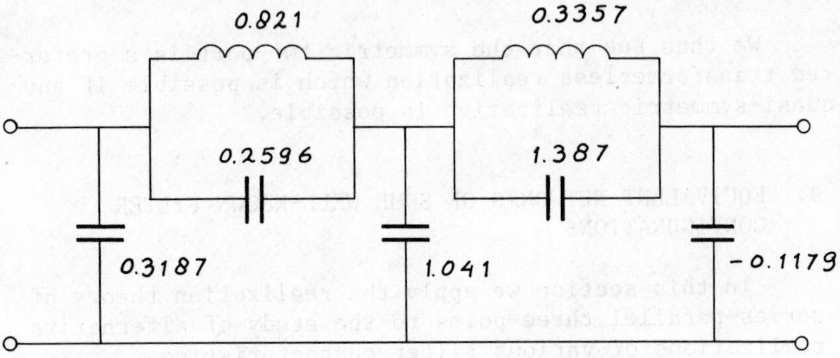

FIG. 5

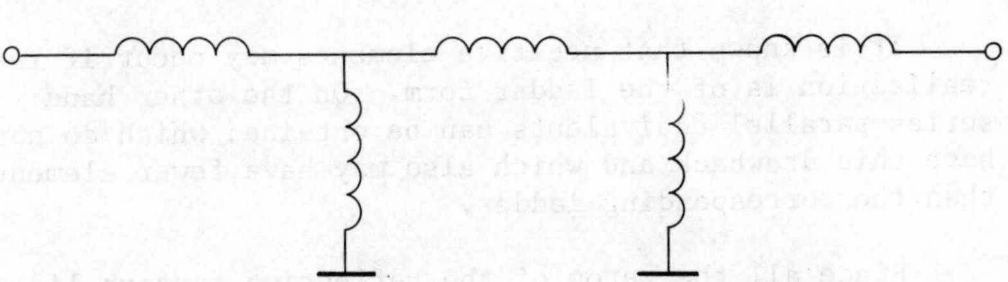

FIG. 6

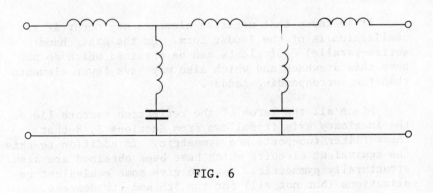

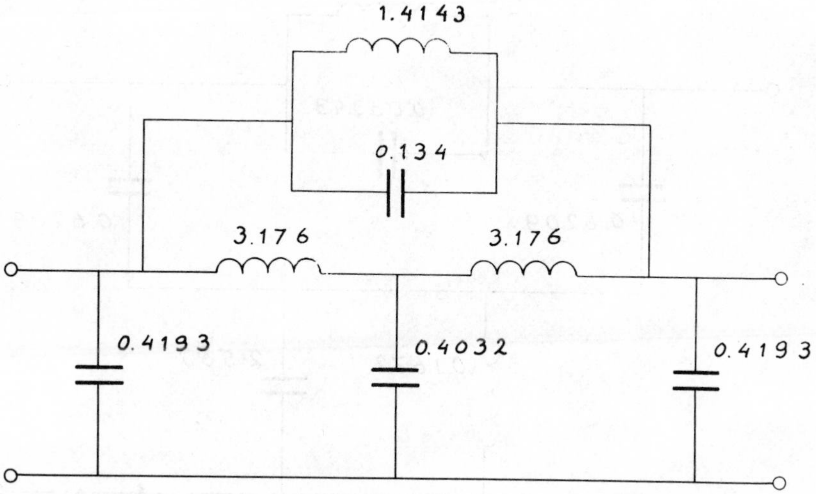
1.4143
0.134
3.176
3.176
0.4193
0.4032
0.4193

FIG. 7

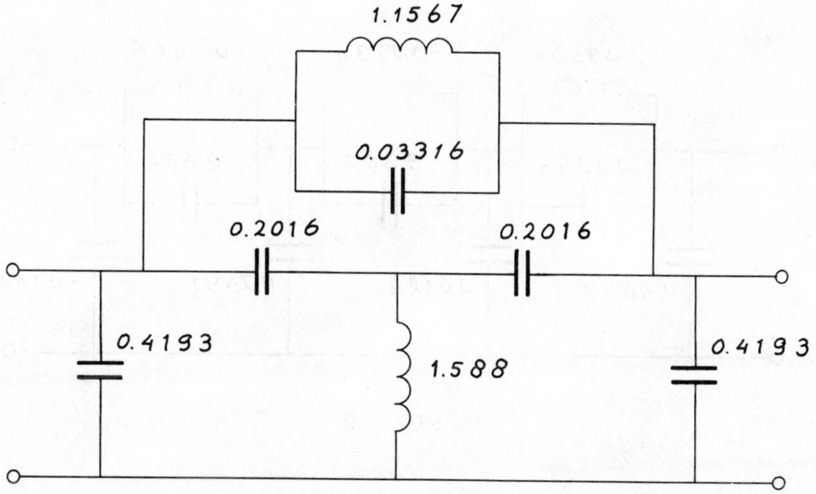
1.1567
0.03316
0.2016
0.2016
0.4193
1.588
0.4193

FIG. 8

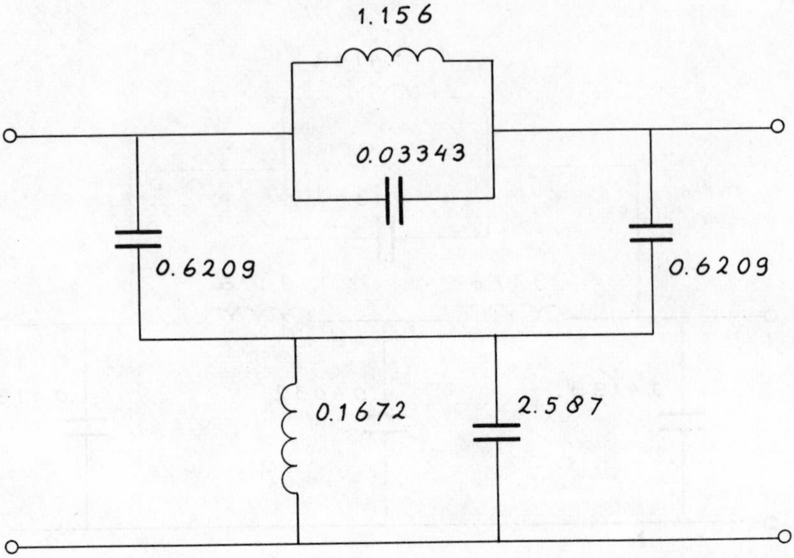
1.156
0.03343
0.6209
0.6209
0.1672
2.587

FIG. 9

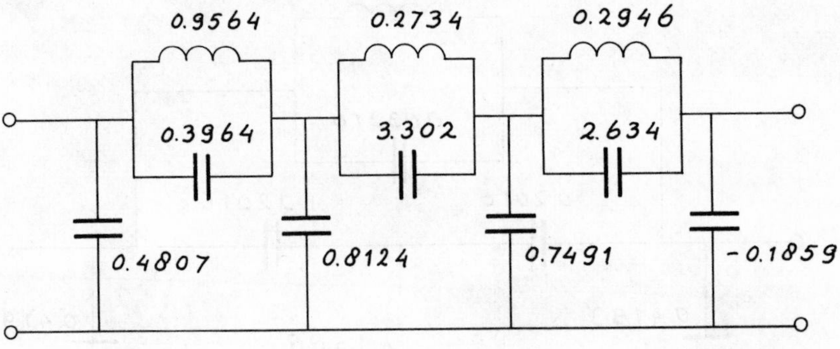
0.9564
0.2734
0.2946
0.3964
3.302
2.634
0.4807
0.8124
0.7491
-0.1859

FIG. 10

It may be stated here that Beccari [19] gave an example in which he found Fig. 8 as an equivalent for Fig. 5. Hoever, he did not mention that in his example (C 0501, θ = 45, Saal), Figs. 7 and 9 are also equivalents; the numerical values of the elements in Figs. 5, 7, 8 and 9 refer to this particular low-pass filter.

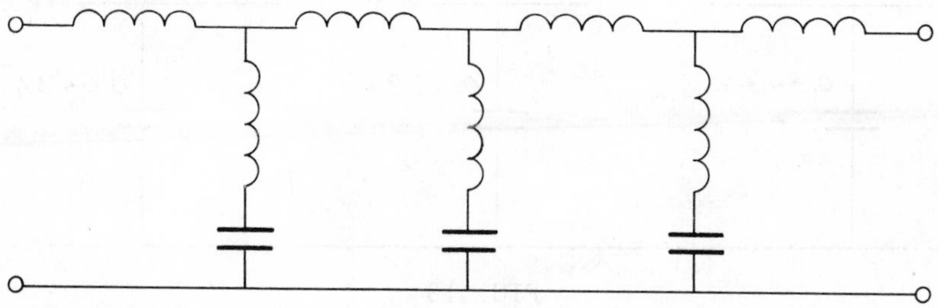

FIG. 11

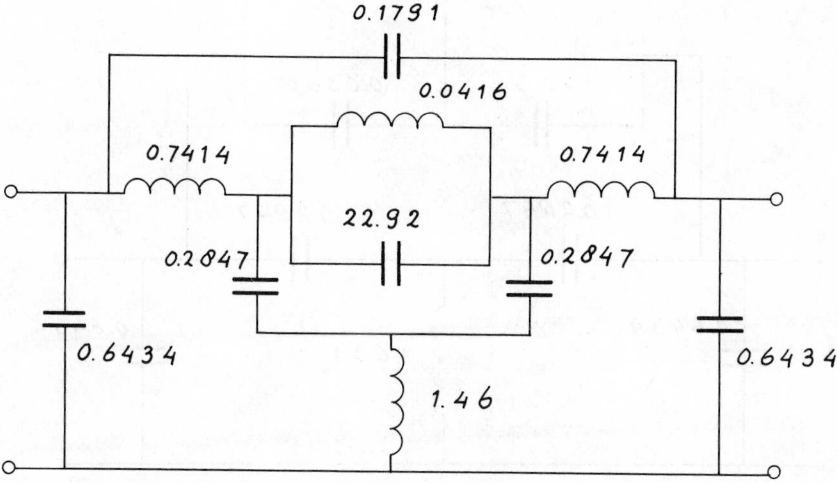

FIG. 12

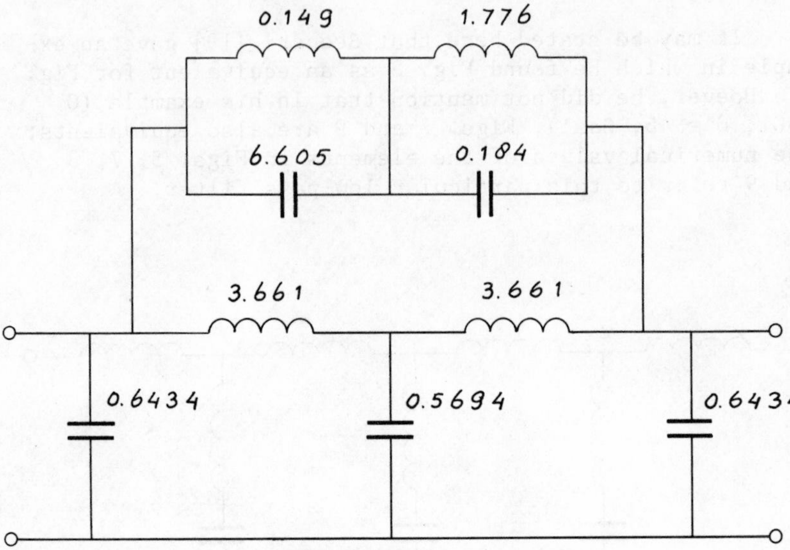
0.149
1.776
6.605
0.184
3.661
3.661
0.6434
0.5694
0.6434

FIG. 13

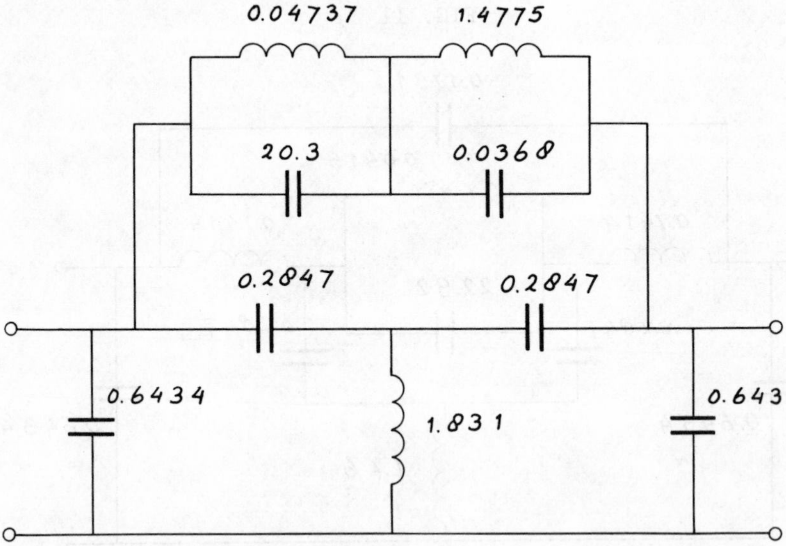
0.04737
1.4775
20.3
0.0368
0.2847
0.2847
0.6434
1.831
0.6434

FIG. 14

7th degree: Figs. 10 and 11 show the ladder realizations and at least one of Figs. 12, 13, 14 and 15 can be used to replace Fig. 10.

The numerical values refer to a low-pass filter (C 0704, θ = 73, Saal) for which each possibility can be used. As can be seen from Figs. 9, 13, 14 and 15, equivalents may be found with fewer elements than the corresponding ladder. This results from the symmetry property. Looking from Fig. 7 to Fig. 13 (or Fig. 8 and 14 or Fig. 9 and 15) we see that increasing the degree by 2 may result in increasing the number of elements by 2 (a parallel resonant circuit) instead of 3 in the ladder case. A fundamental difference between the equivalent circuits described here and the corresponding ladder is that the attenuation poles are no longer "visible" in the circuit and accordingly cannot be adjusted independently. However, series or parallel resonant circuits present in the realizations may be adjusted. The sensitivity of the attenuation for variations in the network elements is at present under investigation.

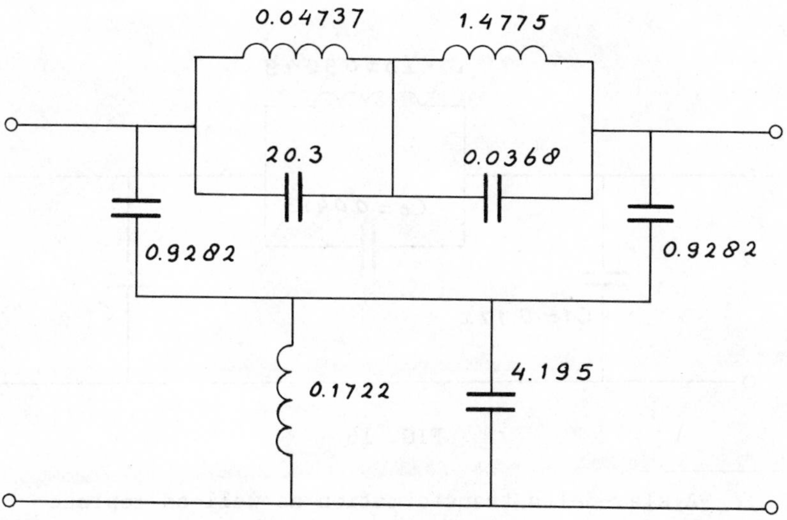

FIG. 15

The possibility of saving elements makes it quite interesting to look for series-parallel realizations in the band-pass and band-stop cases. Some examples are given for the band-stop case, derived from the 3rd and 5th degree low-pass filters by the well-known frequency transformation:

$$p \rightarrow \frac{p}{a(p^2 + 1)} \tag{10}$$

3rd degree: The low-pass ladder is shown in Fig. 16 and the corresponding band-stop ladder in Fig. 17. If the polynomials describing Fig. 16 are transformed by (10) and synthetised, Fig. 18 may result. Taking the dual and applying the transformation $p \rightarrow \frac{1}{p}$ results in Fig. 19*. The numerical values in the Figs. 16, 17, 18 and 19 refer to C 0350, θ = 13 with a = 10.

The elements of Fig. 19 can easily be expressed in

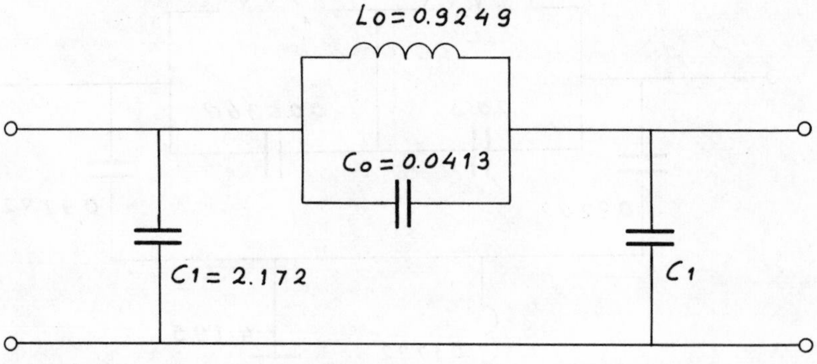

FIG. 16

*A star-delta transformation as well as replacements by Foster equivalents are also necessary.

terms of the elements of Fig. 16 and the parameter a:

$$L_1 = \frac{2C_1}{a} \qquad C_3 = \frac{L_0}{2a}$$

$$C_2 = \frac{2a^2 - L_0C_1}{4aC_1} \qquad C_4 = \frac{2a}{2C_0 + C_1}$$

$$\omega_1^2 = \frac{2a^2}{2a^2 - L_0C_1} \qquad L_2 = \frac{2C_0 = C_1}{2a}$$

$$\omega_2 = 1 \qquad L_3 = \frac{a}{L_0}$$

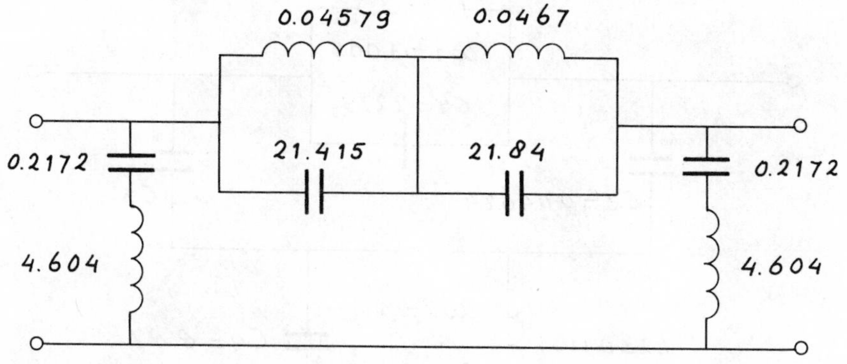

FIG. 17

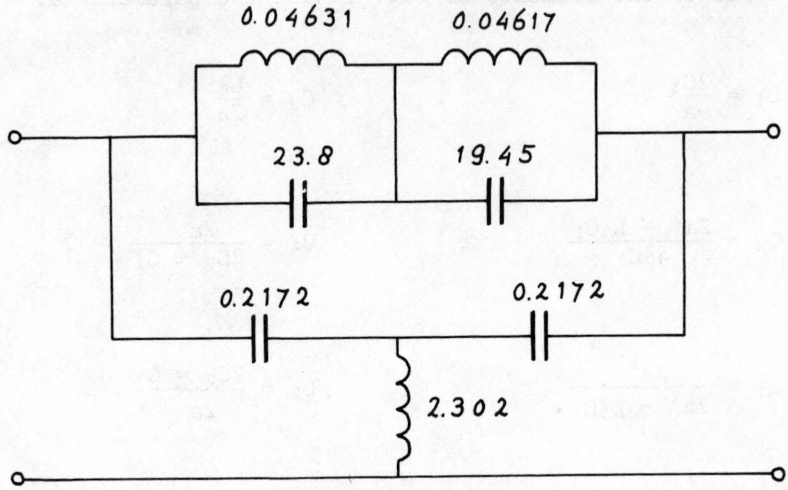
0.04631
0.04617
23.8
19.45
0.2172
0.2172
2.302

FIG. 18

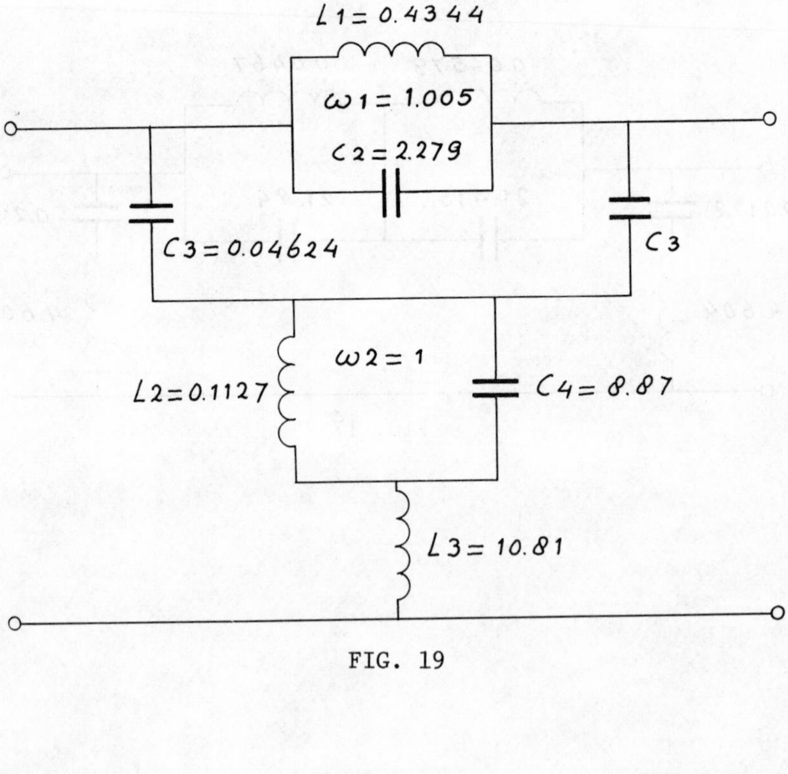
L1= 0.4344
ω1= 1.005
C2= 2.279
C3= 0.04624
C3
ω2= 1
L2=0.1127
C4= 8.87
L3= 10.81

FIG. 19

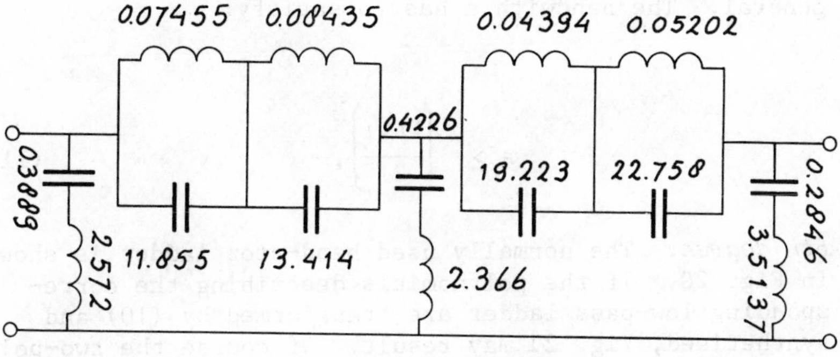
0.07455
0.08435
0.04394
0.05202
0.4226
0.3889
19.223
22.758
0.2846
11.855
13.414
2.572
2.366
3.5137

FIG. 20

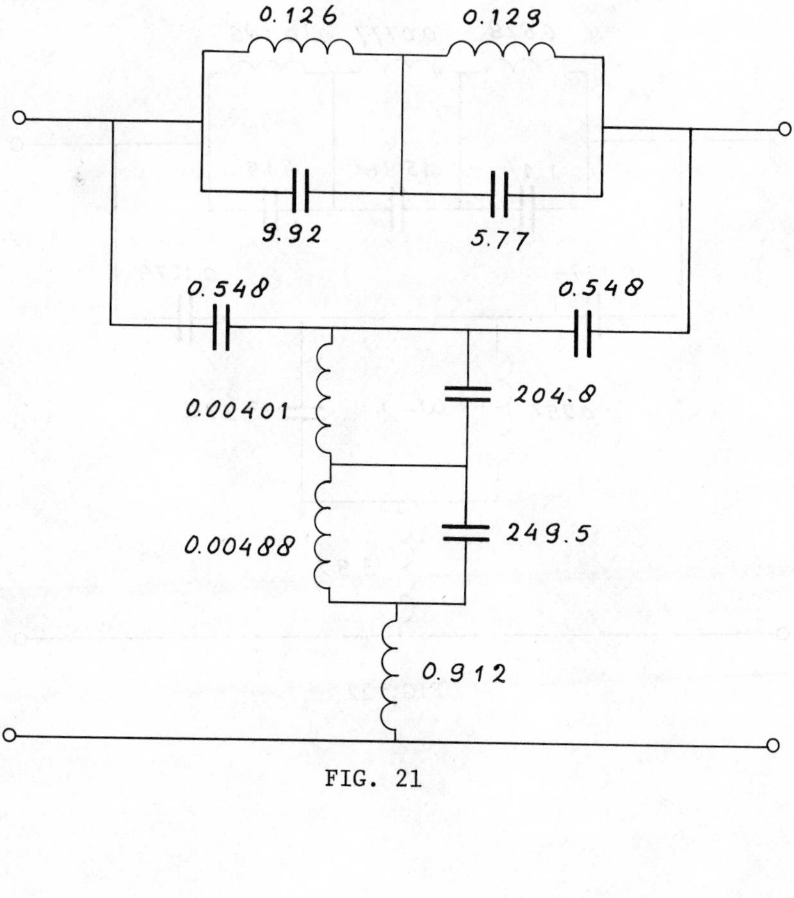
0.126
0.129
9.92
5.77
0.548
0.548
204.8
0.00401
0.00488
249.5
0.912

FIG. 21

From these formulas we see that Figs. 18 and 19 are not general. The bandwith a has to satisfy:

$$a \geq \left(\frac{L_0 C_1}{2}\right)^{\frac{1}{2}}. \tag{11}$$

5th degree: The normally used band-stop ladder is shown in Fig. 20. If the polynomials describing the corresponding low-pass ladder are transformed by (10) and synthetised, Fig. 21 may result. Of course the two-pole present in this bridged-T structure may be replaced by their Forster or Cauer equivalents. Again taking the

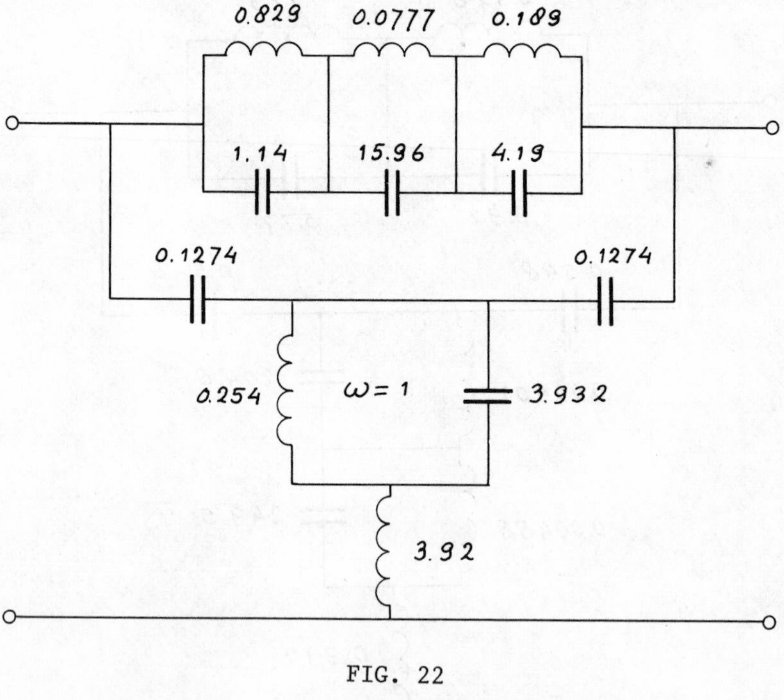

FIG. 22

dual of Fig. 21 and applying $p \to \frac{1}{p}$ results in Fig. 22 (also Foster equivalent two-poles are interchanged and a star-delta transformation has been applied).

The numerical values of Figs. 20, 21 and 22 refer to C 0550, θ = 60, with a = 5. From Figs. 18, 19, 21 and 22 we see that the realizations have the minimum number of inductances and also fewer capacitances than the ladder.

If the inductances have to be simulated by grounded gyrator-capitance combinations the number of gyrators needed in the bridged-T structures will be considerably smaller than in the corresponding ladder. These resulting bridged-T structures can also be obtained by synthetising from the Jaumann structure, which is a canonic structure equivalent to the symmetrical lattice [20] and is shown in Fig. 23 in which Z_a and Z_b are the lattice impedances.

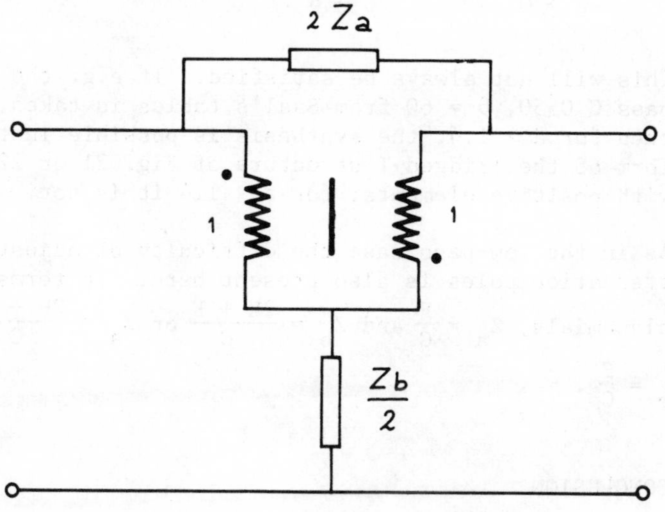

FIG. 23

Two methods of avoiding the ideal transformer are available.

(a) A parallel inductance (L_1) from $\frac{1}{2Z_a}$ and a series-inductance (L_2) from $\frac{Z_b}{2}$ is needed and the relation $4L_2 \geq L_1$ has to be satisfied. Stated otherwise:

$$\lim_{p \to \infty} \frac{Z_b}{p} \geq \lim_{p \to 0} \frac{Z_a}{p} > 0 \,. \tag{12}$$

(b) A parallel capacitance (C_1) from $\frac{1}{2Z_a}$ and a series-capacitance (C_2) from $\frac{Z_b}{2}$ is needed and the relation $4C_1 \geq C_2$ has to be satisfied. Or:

$$\lim_{p \to 0} pZ_b \geq \lim_{p \to \infty} pZ_a \tag{13}$$

This will not always be satisfied. If e.g. the low-pass C 0550, $\theta = 60$ from Saal's tables is taken, then for $a > 1.4$, the synthesis is possible in the form of the bridged-T structure of Fig. 21 or 22 with positive elements; for $a < 1.4$ it is not.

As in the low-pass case the difficulty of adjusting the attenuation poles is also present here. In terms of the polynomials, $Z_a = \frac{F}{C}$ and $Z_b = \frac{2H + F}{C}$ or $Z_a = \frac{2H + F}{C}$ and $Z_b = \frac{F}{C}$.

10. CONCLUSION

In the foregoing we have considered some of the possibilities that are now known for realizing reactance three-pole networks. In particular, the application of this material to filter synthesis has been examined. Conventional filter networks are usually constructed in ladder form, partly because of the relative ease of computation and partly because during the construction of the filter the attenuation poles can be independently adjusted using simple measuring techniques. The general form of series-parallel three-pole does not have these advantages. However, with the availability of computers, the computation problem is not serious and the technique of adjusting the characteristic after the components have been assembled is in conflict with one of the basic tenets of the philosophy of integrated circuits. If for example one is to realize the filter using capacitors and gyrators, then the ideal situation is to produce the various components by a series of diffusion and vacuum deposition processes without a single electrical measurement or adjustment being made. If this stage should finally be attained, the easy adjustment facilities of the ladder network would be superfluous.

Whether filters constructed in this manner can ever provide satisfactory solutions in a given case is another matter. It is well known that many filters in communication systems such as telephony have to satisfy very stringent requirements which can be met only if adjustment facilities are available. Nevertheless, there are many types of filters in electronic engineering for which such demands are not necessary. In any case it would be unwise to limit a theoretical development at this stage because of the limitations of present technology, in particular the realizable tolerances and temperature coefficients of coils and capacitors.

There is, however, another aspect of this problem whereby, apart from the broader class of realizable characteristics, the general series-parallel network may have advantages over the ladder network, It is known that the sensitivity of the zeros of C (open-circuit poles of the complete network) to variations in individual elements is less in the case of a normal passive network - whereby the poles are dependent on all the elements - than in a circuit in which each pole is realized by a damped resonant circuit which is separated from its neighbor by an isolating amplifier. In the passive case, the error due to a deviation of one element is apparently "spread out" over the whole network and has little influence on any one pole. This is not the case if the poles are formed by isolated circuits. It is worth investigating whether similar conclusions apply to the zeros of H, whereby the ladder network is then analogous to the above-mentioned active network.

The influence of the arrival of integrated circuits on theoretical developments has been mentioned. A further important question that now arises is: to what extent is the three-pole still preferable to the general two-port? Or more generally, which form of n-port network is going to be the most useful in practice?

In the theory as such the importance of the non-series-parallel three-poles has been emphasized. It is clear that a lot of exploratory work has yet to be done before one can begin to grasp what is the real significance of the few results that are now available. For

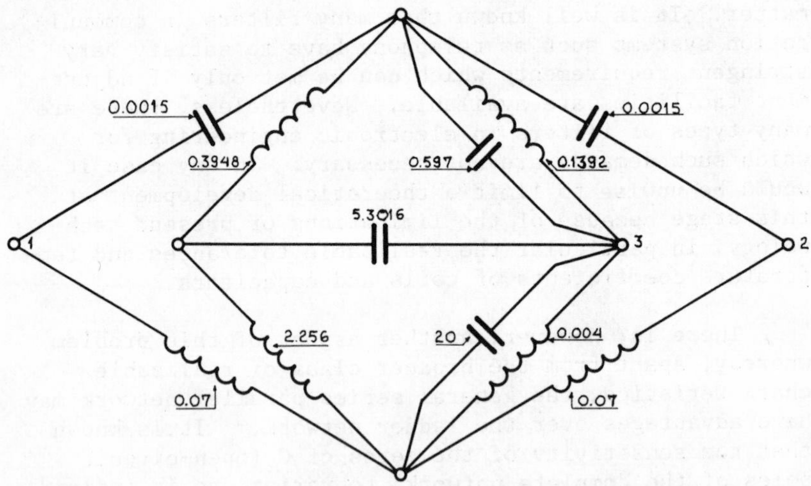

FIG. 24

further progress and also for the applications, it is of prime importance that the existing knowledge should be collated and presented in the simplest and most easily understandable form, not only with respect to the concepts but also to the notation employed. As a very minor contribution to this end we conclude with some remarks on Fialkow's counter-example to Darlington's conjecture which may be suggestive of further work. The example is shown in Fig. 24 where the terminals are numbered 1, 2, 3. The units are farad and henry $^{-1}$.

This network is of the 11th order and supports one mode with zero natural frequency, that is unobservable at the terminals as a result of the loop of inductors. There is also an unobservable mode with frequency unity. The network was specially constructed to exhibit this mode, and on dividing all the polynomials by $\lambda^2 + 1$, negative coefficients appear in H. This is the only other unobservable mode so that the order of the *three-pole* is 8 and the degree is 10.

It is interesting to note in passing the highest

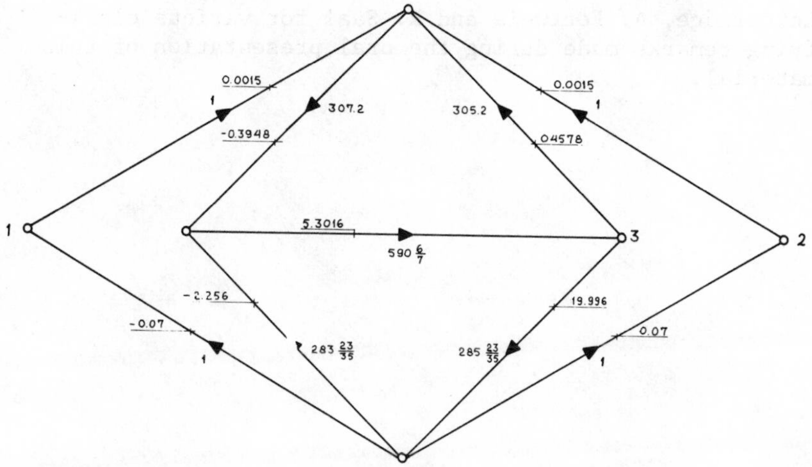

FIG. 25

powers of λ in the remaining polynomials and in brackets the number of zero natural frequencies: D, 9 (3); A, B, 10 (2); F + G, 10 (2). All of these zero frequencies are produced by loops of inductors. It would be worthwhile investigating the possibility of constructing a network with similar properties but containing cut-sets of capacitors. [Note that the complete network consisting of the reactance network and its three port-terminations is non-planar.]

The condition that an unobservable mode of unit frequency exists is equivalent to the linear-dependence conditions given by Fialkow [10]. Figure 25 shows the branch susceptances at unit frequency and the current ratios alongside the reference-direction arrows of the branch currents. It is easily seen from the two balanced bridges present that the voltages between the terminals are zero.

ACKNOWLEDGEMENTS

We are indebted to V. Belevitch, G. Bown, O.P.D. Cutteridge, A. Fettweis and R. Saal for various clarifying remarks made during the oral presentation of this material.

REFERENCES

1. ORCHARD, H.T. (1966) Inductorless filters, *Electron. Lett.*, pp. 224-225.

2. NEIRYNCK, J. and THIRAN, J.P. (1967) Sensitivity analysis of lossless 2-port by uniform variations of the element values, *Electron. Lett. 3*, pp. 74-76.

3. TELLEGEN, B.D.H. (1946) Network synthesis, especially the synthesis of resistanceless four-terminal networks, *Philips Res. Rep. 1*, pp. 169-184.

4. ADAMS, K.M. (1958) On the synthesis of three-terminal networks composed of two kinds of elements, *Philips Res. Rep. 13*, pp. 201-264, p. 209.

5. BELEVITCH, V. (1963) On network analysis by polynomial matrices, in: *Recent Developments in Network Theory* (ed. S.R. Deards), Pergamon Press, pp. 19-30.

6. ADAMS, K.M., ibid., p. 211.

7. ADAMS, K.M., ibid., pp. 226-227; also FIALKOW, A.D. (1966) Series-parallel grounded two-ports, *Quart. Appl. Math. 24*, pp. 195-213.

8. ADAMS, K.M., ibid., pp. 202-236.

9. DARLINGTON, S. (1955) A survey of network realization techniques, *IRE Trans. Circuit Theory CT-2*, pp. 291-297.

10. FIALKOW, A.D. (1968) A limitation of the series-parallel structure, *IEEE Trans. Circuit Theory, CT-15*, pp. 124-132.

11. ADAMS, K.M., ibid., pp. 247-252.

12. ADAMS, K.M. (1963) A review of the synthesis of

linear three-terminal networks composed of two kinds of elements, in : *Recent Developments in Network Theory* (ed. S.R. Deards), Pergamon Press, pp. 65-73.

13. FIALKOW, A.D. (private communication).

14. PAWSEY, D.C. (1968) Adaption of the 7th degree electrically symmetrical ladder filters for unsymmetrical operation, *IEEE Trans.Circuit Theory CT-15*, pp. 499-500.

15. ADAMS, K.M., ibid., pp. 215-222.

16. BELEVITCH, V. (1968) *Classical Network Theory*, Holden-Day, p. 280.

17. BELEVITCH, V., ibid., chap. 8.

18. SAAL, R. (1963) *Der Entwurf von Filtern mit Hilfe des Kataloges normierter Tiefpasse*, Telefunken GmbH.

19. BECCARI, C. (1966) Alcune possibilità nella sintesi di bipoli senza trasformatori, *Alta Frequenza 9, 35*, pp. 688-699.

20. BELEVITCH, V., ibid., p. 208.

CASCADE SYNTHESIS OF LOSSLESS TWO-PORTS BY TRANSFER MATRIX FACTORIZATION

ALFRED FETTWEIS

Department of Electrical Engineering
Ruhr-Universität Bochum, Germany

SUMMARY

The decomposition of the transfer matrix into a product of two simpler transfer matrices of the same type is treated rigorously for all reciprocal and nonreciprocal lumped finite lossless two-ports. The order in which the factorization should be realized can be arbitrarily prescribed; for any given order, an infinite number of solutions exists.

It is shown that the original factorization problem is equivalent to a much simpler problem, which in turn mainly reduces to the problem of solving a set of so-called fundamental equation, consisting of n linear homogeneous equeations in n + 2 unknowns. From a single suitable solution of this set, all solutions of the factorization problem can easily be obtained. A simple procedure for establishing the fundamental equation is given.

The relationship between the transfer matrix factorization and the cascade synthesis of lossless two-ports by means of elementary sections is discussed in terms of the present method.

1. INTRODUCTION

The cascade synthesis of lossless two-ports represents one of the main problems of classical network theory. It has been solved in the past by suitable expansions of the input impedance or input reflectance [1-10] and by factorization of the chain martix [11,12]. It had long been recognized, however, that the transfer matrix T (also called transfer scattering matrix) should constitute a much better tool for achieving the same purpose [13,14]. Indeed, as will also be discussed later, any problem of cascade synthesis amounts to a certain problem of factorizing the transfer matrix, while on the other hand, for lossless two-ports, the transfer matrix can very easily be expressed in terms of the so-called canonic polynomials. This latter advantage is not obtained if the cascade synthesis is based on a factorization of the chain matrix as discussed by Piloty [11] and Talbot [12].

Although several publications have appeared on the problem thus defined [9, 14-16], an exact proof has only recently [17-19] become available. The method by Youla [17] deals with real reciprocal two-ports but is rather complicated. Independent of the Youla's work, an alternate method has been obtained by the author. It is described for general lossless two-ports (which may thus also contain nonreciprocal and constant imaginary elements) in [18], and for real reciprocal two-ports (reactance two-ports) in [19]. In the present paper, the method shall be described under the important assumption that the lossless two-port is real; thus it may not contain imaginary elements, but may be either reciprocal or not. For reasons of simplicity, the term "real" shall, however, henceforth be dropped, i.e., we shall simply speak of "lossless two-ports."

For the sake of readers less familiar with the subject, we shall first give a more detailed description of the origin of the problem than was given in the earlier publications [18,19]. Additional details will also be added at various points. Finally, the cascade synthesis of lossless two-ports by means of the classic elementary sections will be discussed in the light of the present method.

As in the earlier publications [18,19], the proof of the method hardly makes use of previous network results; it thus indirectly proves a large number of further network results which otherwise are obtained by much more lengthy procedures. It will be shown how, for a given order in which the factorization should be realized, all solutions can be obtained in a simple way once a single solution has been found. In section 7, a particularly simple method for establishing the fundamental equations, on which the solution is based, is given. This method can easily serve as basis for practical computation.

The relationship between the computation of the elementary sections and the transfer matrix has been discussed by Neirynck, van Bastelaer, and Leich [16,20]. In section 9, these results are summarized and presented in the light of the present theory. The discussion given there shows clearly that the present method allows very easily to make a systematic search for finding all the elementary sections which can be used in cascade synthesis. In section 10, finally, an extension to networks contain ing unit elements will be discussed.

2. TRANSFER MATRICES OF LOSSLESS TWO-PORTS

2.1. *Basic properties*

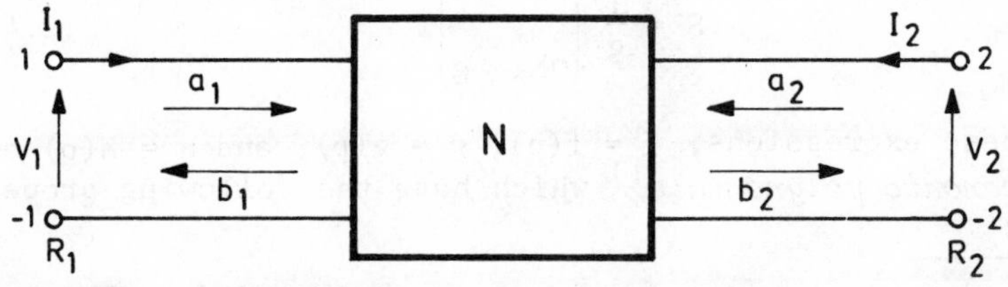

Fig. 2.1 Incident waves a_i and reflected waves b_i for a two-port N.

For a two-port N (Fig. 2.1), the *incident wave* a_i and the *reflected wave* b_i at port i (i = 1 or 2) are defined by

$$a_i = (V_i + R_i I_i)/2\sqrt{R_i}, \quad b_i = (V_i - R_i I_i)/2\sqrt{R_i}, \tag{2.1}$$

where R_i is the characteristic (normalizing) resistance attributed to port i. The *scattering matrix* S and the *transfer matrix* T of N are then defined by

$$\begin{pmatrix} b_1 \\ b_2 \end{pmatrix} = S \begin{pmatrix} a_1 \\ a_2 \end{pmatrix}, \quad \begin{pmatrix} b_1 \\ a_1 \end{pmatrix} = T \begin{pmatrix} a_2 \\ b_2 \end{pmatrix}. \tag{2.2}$$

The relationship between S and T is thus quite similar to the relationship between the impedance matrix and the chain matrix. In particular, if S_{ij} designates the entry of S in row i and column j, the reciprocity condition $S_{12} = S_{21}$ is equivalent to det T = 1.

For a lossless two-port, the *canonic form* of the scattering matrix is given by [9]

$$S = \frac{1}{g} \begin{pmatrix} h & \sigma f_* \\ f & -\sigma h_* \end{pmatrix}, \tag{2.3}$$

and the *canonic form* of the transfer matrix by

$$S = \frac{1}{g} \begin{pmatrix} \sigma g_* & h \\ \sigma h_* & g \end{pmatrix}. \tag{2.4}$$

In these expressions, f = f(p), g = g(p), and h = h(p) are the *canonic polynomials*[1)] which have the following properties:

[1)] For sake of improved elegance in writing the various expressions, we have slightly modified the traditional notation: Our polynomial f corresponds to σf_* in the notation used by Belevitch [9].

1. They are all *real polynomials* in the complex frequency p.

2. g is *Hurwitz*.

3. f is *monic*, i.e., its leading coefficient is equal to unity.[2)]

4. f, g, and h are *related* by

$$gg_* = hh_* + ff_*. \tag{2.5}$$

In all these expressions, the lower asterisk indicates the Hurwitz *conjugate* of the corresponding function; for real functions as we are considering here, this amounts to replacing p by -p, i.e., $f_* = f(-p)$, etc. Furthermore, σ is a constant which is either equal to +1 or to -1, i.e.

$$\sigma = \pm 1; \tag{2.6}$$

we call σ the *parity constant* (it is often related to the parities of the degrees of various polynomials). If we are dealing with a *reactance* two-port, i.e., if the two-port is not only real but also reciprocal, f is either even ($f_* = f$) or odd ($f_* = -f$) and we have

$$\sigma = f_*/f, \tag{2.7}$$

i.e., $\sigma = +1$ for f even and $\sigma = -1$ for f odd.[3)]

From (2.5), a simple relation between the degrees of the canonic polynomials can be derived. Indeed, for $p = j\omega$, we have $ff_* = |f|^2$, etc., and therefore, in particular

---

[2)] In contrast to what is usually done, it is simpler for our purpose to choose f monic rather than g.

[3)] With reference to footnote 1, note that the modified definition of f does not imply any modification in case of reciprocal two-ports.

$$|h| \leq |g|, \; |f| \leq |g|. \tag{2.8}$$

Hence, using the notation "deg f" for "degree of f," etc., we have

$$\deg h \leq n, \; \deg f \leq n, \tag{2.9}$$

where n is defined by

$$n = \deg g. \tag{2.10}$$

The non-negative number m defined by

$$m = n - \deg f \tag{2.11}$$

indicates the number of transmission zeros at infinity. These, together with the zeros of f form the poles of T.

In section 3, we shall make use of the inverse T^{-1} of a transfer matrix T. The following expression is easily obtained from (2.4) and (2.5)

$$T^{-1} = \frac{1}{\sigma f_*} \begin{pmatrix} g & -h \\ -\sigma h_* & \sigma g_* \end{pmatrix}. \tag{2.12}$$

Clearly, T and T^{-1} only exist if f is not identically zero. Henceforth, it will always implicitly be assumed that this requirement is fulfilled. If $f \equiv 0$, the synthesis of the two-port defined by (2.3) becomes trivial and a problem of cascade synthesis does not arise. Indeed, the two-port then corresponds to two isolated impedances z_1 and z_2 of reflectances

$$(z_1 - 1)/(z_1 + 1) = h/g^- \text{ and } (z_2 - 1)/(z_2 + 1) = -\sigma h_*/g,$$

respectively (Fig. 2.2).

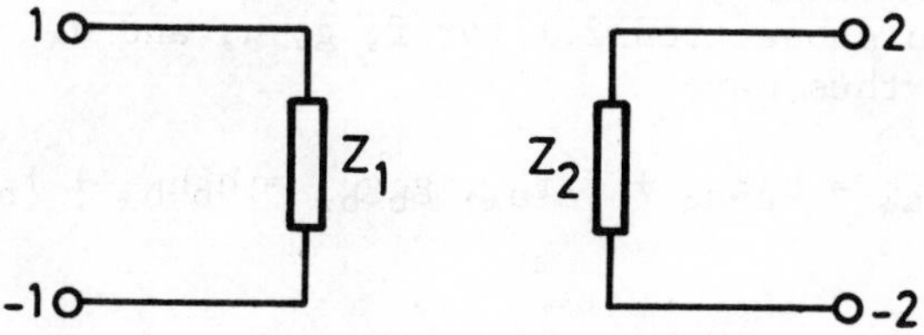

Fig. 2.2 Two-port for which $f \equiv 0$.

2.2. *Transfer matrix of cascaded lossless two-ports*

If two two-ports N_a and N_b, with ports 1, 2, and 3, 4, respectively are connected in cascade at ports 2 and 3 (Fig. 2.3), and if we choose $R_2 = R_3$, we have $a_3 = b_2$ and

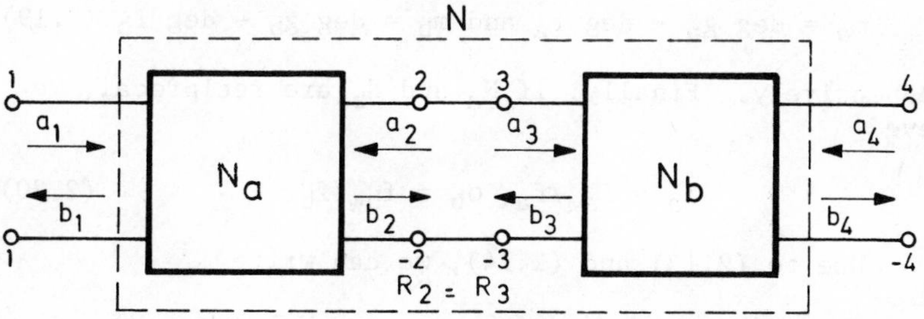

Fig. 2.3 Cascade connection of two two-ports N_1 and N_2.

$a_3 = b_2$. Hence, the transfer matrix T of the combined two-port N is related to the transfer matrices T_a and T_b of N_a and N_b by

$$T = T_a T_b. \tag{2.13}$$

Assume now again that N_a and N_b are lossless, i.e.

$$T_a = \frac{1}{f_a}\begin{pmatrix}\sigma_a g_{a*} & h_a \\ \sigma_a h_{a*} & g_a\end{pmatrix}, \quad T_b = \frac{1}{f_b}\begin{pmatrix}\sigma_b g_{b*} & h_b \\ \sigma_b h_{b*} & g_b\end{pmatrix} \tag{2.14}$$

where f_a, g_a, h_a and σ_a on the one hand, and f_b, g_b, h_b and σ_b on the other, have the same properties as those discussed in subsection 2.1 for f, g, h, and σ. In particular, we thus have

$$g_a g_{a*} = h_a h_{a*} + f_a f_{a*}, \quad g_b g_{b*} = h_b h_{b*} + f_b f_{b*}, \tag{2.15}$$

$$\sigma_a = \pm 1, \qquad \sigma_b = \pm 1, \tag{2.16}$$

$$\deg h_a \leq \deg g_a, \qquad \deg h_b \leq \deg g_b, \tag{2.17}$$

and, for $p = j\omega$,

$$|h_a| \leq |g_a|, \ |h_b| \leq |g_b|. \tag{2.18}$$

The numbers of transmission zeros at infinity are given by the non-negative integers

$$m_a = \deg g_a - \deg f_a \text{ and } m_b = \deg g_b - \deg f_b \tag{2.19}$$

respectively. Finally, if N_a and N_b are reciprocal, we have

$$\sigma_a = f_{a*}/f_a, \ \sigma_b = f_{b*}/f_b. \tag{2.20}$$

Due to (2.13) and (2.14), we can write

$$T = \frac{1}{f'} \begin{pmatrix} \sigma' g'_* & h' \\ \sigma' h'_* & g' \end{pmatrix} \tag{2.21}$$

where

$$f' = f_a f_b, \ \sigma' = \sigma_a \sigma_b \tag{2.22}$$

$$g' = g_a g_b + \sigma_a h_{a*} h_b, \ h' = h_a g_b + \sigma_a g_{a*} h_b. \tag{2.23}$$

As can be checked, f', g', and h' have the same properties 1, 3, and 4 as those mentioned in subsection 2.1 for f, g, and h; thus, in particular,

$$g' g'_* = h' h'_* + f' f'_*. \tag{2.24}$$

We also still have $\sigma' = \pm 1$, and furthermore, if (2.20) is fulfilled, $\sigma' = f'_*/f'$. However, g' does, in general, not have to be Hurwitz, i.e., (2.21) does not have to be the canonic form of T. Furthermore, the degree of g' may be lower than the sum of the degrees of g_a and g_b.

The possibility that such anomalies can actually arise can easily be seen from the simple example shown in Fig. 2.4, where the two series arms at the cascaded ports

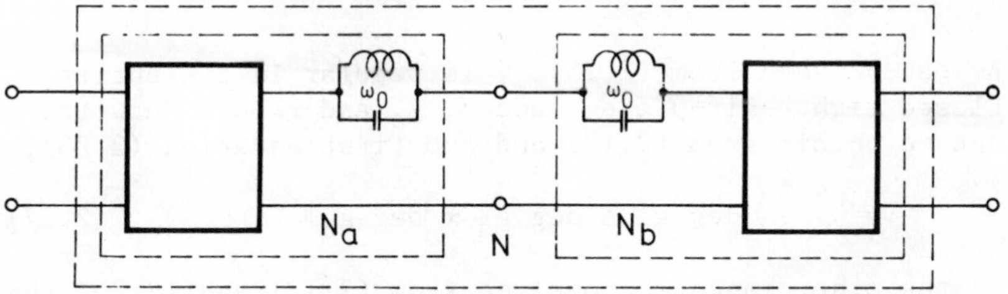

Fig. 2.4 Example of a two-port N for which f is of lower degree than the product $f_a f_b$.

have the same parallel resonant frequency ω_0. In this case, the two series arms contribute a factor $(p^2 + \omega_0^2)^2$ to $f' = f_a f_b$, but together they may only contribute a factor $p^2 + \omega_0^2$ to the canonic polynomial f of the combined two-port N. Hence, it must be possible in this case to cancel a factor $p^2 + \omega_0^2$ from all polynomials in (2.21). Similarly, if the two parallel resonant circuits in Fig. 2.4 are replaced by simple capacitances, it must be possible to cancel a factor p. Finally, if these two parallel resonant circuits are replaced by simple inductances, these can contribute together only a simple pole at infinity. This requires that a cancellation occurs among the leading coefficients in the right-hand member of the first equation (2.23). Equally simple examples can be obtained by assuming shunt arms rather than series arms at the common port of N_a and N_b.

All this shows that we should examine more closely the general mathematical nature of all cancellations which can occur. For doing this, we consider the function Φ defined by

$$\Phi = g'/g_a g_b \tag{2.25}$$

which, due to the first equation (2.23), can also be written

$$\Phi = 1 + \frac{\sigma_a h_a h_{a*}}{g_a} \frac{h_b}{g_b} \tag{2.26}$$

As can be seen from (2.25), Φ is regular in the entire closed right half-plane; indeed, g_a and g_b are Hurwitz, and we obtain from (2.17) and the first equation (2.23),

$$\deg g' \leq \deg g_a + \deg g_b. \tag{2.27}$$

On the other hand, we conclude from (2.18) and (2.26) that Re $\Phi \geq 0$ for $p = j\omega$. Hence, Φ is a positive function; it thus has no zeros in the right half-plane, while zeros on the $j\omega$-axis are simple. Consequently, since possible common factors of g' and $g_a g_b$ are Hurwitz, g' is semi-Hurwitz (i.e., such that it has no zeros in the right half-plane and that zeros on the $j\omega$-axis are simple), and we can write $g' = gc$ where g is Hurwitz and c is a real polynomial grouping those zeros of g' which are located on the $j\omega$-axis. Clearly, c is either even or odd, and its zeros are all simple. Another consequence of the positive reality of Φ is

$$\deg g_a + \deg g_b - 1 \leq \deg g'. \tag{2.28}$$

For $p = j\omega$, we obtain from (2.24),

$$|g'|^2 = |h'|^2 + |f'|^2.$$

This shows, that for $c = 0$, we also have $h' = f' = 0$. Thus, as the zeros of c are simple, c also divides h' and f', i.e., $h = h'/c$ and $f = f'/c = f_a f_b/c$ are polynomials. This last conclusion can even be made more precise. Indeed, eliminating first h_b and then h_a between the two equations (2.23), and using (2.15), one obtains

$$f_a f_{a*} = (g' g_{a*} - h' h_{a*})/g_b \qquad (2.29)$$

$$f_b f_{b*} = (g' g_{b*} - \sigma_a h'_* h_b)/g_a. \qquad (2.30)$$

Hence, as c divides g', h', and h'_* (c is even or odd!), but certainly not the Hurwitz polynomials g_a and g_b, it must divide $f_a f_{a*}$ and $f_b f_{b*}$, and therefore f_a as well as f_b (any polynomial and its Hurwitz conjugate have the same zeros on the $j\omega$-axis!). Consequently, the zeros of c are transmission zeros of N_a as well as of N_b.

Clearly, the polynomials f, g, and h obtained this way also satisfy (2.5) and thus have all the properties of canonic polynomials. The scattering martix S and the transfer matrix T of the combined two-port N can be written in the form (2.3) and (2.4), respectively, with $\sigma = \sigma' c_*/c$, i.e., $\sigma = \sigma'$ if c is even, and $\sigma = -\sigma'$ if c is odd. Clearly, this factor c covers all possible cancellations corresponding to transmission zeros at finite frequencies.

Let us now examine more closely the possibility of a cancellation due to transmission zeros at $p = \infty$. Due to $g = g'/c$ and $f = f_a f_b/c$, we obtain, using (2.27) and (2.28) together with $n = \deg g$,

$$\deg g_a + \deg g_b - \deg c - 1 \le n \le \deg g_a + \deg g_b - \deg c, \qquad (2.31)$$

and for the number of transmission zeros at infinity, m, using (2.11), (2.19), and (2.31),

$$m_a + m_b - 1 \le m \le m_a + m_b. \qquad (2.32)$$

However, the first equality sign in (2.32) can only occur if we have simultaneously $m_a > 0$ and $m_b > 0$. Indeed, due to (2.17) and the relation $\deg h' \le \deg g'$, which can be derived from (2.24), one can derive from (2.29) and (2.30) the two relations

$$2 \deg f_a \le \deg g' + \deg g_a - \deg g_b,$$

$$2 \deg f_b \le \deg g' + \deg g_b - \deg g_a$$

whence, using (2.19),

$$\deg g' \geq \deg g_a + \deg g_b - 2\, m_a,$$

$$\deg g' \geq \deg g_a + \deg g_b - 2\, m_b.$$

Hence, due to (2.27), one has $\deg g' = \deg g_a + \deg g_b$ as soon as $m_a = 0$ or $m_b = 0$.

Summarizing, we thus conclude that the only cancellations possible are those due to transmission zeros on the $j\omega$-axis as well as at $p = \infty$. In all cases, the cancellation cannot decrease the order of the transmission zero by more than one, and a cancellation is only possible if N_a as well as N_b have a transmission zero at the frequency in question. All these conditions are clearly fulfilled for the examples discussed in relation with Fig. 2.4.

3. STATEMENT OF THE FACTORIZATION PROBLEM

Assume that the canonic polynomials f, g, and h as well as the parity constant σ of a lossless two-port N are given. We wish to decompose N into a cascade of two lossless two-ports N_a and N_b. This amounts to decomposing the transfer matrix T into a product T_aT_b as given by (2.13) where the matrix factors T_a and T_b have the form (2.14), f_a, g_a, and h_a being the canonic polynomials and σ_a the parity constant of N_a, and f_b, g_b, and h_b being the canonic polynomials and σ_b the parity constant of N_b.

In the present paper, we are only interested in minimal decompositions, i.e., in factorizations of T for which the excess polynomial c is identically equal to unity and for which no cancellation of leading coefficients in the right-hand member of the first equation (2.23) can occur. Referring to the notation used in subsection 2.2, we are thus only interested in such factorizations of T for which $f = f'$, $g = g'$, $h = h'$, $\sigma = \sigma'$, and $\deg g = \deg g_a + \deg g_b$. For the rest, we would like to be able to divide the zeroes of f and the transmission zeros at infinity in an arbitrary way among N_a and N_b, except of course for the obvious requirement that f_a and f_b have to be real. Finally, we also have to make some decision concerning the parity constants σ_a and σ_b. For reciprocal two-ports, we have to choose them according to (2.20), but in the general case, no such requirement exists. Hence, we may want to be able to choose them in any of the two possible ways compatible with $\sigma = \sigma_a\sigma_b$.

In order to derive the relations required as basis for our further analysis, it is simpler not to start from $T = T_aT_b$ but from one of the two equivalent relations

$$T_b = T_a^{-1}T \quad \text{or} \quad T_a = TT_b^{-1}. \tag{3.1}$$

T_a^{-1} and T_b^{-1} can easily be written down by comparing (2.4), (2.12), and (2.14). We use here the first of the two expressions (3.1). Taking all previous conditions into account, one finds this way that our factorization problem is equivalent to the following

Problem 1

Let there be *given*:

1. Three real polynomials $f = f(p)$, $g = g(p)$, and $h = h(p)$, of which f is monic and g Hurwitz, and for which

$$\Delta = (gg_* - hh_*)/ff_* = 1. \tag{3.2}$$

2. A constant σ which is either equal to +1 or to -1.

3. Two monic real polynomials $f_a = f_a(p)$ and $f_b = f_b(p)$ such that

$$f_af_b = f. \tag{3.3}$$

4. Two integers m_a and m_b such that

$$m_a \geq 0,\ m_b \geq 0,\ m_a + m_b = m \tag{3.4}$$

where m is a non-negative integer defined by

$$m = n - \deg f,\ n = \deg g \tag{3.5}$$

5. Two integers σ_a and σ_b such that

$$\sigma_a = \pm 1, \sigma_b = \pm 1, \sigma_a\sigma_b = \sigma \qquad (3.6)$$

In addition to the quantities thus given we *define* two further integers n_a and n_b by

$$n_a = m_a + \deg f_a, n_b = m_b + \deg f_b. \qquad (3.7)$$

For these, we obtain by (3.3), the last expression (3.4), and the first expression (3.5),

$$n_a + n_b = n \qquad (3.8)$$

Using these various quantities, we can now state the task of problem 1 in the following way:

Find two real Hurwitz polynomials $g_a = g_a(p)$ and $g_b = g_b(p)$, and two further real polynomials $h_a = h_a(p)$ and $h_b = h_b(p)$ such that

1. $$h_b = \sigma_a(hg_a - gh_a)/f_af_{a*}; \qquad (3.9)$$

2. $$g_{b*} = (g_*g_a - h_*h_a)/f_af_{a*}; \qquad (3.10)$$

3. $$\deg g_a = n_a, \quad \deg g_b = n_b; \qquad (3.11)$$

4. $$\Delta_a = 1, \quad \Delta_b = 1 \qquad (3.12)$$

where Δ_a and Δ_b are defined by

$$\Delta_a = (g_ag_{a*} - h_ah_{a*})/f_af_{a*} \qquad (3.13)$$

$$\Delta_b = (g_bg_{b*} - h_bh_{b*})/f_bf_{b*} \qquad (3.14)$$

In case of a reciprocal two-port N, problem 1 thus defined is compatible with the request for two-ports N_a and N_b which are also reciprocal. In this case, indeed, (2.7) is fulfilled and we have to satisfy (2.20). Hence, the requirement $f = f_af_b$ then is compatible with $\sigma = \sigma_a\sigma_b$.

4. REDUCTION OF PROBLEM 1 TO A SIMPLER PROBLEM

As problem 1 is rather complex, it would be difficult to solve it directly. We shall, therefore, now consider the following problem 2. There is some advantage, however, not to formulate problem 2 as a whole but to split it into two partial problems 2a and 2b.

Problem 2a

Let there again be *given* the real polynomials f, g, h, f_a, and f_b as well as the integers m_a, m_b, n_a, n_b, σ, σ_a, and σ_b as in problem 1.

Find two polynomials g_a and h_b such that

1. $$\deg g_a \leq n_a, \ \deg h_b \leq n_b, \qquad (4.1)$$

2. The functions h_a and g_b defined by

$$h_a = (hg_a - \sigma_a f_a f_{a*} h_b)/g, \qquad (4.2)$$

$$g_{b*} = (\sigma_a h_* h_b + f_b f_{b*} g_a)/g \qquad (4.3)$$

are real polynomials.

Problem 2b

Consider the rational function Δ_0 defined by

$$\Delta_0 = (g_a g_b + \sigma_a h_{a*} h_b)/g. \qquad (4.4)$$

Among the solutions of problem 2a, *find* one for which

$$\Delta_0 = 1 \text{ for at least } one \text{ value of } p. \qquad (4.5)$$

The equivalence of problems 1 and 2 will now be established through the following theorems 1a, 1b, and 1c.

Theorem 1a

Every solution of problem 1 is a solution of problem 2.

Proof

1. Using (3.2), (3.3), and (3.6), one can verify easily that the two equations (3.9) and (3.10) are equivalent with (4.2) and (4.3).

2. Expressing h_b and g_b in (4.4) by means of (3.9) and (3.10), one obtains $\Delta_0 = \Delta_a$. Hence, (4.5) is certainly fulfilled as soon as we have $\Delta_a = 1$.

3. The first inequality (4.1) is satisfied by the first equality (3.11).

4. From $\Delta_b = 1$, we obtain $\deg h_b \leq \deg g_b$ (cf. (2.9)). Hence, the second inequality (4.1) is satisfied by the second equality (3.11).

Theorem 1b

For every solution of problem 2a, the two equations (3.9) and (3.10) are fulfilled, and we have

$$\deg h_a \leq n_a, \ \deg g_b \leq n_b, \tag{4.6}$$

as well as

$$\Delta_a = \Delta_b = \Delta_0. \tag{4.7}$$

Furthermore, Δ_0 turns out to be a real constant.

Proof

1. As we have seen under step 1 of the proof of Theorem 1a, the equations (4.2) and (4.3) are equivalent with (3.9) and (3.10). Hence, the result $\Delta_0 = \Delta_a$ mentioned under step 2 of the same proof is also valid for every solution of problem 2a. Substituting (3.9) and (3.10) in (3.14), and taking into account (3.3) and (3.6), we also obtain $\Delta_a = \Delta_b$, whence (4.7) follows.

2. Using the earlier definition $n = \deg g$, we conclude from (4.2) that the degree of h_a cannot exceed the larger of the two numbers

$$\deg h + \deg g_a - n \text{ and } 2\deg f_a + \deg h_b - n. \tag{4.8}$$

Hence, taking into account the first expression (2.9) (which follows from (3.2)), the first expression (3.7), (3.8), and (4.1), we see that deg h_a exceeds neither n_a nor $n_a - 2m_a$, whence the first inequality (4.6) follows due to $m_a \geq 0$. The second inequality (4.6) is obtained in a similar way by starting from (4.3) rather than (4.2), and using the second expression (3.7) rather than the first.

3. The function Δ_0 defined by (4.4) has no pole at finite values of p. Indeed, according to (4.4), such poles would have to be located in the left half-plane; on the other hand, (3.13) shows that Δ_0, which has already been shown to be equal to Δ_a, is an even function of p and can thus not have poles in Re $p < 0$ without having poles in Re $p > 0$.

4. As can be concluded from the second expression (3.5), (3.8), (4.1) and (4.6), Δ_0 has no pole at $p = \infty$ either. Hence, it is a constant, which is necessarily real since (4.4) only involves real polynomials.

Theorem 1c

Every solution of problem 2b is a solution of problem 1.

Proof

1. According to Theorem 1b, Δ_0 is a constant. Thus, due to (4.7), (3.12) holds for all p as soon as (4.5) is fulfilled. We then obtain by (4.4),

$$g = g_a g_b + \sigma_a h_{a*} h_b. \qquad (4.9)$$

2. As $\Delta_a = \Delta_b = 1$, we obtain from (3.13) and (3.14), for $p = j\omega$,

$$|g_a|^2 = |h_a|^2 + |f_a|^2, \quad |g_b|^2 = |h_b|^2 + |f_b|^2. \qquad (4.10)$$

Hence, since due to (3.3), f_a and f_b cannot vanish identically, the same is also true for g_a and g_b. We also derive from (4.10),

$$\deg h_a \leq \deg g_a, \ \deg h_b \leq \deg g_b,$$

and thus, from (4.9)

$$n = \deg g \leq \deg g_a + \deg g_b, \tag{4.11}$$

whence (3.11) follows if we take into account (3.8), the first inequality (4.1), and the second inequality (4.6). (4.11) can now be written more precisely,

$$n = \deg g = \deg g_a + \deg g_b. \tag{4.12}$$

3. In order to show that g_a and g_b are Hurwitz polynomials, we consider the function Φ defined by

$$\Phi = g_a g_b / g \tag{4.13}$$

which, according to (4.9), is also given by

$$1/\Phi = 1 + \sigma_a \frac{h_{a*}}{g_a} \frac{h_b}{g_b} \tag{4.14}$$

As g is Hurwitz, all possible common zeros of g and $g_a g_b$ are necessarily located inside of the left half-plane. Due to (4.13), all further zeros of $g_a g_b$ are necessarily zeros of Φ. Consequently, we have to show that Φ has no zero for Re $p \geq 0$.

In order to show this[4)], we observe first that Φ has neither pole nor zero on the closed $j\omega$-axis (i.e., neither for $p = j\omega$ nor for $p = \infty$) and that it is regular inside of the right half-plane. Indeed, due to (4.13), Φ is regular in Re $p \geq 0$ (g is Hurwitz) and has no pole or zero at $p = \infty$ (see (4.12)); furthermore, (4.10) shows that we have, for $p = j\omega$,

$$|h_a| \leq |g_a|, \ |h_b| \leq |g_b|, \tag{4.15}$$

whence we conclude from (4.14) that $1/\Phi$ has no pole, and

[4)] The proof used hereafter, which is somewhat modified with respect to the one used in [18], follows a suggestion by G. Verkroost (private communication).

Φ thus no zero, on the $j\omega$-axis.

From (4.14) and (4.15), we also conclude that $\mathrm{Re}(1/\Phi) \geq 0$, and thus $\mathrm{Re}\ \Phi \geq 0$, for $p = j\omega$. This property and the regularity in the closed right half-plane show that Φ is a positive function and that it thus has no zero inside of the right half-plane.

5. GENERAL PROPERTIES OF THE SOLUTIONS OF PROBLEM 2

5.1 *Derivation of further solutions of problem 2a*

We want to show how a first solution of problem 2a can be used to derive further ones. In order to do this, we shall designate the more general solution by the notation used so far, i.e., by g_a, h_b, h_a, and g_b, while the solution obtained first shall be designated by g_a', h_b', h_a', and g_b'. These latter polynomials thus satisfy in particular the relations

$$\deg g_a' \leq n_a, \ \deg h_b' \leq n_b, \tag{5.1}$$

$$\deg h_a' \leq n_a, \ \deg g_b' \leq n_b, \tag{5.2}$$

$$h_b' = \sigma_a(h_a' - gh_a')/f_a f_{a*}, \tag{5.3}$$

$$g_{b*}' = (g_* g_a' - h_* h_a')/f_a f_{a*}, \tag{5.4}$$

where we have already partially made use of Theorem 1b by writing (5.3) and (5.4) in a form corresponding to (3.9) and (3.10) rather than (4.2) and (4.3). The result to be given can best be stated in form of the two theorems 2a and 2b which will now be discussed.

Theorem 2a

Let the polynomials g_a', h_b', h_a', and g_b' form a solution of problem 2a, and let α and β be two arbitrary real constants. In this case, the polynomials g_a, h_b, h_a, and g_b defined by

$$g_a = \alpha g_a' + \beta\sigma_a h_{a*}', \ h_b = \alpha h_b' - \beta g_b' \tag{5.5}$$

$$h_a = \alpha h'_a + \beta\sigma_a g'_{a*},\quad g_b = \alpha g'_b - \beta h'_b, \tag{5.6}$$

also form a solution of problem 2a, and we have

$$\Delta_0 = (\alpha^2 - \beta^2)\,\Delta'_0 \tag{5.7}$$

where Δ_0 is defined by (4.4), and Δ'_0 by

$$\Delta'_0 = (g'_a g'_b + \sigma_a h'_{a*}\, h_b)/g. \tag{5.8}$$

Proof

1. By means of (5.1) and (5.2), one checks that the polynomials g_a and h_b defined by (5.5) satisfy (4.1).

2. Substituting (5.5) and (5.6) in (3.9) and (3.10), one checks easily that these latter two equations are automatically fulfilled due to (5.3) and (5.4).

3. Substitution of (5.5) and (5.6) in (4.4) leads to (5.7). Q.E.D.

We shall now assume that we are dealing with a first solution which not only satisfies problem 2a but also problem 2b, i.e., that we also have $\Delta'_0 = 1$. The two lossless two-ports to which this first solution corresponds shall be designated by N'_a and N'_b, respectively. This allows us to state Theorem 2b in the following way:

Theorem 2b

If the polynomials g'_a, h'_b, h'_a, and g'_b are a solution of problem 2b and if the real constants α and β are such that

$$\alpha^2 - \beta^2 = 1, \tag{5.9}$$

the new polynomials defined by (5.5) and (5.6) also form a solution of problem 2b. Physically, this new solution corresponds to inserting between N'_a and N'_b two constant two-ports in cascade having transfer matrices

$$T_1 = \begin{pmatrix} \alpha & \beta \\ \beta & \alpha \end{pmatrix} \text{ and } T_2 = \begin{pmatrix} \alpha & -\beta \\ -\beta & \alpha \end{pmatrix} \tag{5.10}$$

respectively. These two constant two-ports correspond to ideal transformers of transformation ratios $(\alpha + \beta)/1$ and $1/(\alpha + \beta) = (\alpha - \beta)/1$, respectively (Fig. 5.1)[5].

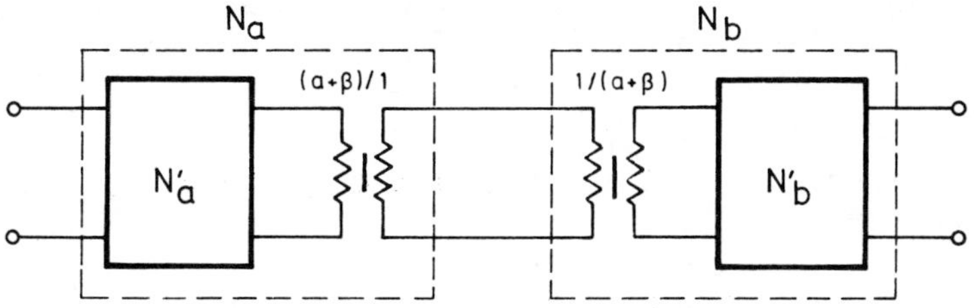

Fig. 5.1 Derivation of a new solution (N_a, N_b) of the factorization problem by inserting two ideal transformers of opposite transformation ratios between the two-ports N_a' and N_b' originally obtained.

Proof

1. From (5.7) and (5.9), one obtains $\Delta_0 = 1$ for $\Delta_0' = 1$.

2. From (5.9) and (5.10), one obtains $T_2 = T_1^{-1}$. Let us designate the transfer matrices corresponding to N_a' and N_b' by T_a' and T_b', respectively. We have

$$T = T_a'T_b' = T_a'T_1T_2T_b' = T_aT_b,$$

where we have written $T_a = T_a'T_1$ and $T_b = T_2T_b'$. As can easily be checked, these matrices T_a and T_b indeed correspond to the new polynomials defined by (5.5) and (5.6).

3. For the interpretation of (5.10) in terms of ideal transformers, see subsection 9.2.

[5] As can be checked, inserting two constant lossless nonreciprocal two-ports (i.e. two gyrators) would amount to changing σ_a to $-\sigma_a$. The new solution obtained this way would thus not correspond to the same given quantities.

5.2. GENERAL PROPERTIES OF THE SOLUTIONS OF PROBLEM 2a

Clearly, problem 2a always has the *trivial* solution $g_a \equiv h_b \equiv 0$, for which we also have, due to (4.2) and (4.3), $h_a \equiv g_b \equiv 0$. All other solutions shall be called *nontrivial*. For these, the following lemmas hold.

Lemma 1

For every nontrivial solution of problem 2a, for which $\Delta_0 = 0$, we have

1. $$\deg g_a + \deg g_b \geq n - 1 \tag{5.11}$$

2. g_a and g_b cannot both be zero for $p = 0$.

Proof

1. Note first that we obtain from (3.14) and (4.7), for $\Delta_0 = 0$,

$$g_b g_{b*} = h_b h_{b*}. \tag{5.12}$$

This shows already that we cannot have $g_b \equiv 0$ since otherwise, we would also have $h_b \equiv 0$, and thus, by (4.3), $g_a \equiv 0$.

2. Consider again the function $\Phi = g_a g_b / g$ already defined by (4.13). This time, however, we may not make use of (4.9), which is valid only for $\Delta_0 = 1$, but we can solve (4.3) for g_a and substitute the result in the defining expression of Φ. This gives

$$\Phi = \frac{g_b g_{b*}}{f_b f_{b*}} \left(1 - \sigma_a \frac{h_*}{g} \frac{h_b}{g_{b*}}\right) \tag{5.13}$$

Due to (3.8), the first equation (4.1), and the second equation (4.6), we certainly have

$$\deg g_a + \deg g_b \leq n.$$

Thus, as g is Hurwitz, Φ has no pole in the entire closed right half-plane (i.e., neither in Re $p \geq 0$ nor at $p = \infty$).

Furthermore, for $p = j\omega$, we have

$$g_b g_{b*} \geq 0, \quad f_b f_{b*} \geq 0,$$

and, due to (3.2) and (5.12),

$$|h_*/g| \leq 1, \quad |h_b/g_{b*}| = 1.$$

Hence, (5.13) shows that Re $\Phi \geq 0$ for $p = j\omega$. Consequently Φ is again a positive function. It can thus at most have simple zeros at $p = \infty$ and $p = 0$. This proves the two statements of Lemma 1.

Theorem 3

Problem 2a does not have any nontrivial solutions for which we have simultaneously

$$\deg g_a \leq n_a - 1, \quad \deg h_b \leq n_b - 1, \tag{5.14}$$

or simultaneously

$$g_a = h_b = 0 \text{ for } p = 0. \tag{5.15}$$

Proof

1. Assume that (5.14) or (5.15) is fulfilled. In the former case, (3.8), (4.4) and (4.6) show that we would have $\Delta_0 = 0$ for $p = \infty$, while in the latter case, the conclusion $\Delta_0 = 0$ for $p = 0$ would immediately be obtained from (4.4). Thus, as due to Theorem 1b, Δ_0 is a constant, we would in both cases have $\Delta_0 = 0$ for all p, so that Lemma 1 is applicable.

2. By the reasoning used in step 2 of the proof of Theorem 1b, one can check that due to (5.14), the second inequality (4.6) could now be replaced by $\deg g_b \leq n_b - 1$ so that we would have deg. $g_a + \deg g_b \leq n - 2$. This, however, would contradict statement 1 of Lemma 1.

3. Due to (4.3), (5.15) would imply $g_b = 0$ for $p = 0$. Thus statement 2 of Lemma 1 would be contradicted.

Theorem 4

1. If g and h are relatively prime, or, more generally, if g, h, and $f_a f_{a*}$ are relatively prime, the requirement in problem 2a that g_b defined by (4.3) is a polynomial, is superfluous

2. If g and h_* are relatively prime, the requirement in problem 2a that h_a defined by (4.2) is a polynomial, is superfluous.

Proof

1. Eliminating f and f_a between (3.2), (3.3), and (4.2), we obtain

$$g(\sigma_a g_* h_b + f_b f_{b*} h_a) = h(\sigma_a h_* h_b + f_b f_{b*} g_a). \quad (5.16)$$

Hence, if g and h are relatively prime, g must divide the expression between parentheses in the right-hand side of (5.16), i.e., g_b defined by (4.3) then is automatically a polynomial. The same conclusion can be drawn from Lemma 2, which will be discussed below, if g, h, and $f_a f_{a*}$ are relatively prime.

2. Eliminating f and f_b between (3.2), (3.3), and (4.3), we obtain

$$g(g_a g_* - f_a f_{a*} g_{b*}) = h_*(g_a h - \sigma_a f_a f_{a*} h_b). \quad (5.17)$$

Hence, if g and h_* are relatively prime, g must divide the expression between parentheses in the right-hand side of (5.17), i.e., h_b defined by (4.2) then is automatically a polynomial. The same conclusion can be drawn from Lemma 3, which will be discussed below, if g, h_*, and $f_b f_{b*}$ are relatively prime.

Lemma 2

Let c_a be the g.c.d. (greatest common divisor) or the polynomials g, h, and $f_a f_{a*}$, and let g_a, h_a, and h_b be polynomials satisfying (4.2). Then, the function y_b defined by

$$y_b = c_a(\sigma_a h_* h_b + f_b f_{b*} g_a)/g \quad (5.18)$$

is a polynomial.

Proof

1. The g.c.d. of g and h can be written as $c_a d_b$, where d_b is a polynomial. Thus, $g_0 = g/c_a d_b$ and $h_0 = h/c_a d_b$ are polynomials, and d_b is relatively prime with the polynomial $\phi_a = f_a f_{a*}/c_a$. Hence, due to (3.2) and (3.3), $\psi_b = f_b f_{b*}/d_b$ is also a polynomial; furthermore, the expression

$$h_b = d_b(h_0 g_a - g_0 h_a)/\sigma_a \phi_a,$$

derived from (4.2), shows that $h_{b0} = h_b/d_b$ is a polynomial.

2. Using these results, one obtains from (5.18),

$$y_b = (\sigma_a h_* h_{b0} + \psi_b g_a)g_0, \tag{5.19}$$

and from (5.16)

$$g_0(\sigma_a g_* h_{b0} + \phi_b h_a) = h_0(\sigma_a h_* h_{b0} + \phi_b g_a). \tag{5.20}$$

This shows that g_0 must divide the term between parenthesis in the right-hand side of (5.20), i.e., the numerator in the right-hand side of (5.19).

Lemma 3

Let c_b be the g.c.d. of g, h_*, and $f_b f_{b*}$, and let g_a, g_b, and h_b be polynomials satisfying (4.3). Then, the function y_a defined by

$$y_a = c_b(hg_a - \sigma_a f_a f_{a*} h_b)/g \tag{5.21}$$

is a polynomial.

Proof

The proof is very similar to the one of Lemma 2. We now can write the g.c.d. of g and h_* as $c_b d_a$; $g/c_b d_a$ and $h_*/c_b d_a$ are polynomials while d_a is relatively prime with $f_b f_{b*}/c_b$, and $f_a f_{a*}/d_a$ as well as g_a/d_a turn out to be polynomials. The proof is completed by deriving from

(5.17) an expression replacing (5.20).

6. SOLUTION OF PROBLEM 2

6.1 *Solution of problem 2a*

We satisfy the conditions (4.1) by writing g_a as a polynomial of degree n_a with arbitrary coefficients A_i, $i = 1, 2, \ldots, n_a$, and h_b as polynomial of degree n_b with arbitrary coefficients B_i, $i = 1, 2, \ldots, n_b$, i.e.,

$$g_a = A_{n_a} p^{n_a} + A_{n_a-1} p^{n_a-1} + \ldots + A_0, \qquad (6.1)$$

$$h_b = B_{n_b} p^{n_b} + B_{n_b-1} p^{n_b-1} + \ldots + B_0. \qquad (6.2)$$

The total number of coefficients to be determined is thus $n_a + 1 + n_b + 1 = n + 2$.

Assume first that g, h, and $f_a f_{a*}$ are relatively prime. Hence, due to statement 1 of Theorem 4, we only have to take into account equation (4.2). We first substitute (6.1) and (6.2) in the expression

$$hg_a - \sigma_a f_a f_{a*} h_b$$

and divide the result formally by g. This gives as remainder a polynomial r of degree not larger than n - 1, whose coefficients are linear combinations of the A_i and B_i. The condition that h_a defined by (4.2) is a polynomial is then equivalent to the condition that r is identically zero, and this in turn leads to a set of at most n linear homogeneous equations, henceforth called *fundamental equations*, in the n + 2 undetermined coefficients A_i and B_i.

If g, h, and $f_a f_{a*}$ are not relatively prime but have a g.c.d., c_a, of degree $\deg c_a > 0$, (4.2) can be written

$$h_a = (h_1 g_a - \sigma_a \phi_a h_b)/g_1$$

where $h_1 = h/c_a$, $\phi_a = f_a f_{a*}/c_a$, and $g_1 = g/c_a$. Hence the requirement for h_a to be a polynomial now gives only at most n - deg c_a conditions; all these are again linear homogeneous equations in the A_i and B_i. On the other hand, due to (4.3) and equation (5.18) of Lemma 2, we now have

$$g_{b*} = y_b/c_a$$

where y_b is a polynomial. Thus, the requirement for g_b to be a polynomial also amounts to at most deg c_a conditions.

These latter conditions can be obtained explicitly as linear homogeneous equations of the A_i and B_i by first substituting (6.1) and (6.2) in the expression between parentheses in (5.18) and dividing the result formally by $g_1 = g/c_a$. From this division, only the quotient, which we have designated by y_b, has to be retained since the remainder will automatically be zero once the aforementioned n - deg c_a conditions are satisfied; the coefficients of the quotient y_b are linear combinations of the A_i and B_i. The required additional conditions are then obtained by dividing in turn the quotient y_b formally by c_a and writing that the resulting remainder must vanish. Altogether, we thus again obtain a system of at most n - deg c_a + deg c_a = n linear homogeneous fundamental equations in the n + 2 unknowns (also see appendix).

It should be clear that equation (4.3) could be used as a basis for deriving the fundamental equations just as well as (4.2). According to Lemma 3, (4.3) alone suffices if g, h_*, and $f_b f_{b*}$ are relatively prime, while (4.3) together with

$$h_a = y_a/c_b,$$

which can be derived from (4.2) and (5.21), allows us to attain our goal if c_b is the g.c.d. of g, h_*, and $f_b f_{b*}$.

The result obtained so far will now be made more precise by means of the following

Theorem 5

The matrix of the set of fundamental equations is of rank $n = \deg g$; in particular, its submatrices of order n obtained by deleting either the two columns corresponding to A_{n_a} and B_{n_b} or those corresponding to A_0 and B_0, are nonsingular.

Proof

Assume that the theorem is not true. If the submatrix obtained by deleting the columns corresponding to A_{n_a} and B_{n_b} is singular, the fundamental equations have a nontrivial solution for which $A_{n_a} = B_{n_b} = 0$, i.e., for which g_a and h_b satisfy (5.14). On the other hand, if the submatrix obtained by deleting the columns corresponding to A_0 and B_0 is singular, the fundamental equations have a nontrivial solution for which $A_0 = B_0 = 0$, i.e., for which g_a and h_b satisfy (5.15). Both these conclusions are excluded by Theorem 3.

Corollary

For every arbitrary A_{n_a} and B_{n_b} or for every arbitrary A_0 and B_0, the fundamental equations have a unique solution.

6.2 *Solution of problem 2b*

Hereafter, we shall sometimes make use of certain of the coefficients of g, h, f_a, and f_b. These are defined by

$$g = G_n p^n + G_{n-1} p^{n-1} + \ldots + G_0 \tag{6.3}$$

$$h = H_n p^n + H_{n-1} p^{n-1} + \ldots + H_0 \tag{6.4}$$

$$f_a = M_{n_a} p^{n_a} + M_{n_a-1} p^{n_a-1} + \ldots + M_0 \tag{6.5}$$

$$f_b = N_{n_b} p^{n_b} + N_{n_b-1} p^{n_b-1} + \ldots + N_0 \tag{6.6}$$

Due to our normalization, we have $M_{n_a} = 1$ as soon as $\deg f_a = n_a$; if $M_{n_a} = 0$, the first coefficient of f_a which is not zero is equal to 1. Similarly, we have $M_{n_b} = 1$ as soon as $\deg f_b = n_b$; if $M_{n_b} = 0$, the first coefficient of f_b which is not zero is equal to 1.

Theorem 6

Problem 2a always has solutions for which $\Delta_0 \neq 0$.

Proof

Substituting (4.2) and (4.3) in (4.4), we obtain, using (3.6),

$$\Delta_0 = (\sigma_a g_a h_{b*} h + \sigma_a g_{a*} h_b h_* + g_a g_{a*} f_b f_{b*} - h_b h_{b*} f_a f_{a*})/gg_* \tag{6.7}$$

where the expression between paranthesis is a polynomial of degree $\leq 2n$. By Theorem 1b, we already know that Δ_0 is a constant. We can thus calculate it by taking the limit of expression (6.7) for $p \to 0$; this gives, using (6.1) - (6.6),

$$\Delta_0 = (2\sigma_a A_0 B_0 H_0 + A_0^2 N_0^2 - B_0^2 M_0^2)/G^2 \tag{6.8}$$

Now, according to the corollary to Theorem 5, A_0 and B_0 may be chosen arbitrarily. Thus, according to (6.8), Δ_0 can always be made different from zero except if we would have simultaneously $H_0 = M_0 = N_0 = 0$. This, however, is impossible, since by (3.3), $M_0 = 0$ and/or $N_0 = 0$ implies $f(0) = 0$, and thus, by (3.2), $H_0 \neq 0$ (as g is Hurwitz, we certainly have $G_0 \neq 0$). Q.E.D.

Note that the same conclusion could be reached if we take the limit of (6.7) for $p \to \infty$, in which case we find[6)]

---

6) Note that results for $p = \infty$ can also be derived from the corresponding results for $p = 0$, and vice versa, through simply replacing p by 1/p in the original factor-

$$\Delta_0 = [(-1)^{n_a} 2\sigma_a A_{n_a} B_{n_b} H_n + \nu A_{n_a}^2 - \mu B_{n_b}^2]/G_n^2$$

where

$$\mu = 0 \text{ if } m_a > 0, \quad \mu = 1 \text{ if } m_a = 0,$$

$$\nu = 0 \text{ if } m_b > 0, \quad \nu = 1 \text{ if } m_b = 0. \text{ Q.E.D.}$$

The *solution* of problem 2b is now immediate. Suppose, indeed, that we start from a first solution g_a', h_b', with corresponding polynomials h_a', g_b', for which $\Delta_0' \neq 0$, Δ_0' being defined by (5.8). By means of Theorem 2a, it is then always possible to choose a new solution in such a way that $\alpha^2 - \beta^2 = 1/\Delta_0'$. By (5.7) we then have $\Delta_0 = 1$.

6.3 *Determination of all solutions of problem 2*

As is well-known, a system of n linear homogeneous algebraic equations in n + k unknowns, whose matrix has rank n, has precisely k linearly independent solutions. All other solutions can be obtained as linear combinations of a set of k independent ones.

If we apply this to our present situation, we conclude from Theorem 5 that the fundamental equations have precisely two independent solutions. Each of these is, in principle, a (n + 2)-dimensional vector in the coefficients of g_a and h_b. For our present purpose, however, it is more advantageous to consider instead the corresponding vectors expressed in terms of the solution polynomials themselves. It is then obvious that for problem 2a we can find two linearly independent solutions

---

ization problem; such a substitution allows us to replace the transfer matrices T, T_a, and T_b by corresponding transfer matrices, which again correspond to lossless two-ports, but for which the behavior at p = 0 corresponds to the original behavior at p = ∞, and the behavior at p = ∞ to the behavior at p = 0. This remark can, of course, also be used in all past and future situations, in which the behavior at p = 0 is considered separately from the behavior at p = ∞.

$$\begin{pmatrix} g_a' \\ h_b' \end{pmatrix} \quad \text{and} \quad \begin{pmatrix} g_a'' \\ h_b'' \end{pmatrix} \tag{6.9}$$

and that the general solution can be written as a linear combination of these, i.e.

$$\begin{pmatrix} g_a \\ h_b \end{pmatrix} = \alpha \begin{pmatrix} g_a' \\ h_b' \end{pmatrix} + \beta \begin{pmatrix} g_a'' \\ g_b'' \end{pmatrix} \tag{6.10}$$

where α and β are real constants. This leads us to the following.

Theorem 7

If $\Delta_0' \neq 0$, the methods described in Theorems 2a and 2b furnish all the solutions of problems 2a and 2b respectively.

Proof

1. If we compare (5.5) with (6.10), we see immediately that (5.5) can be written in the form (6.10) if we define g_a'' and h_b'' by

$$g_a'' = \sigma_a h_{a*}', \quad h_b'' = -g_b'. \tag{6.11}$$

As there only exist two independent solution vectors, it is thus sufficient to prove that the second vector (6.9) is independent of the first one if g_a'' and h_b'' are defined by (6.11), and if furthermore $\Delta_0' \neq 0$, where Δ_0' is defined by (5.8).

In order to show this, let us assume e.g. that these two vectors are not independent, i.e., that there exist two constants α and β, which are not both zero, such that the relations

$$\alpha g_a' + \beta g_a'' = 0 \quad \text{and} \quad \alpha h_b' + \beta_b'' = 0$$

are both identically fulfilled. Using (5.8), one easily checks that this always implies $\Delta_0' = 0$.

6.4 *Additional properties*

We shall now also make use of some of the coefficients of h_a and g_b. These are defined by

$$h_a = C_{n_a}p^{n_a} + C_{n_a-1}p^{n_a-1} + \ldots + C_0 \tag{6.12}$$

$$g_b = D_{n_b}p^{n_b} + D_{n_b-1}p^{n_b-1} + \ldots + C_0 \tag{6.13}$$

Theorem 8

1. If $m_a = 0$, i.e. if deg $f_a = n_a$, the factorization problem has a solution for which $A_{n_a} = 1$ and $C_{n_a} = 0$.

2. If $f_a(0) \neq 0$, the factorization problem has a solution for which $A_0 = M_0$ and $C_0 = 0$.

Proof

1. Assume that we have obtained a first solution of the factorization problem. Further solutions can be obtained by the method of subsection 5.1. We shall add primes (') in order to distinguish quantities related to g_a' and h_a' from the corresponding quantities related to g_a and h_a. Due to the first equation (5.5), we can write

$$A_{n_a} = \alpha A_{n_a}' + \beta(-1)^{n_a}\sigma_a C_{n_a}' \tag{6.14}$$

$$A_0 = \alpha A_0' + \beta\sigma_a C_0' \tag{6.15}$$

2. If $m_a = 0$, we have $M_{n_a} = 1$ (see the discussion following (6.6)). Due to $\Delta_a' = \Delta_a = 1$ (cf. (3.13)), together with (6.1), (6.5) and (6.12) we thus have

$$A_{n_a}'^2 - C_{n_a}'^2 = 1, \; A_{n_a}^2 - C_{n_a}^2 = 1.$$

Hence, due to (6.14) we have $A_{n_a} = 1$, and thus $C_{n_a} = 0$, for

$$\alpha = A_{n_a}', \; \beta = -(-1)^{n_a}\sigma_a C_{n_a}'$$

for which (5.9) is also satisfied. Note that by applying the method of subsection 5.1 once again to the result just obtained, one checks easily that the solution with $A_{n_a} = 1$ is unique.

3. Due to $\Delta_a' = \Delta_a = 1$ together with (6.1), (6.5), and (6.12), we also have

$$A_0'^2 - C_0'^2 = M_0^2, \; A_0^2 = C_0^2 = M_0^2,$$

where $M_0 \neq 0$ if $f_a(0) \neq 0$. Hence, due to (6.15), we have $A_0 = M_0$, and thus $C_0 = 0$, if

$$\alpha = A_0'/M_0, \; \beta = -\sigma_a C_0'/M_0,$$

for which (5.9) is again satisfied. Here, too, the solution for which $A_0 = M_0$ is unique.

Theorem 9

1. If $f_a(0) = 0$ and/or $f_b(0) = 0$, we have $H_0/G_0 = \pm 1$. Similarly, if $m_a > 0$ and/or $m_b > 0$, we have $H_n/G_n = \pm 1$.

2. The following relations hold:

$$\text{If } f_a(0) = 0, \qquad \text{then } C_0/A_0 = H_0/G_0 \tag{6.16}$$

$$\text{If } f_b(0) = 0, \qquad \text{then } \frac{B_0}{D_0}\frac{H_0}{G_0} = \sigma_a \tag{6.17}$$

$$\text{If } m_a > 0, \qquad \text{then } C_{n_a}/A_{n_a} = H_n/G_n \tag{6.18}$$

$$\text{If } m_b > 0, \qquad \text{then } \frac{B_{n_b}}{D_{n_b}}\frac{H_n}{G_n} = (-1)^{n_a}\sigma_a \tag{6.19}$$

Proof

The statement 1 follows immediately from (3.2) - (3.5) together with (6.3) and (6.4). Consider thus statement 2. If we compute (4.2) for $p = 0$ and assume $f_a(0) = 0$, we find, using (6.1) - (6.4), $C_0 = H_0A_0/G_0$. But $A_0 \neq 0$

since g_a is Hurwitz, whence (6.16) follows. On the other hand, (6.18) is obtained by computing (4.2) for $p = \infty$. Indeed, $A_{n_a} \neq 0$ since deg $g_a = n_a$, and, for $m_a > 0$, the degree of $f_a f_{a*} h_b$ is smaller than the degree of hg_a. In a similar way, (6.17) and (6.19) can be obtained from (4.3).

7. A SIMPLE PROCEDURE FOR ESTABLISING THE FUNDAMENTAL EQUATIONS

For the method to be described here, we shall omit considering equation (4.3). As we know from Theorem 4 as well as the appendix, this is almost always permitted in practical situations. It should be obvious that a similar method is applicable if we omit considering equation (4.2) rather than (4.3).

We first divide g into $p^i h$ and $p^i f_a f_{a*}$, with $i = 0, 1, 2, \ldots, n_a$ and $i = 0, 1, 2, \ldots, n_b$, respectively. The result can be written in the form

$$\begin{aligned} h &= x_0 g + u_0, & \sigma_a f_a f_{a*} &= y_0 g + v_0, \\ ph &= x_1 g + u_1, & \sigma_a p f_a f_{a*} &= y_1 g + v_1 \\ &\ldots & &\ldots \\ p^{n_a} h &= x_{n_a} g + u_{n_a}, & \sigma_a p^{n_b} f_a f_{a*} &= y_{n_b} g + v_{n_b}, \end{aligned} \tag{7.1}$$

where the x_i, y_i, u_i, and v_i are polynomials, with

$$\deg u_i < n, \quad \deg v_i < n. \tag{7.2}$$

If we now substitute (6.1) and (6.2) in (4.2), and take into account (7.1), we find that h_a is a polynomial if the condition

$$A_0 u_0 + A_1 u_1 + \ldots + A_{n_a} u_{n_a} = B_0 v_0 + B_1 v_1 + \ldots + B_{n_b} v_{n_b} \tag{7.3}$$

is identically fulfilled. This gives immediately the required n linear homogeneous equations in the n + 2 unknowns A_i and B_i, while h_a then follows from

$$h_a = A_0x_0+A_1x_1+\dots+A_{n_a}x_{n_a}-B_0y_0-B_1y_1-\dots-B_{n_b}y_{n_b}. \tag{7.4}$$

In order to obtain for g_b an expression similar to (7.4), we still have to divide g into p^ih_* and $p^if_bf_{b*}$, with i = 0, 1, 2, ..., n_b and i = 0, 1, 2, ..., n_a, respectively. We have,

$$\sigma_ap^ih_* = x_i'g + u_i', \quad p^if_bf_{b*} = y_i'g + v_i', \tag{7.5}$$

with

$$\deg u_i' < n, \quad \deg v_i' < n. \tag{7.6}$$

with these definitions, (4.3) gives

$$g_{b*} = B_0x_0'+B_1x_1'+\dots+B_{n_b}x_{n_b}'+A_0y_0'+A_1y_1'+\dots+A_{n_a}y_{n_a}'. \tag{7.7}$$

From the procedure just described, one should not conclude that it is necessary to carry out individually all the divisions defined by (7.1) and (7.2) as well as by (7.5) and (7.6). Assume, e.g., that we start by determining the first equation in the first column of (7.1). Due to deg h ≤ deg g, x_0 is a constant. We then divide g into pu_0; this gives

$$pu_0 = X_1g + u_1$$

where X_1 is again a constant, and we have

$$x_1 = px_0 + X_1.$$

This procedure can now be continued in the same way until x_{n_a} and u_{n_a} have been found. The same method can also be applied to the equations in the second column of (7.1) as well as those of (7.5), although either y_0 or y_0' may not be a constant.

We still wish to draw attention to the following fact: Any of the elementary sections which appear in the canonic cascade connection of Darlington [1] and Piloty [11], and which will be discussed in section 9, can be obtained by at most two computing steps. Indeed, it is always possible to split a two-port first into a cascade of two-ports N_a and N_b, and then split, e.g., N_b into a cascade of N_c and N_d, where N_c is the section to be determined (Fig. 7.1). This reduces the possibility of accumulation of errors in the calculation of more complicated structures.

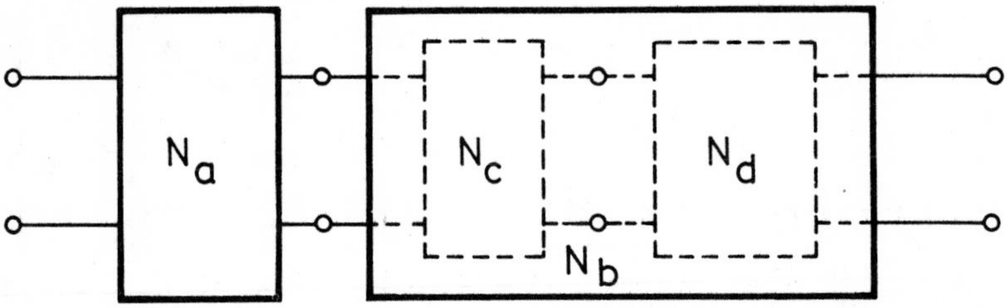

Fig. 7.1 Determination of any elementary section N_c in only two computing steps.

8. PROBLEMS EQUIVALENT TO PROBLEM 2

Instead of problem 2, various other equivalent problems can be used. Of these, we want to mention here only two, which we designate as problem 3 and 4, respectively. We shall assume as given the same quantities as for problems 1 and 2.

Problem 3

Find two real polynomials g_a and h_a such that

1. $$\deg g_a = n_a, \quad \deg h_a \leq n_a,$$

2. the functions g_b and h_b defined by (3.9) and (3.10), respectively, are polynomials of degree

$$\deg g_b \le n_b, \quad \deg h_b \le n_b.$$

3. $\Delta_a = 1$ for at least one value of p, where Δ_a is defined by (3.13).

Problem 4

Find two polynomials g_a and g_b such that

1. $$\deg g_a \le n_a, \quad \deg g_b \le n_b,$$

2. the functions h_a and h_b defined by

$$h_a = (g_* g_a - f_a f_{a*} g_{b*})/h_*$$

$$h_b = \sigma_a(g g_{b*} - f_b f_{b*} g_a)/h_*$$

are polynomials of degree

$$\deg h_a \le n_a, \quad \deg h_b \le n_b;$$

3. $\Delta_0 = 1$ for at least one value of p, where Δ_0 is defined by

$$\Delta_0 = (h_a g_b + \sigma_a g_{a*} h_b)/h.$$

The equivalence of these problems with the earlier problem 2 can be checked without great difficulty. Both problems 3 and 4, however, are less advantageous for establishing the complete theory than problem 2. For problem 3, this is due to the fact that the number of transmission zeros at infinity, m_a, can only be fixed by the equality $\deg g_a = n_a$ and not by the inequalities (4.1). For problem 4, it is due to the fact that deg h may be smaller than n = deg g.

Nevertheless, for practical computations, problem 3 may be simpler. Indeed, $f_a f_{a*}$ may be of rather low degree so that a division by this polynomial may be simpler than a division by g. Furthermore, a method similar to the one described in section 7 only requires dividing $f_a f_{a*}$ into g and h, even for the determination of g_b and h_b.

9. CASCADE SYNTHESIS BY ELEMENTARY SECTION

If we want to synthesize a given two-port N as a cascade of elementary sections, the method of factorization may be applied successively until the resulting individual sections are of the lowest possible degree[7] compatible with the general requirements imposed. As all polynomials must be real, this leads us to consider hereafter sections of degree 0, 1, and 2 in the nonreciprocal case. For the synthesis of reciprocal two-ports by means of reciprocal sections, we must in addition consider the case where the polynomials f and g are both of degree 4, with f being even.

In the following subsections, we always use the notation defined for N rather than for N_a or N_b. This must be taken into account in particular when we refer e.g., to Theorem 8, which has been stated with reference to N_a.

For each of the elementary sections to be discussed in subsections 9.1 to 9.4, a certain number of inequalities are derived in the beginning. Although this is never mentioned explicitly, it can easily be checked by means of these inequalities that the elements appearing in the corresponding elementary sections always turn out to be positive. Note that these inequalities essentially hold due to the fact that g is Hurwitz.

9.1 *Impedance, admittance, and chain matrices*

The normalized impedance matrix Z and the normalized admittance matrix Y of a two-port are known to be related to the scattering matrix S by

$$Z = 2(I - S)^{-1} - I, \quad Y = 2(I + S)^{-1} - I$$

where I is the unit matrix. This way, one obtains from (2.3) the expressions given below for Z, Y, and the chain matrix K. In these expressions, which are different de-

---

[7] The degree of a lossless two-port can be defined as the degree of its canonic polynomial g.

pending on whether $\sigma = +1$ or -1, the subscripts e and o designate the even and odd parts of the corresponding polynomials, i.e.,

$$g_e = (g + g_*)/2, \quad h_e = h + h_*)/2,$$

$$g_o = (g - g_*)/2, \quad h_o = h - h_*)/2.$$

We have,

for $\sigma = 1$

$$Z = \frac{1}{g_o - h_o}\begin{pmatrix} g_e + h_e & f_* \\ f & g_e - h_e \end{pmatrix}$$

$$Y = \frac{1}{g_o + h_o}\begin{pmatrix} g_e - h_e & -f_* \\ -f & g_e + h_e \end{pmatrix}$$

$$K = \frac{1}{f}\begin{pmatrix} g_e + h_e & g_o + h_o \\ g_o - h_o & g_e - g_e \end{pmatrix}$$

for $\sigma = -1$

$$Z = \frac{1}{g_e - h_e}\begin{pmatrix} g_o + h_o & -f_* \\ f & g_o - h_o \end{pmatrix}$$

$$Y = \frac{1}{g_e + h_e}\begin{pmatrix} g_o - h_o & f_* \\ -f & g_o + h_o \end{pmatrix}$$

$$K = \frac{1}{f}\begin{pmatrix} g_o + h_o & g_e + h_e \\ g_e - h_e & g_o - h_o \end{pmatrix}$$

9.2 *Sections of degree zero*

If T is constant, we can write $f = 1$, $g = \alpha$, $h = \beta$, where α and β are constants. Due to (2.5), we have

$$\alpha^2 - \beta^2 = 1$$

and thus, by (2.4), $\det T = \sigma = \pm 1$. We have, for $\sigma = 1$,

$$K = \begin{pmatrix} \alpha - \beta & 0 \\ 0 & \alpha - \beta \end{pmatrix},$$

which corresponds to a transformer (Fig. 9.1a) of ratio $k = \alpha + \beta = 1/(\alpha - \beta)$.

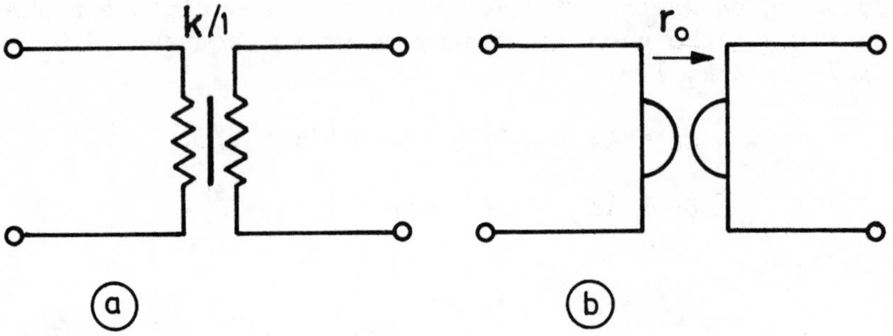

FIG. 9.1 Sections of degree zero: transformer (a) and gyrator (b).

For $\sigma = -1$, we have

$$K = \begin{pmatrix} 0 & \alpha + \beta \\ \alpha - \beta & 0 \end{pmatrix},$$

and this corresponds to a gyrator (Fig. 9.1b[8]) of normalized gyration ratio $r_0 = \alpha + \beta = 1/(\alpha - \beta)$.

9.3 *Sections of degree one*

a) $f = p$, $m = 0$. (m defined by (2.11)). As f is odd, we may choose a reciprocal realization, i.e., $\sigma = -1$. By making use of statement 1 of Theorem 8 (subsection 6.4), we can write,

$$g = p + G_0, \text{ with } G_0 > 0, \; h = H_0.$$

Due to (2.5), we have $H_0 = \pm G_0$. For $H_0 = G_0$, we obtain

$$K = \begin{pmatrix} 1 & 1/pc \\ 0 & 1 \end{pmatrix}, \quad c = 1/2G_0,$$

---

[8] The arrow indicates the direction for which for a positive value of r_0 a positive input current causes a positive output voltage.

which corresponds to the section of Fig. 9.2a. On the other hand, for $H_0 = -G_0$, we obtain

$$K = \begin{pmatrix} 1 & 0 \\ 1/lp & 1 \end{pmatrix}, \quad l = 1/2G_0,$$

which corresponds to the section of Fig. 9.2b.

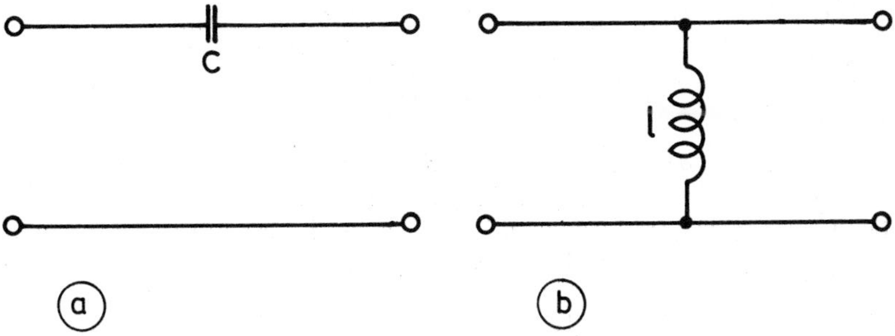

FIG. 9.2 Sections of degree 1, with $f = p$.

b) $f = 1$, $m = 1$. As f is even, we may choose a reciprocal realization, i.e., $\sigma = 1$. By making use of statement 2 of Theorem 8, we can write

$$g = G_1 p + 1, \quad G_1 > 0, \quad h = H_1 p.$$

Due to (2.5), we have $H_1 = \pm G_1$. For $H_1 = G_1$, we obtain

$$K = \begin{pmatrix} 1 & lp \\ 0 & 1 \end{pmatrix}, \quad l = 2G_1,$$

which corresponds to Fig. 9.3a, and for $H_1 = -G_1$,

$$K = \begin{pmatrix} 1 & 0 \\ pc & 1 \end{pmatrix}, \quad c = 2G_1,$$

which corresponds to Fig. 9.3b.

c) $f = p + F_0$, $F_0 \neq 0$, $m = 0$. No reciprocal realiz-

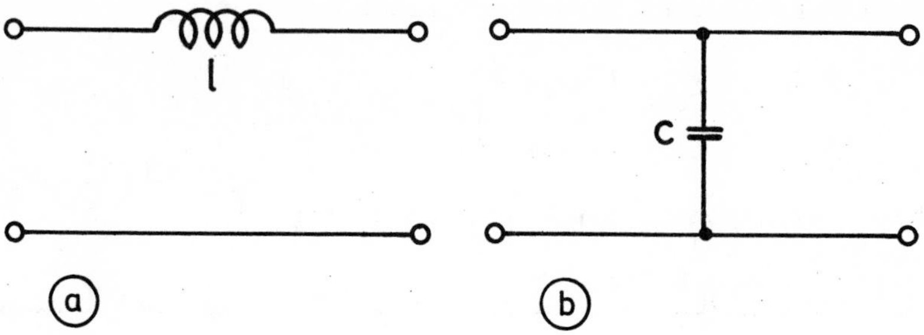

FIG. 9.3 Sections of degree 1, with $f = 1$.

ation is possible. We *first* choose $\sigma = -1$; by making use of statement 1 of Theorem 8, we can write

$$g = p + G_0, \quad G_0 > 0, \quad h = H_0,$$

whence, due to (2.5),

$$(G_0 - H_0)(G_0 + H_0) = F_0^2,$$

and therefore,

$$G_0 - H_0 > 0, \quad G_0 + H_0 > 0.$$

We obtain

$$Z = \frac{1}{G_0 - H_0}\begin{pmatrix} p & p-F_0 \\ p+F_0 & p \end{pmatrix} = \frac{p}{G_0-H_0}\begin{pmatrix} 1 & 1 \\ 1 & 1 \end{pmatrix} + \frac{F_0}{G_0-H_0}\begin{pmatrix} 0 & -1 \\ 1 & 0 \end{pmatrix}$$

$$Y = \frac{1}{G_0 + H_0}\begin{pmatrix} p & -p+F_0 \\ -p-F_0 & p \end{pmatrix} = \frac{p}{G_0+H_0}\begin{pmatrix} 1 & -1 \\ -1 & 1 \end{pmatrix} + \frac{F_0}{G_0+H_0}\begin{pmatrix} 0 & 1 \\ -1 & 0 \end{pmatrix}$$

The expression for Z leads to the realization of Fig. 9.4a, and the expression of Y leads to Fig. 9.4b, where

$$r_0 = F_0/(G_0 - H_0) = (G_0 + H_0)F_0$$

$$l = 1/(G_0 - H_0), \quad c = 1/(G_0 + H_0),$$

with $l/c = r_0^2$.

Next, we choose $\sigma = 1$ and make use of statement 2 of Theorem 8; we can write,

$$g = G_1 p + F_0, \quad G_1/F_0 > 0, \quad h = H_1 p,$$

whence, due to (2.5),

$$(G_1 - H_1)(G_1 + H_1) = 1,$$

and therefore,

$$(G_1 - H_1)/F_0 > 0, \quad (G_1 + H_1)/F_0 > 0.$$

We obtain

$$Z = \frac{1}{(G_1-H_1)p}\begin{pmatrix} F_0 & -p+F_0 \\ p+F_0 & F_0 \end{pmatrix} = \frac{F_0}{(G_1-H_1)p}\begin{pmatrix} 1 & 1 \\ 1 & 1 \end{pmatrix} + \frac{1}{G_1-H_1}\begin{pmatrix} 0 & -1 \\ 1 & 0 \end{pmatrix},$$

$$Y = \frac{1}{(G_1+H_1)p}\begin{pmatrix} F_0 & p-F_0 \\ -p-F_0 & F_0 \end{pmatrix} = \frac{F_0}{(G_1+H_1)p}\begin{pmatrix} 1 & -1 \\ -1 & 1 \end{pmatrix} + \frac{1}{G_1+H_1}\begin{pmatrix} 0 & 1 \\ -1 & 0 \end{pmatrix}.$$

The expression for Z leads to the realization of Fig. 9.4c, and the expression of Y to (9.4d), where

$$r_0' = 1/(G_1 - H_1) = G_1 + H_1,$$

$$c' = (G_1 - H_1)/F_0, \quad l' = (G_1 + H_1)/F_0,$$

with $l'/c' = r_0'^2$. Note that the sections of Fig. 9.4a

and b can be transformed into those of Fig. 9.4c and d by connecting a suitable gyrator in cascade.

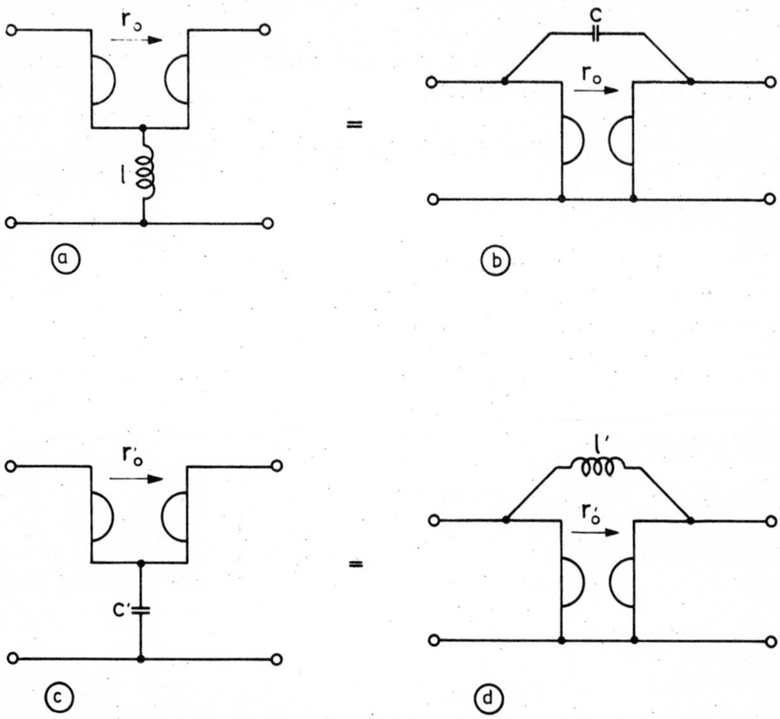

FIG. 9.4 Sections of degree 1 with $f = p + F_0$. The equality signs between two sections indicate that these sections are equivalent.

9.4 *Sections of degree two*

The following cases either cannot be reduced to first degree sections at all, or at least not to reciprocal first degree sections.

a) $f = p^2 + F_0$, $F_0 \neq 0$, $m = 0$. As f is even, we may choose a reciprocal realization i.e., $\sigma = 1$. By making *first* use of statement 1 of Theorem 8, we can write,

$$g = p^2 + G_1p + G_0, \quad G_1 > 0, \quad G_0 > 0, \quad h = H_1p + H_0. \quad (9.1)$$

Due to (2.5), we have

$$G_0^2 = H_0^2 + F_0^2, \quad G_1^2 = H_1^2 + 2(G_0 - F_0), \quad (9.2)$$

whence

$$G_0 > |H_0|, \quad G_0 \geq |F_0|, \quad G_1 \geq |H_1|. \quad (9.3)$$

For $H_1 \neq G_1$,

$$Z = \frac{1}{(G_1 - H_1)p} \begin{pmatrix} p^2 + G_0 + H_0 & p^2 + F_0 \\ p^2 + F_0 & p^2 + G_0 - H_0 \end{pmatrix},$$

and for $H_1 \neq -G_1$,

$$Y = \frac{1}{(G_1 + H_1)p} \begin{pmatrix} p^2 + G_0 - H_0 & -p^2 - F_0 \\ -p^2 - F_0 & p^2 + G_0 + H_0 \end{pmatrix}$$

For $H_1 \neq G_1$, the expression for Z leads to Fig. 9.5a, with

$$l_1 = 1/(G_1 - H_1), \quad c_1 = (G_1 - H_1)/(G_0 + H_0),$$
$$n_1 = l_1c_1F_0. \quad (9.4)$$

For $H_1 \neq -G_1$, the expression for Y leads to Fig. 9.5b, with

$$c_2 = 1/(G_1 + H_1), \quad l_2 = (G_1 + H_1)/(G_0 - H_0),$$
$$n_2 = l_2c_2F_0 = 1/n_1. \quad (9.5)$$

Next, we make use of statement 2 of Theorem 8; we can write

$$g = G_2p^2 + G_1p + F_0, \quad G_2/F_0 > 0, \quad G_1/F_0 > 0,$$
$$h = H_2p^2 + H_1p. \quad (9.6)$$

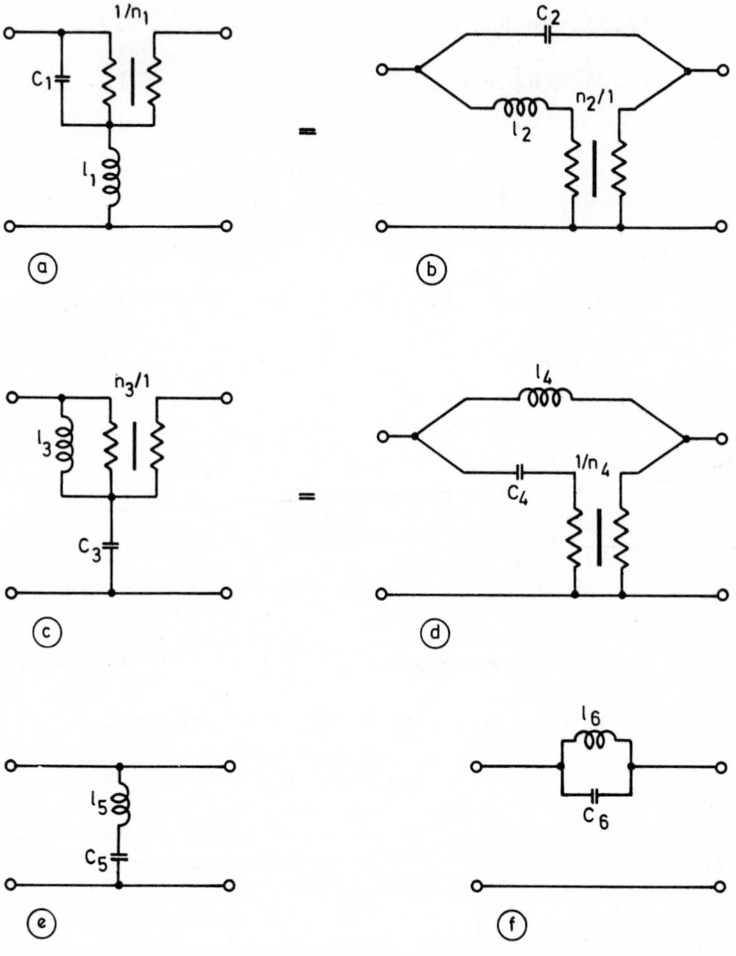

FIG. 9.5 Reciprocal sections of degree 2. The equality signs between two sections indicate that these sections are equivalent as soon as they both exist.

Due to (2.5), we have

$$G_2^2 = H_2^2 + 1, \quad G_1^2 = H_1^2 + 2(G_2 - 1)F_0, \tag{9.7}$$

whence

$$G_2/F_0 > |H_2/F_0|, \quad G_2/F_0 \geq |1/F_0|, \quad G_1/F_0 \geq |H_1/F_0|. \tag{9.8}$$

For $H_1 \neq G_1$,

$$Z = \frac{1}{(G_1 - H_1)p} \begin{bmatrix} (G_2 + H_2)p^2 + F_0 & p^2 + F_0 \\ p^2 + F_0 & (G_2 - H_2)p^2 + F_0 \end{bmatrix},$$

and for $H_1 \neq G_1$,

$$Y = \frac{1}{(G_1 + H_1)p} \begin{bmatrix} (G_2 - H_2)p^2 + F_0 & -p^2 - F_0 \\ -p^2 - F_0 & (G_2 + H_2)p^2 + F_0 \end{bmatrix}.$$

For $H_1 \neq G_1$, we obtain the section of Fig. 9.5c, with

$$c_3 = (G_1 - H_1)/F_0, \quad l_3 = (G_2 + H_2)/(G_1 - H_1),$$
$$n_3 = l_3 c_3 F_0, \tag{9.9}$$

and for $H_1 \neq G_1$, we obtain the section of Fig. 9.5d, with

$$l_4 = (G_1 + H_1)/F_0, \quad c_4 = (G_2 - H_2)/(G_1 + H_1),$$
$$n_4 = l_4 c_4 F_0 = l/n_3. \tag{9.10}$$

A *particular* situation arises if we are simultaneously in the two situations which we have just discussed, i.e., if we have simultaneously $G_2 = 1$ and $G_0 = F_0$, in which case $F_0 > 0$. We then obtain from (9.2) or (9.7), $H_2 = H_0 = 0$ and $H_1 = \pm G_1$. For $H_1 = -G_1$, one finds this way the section of Fig. 9.5e, with

$$l_5 = 1/2G_1, \quad c_5 = 2G_1/F_0, \tag{9.11}$$

and for $H_1 = G_1$, one finds the section of Fig. 9.5f, with

$$l_6 = 2G_1/F_0, \quad c_6 = 1/2G_1. \tag{9.12}$$

Note that the sections of Fig. 9.5a and b can be transformed into those of Fig. 9.5c and d by connecting a suitable transformer in cascade.

b) $f = p^2 + F_1 p + F_0$, $F_1 \neq 0$, $m = 0$. No reciprocal realization is possible. We choose $\sigma = 1$ throughout. All calculations are very similar to those for $F_1 = 0$.

First, we consider the case defined by (9.1); (9.2) and (9.3) now have to be replaced by

$$G_0^2 = H_0^2 + F_0^2, \quad G_1^2 = H_1^2 + F_1^2 + 2(G_0 - F_0),$$

$$G_0 > |H_0|, \quad G_0 \geq |F_0|, \quad G_1 > |H_1|. \tag{9.13}$$

We then obtain the sections of Fig. 9.6a and b, where l_1, c_1, and n_1 are still given by (9.4), and l_2, c_2, and n_2 by (9.5), but where

$$r_1 = F_1/(G_1 - H_1), \quad r_2 = (G_1 + H_1)/F_1. \tag{9.14}$$

Next, we consider the case defined by (9.6), where (9.7) and (9.8) have now to be replaced by

$$G_2^2 = H_2^2 + 1, \quad G_1^2 = H_1^2 + F_1^2 + 2F_0(G_2 - 1),$$

$$G_2/F_0 > |H_2/F_0|, \quad G_2/F_0 \geq |1/F_0|, \quad G_1/F_0 > |H_1/F_0|. \tag{9.15}$$

We then obtain the sections of Fig. 9.6c and d, where l_3, c_3, and n_3 are still given by (9.9), and l_4, c_4, and n_4 by (9.10), but where r_1 and r_2 are given by (9.14).

Note that in both cases the possibility $G_1 = |H_1|$ can never arise due to $F_0 \neq 0$ and $F_1 \neq 0$.

A *particular* situation arises again if we are simultaneously in the two situations just discussed, i.e., if we have simultaneously $G_2 = 1$ and $G_0 = F_0$, in which case $F_0 > 0$ and

$$G_1^2 = H_1^2 + F_1^2.$$

We obtain the sections of Fig. 9.6e and f, which are now

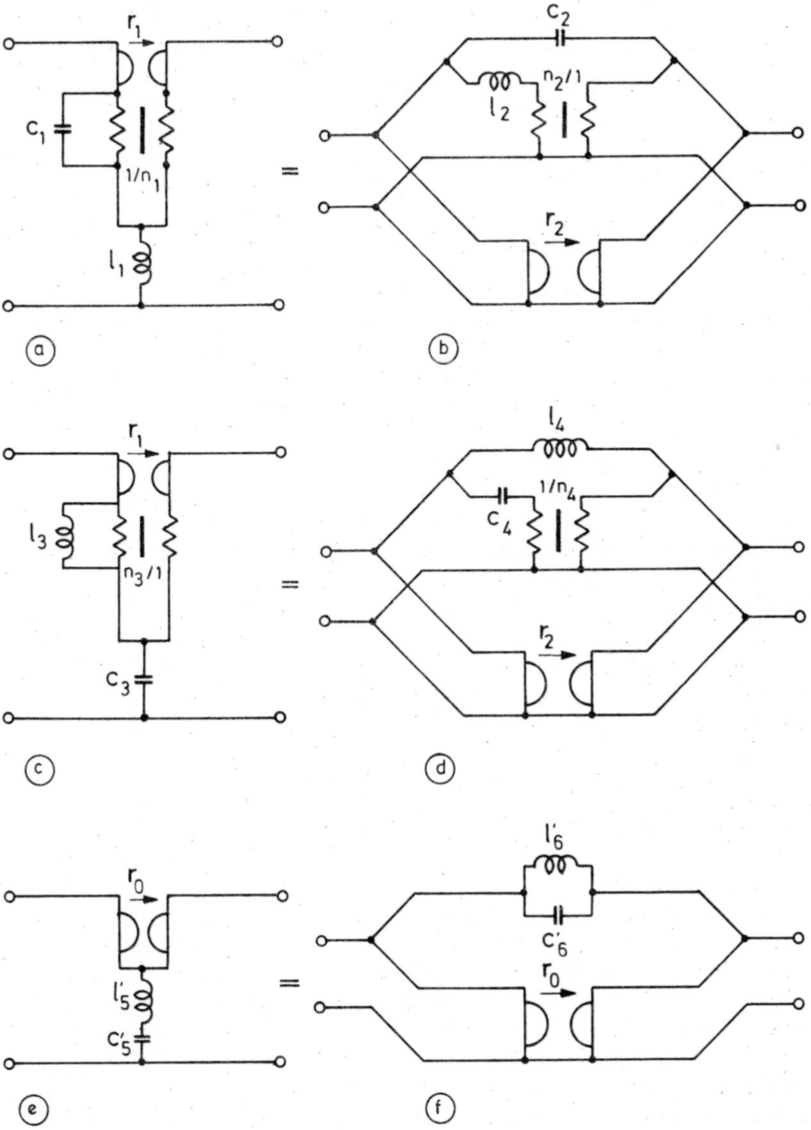

FIG. 9.6 Nonreciprocal sections of degree 2. The equality signs between two sections indicate that these sections are equivalent.

equivalent, and for which

$$l_5' = 1/(G_1 - H_1), \quad c_5' = (G_1 - H_1)/F_0$$

$$l_6' = (G_1 + H_1)/F_0, \quad c_6' = 1/(G_1 + H_1)$$

$$r_0 = F_1/(G_1 - H_1) = (G_1 + H_1)/F_1.$$

9.5 *Sections of degree four*

If a reciprocal realization is required, the only case which cannot be reduced to sections of lower degree is the following:

$$f = p^4 + F_2p^2 + F_0, \quad F_0 > F_2^2/4. \tag{9.16}$$

In this case, the zeros of f cannot be located on the real or imaginary p-axis. For a reciprocal realization, we must choose $\sigma = 1$. With regard to the limited practical importance of the present case, we consider only a single one of the many possible realizations, viz. Darlington's type D section.

We can write, using statement 2 of Theorem 8, and splitting g and h immediately into their even and odd parts,

$$g_e = G_4p^4 + G_2p^2 + F_0, \quad g_o = p(G_3p^2 + G_1) \tag{9.17}$$

$$h_e = H_4p^4 + H_2p^2, \quad h_o = p(H_3p^2 + H_1) \tag{9.18}$$

As the polynomials involved are already of rather high degree, a direct analysis of the relative magnitudes of their coefficients, as was done in all previous cases, is rather intricate. We, therefore, use a more general approach instead. It is obvious that we could have used similar methods in our earlier discussions. The present method corresponds in principle to the one already used by Darlington [1] for showing that the impedance matrix obtained by the insertion loss method always satisfies Cauer's residue conditions.

We note first that the polynomial $g + h_*$ is a Hurwitz

polynomial of degree 4. Indeed, as is well-known, it is certainly semi-Hurwitz (no zeros in Re $p > 0$ and zeros on the $j\omega$-axis being simple) since g is Hurwitz and $|h_*/g| < 1$ for Re $p > 0$. Furthermore, if $g + h_*$ had a zero for a certain $p = j\omega_1$, we would also have, by (2.5), $f(j\omega_1) = 0$, which is excluded by hypothesis. In a similar way, one checks by means of (2.5) that the degree of $f + h_*$ cannot be lower than the degree of f.

Consider now the entries z_{ij} of the impedance matrix Z given in subsection 9.1, with k, j = 1, 2, and $z_{ij} = z_{ji}$. We have, e.g.,

$$z_{11} = (g_e + h_e)/(g_o - h_o),$$

which is the ratio of the even to the odd part of $g + h*$. Consequently, z_{11} is a reactance function of degree 4, having a pole at infinity as well as at zero. By inspection of the expression for Z, we can write for all z_{ij}, using (9.17) and (9.18),

$$z_{ij} = a_{ij}p + b_{ij}/p + c_{ij}p/(p^2 + \omega_0^2) \qquad (9.19)$$

where ω_0 as well as the a_{ij}, b_{ij}, and c_{ij} can easily be expressed in terms of the coefficients of f, g, and h, and where

$$a_{11} > 0, \quad b_{11} > 0, \quad c_{11} > 0. \qquad (9.20)$$

In particular, as $g_e + h_e$ and $g_e - h_e$ have the same constant term as f, viz. F_0, we have

$$b_{11} = b_{12} = b_{22}. \qquad (9.21)$$

From the expression for Z (subsection 9.1) together with (2.5), we derive

$$\det Z = z_{11}z_{22} - z_{12}^2 = (g_o + h_o)/(g_o - h_o), \qquad (9.22)$$

while we obtain from (9.19)

$$a_{11}a_{22} - a_{12}^2 = \lim_{p\to\infty} [(\det Z)/p^2], \qquad (9.23)$$

$$c_{11}c_{22} - c_{12}^2 = \lim_{p^2 \to -\omega_0^2} [(\det Z)(p^2 + \omega_0^2)^2/p^2] \quad (9.24)$$

Thus, comparing (9.23) and (9.24) with (9.22), we see that

$$a_{11}a_{22} - a_{12}^2 = c_{11}c_{22} - c_{12}^2 = 0. \quad (9.25)$$

Finally, using (9.19, (9.20), (9.21) and (9.25), we

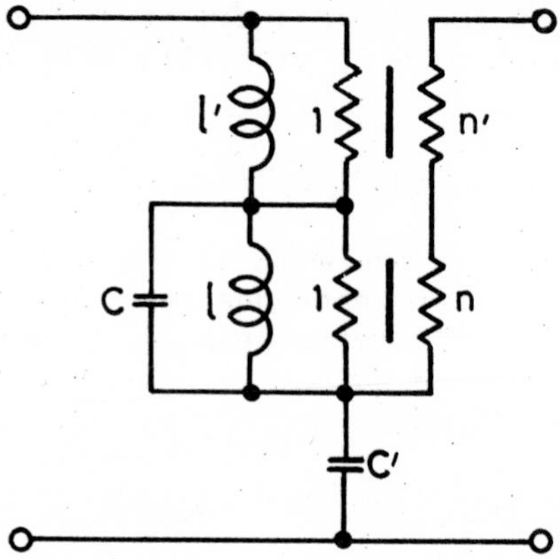

FIG. 9.7 Reciprocal fourth degree section (Darlington's type D section).

see that a possible realization is as shown in Fig. 9.7, where

$$l' = a_{11}, \quad c' = 1/b_{11}, \quad c = 1/c_{11}, \quad l = 1/\omega_0^2 c,$$

$$n' = a_{12}/a_{11}, \quad n = c_{12}/c_{11}.$$

9.6 *Alternation of sections in case of multiple transmission zeros at zero and infinity*

In subsection 9.3, we have seen that there exist two essentially different sections for the case $f = p$, $m = 0$, and that the same is true for $f = 1$, $m = 1$. For physical reasons, it is obvious that for reciprocal realizations of reciprocal two-ports, the two sections of Fig. 9.2 must alternately be used in case of multiple transmission zeros at $p = 0$, and the two sections of Fig. 9.3 alternately in case of multiple transmission zeros at $p = \infty$. It is useful to examine how this is confirmed by our mathematical expressions.

Care must be taken in interpreting our notation since the sections of Figs. 9.2 and 9.3 may now correspond to the two-port N_a of the factorization problem. In this sense, the results of subsection 9.3 may be interpreted in the following way: For $f = p$, $m = 0$, the section of Fig. 9.2a has to be used if $C_0/A_0 = 1$, and the section of Fig. 9.2b if $C_0/A_0 = -1$. Similarly, for $f = 1$, $m = 1$, the section of Fig. 9.3a has to be used if $C_1/A_1 = 1$, and the section of Fig. 9.3b if $C_1/A_1 = -1$. The basis for the following discussion is Theorem 9 proved in subsection 6.4.

Consider first the case of transmission zeros at $p = 0$. If a transmission zero at $p = 0$ should be realized by the two-port N_a, then (6.16) holds, i.e., C_0/A_0 is simply equal to H_0/G_0. Assume next that the two-port N_a should be reciprocal, i.e., f_a even or odd, with $\sigma_a = f_{a*}/f_a$, and that f_a has a zero of order k at $p = 0$, plus, possibly, further zeros at points $p \neq 0$; in this case $\sigma_a = (-1)^k$. Assume furthermore that f_b still has at least one zero at $p = 0$. In this case, (6.17) shows that B_0/D_0 has the same sign as H_0/G_0 if k is even (including $k = 0$), and that it has the opposite sign of H_0/G_0 if k is odd. The alternate appearance of the two sections of Fig. 9.2 is now obvious if we note, by comparing (6.3) and (6.4) to (6.2) and (6.13), that B_0/D_0 plays the same role for the subsequent realization of two-port N_b as H_0/G_0 did for the original two-port N.

The alternate appearance of the two sections of Fig. 9.3 can be shown in exactly the same way by making use of (6.18) and (6.19) rather than (6.16) and (6.17). Note in this respect that for a reciprocal two-port N_a, one always has, due to $\sigma_a = f_{a*}/fa$ and the first expression (3.7), $\sigma_a = (-1)^{n_a - m_a}$.

9.7 *Realization by means of grounded gyrators*

As has been pointed out by Leich and van Bastelaer [20], only grounded gyrators are required in the synthesis of lossless two-ports. In order to show this, note first that no sections of degree higher than two are needed if gyrators are available. If we have at our disposition gyrators in addition to all conventional elements and if gyrators should only be used for realizing nonreciprocal sections, the objective can e.g., be met through using the sections b and d among those of Fig. 9.4, and the sections b, d, and f among those of Fig. 9.6. Finally, if inductances and transformers have to be proscribed, the section of Fig. 9.5b, which also appears in Fig. 9.6b, can be modified appropriately by making use of the simple equivalence shown in Fig. 9.8, where the gyration constants r_1 and r_2 have to be such that $n = r_1 r_2$ and where $c = l/r_1^2$. Note that this equivalence includes the case of floating inductances for $n = 1$.

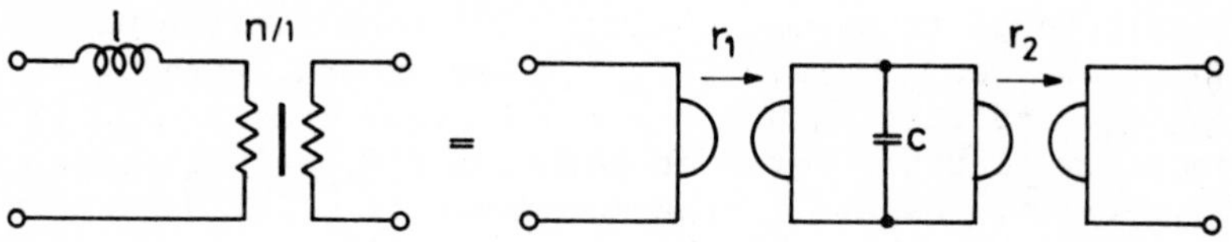

FIG. 9.8 A simple two-port equivalence.

10. NETWORKS CONTAINING UNIT ELEMENTS

In the theory of networks containing unit elements, one is led to consider scattering and transfer matrices which have all the properties of those discussed in section 2 except for the following detail: The function $f = f(p)$ does no longer have to be a polynomial but may be of the form

$$f = (1 - p^2)^{k/2} f_0$$

where k is a nonnegative integer and $f_0 = f_0(p)$ a real monic polynomial [19], with

$$f_* = (1 - p^2)^{k/2}\ f_{0*}$$

It is obvious, that the problem of factorizing T into a product of two simpler matrices T_a and T_b of the same type as T, can then be formulated in exactly the same way as before, except that f_a and f_b may now be of the same general form as f itself. All proofs given in sections 4 to 8 remain valid without any modification. This is due to the fact that f, f_a, and f_b occur exclusively in form of the combinations ff_*, $f_a f_{a*}$, and $f_b f_{b*}$, respectively, which obviously always are polynomials.

As for the cascade synthesis by elementary sections, all resuls of section 9 remain valid too but we now have to examine in addition a section of degree one for which

$$f = \sqrt{1 - p^2}, \text{ and thus } m = 0.$$

We shall use the same notation as in section 9. By means of a similar reasoning as for Theorem 8, one can verify that the factorization can be carried out in such a way that $G_0 = 1$ and thus $H_0 = 0$, i.e., such that

$$g = G_1 p + 1,\ h = H_1 p,\quad G_1 > 0,$$

whence, due to (2.5),

$$(G_1 - H_1)(G_1 + H_1) = 1.$$

This implies, in particular, $G_1 > |H_1|$.

For a reciprocal realization, we have to choose $\sigma = f_*/f = 1$. Thus, using the general expression for the chain matrix given in subsection 9.1, we obtain

$$K = \frac{1}{\sqrt{1 - p^2}} \begin{pmatrix} 1 & Rp \\ p/R & 1 \end{pmatrix}$$

where R is a positive constant given by

$$R = G_1 + H_1 = 1/(G_1 - H_1).$$

Clearly, (10.1) corresponds to a unit element of characteristic impedance R (Fig. 10.1). This shows that for a cascade realization of the more general type of transfer matrix, which we are considering at present, the only elementary section required in addition to those discussed in section 9 is a unit element.

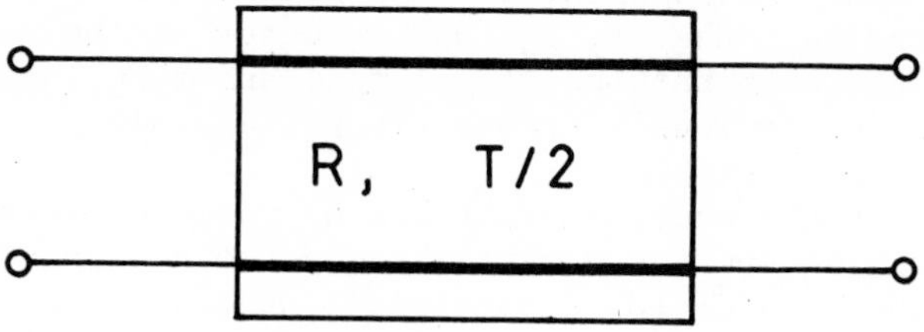

FIG. 10.1 Unit element of characteristic impedance R and delay T/2.

APPENDIX

Elimination of all-pass factors

We have seen that all general results remain valid even if g and h, and/or g and h_* are not relatively prime. Nevertheless, certain complications may arise under such conditions, and in particular, it may become more difficult to establish the fundamental equations. As has already been pointed out by Belevitch [9], however, common factors between g and h, and/or g and h_*, may always be eliminated by first extracting suitable all-pass sections at the terminating ports. We shall briefly show how this can be done.

Let v be a common divisor of g and h. Due to (2.5), v divides ff_*. We can thus certainly split v into a product of two factors, $v = \gamma\delta$, where γ divides f, and δ divides f_*. Now, γ and δ_* are relatively prime since γ and δ are Hurwitz (they divide g). Thus, f is divisible not only by γ and δ_* but also by $\gamma\delta_*$. Altogether, we can thus define new polynomials f', g', and h' by

$$f' = f/\gamma\delta_*, \quad g' = g/\gamma\delta, \quad h' = h/\gamma\delta. \tag{A-1}$$

This leads to the following

Theorem A-1

If the canonic polynomials g and h of a lossless two-port N have a common factor v and if σ' and σ_v are such that

$$\sigma' = \pm 1, \quad \sigma_v = \pm 1, \quad \sigma'\sigma_v = \sigma, \tag{A-2}$$

then there always exist two polynomials γ and δ, with $\gamma\delta = v$, such that the transfer matrix T of N can be written as $T = T'T_v$, where

$$T' = \frac{1}{f'}\begin{pmatrix} \sigma' g'_* & h' \\ \sigma' h'_* & g' \end{pmatrix}, \quad T_v = \frac{1}{\gamma\delta_*}\begin{pmatrix} \sigma_v \gamma_* \delta_* & 0 \\ 0 & \gamma\delta \end{pmatrix}, \tag{A-3}$$

f', g', and h' being polynomials such that (A-1) holds. T_v, given by the second Eq. (A-3), corresponds to the canonic form of the transfer matrix of an all-pass two-port[9]. T', given by the first Eq. (A-3), corresponds to the canonic form of the transfer matrix of a lossless two-port, i.e., g' is Hurwitz and we have

$$g'g'_* = h'h'_* + f'f'_*; \tag{A-4}$$

γ and δ may be chosen such that f' is monic.

Proof

The equalities $T = T'T_v$ and (A-4) are easily established by using (2.4), (2.5), and (A-1) - (A-3). The rest of the proof is elementary.

Assume next that u is a common divisor of g and h_*.

---

[9] An all-pass two-port is a lossless two-port whose canonic polynomial h is identically zero.

One can show now that u can be split into two polynomials ε and η such that f", g", and h" given by

$$f'' = f/\varepsilon\eta_*, \quad g'' = g/\varepsilon\eta, \quad h'' = \sigma_u h/\varepsilon_*\eta_* \qquad \text{(A-5)}$$

are polynomials. More precisely, we have the following

Theorem A-2

If the canonic polynomials g and h of a lossless two-port are such that g and h_* have a common factor u, and if σ'' and σ_u are such that

$$\sigma'' = \pm 1, \quad \sigma_u = \pm 1, \quad \sigma''\sigma_u = \sigma,$$

then there always exists two polynomials ε and η, with $\varepsilon\eta = u$, such that the transfer matrix T of N can be written as $T = T_u T''$, where

$$T_u = \frac{1}{\varepsilon\eta_*}\begin{pmatrix} \sigma_u\varepsilon_*\eta_* & 0 \\ 0 & \varepsilon\eta \end{pmatrix}, \quad T'' = \frac{1}{f''}\begin{pmatrix} \sigma'' g''_* & h'' \\ \sigma'' h''_* & g'' \end{pmatrix}, \qquad \text{(A-6)}$$

f", g", and h" being polynomials such that (A-5) holds. T_u given by the first equation (A-6) corresponds to the canonic form of an all-pass two-port. T" given by the second equation (A-6) corresponds to the canonic form of the transfer matrix of a lossless two-port, i.e., g" is Hurwitz and

$$g''g''_* = h''h''_* + f''f''_*;$$

ε and η may be chosen such that f" is monic.

CONCLUSIONS

1. We can always ensure that g is relatively prime with h and h_* by extracting suitable all-pass sections at both ports. As can easily be checked, the extraction of an all-pass at one port cannot create new common factors corresponding to the other port. Hence, the sum of the degrees of the two all-pass sections to be extracted can-

not exceed the degree of the g.c.d. of g and h plus the degree of the g.c.d. of g and h_*. It may, however, be smaller, as under special situations, the elimination of a common zero between g and h may automatically eliminate a common zero between g and h_*, and vice versa.

2. If the transfer matrices T_u and T_v do not yet correspond to elementary all-pass sections, they can easily be factorized further until a realization by means of sections of degree 1 or 2 becomes possible. In the first case, the sections of Fig. 9.4 can, of course, be used. In the second case, the discussion given in subsection 9.4b shows that the sections of Fig. 9.6e and f are applicable. Indeed, we are then in a situation in which the polynomial h is identically zero, i.e., in which all the H_i are zero. Hence, for $G_2 = 1$, we have $G_0 = |F_0|$, i.e., if $F_0 > 0$ (which is always the case if the zeros of f are complex) we have $G_0 = F_0$.

3. The all-pass extraction cannot always be carried out by means of reciprocal sections, not even if N is reciprocal, i.e., if $\sigma = f_*/f$. Indeed, in order to obtain, e.g., det $T_v = 1$, we must have $\gamma\delta_* = \sigma_v\gamma_*\delta_*$. As γ and δ are Hurwitz, this requires $\sigma_v = 1$ and $\delta = k\gamma$, where k is a constant. We thus have $v = k\gamma^2$, whence

$$T_v = \begin{pmatrix} \gamma_*/\gamma & 0 \\ 0 & \gamma/\gamma_* \end{pmatrix}$$

This corresponds to a reciprocal all-pass. Hence, the necessary and sufficient conditions for the all-pass T_v to be reciprocal is that v is equal to ± a perfect square. The realization of reciprocal all-pass two-ports is well-known; it can, in principle, always be achieved by the sections shown in Figs. 9.5a to d and 9.7.

REFERENCES

1. DARLINGTON, S. (September 1939) Synthesis of reactance 4-poles, *J. Math. and Phys. 18*, pp. 257-353.

2. UNBEHAUEN, R. (1957) Neuartige Verwirklichung von Zweipolfunktionen durch kanonische oder durch kopplungsfreie Schaltungen (Dissertation), Technische Hochschule Stuttgart, Germany.

3. UNBEHAUEN, R. (December 1958) Zur Ermittlung kanonischer Reaktanzvierpole mit vorgeschriebene Kettenmatrix, *Archiv der Elektr. Übertr. 12*: (12), pp. 545-556.

4. HAZONY, D. (June 1961) Zero cancellation synthesis using impedance operators, *IRE Trans. on Circuit Theory PGCT-8*: (2) pp. 114-120.

5. HAZONY, D. (September 1961) Two extensions of the Darlington synthesis procedure, *IRE Trans. on Circuit Theory PGCT-8*: (3), pp. 284-288.

6. YOULA, D.C. A new theory of cascade synthesis, *IRE Trans. on Circuit Theory PGCT-8*: (2), pp. 244-260; (September 1961). *PGCT-9*: (2), p. 195; (June 1962). *CT-13*: (1), p. 90; (March 1966).

7. RUBIN, W.L. and CARLIN, H.J. (March 1962) Cascade synthesis of nonreciprocal lossless 2-ports, *IRE Trans. Circuit Theory CT-9*: (1), pp. 48-55.

8. BELEVITCH, V. (August 1963) Factorization of scattering matrices with applications to passive-network synthesis, *Philips Research Reports 18*: (4), pp. 275-317.

9. BELEVITCH, V. (1968) *Classical Network Theory*, San Francisco, Holden-Day.

10. KAO, C. and TOKAD, Y. (July 1968) New method for cascade synthesis of 1-port passive networks, *Proc. IEE 155*, No. 7, pp. 937-944.

11. PILOTY, H. (Sept., Oct., Nov. 1940) Kanonische Kettenschaltungen für Reaktanzvierpole mit vorgeschriebenen Betriebseigenschaften, *Telegr.-, Ferspr.-, Funk und Fernsehtechn. 29*, pp. 249-258, 279-290, 320-325.

12. TALBOT, A. (October 1953) *A new method of synthesis of reactance 4-poles*, IEEE Monograph No. 77.

13. BAUER, F.L. (December 1955) Die Betriebs-Kettenmatrix von Vierpolen, *Archiv d. Elektr. Übertragung 9*: (12), pp. 559-560.

14. BELEVITCH, V. (June 1956) Four-dimensional transformations of 4-pole matrices with applications to the synthesis of reactance 4-poles, *IEEE Trans. Circuit Theory CT-3*: (2), pp. 105-111.

15. BOSSE, G. (1963) *Einführung in die Synthese elektrischer Siebschaltungen mit forgeschriebenen Eigenschaften*, Stuttgart, Germany, S. Hirzel Verlag.

16. NEIRYNCK. J. and VAN BASTELAER, Ph. (1967) La synthèse des filtres par factorisation de la matrice de transfert, *Revue MBLE 10*: (1), pp. 5-32.

17. YOULA, D.C (June 1967) On the cascade decomposition of lossless 2-ports, Polytechnic Institute of Brooklyn, Electrophysics Department, Electrophysics Memo PIBMRI-1377-67.

18. FETTWEIS, A. (February 1970) On the factorization of the transfer matrix of lossless two-ports, (scheduled for publication in the *IEEE Trans. on Circuit Theory CT-17*: (1), pp. 86-94.

19. FETTWEIS, A. Zur Faktorzerlegung der Betriebs-Kettenmatrix von Reaktanz-Zweitoren, *Archiv. Elektr. Übertr.* (in print).

20. LEICH, H. and VAN BASTELAER, Ph. (1968) La synthèse des quadripoles non dissipatifs au moyen de capacités et de gyrateurs déséquilibrés, *Revue MBLE XI*: (2), pp. 31-48.

21. FRAITURE, L. and NEIRYNCK, J. (April 1967) Theory of unit-elements filters, MBLE-Research Laboratory Report R-62.

TRANSFORMATIONS OF 4-TERMINAL NETWORKS USING CAPACITIVE BRIDGES

J.E. COLIN

Société Anonyme de Télécommunications,
41, Rue Cantagrel, Paris 13

ABSTRACT

The paper discusses a number of equivalence relations between a ladder network and a second network formed by adding a capacitor - either in series or in parallel - to a different ladder network. This capacitor adds a degree of freedom to the network, enabling one to modify the values of the network elements in order to make them easier to realize. It is sometimes even possible to eliminate the negative capacitance which can appear when designing a particular filter; this can then be effected without resorting to lattice networks or to tappings on the inductance coils. The additional capacitor can also be used to eliminate the effects of parasitic capacitors occurring in the actual construction of a designed network. It indeed becomes possible to incorporate these parasitic elements into the capacitors provided by the design.

1. INTRODUCTION

Fig. 1 represents one of the network structures which is currently handled by the technical literature. This network is equivalent to the classical network of Fig. 2, where the elements are given different values. This means therefore that there exists a two-port transformation enabling one to go over from the network of Fig. 2 to the network of Fig. 1 and conversely. In this transformation appear two new elements, denoted here by T and K, which behave like additional degrees of freedom. An appropriate choice of these elements enables one to obtain some interesting results.

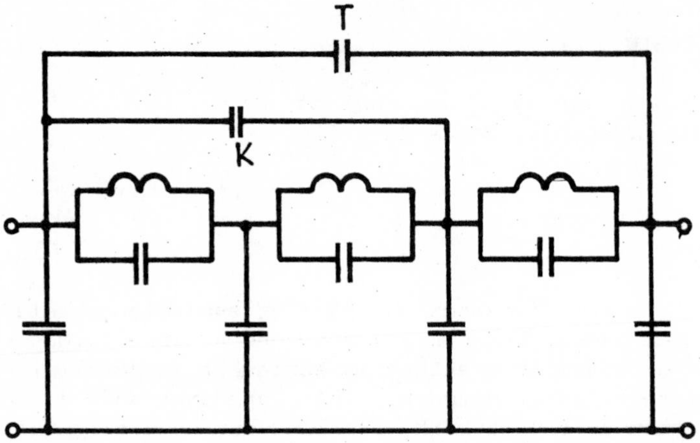

FIG. 1. Example of a network with capacitive bridges

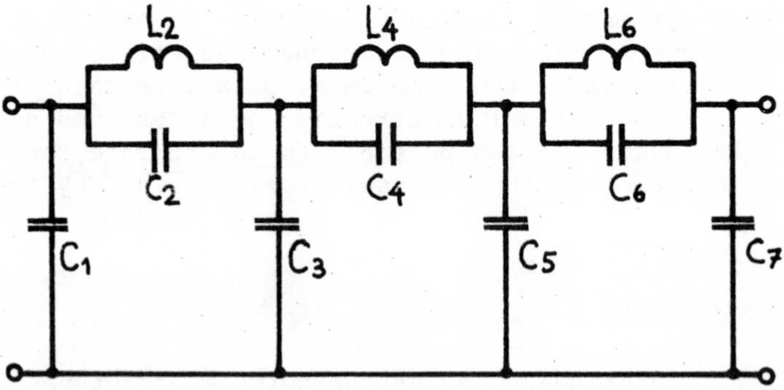

FIG. 2. Classical network

The purpose of this paper is to try to show some of these results and to supply a few useful transformations.

The network of Fig. 1 contains specifically two capacitors K and T bridging a certain two-port or part of it; this two-port is topologically identical to the initial two-port of Fig. 2, whence the name "capacitive bridges" given to such networks.

2. SURVEY OF BASIC NOTIONS ON TWO-PORTS

It is sometimes quite difficult to compute the coefficients of the chain-matrix of a network, given the coefficients of the chain-matrices of each of the two-ports which constitute this network: this is usually a consequence of the complexity of the whole network. The method described here gives a solution to this problem when the elementary two-ports are all reciprocal and passive. This method is based upon the equivalence relations described in Fig. 3 and Fig. 4 and upon the classical Y-Δ transformation.

Fig. 3 illustrates the fact that any reciprocal passive two-port (where A,B,C and D are the coefficients of the chain-matrix) can be replaced by a network with 3 impedances. Fig. 4 describes another well-known equivalence relation.

Now consider, for example, the network of Fig. 5a: this network consists of five reciprocal and passive two-ports. It is quite difficult to compute the coefficients A,B,C,D directly. However, if one replaces each two-port by its equivalent network with 3 impedances (in triangle), one obtains the network of Fig. 5b, which consists of one-ports only. By a number of additional Y-Δ transformations, the network of Fig. 5b is reduced to that of Fig. 5c. Fig. 3 then shows how to compute the coefficients of the chain-matrix for the resulting network of Fig. 5c.

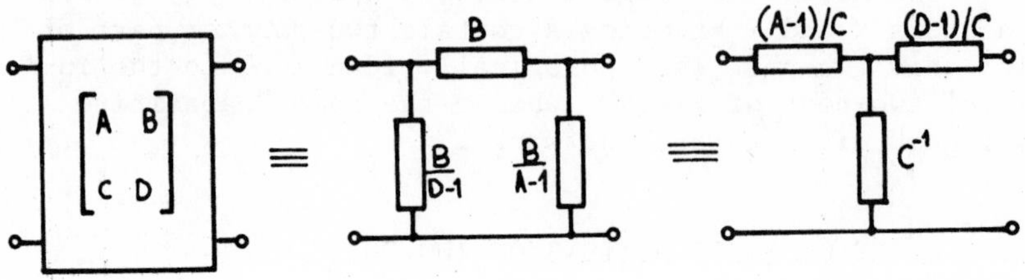

FIG. 3. Equivalence relations for a reciprocal and passive network

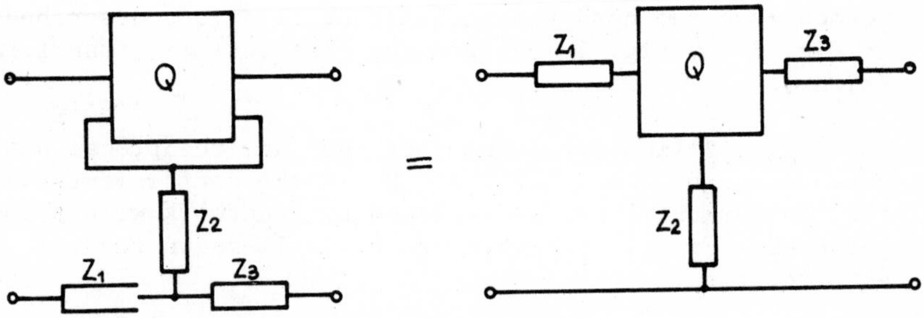

FIG. 4. Equivalence of two networks

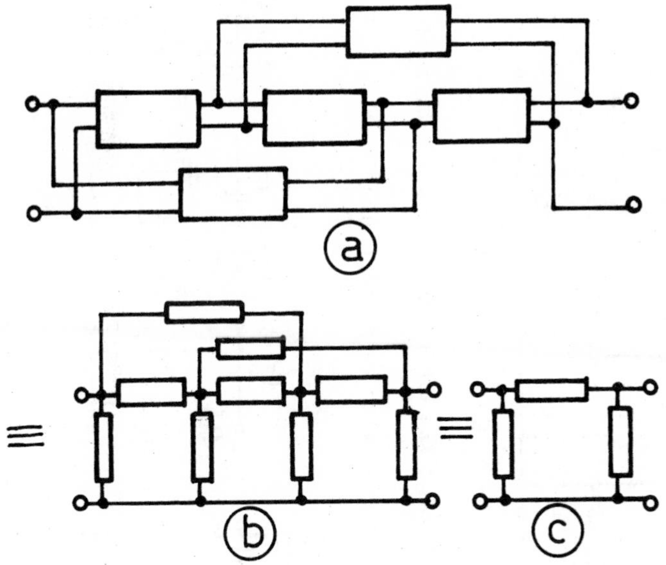

FIG. 5. Example of equivalent networks

3. TRANSFORMATION OF A NETWORK WITH ONE INDUCTANCE

3.1. *Description of the transformation*

Going over from the network of Fig. 6a to that of Fig. 6b is a good example of a transformation of a two-port through the use of a single capacitive bridge. By application of the duality principle, the same transformation enables one to proceed from the network of Fig. 6c to that of Fig. 6d. Two kinds of capacitive bridges thus appear: one is obtained through the capacitor of value G in Fig. 6b, the other through the capacitor of value 1/G in Fig. 6d.

An additional degree of freedom t appears in the transformation; if the elements of the network of Fig. 6a are positive, those of Fig. 6b will be positive only if

$$t_1 < t < t_0$$

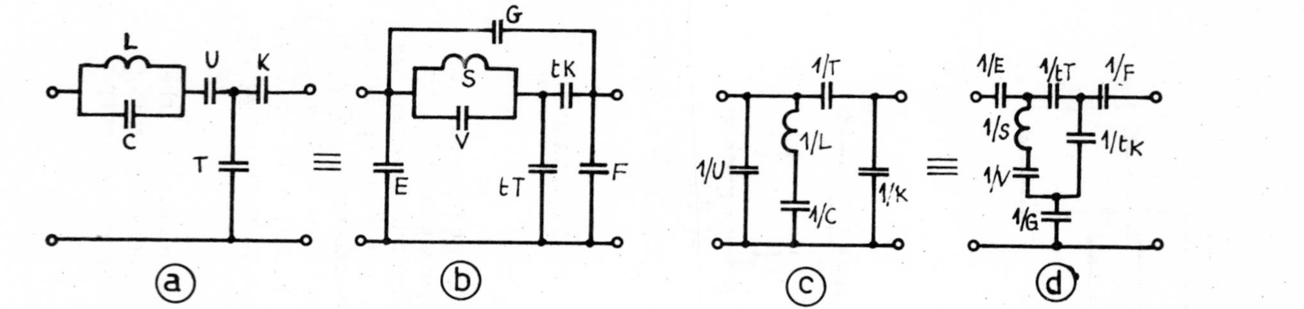

$$E/T = G/K = t_0 - t\ ;\ F = KT/(K+T+U)\ ;\ S = L\,t_0^2/t^2\ ;\ V = t(K+T)(t-t_1)/t_1\ ;\ t_0 = U/(T+K+U)\ ;\ t_1 = U t_0(1-t_0)/[U(1-t_0)+C]$$

FIG. 6. Transformation of a network with one inductance

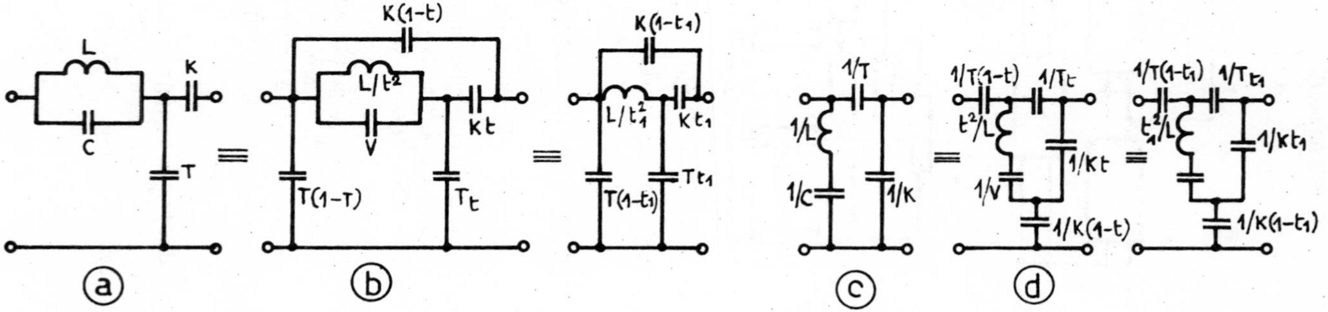

$$V = t(K+T)(t-t_1)/t_1 \quad ; \quad t_1 = (K+T)/(C+K+T)$$

FIG. 7. Transformation of Fig. 6 when 1/U = 0

For $t = t_0$ the transformation disappears, since the two-ports of Fig. 6a and 6b become identical, up to a Y-Δ transformation on the capacitors U,K,T.

For $t = t_1$, the capacitor V disappears.

Note that the transformation is not applicable if $C = 0$ and it is trivial if $T = 0$ or $1/K = 0$. On the other hand, if $1/U = 0$, the transformation becomes identical to that of Fig. 7.

3.2. *Applications of the transformation*

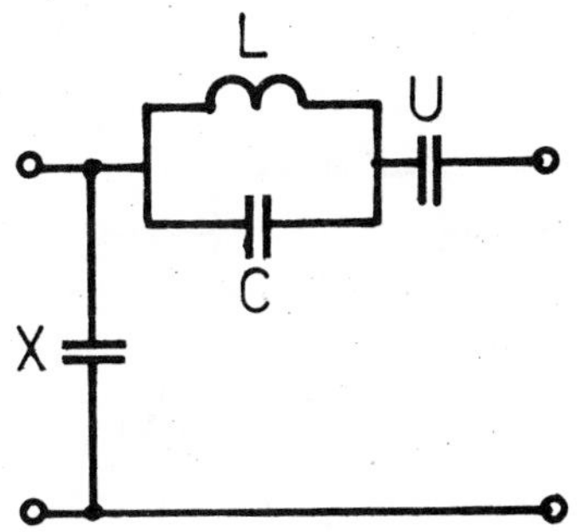

FIG. 8. A filter terminated by $X < 0$

When designing a filter it sometimes happens that a capacitor X with negative value appears at one of the terminals of the network. The physical realization of the filter requires a solution to this problem. A common solution, based on the autotransformer principle, uses a tapping on the inductance coil.

A transformation like the one which enabled us to go from the network of Fig. 6a to that of Fig. 6b can also be used to solve the problem, since a positive capacitor E appears in Fig. 6b in parallel with the capacitor X. It is of course sufficient to make $E + X > 0$ to permit physical realization of the filter. If one cannot satisfy this relation, one has to resort to transformations with more than one inductance. The Appendix gives a number of transformations for networks with two inductances. On the other hand, when the relation E + X > 0 is true, the presence of the capacitor G (that is, the capacitive bridge) allows a physical realization without tappings on the inductance coil.

Another application arises when in a network like that of Fig. 6a the value of L is found to be very small, which makes physical realization difficult. One sees, however, that the inductance S appearing in the transformed network of Fig. 6b has a value $S = L.t_0{}^2/t^2$. S is therefore always greater than L, especially as C is made bigger.

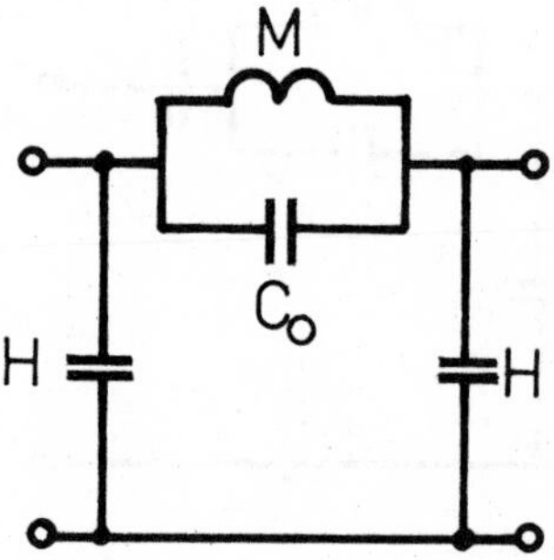

FIG. 9. Ferrite core inductance

Yet another application of these transformations occurs when parasitic capacitances hinder the physical realization of designed filters. Fig. 9 represents the equivalent network of an inductance of value M, wound on a ferrite-pot. This network contains a distributed capacitance C_0 and two capacitors H between the terminals of M and the ground. Figs. 6b and 7b show how to eliminate the effects of C_0 and H by decreasing the values of E and tT by H and the value V by C_0.

4. EXAMPLE OF APPLICATION: A LOW-PASS FILTER

It is well known that two-ports transformations are more easily applied to pass-band filters than to low-pass filters. It therefore seems interesting to give an example of a two-ports transformation using a capacitive bridge for the latter case.

Consider, for example, a low-pass filter of the kind illustrated in Fig. 2. The attenuation in the passband has a Chebyshev behavior and the maximum ripple does not exceed 10%. The values of the elements are as follows:

$C_1 = 0.6286$; $C_2 = 0.5703$; $L_2 = 0.8788$; $C_3 = 0.7847$;

$C_4 = 6.220$; $L_4 = 0.1549$; $C_5 = 0.6377$; $C_6 = 3.893$;

$L_6 = 0.226$; $C_7 = -0.09032$.

The capacitance C_7 is thus negative. Through application of the transformation of Fig. A_1 (see Appendix), one can realize the filter with

$L = L_4$; $C = C_4$; $T = C_5$; $M = L_6$; $K = C_6$.

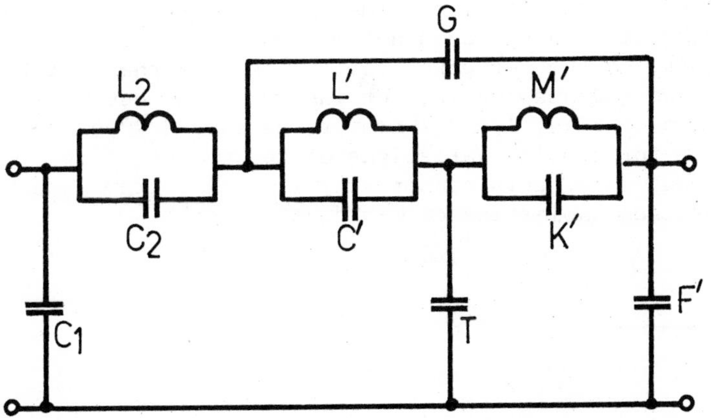

FIG. 10. Transformed filter (type C0710 80°)

In order not to increase the number of condensers, the parameter u is chosen so that $E + C_3 = C$, yielding $u = 0.3253$. Through the application of the transformation, the network of Fig. 2 is reduced to that of Fig. 10, with the following values for the elements:

$C_1 = 0.6286$; $C_2 = 0.5703$; $L_2 = 0.8788$ (these values are unchanged)

$L' = 0.02576$; $C' = 38.04$; $T' = 1.247$; $M' = 0.3551$;

$X' = 1.845$; $G = 0.5458$; $F' = F + C_7 = 0.08465$.

All these elements are therefore physically realizable.

5. CONCLUSION

The present paper has shown that the use of capacitive bridges solves certain problems which may appear in designing filters, without resorting either to lattice networks, T-bridge networks, or tappings on the inductance coils.

It therefore seems useful to introduce this new variety of filters between the ladder networks and the lattice filters. The subject is indeed capable of further development. It also seems that the presence of these capacitive shunts does not seriously affect the sensitivity of the network. This means that the variations of the attenuation in, for instance, the equivalent networks of Figs. 1 and 2 are of the same degree of magnitude for similar variations of the values of the elements. At any rate, this sensitivity is much better than that obtained in lattice or T-brdige networks.

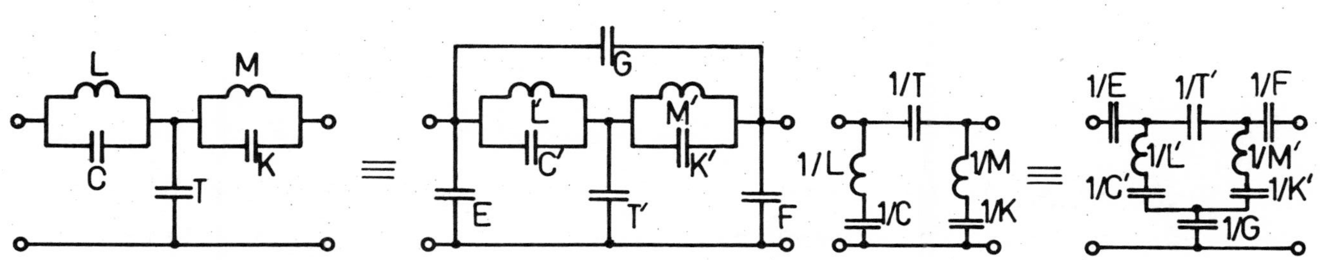

u = parameter ; $g = uL[LC - M(K+T)] + M[L(C+T) - MK]$; $\lambda = M(C+K+T) + u[LC - M(K+T)]$;

$\mu = uL(C+K+T) + MK - L(C+T)$

$E = MT(u-1)/u(L+M)$; $F = LT(1-u)/(L+M)$; $G = g(1-u)/u(L+M)^2$; $T' = T(u^2L+M)/u(L+M)$; $E+F+T' = T$

$L' = Lu^2(L+M)/(u^2L+M)$; $C' = \lambda(u^2L+M)/u^2(L+M)^2$; $M' = M(L+M)/(u^2L+M)$; $K' = \mu(u^2L+M)/u(L+M)^2$; $L'+M' = L+M$

FIG. A1. Transformation 1

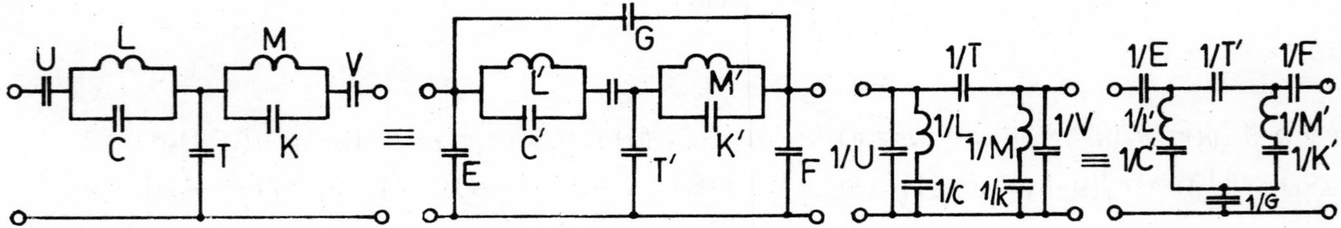

$$a = LU^2 + MV^2;\ b = M(V+K) - L(U+C)\ ;\ d = MV^2 - LU(V+T)\ ;\ E = UT/(U+V+T)\ ;\ F = T[MKV - L(U^2+UV+VC)]/b(U+V+T)$$

$$G = U[MKV + L(UT - VC)]/b(U+V+T)\ ;\ U' = dU/b(U+V+T)\ ;\ T' = aT/(U+V+T)b\ ;\ L' = b^2 LMV^2(U+V+T)^2/ad^2$$

$$C' = ad[CV^2 - KU(V+T) - UVT]/b^2V^2(U+V+T)^2;\ M' = b^2/a\ ;\ K' = a[MKV + LC(U+T) + LUT]/b^2(U+V+T)^2$$

FIG. A2. Transformation 2

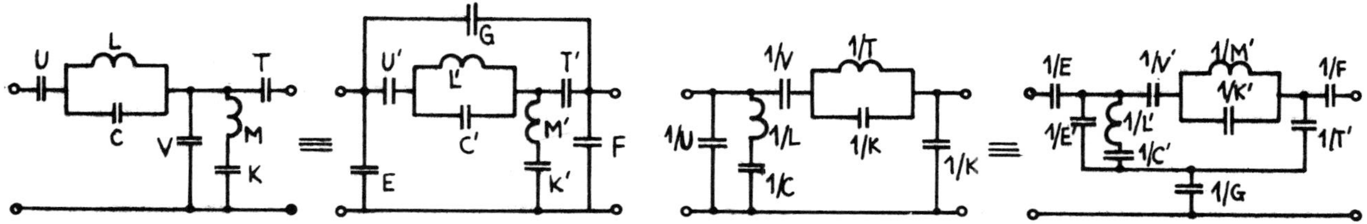

$a = CK(U+T)+U(T+V)(K+V)$; $b = MKV + L(UC+UT+CT)$; $d = UC+(U+C)(T+V)$; $f = K(UC+UT+CT)-U^2V$

$e = L(K+V)(UC+UT+CT) - MKV(U+T)$; $g = L(UC+UT+CT)^2 + MU^2V^2$; $E = UVC/d$; $F = TV(U+C)/d$; $T' = TKg/dc$

$G = UVT\{LC(UC+UT+CT) - MK[C(U+T)+U(T+V)]\}/de$; $U' = KUag/def$; $L' = d^2e^2/a^2g$; $M' = LMd^2/g$

$C' = agK[LC(UC+UT+CT) + MUV(T+V)]/d^2e^2$; $K' = Kg/bd$

FIG. A3. Transformation 3

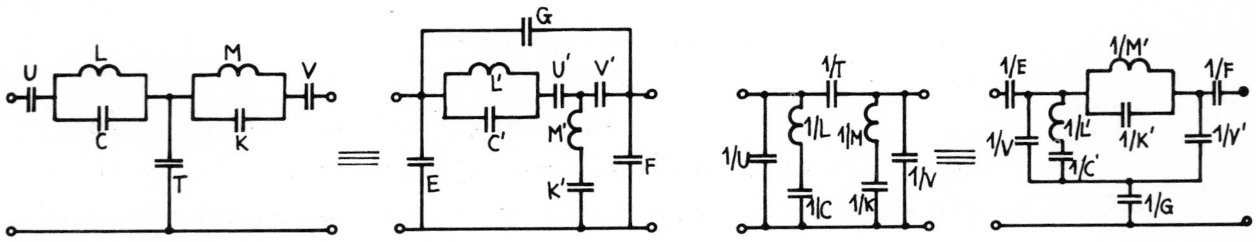

$a = U(VK+TV+TK) - CV^2$; $b = V(UC+UT+CT) - KU^2$; $d = UC(V+K) + VK(U+C) + T(V+K)(U+C)$

$e = MU^2(V+K)^2 + LV^2(U+C)^2$; $f = M(V+K)(U+T) + LV(U+C)$; $g = L(U+C) - M(V+K)$

$E = UTC(V+K)/d$; $F = VTK(U+C)/d$; $G = U(V+K)\{LVC(U+C) - M[TU(V+K) + VK(U+C)]\}/gd$

$V' = eT/dg$; $U' = aUV'/Vb$; $L' = V^2 g^2 d^3/a^2 e$; $C' = ae\{MU(V+K)(VK+TV+TK) - LCV^2(U+C)\}/V^2 d^2 g^2$

$M' = LMd^2/eT^2$; $K' = eT^2/df$.

FIG. A4. Transformation 4

RESPONSE OPTIMIZATION OF LOSSY TWO-PORT NETWORKS

J. K. SKWIRZYNSKI

Research Division, Marconi Co. Ltd.
Gt. Baddow, Essex, England

INTRODUCTION AND MOTIVATION

Optimization techniques can be successfully applied to overcome the limitations imposed by classical methods of network synthesis. Since the analysis of quite large networks is feasible with a computer, an attractive method of design is the successive adjustment of the circuit parameters of such a network in order to achieve the desired response. We shall concentrate here on applications of some well-known optimization algorithms to the improvement of responses of two-port networks and in particular of electrical frequency filters. This seems to be an attractive proposition for several reasons:

1) Two-port networks can be speedily analyzed by using the chain matrix methods. It is therefore possible to study and evaluate relative merits of various search techniques for locating minima of many-variable functions, by applying them to eminently practical problems.

2) It can be shown that the chain matrix analysis allows for equally speedy calculation of response derivatives.

3) Constraints of component values can be readily handled by substitution or with the aid of penalty or barrier function methods.

4) It is interesting to compare optimization of two kinds of network problems. On the one hand ladder filter networks designed by synthesis provide examples of systems in which effects of components are highly interactive, that is their variations are reflected on responses almost everywhere in the frequency band. This provides a practical field for experimentation with optimization algorithms and with differently formulated error criteria for complicated many-dimensional functions, with many minima, banana-shaped valleys, ridges etc. which are so dear to specialists in this field of non-linear programming. On the other hand, there are important circuits and types of design techniques (such as constant-resistance amplitude equalizers, or all-pass group delay equalizers, or location of frequencies of zero transmission in filters by the template methods), where there is a minimal interaction between groups of parameters of the optimized response function. This function is then a sum of independent contributions due to network sections (or terms), connected in tandem, each section contributing in a limited part of the frequency band. The 'weakly-interacting parameter' problem will not be considered here as there are several reports already published in this field [1].

5) Finally, the so-called insertion loss method of filter design has severe practical limitations. The method is applicable to reactive networks built in a ladder structure and operating between resistive loads, whereas practical components (e.g. inductances) are associated with stray effects such as dissipation, with serious consequences on the relation between theoretical and practical responses of selective filter networks. The effects of strays can be readily modelled for analysis and therefore can be included in the construction of objective functions for optimization purposes. Also, the insertion-loss method consists of two distinct steps (approximation and realization) and a choice in the first step necessarily curtails the scope of success in the second. The use of optimization algorithms seems to be a good strategy to accomplish designs which are independent of limitations of the network theory, particularly since a theoretical response provides a good guess to start searching for optimal conditions.

CONSTRUCTION OF OBJECTIVE FUNCTIONS FOR LADDER FILTERS

Practical filters have a general ladder structure (including twin-T and lattice sections). The voltages and currents at any ladder node are here denoted by vectors $(V_i, I_i)^t = a_i$, where t indicates the transpose. Then

$$a_i = T_i \, a_{i+1} \tag{1}$$

where T_i is the chain matrix of the network section between nodes i and (i+1). The elements of T_i are rational functions of the complex frequency p, where Im(p) = 2πf and f is the frequency. For reciprocal networks $\det(T_i) = 1$. Then, if a_0 is the generator vector and a_n is the load vector:

$$a_0 = T_0 \, T_1 \ldots T_n \, a_n = \left(\prod_{s=0}^{n} T_s \right) a_n = T \, a_n \tag{2}$$

In particular, for a series generator resistance (normalized to unity) and a shunt load conductance (normalized), we have:

$$T_0 = \begin{pmatrix} 1 & 1 \\ 0 & 1 \end{pmatrix} ; \quad T_n \begin{pmatrix} 1 & 0 \\ g & 1 \end{pmatrix} \tag{3}$$

and

$$\prod_{s=2}^{n-1} T_s = \prod_{s=2}^{n-1} \begin{pmatrix} A_s & B_s \\ C_s & D_s \end{pmatrix} = \begin{pmatrix} A & B \\ C & D \end{pmatrix} \tag{4}$$

where A, B, C and D are rational functions of p and bilinear functions of any network element value. Therefore

$$a_0 = \begin{pmatrix} A+C+g(B+D) & B+D \\ C+gD & D \end{pmatrix} a_n = T \, a_n \tag{5}$$

The conventional effective transmission loss is defined as 20 log S (dB) where

$$S = \frac{1}{2(g)^{\frac{1}{2}}}[A + C + g(B + D)] \tag{6}$$

User-orientated program specifications for the transmission loss are given in dB values, but it is advisable inside the program to use an objective function in terms of $F = |S|^{-1}$. The inverted function is used because in the stop-bands of filter responses the transmission loss may vary from values as low as 20 dB to infinite values at poles of transmission and it is better to optimize a function confined to the range (0,1). Other functions of S may be constructed to optimize phase or group delay responses, or to achieve optimal impedance matching at the input and/or output ports. One can also optimize a suitably weighted sum of these.

An objective function is chosen to decide which of two given response curves, plotted against frequency, is better with respect to some requirement. One defines a scalar measure of the error which is a function of m points on the curve and of k values of network elements (note that $k \geq n$, since there may be more than one circuit component in a given network section). Rather than using a single value of specification at each frequency point, it is more practical to replace this by upper and lower bounds of the specification. Two alternative error criteria are generally used: the sum of squares (or other even powers), or maximum modulus of error. The first is readily adapted to optimization algorithms which are based on the Taylor expansion: the second is acceptable for direct search, pattern search [2], or simplex [3] procedures, since these do not require the existence of first and second derivatives of the objective function with respect to element values in a filter, neither of which can be defined for the maximum modulus error.

Let $F_r = F(f_r;e_1,e_2,\ldots,e_k) = F(f_r;\mathbf{e})$ be a response calculated from equs. (1-6) at a sample frequency f_r for a set $\mathbf{e}$ of k network elements, which we require to vary. In a well organized optimization program a facility should be available to vary only a designated number of network elements. Such a facility is particularly useful for designing a matching network, when the latter is to be optimized in the presence of a fixed transducer and the response of both networks has to be calculated at each stage of optimization.

In general it is not easy to specify an exact target at a sample frequency point, but acceptable limits for the result are usually known. In this case the sum of squares of errors may be modified to have zero contribution to the overall error function from a quantity which is within the prescribed limits, and a contribution proportional to the square of the amount by which the quantity exceeds the limits where this is the case. In other words, if the response at any sample point is within the prescribed limits, the specification is taken as that value of the response; when the limits are not satisfied, the specification is taken to be the value of the nearest limit. With this type of policy differentials have only a meaning when associated with one particular evaluation of the error function, since the specification may be different in the following evaluation. Such a policy can be used only with methods of optimization which do not require continuous differentials. It has been found successful when used with pattern search and simplex algorithms in the 'opening game' of a general search (see below).

An error function which can be used with any algorithm is constructed as follows:

Let $F_{min,r}$ and $F_{max,r}$ be the maximum and minimum values of the acceptable specification limits of a filter response at the r-th frequency sample point. Let also:

$$F_{av,r} = \frac{1}{2}(F_{max,r} + F_{min,r})$$

be the corresponding mid-point between these limits. The situation is illustrated in Fig. 1. Consider the error function:

$$K_r = \frac{2F_r - F_{max,r} - F_{min,r}}{F_{max,r} - F_{min,r}} = \frac{F_r - F_{av,r}}{F_{max,r} - F_{av,r}} \qquad (7)$$

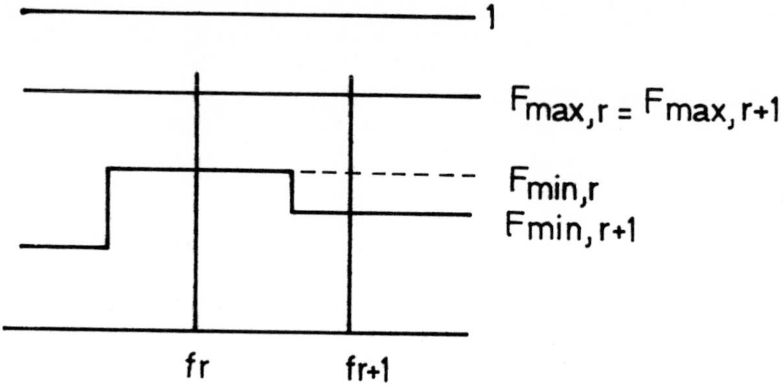

FIG. 1

A term K_r varies between -1 and +1 when the response F_r lies between the limits $F_{min,r}$ and $F_{max,r}$ of the specification and increases linearly when F_r lies outside these limits. A suitable error function used as the objective function in an optimization algorithm is:

$$P(\mathbf{e}) = \sum_{r=1}^{m} w_r (K_r)^{\alpha_r} \tag{8}$$

where w_r = weighting factor,
α_r = even integer, which may be varied from sample to sample to affect relative weighting (for least square error functions $\alpha_r = 2$).

Therefore the objective function $P(\mathbf{e})$ is continuous and has continuous differentials. It can effect arbitrarily heavy penalization of sampled responses which lie outside limits of specification. We usually specify $F_{av,r}$ by taking into account a constant transmission loss due to dissipation with practical components. A filter is first designed by the synthesis method, or a rough model is provided using image parameter techniques. Then practical components are modelled (for strays and dissipation) and the resultant circuit is analyzed to give reliable values for $F_{av,r}$. Care is taken in critical cases to share these values between the filter and its amplitude equalizer, in order to avoid sensitive conditions of pre-distorted filters.

When frequency sample points are chosen, particularly in the pass-band of a filter, proper account should be taken of usual distribution of Hurwitz roots in the complex p-plane and of the number of possible amplitude ripples which is governed by the complexity of the network. This is important as a success of given optimization game may crucially depend on the choice of sample points. Obvious benefits of a well-chosen objective function can be severely lessened by an injudicious choice of sample points. This is probably most clearly illustrated by the exaggerated case shown in Fig. 2. If the error function involves only points B_1 and B_2, result ① would produce a lower value for an objective function than result ② However, if points A_1, A_2, A_3 and A_4 are chosen, a more realistic estimate of the value of the two results can be made.

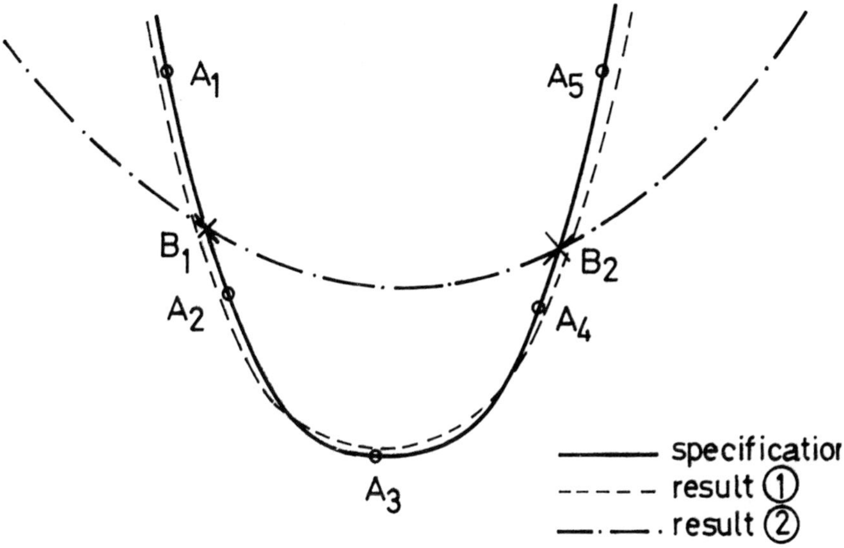

FIG. 2

CALCULATION OF AN OBJECTIVE FUNCTION AND OF DERIVATIVES

The objective function $P(\mathbf{e})$ in Eq. (8) is a sum of generalized squares of non-linear functions K_r of the independent variable $\mathbf{e} = (e_1, e_2, \ldots, e_k)^t$. It is therefore suitable for using a Newton method of residual minimization. One of the major problems with optimization is the amount of computer time it can use. This time is almost entirely taken by repetitive network analysis. It is therefore important to organize the analysis in the most economical but also sufficiently general fashion.

Since there is a continuous need to experiment with new or improved optimization algorithms, the analysis of networks has to be arranged in such a way that as far as possible the presentation of data to the program has a common format which can be readily extended. This means that there is a standard set of network branches and sections which can be used in the design of two-port networks by optimization. A part of this set (for ladder branches of selective filters) is shown in an ordered fashion in Table 1. The optimizable variables are the capacitor, inductor and resistor values, when these are discrete, but in the branches where an inductor and a capacitor are tuned (to produce a pole of attenuation), the optimizable variables may become optionally the inductor and the tuning frequency. The series and parallel resistive losses are simulated for variation of Q-factors of inductors with frequency, particularly with wide-band filters. It is possible to fix certain components or to relate them in a constrained fashion (see below). A suitable, user-oriented code has been devised for this purpose.

In several algorithms for optimization it is necessary to find the differentials of $P(\mathbf{e})$ with respect to each variable element e_j at each sample frequency f_r. There are two ways in which this may be done. We shall concentrate on differentials of F since the rest is straightforward. The first way is to find an approximation to the differential of F with respect to e_j by forming:

$$\left.\frac{\partial F}{\partial e_j}\right|_{f=f_r} = F_r(e_1,e_2,\ldots,e_j+\Delta e_j,\ldots,e_k) - F_r(e_1,e_2,\ldots,e_k)$$

The relative increment is then usually chosen to be of

TABLE 1

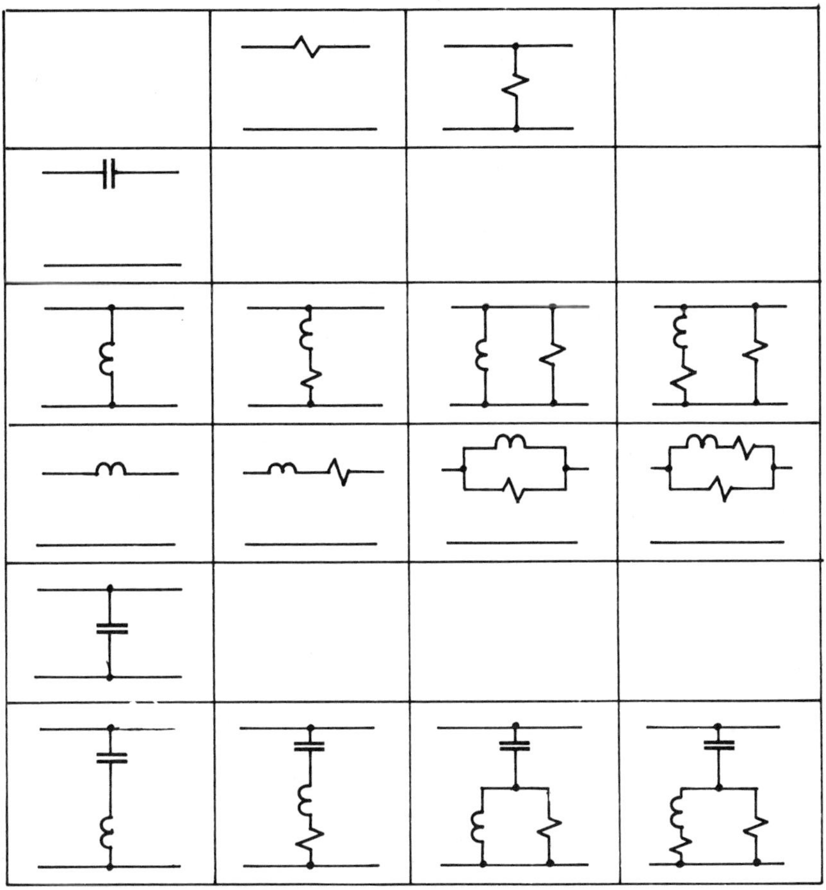

the order of 10^{-5}. An optional modification is to replace the gradient derivatives with normalized derivatives [1]:

$$\frac{\partial F}{\partial \ln(e_j)} = \frac{\partial F}{\partial e_j} \cdot e_j$$

and thus remove the effect of scale.

The other way is to utilize the fact that the differentials of S may be explicitly defined:

$$\frac{\partial F_r}{\partial e_j} = - |S|^{-3} \left\{ Re(S) \frac{\partial}{\partial e_j} [Re(S)] + Im(S) \frac{\partial}{\partial e_j} [Im(S)] \right\} \Bigg|_{f=f_r} \tag{9}$$

The efficient formulation of these differentials is carried out by first forming the chain matrix in Eq. (2) at $f = f_r$. S and $|S|$ can then be found by substitution in Eq. (6). The n sets of values of chain matrix elements are stored as they are formed throughout this part of computation. Now let us assume that the element e_j is contained in the network section represented by the chain matrix T_q ($e_j \in T_q$). Note that e_j is defined as a 'component' in the general sense; it may be an element value, or frequency value, or a relation constraint (see below). Let

$$T_{q,e_j} = T_{q,e_j}(f_r) = \frac{\partial T_q}{\partial e_j} \Bigg|_{f=f_r} \qquad (e_j \in T_q) \tag{10}$$

Then from Eq. (2):

$$T_{ej} = \left(\prod_{s=1}^{q-1} T_s \right) \left(T_{q,e_j} \right) \left(\prod_{s=q+1}^{n} T_s \right) \tag{11}$$

After evaluation of S we put the product in the first bracket in Eq. (11) equal to the unit matrix, and the product in the third bracket in Eq. (11) equal to T. Increasing the first product and decreasing the third product by successive matrix elements T_q ($q = 2,3,\ldots n-1$), the values of which are available, the differentials can be evaluated, either approximately or explicitly, with the minimum amount of matrix manipulation.

THE GENERAL OPTIMIZATION GAME

A general optimization game consists of three stages; the *opening game* (when the contours of objective function are well dispersed and a simplex method makes a rapid progress to encounter a valley; this stage is used when a design is started from a mad guess or from arbitrarily selected values in an arbitrary network topology; we shall not consider it here, as it is assumed that the synthesis method will always produce a good starting guess), the *middle game* (which uses a variant of Newton-Gauss procedure [4]; this stage is the most commonly used with best reported results in everyday filter design practice; it is described below) and *end game* (when convergence in the middle game slows down and it is necessary to switch over to the procedure suggested by Davidon [5] for improvement of the Hessian; this optimization algorithm is well documented and will not be described here).

A major drawback of adopting a two- or three-stage approach is the difficulty of getting the computer to recognize the change-over and the size of the program involved. In practice therefore the change is usually effected by the use of two or three separate programs and the applied judgment of the designer. Therefore it is practical to concentrate on the middle game algorithm which we have found to be almost invariably successful for filter design applications.

THE DAMPED LEAST SQUARE METHOD [4]

We wish to minimize $P(\mathbf{e})$, where for this purpose we assume that $\alpha_r = 2$ for all r and furthermore, for notational simplicity, we neglect in what follows the weighting factors by assuming $w_r = 1$ for all r. Then in Eq. (8):

$$P = \sum_{r=1}^{m} K_r^2 \qquad (12)$$

where K_r are non-linear functions of the independent variables e_j. The K_r are then residuals, i.e., the deviations between the actual values of filter responses and the specified values. We know that when P is zero it is also a minimum.

If the variables $e_1, e_2, \ldots, e_k$ are trial solutions, we write $e_j = e_{j0}$ and replace Δe_j for $(e_j - e_{j0})$ and K_{r0} for $K_r(e_{10}, e_{20}, \ldots, e_{k0})$. Then

$$P \simeq P_{app} = \sum_{r=1}^{m} (K_{r0} + k_{r1}\Delta e_1 + \ldots . + k_{rk}\Delta e_k)^2 \qquad (13)$$

where k_{ri} are the appropriate derivatives discussed above. Thus

$$P_{app} = b_0 + 2 \sum_{j=1}^{k} b_j \Delta e_j + \sum_{j=1}^{k} \sum_{i=1}^{k} b_{ij} \Delta e_j \Delta e_i \qquad (14)$$

$$\text{where } b_0 = \sum_{r=1}^{m} K_{r0}^2 ; \; b_j = \sum_{r=1}^{m} K_{r0} k_{rj} \quad (j \neq 0)$$

$$b_{ij} = \sum_{r=1}^{m} k_{ri} k_{rj} \qquad (15)$$

At a stationary value of the error function P the derivatives of this function with respect to each variable Δe_j will be zero. We assume that the values $(e_{j0} + \Delta e_j)$, obtained at the end of an iteration, define the stationary point. The condition is expressed by the k simultaneous equations:

$$\frac{\partial P}{\partial \Delta e_j} = 2b_j + 2 \sum_{i=1}^{k} b_{ij} \Delta e_i = 0 \qquad (16)$$

It should be noted that the quadratic approximation P_{app} is not precisely the one that one would obtain by computing the first and second derivatives of P. The expression in Eq. (14) is obtained by squaring the linear expansion of K_r, rather than by expanding, up to the second derivative, the expressions for K_r^2. The 'true' expansion obtains a modified expression for b_{ij} in Eq. (15):

$$b^*_{ij} = b_{ij} + 2 \sum_{r=1}^{m} K_{r0} k_{rij} \tag{17}$$

where k_{rij} is the second derivative of K_r with respect to e_i and e_j. The quantities b^*_{ij} are elements of the Hessian matrix. Nevertheless it is advisable to use expression in Eq. (16) for optimization, for there are certain advantages in using b_{ij} : for example the matrix (b_{ij}) is positive definite, so that the solution vector $(\Delta e_j)^t$ definitely tends to lower values of P. This is a useful property not necessarily shared by the matrix (b^*_{ij}). Furthermore, in the vicinity of minimum K_{r0} are small and $b^*_{ij} \to b_{ij}$.

In practical network optimization the method described so far is very prone to overshoot and oscillate. The remedy is to use the Levenberg-Marquardt method [6] by minimizing not P_{app} but

$$P_{mod} = P_{app} + \lambda \sum_{j=1}^{k} (\Delta e_j)^2 \tag{18}$$

where λ is known as the damping factor or Lagrange multiplier. The modified form of Eq. (16) now becomes:

$$b_j + 2 \sum_{i=1}^{k} (b_{ij} + \lambda\delta_{ij}) \Delta e_j = 0 \tag{19}$$

where δ_{ij} is the Kronecker delta. The new terms appear as additions to diagonal terms of the matrix (b_{ij}), which remains symmetrical. These terms have the effect of reducing the values of Δe_j produced, especially for the variables where these changes would otherwise be largest, and hence they reduce overshoot.

When λ is small the method behaves as for the undamped case, giving good convergence when very close to a minimum, but generally shooting away from the direction to a minimum. When λ is large, the method reduces to that known as 'steepest descent', giving values of Δe_j proportional to individual gradients b_j with a step size inversely proportional to λ. It is therefore advisable to begin optimization with a large λ (usually unity) and to reduce it (by halving) as optimization proceeds. In general, one only changes λ between iterations on the basis of the rate of improvement of the error function. Very slow convergence indicates too high a value of λ, while divergence indicates too low a value of λ.

CONSTRAINTS ON NETWORK COMPONENTS

There are two main types of constraints on element values in filter networks:

1) 'Value Brackets'. The cost of networks is considerably reduced when one uses manufacturer's 'preferred values' of components, particularly capacitors and crystals. Then the value of a component is constrained by the inequality

$$e_{i-} < e_i < e_{i+}$$

This is replaced by [7] : $e_i = e_{i-} + (e_{i+} - e_{i-})\sin^2\theta$ and θ is now the variable parameter while bounds of e_j are fixed.

2) 'Limiting Ratios'. This constraint is particularly important during the design of crystal filters, high frequency filters and filters operating with high power. In the first case it is necessary to maintain realistic ratios between the holder capacity (C_0) and the tuning capacity (C_i) of a simple equivalent circuit of a crystal. Thus

$$C_i C_0^{-1} \leq \tau < 1$$

is replaced by $\tau C_i = C_0 \sin^2\theta$ and C_0, θ are the variable parameters.

Penalty and barrier function techniques for constrained optimization require sequential improvements of Lagrange multipliers and the extra complication is not warranted by simple constraint inequalities encountered in practical filter design.

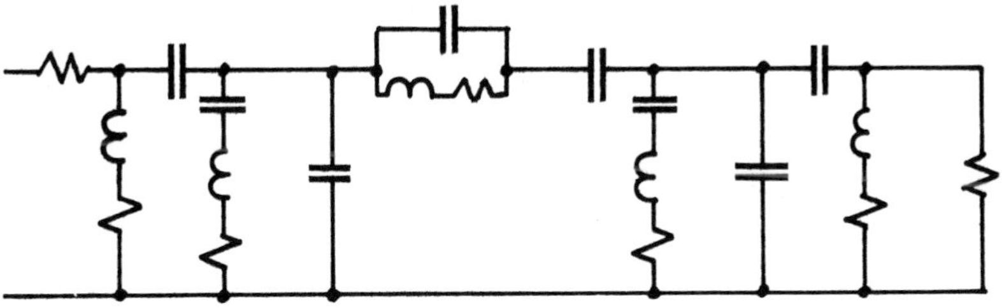

FIG. 3

EXAMPLE AND CONCLUSIONS

The network configuration is shown in Fig. 3. This is a band-pass filter with a pass-band in the KHz range. It was first designed by the synthesis method and analyzed; the dissipation to be expected from practical inductances (Q $\sim$ 200) is simulated by series resistances. The resultant response is shown in Fig. 4 (continuous curve). The value of attenuation at the band-pass edges (about 1.5 dB) was chosen for F_{av} at the sampled reference frequency of 28.8 KHz. The specification is drawn and the response should lie within the dB range limited by broken lines. The stop-band attenuation presented no problems and is not shown. Several optimization programs, for approximately equal lengths of computer run, were tried. The simplex method produced a result shown by the dotted curve. The damped least square method was then tried and produced an excellent result (shown by broken curve) even with slightly tightened specification (shown by the shaded portions).

The difference in the number of evaluations quoted in Fig. 4 may appear surprising, but it must be remembered

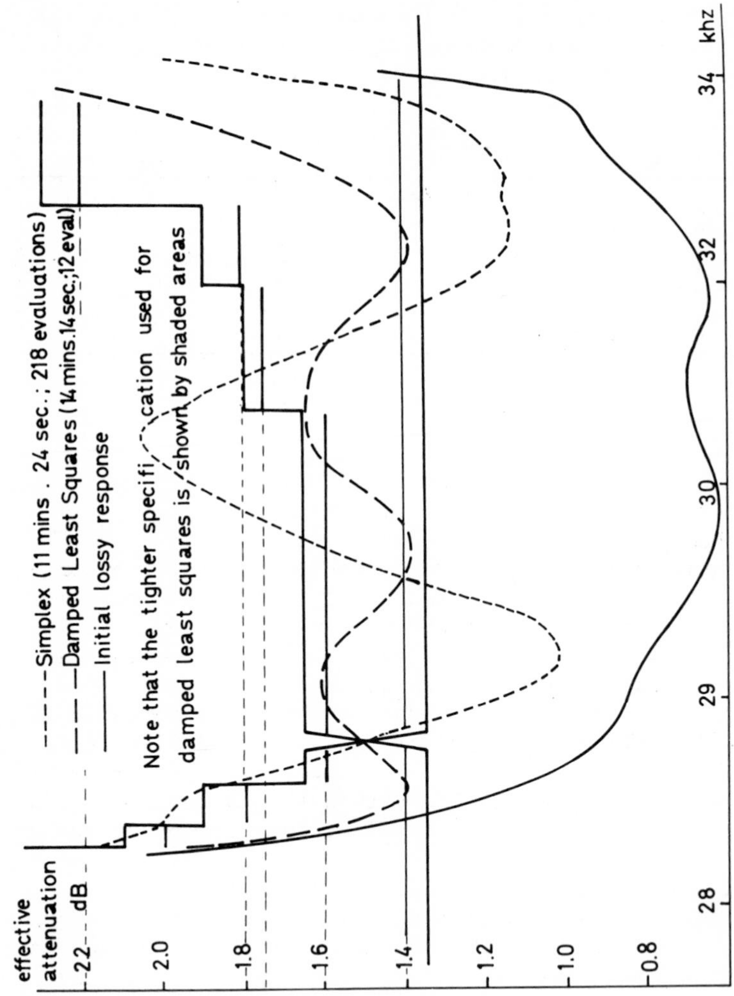
effective attenuation dB
Simplex (11 mins . 24 sec.; 218 evaluations)
Damped Least Squares (14 mins.14 sec.;12 eval)
Initial lossy response
Note that the tighter specifi cation used for
damped least squares is shown by shaded areas
2.2
2.0
1.8
1.6
1.4
1.2
1.0
0.8
28
29
30
32
34
khz

FIG. 4

that evaluation in the simplex method involves finding only the responses. In the least squares method it involves finding the response and its differentials with respect to each variable at each sample frequency. This problem involved 14 variables and 23 sampled frequencies.

The program package described here has been used extensively for design of complex filters (up to 30 components), crystal filters and transducer matching networks. A similar program was devised for design of microwave filters (waveguide with rods, interdigital etc.).

It is hoped that this sketchy outline gives some idea how an application of simple optimization algorithms can expedite difficult design problems.

ACKNOWLEDGMENTS

I wish to thank the Director of Research of the Marconi Company for permission to publish this article. Miss M. Sadler and Mr. J.T.B. Musson are specially thanked for preparation of the program, for devising the policies of optimization and for preparing an exhaustive report which has been used as the basis of this contribution.

REFERENCES

1. SCHEERER, W.G. (Feb. 1967) Optimization with Computers, in: *Circuit Design by Computer Symposium*, New York University, School of Engineering and Science, Department of Electrical Engineering, University Heights, Bronx, New York 10453.

2. HOOKE, R. and JEEVES, F.A. (1959) Direct Search Solution of Numerical and Statistical Problems, *J.A.C.M. 8*, p. 212.

3. NELDER, J.A. and MEAD, R. (1965) A Simplex Method for Function Minimization, *The Computer Journal 7*, p. 308.

4. BEALE, E.M.L. (1967) Numerical Methods, in: *Nonlinear Programming*, (Editor J. Abadie), Amsterdam: North-Holland Publishing Company.

5. DAVIDON, W.C. (1968) Variants Algorithm for Minimization, *Computer Journal 10*, p. 406.

6. MARQUARDT, D.W. (1963) An Algorithm for Least-Square Estimation of Nonlinear Parameters, *Journ. S.I.A.M. 11*, p. 431.

7. POWELL, M.J.D. Private communication.

SPARSE NETWORK ANALYSIS

M.L. BLOSTEIN AND D.J. MILLAR

Department of Electrical Engineering,
McGill University, Montreal, Canada

1. THE FREQUENCY ANALYSIS OF ELECTRICAL NETWORKS

The rapid growth of computer-oriented design techniques has forced network analysts to evaluate more critically the various methods of performing circuit analysis. Designers have come to recognize the power of simulation, mathematical programming and optimization theory as design tools, particularly in conjunction with integrated circuitry. The practical implementation of these sophisticated techniques, however, relies heavily on very efficient algorithms to generate not only network solutions, but also gradients of these solutions with respect to design parameters [1]. In these lectures we shall review briefly a number of existing techniques for generating the frequency response of linear circuits and then introduce a new technique which appears very promising in relation to current needs.

The most direct method of generating the frequency response of a linear circuit is to solve a set of loop or nodal equilibrium equations at each frequency point of interest. The solutions may be obtained by either matrix inversion or triangular decomposition; however, there are significant differences in the computational effort required by each technique [2]. The inversion of an arbitrary $n \times n$ matrix requires approximately n^3 operations (an operation is defined loosely here as a multiply-add) while generating the solution with a full righthand side requires a further n^2 operations. On the other hand, triangular decomposition techniques require only $n^3/3$ operations for triangularization and $n^2/2$ operations each for forward and backward elimination. Clearly, triangular decomposition techniques require fewer operations irrespective of how many different righthand sides are used.

From a network standpoint, the computational advantages of triangular decomposition methods are hardly surprising, since the inverse method generates more information about a network. A nodal impedance matrix permits the direct computation of transfer functions between *any* two node-pairs in a network, while the triangularization process provides data for the direct computation of a single transfer function. Consequently it requires about three times as much work to evaluate all possible transmissions in a network than to evaluate only one. Carrying out the forward and backward elimination steps enables the triangular decomposition technique to compute the transmission between a specific input and any other node-pair. Such eliminations, each with an appropriate unit vector on the righthand side, are required to generate a matrix inverse and the capacity to calculate directly the transmission between arbitrary sets of node-pairs.

The preceding discussion is very pertinent to the computation of sensitivity coefficients (derivatives) of network functions with respect to circuit elements. The following theorem is basic to the calculation of sensitivity coefficients:

Let k represent a circuit element which at a branch level relates network variables (current or voltage) α and β by the simple formula $\alpha = k\beta$. Let E and R represent excitation and response variables, respectively. Then, $F = R/E$ is a bilinear function of k and $dF/dk = -F_{\beta E} F_{R\alpha}$ where $F_{\beta E}$ is the transmission from the excitation port to the input port of k, and $F_{R\alpha}$ is the transmission from the output port of k to the response port. $F_{\beta E} = \beta/E$ and $F_{R\alpha} = R/\alpha'$, where α' is an excitation of the same type as α but applied in either series or shunt with the output port of k, depending whether α is a voltage or current, respectively.

Thus computing a derivative of a network function with respect to a circuit element is equivalent to computing two network transmissions between appropriate node-pairs. The theorem is not new and can be proved very easily by means of signal flowgraphs. It is important, however, as it shows that a nodal inverse contains the necessary information to compute all sensitivity coefficients with respect to any transfer function [3]. On the other hand, a single triangular decomposition with two back eliminations can provide the same data for a specific transfer function [4]. To see this, note that all functions of the type $F_{\beta E}$ can be evaluated using one back elimination since they all have a common input port. Functions of the type $F_{R\alpha}$ do not share this property, but they do have a common output port. Thus, if we define an "adjoint" network [5] as one having a nodal inverse which is the transpose of the original inverse, functions such as $F_{R\alpha}$ become transmissions from the input port of the adjoint network to the input ports of the various elements in the adjoint network, and may be computed by a single back elimination. If LU represents the triangular decomposition of the nodal admittance matrix of the original network, where L and U are lower and upper triangular matrices, respectively, then U^tL^t represents the decomposition of the nodal matrix of the adjoint network, so that only a single triangular decomposition is necessary. Therefore triangular decomposition is more efficient than inversion in both obtaining a solution vector and performing gradient calculations for a given transmission function. The additional computation required for inversion provides the facility to compute the gradient of *every* transfer function with respect to *every* parameter. Often, this added facility is not required.

Klyuyev and Kokovkin-Sheherbak [6] have shown that no general system of linear algebraic equations can be solved in fewer operations than required by triangular decomposition. Therefore any attempt to improve the computational efficiency of a solution scheme must depend on the special character of the coefficient matrix.

There are many practical situations where network matrices are sparse, i.e., they have a large majority of zero elements, and it is reasonable to suspect that such systems can be solved in fewer than $n^3/3 + n^2$ operations. In fact, it is possible to show that the number of operatins required to solve a tridiagonal system is proportional to n rather than n^3. Tinney and Walker [7] have studied the situation where the sparsity structure is arbitrary and have reported very significant increases in computational efficiency by processing the rows in an order which tends to keep the triangular matrices, L and U, sparse. The ordering algorithm may require as much time as a single solution; however, this cost is negligible if the same sparsity structure is solved repeatedly.

Our discussion has concentrated on generating frequency response characteristics by repeatedly solving a system of loop or nodal equations. A variety of other methods have been proposed, but the most promising derive from state variable methods of analysis [8-10]. The methods fall into two categories: (1) those based on eigenvalue determinations of the state matrix, and (2) those based on computing $(j\omega I - A)^{-1}$ where A is the state matrix, I is the identity matrix and ω represents frequency. Kinariwala [8] has proposed to achieve the second approach iteratively by inverting the matrix at one frequency and finding the inverse at other frequencies by relaxation, using the response at the last point as a starting guess. Some efficiency can be obtained since only diagonal elements change from point to point. At this time the effectiveness of the method cannot be evaluated. A program based on the eigenvalue approach has been prepared and distributed widely by Pottle [9]. Preliminary results show that for large networks, unless very many frequency points are required, sparse network solution techniques applied to the conventional nodal equations are more efficient and require considerably less fast-access memory. Undoubtedly the method would be more attractive if very efficient techniques could be found to determine the eigenvalues of sparse matrices. At the present time neither of the methods is attractive from the viewpoint of calculating gradients with respect to circuit elements.

2. AN EFFICIENT ALGORITHM FOR SPARSE NETWORKS

(A) *The Link-at-a-time Algorithm*

A completely different technique of evaluating network functions using nodal equations was described recently by Pinel and Blostein [3]. In this technique, a network is torn apart into a tree with uncoupled two-terminal elements, uncoupled link branches, and voltage-controlled current sources; the nodal inverse of the tree network is then found and modified recursively as the links and controlled sources are restored to the network sequentially.

Let Z represent a nodal impedance matrix and assume that a dependent source is added to the network in which the current in a branch between nodes p and q is controlled by a voltage across nodes m and r. It may be shown [3] that the modified nodal impedance matrix $\tilde{Z}$ can be written as

$$\tilde{Z} = Z - (1/d)\beta\theta'$$

with

$$d = 1/y + z_{mp} + z_{rq} - z_{rp} - z_{mq}$$

$$\beta' = (z_{1p} - z_{1p}, z_{2p} - z_{2p}, \ldots\ldots, z_{np} - z_{nq})$$

$$\theta' = (z_{m_1} - z_{r_1}, z_{m_2} - z_{r_2}, \ldots\ldots, z_{mn} - z_{rn})$$

where y is the transadmittance of the source. If a link of admittance y is added to the network between nodes p and q, the modification is similar except that m = p and r = q.

A simple algorithm involving no multiplication enables the tree inverse to be evaluated in a time which is negligible in comparison with that required for the link addition procedure. Since Z is symmetrical until controlled sources are introduced, the links are restored prior to the controlled sources and the total number of matrix updates necessary for generating the inverse is approximately $n^2(s + \lambda/2)$. This must be compared with the n^3 operations required by inversion schemes based on Gaussian elimination. λ and s represent the number of links and sources, respectively.

Two significant extensions to the link-at-a-time algorithm are described in Sections 2(B) and 2(D), respectively.

(B) *The Branch-at-a-time Algorithm*

The efficiency of computing the nodal impedance matrix can be improved markedly in sparse networks by building up the solution one branch-at-a-time rather than a link-at-a-time, i.e., there is no reason why link branches must always be processed *after all* tree branches have been added. Suppose, for example, we know the nodal impedance matrix of an m-node symmetrical network and we wish to find the new impedance matrix when an open branch with impedance z is added to the pth node (see Fig. 1). The new matrix is simply the old matrix bordered by a row and column in which [11]

$$z_{m+1,m+1} = z_{pp} + z$$

$$z_{i,m+1} = z_{ip} \qquad i = 1, 2, \ldots, m$$

Hence, we require only one add and a "column shift" operation. The bordered row, of course, is not recorded since matrix symmetry is maintained in this process. On the other hand, if an impedance z is added between nodes p and q of the network (the addition of a closed branch), the new matrix can be found from the old in a manner analogous to the link addition described earlier. The "cost" of this addition is proportional to the square of the number of independent nodes in the network, whereas the cost of adding an open branch can be ignored. Since every branch in a network falls in either one of these categories, we have the ingredients of a recursive process to build up the nodal inverse a "branch-at-a-time". The computational cost of the process is difficult to ascertain since it equals the sum of the costs of adding each closed branch or link (open branches or tree branches are free, as remarked earlier), but in contrast to the link-at-a-time algorithm where the cost of each link is proportional to n^2, the cost of each link is now proportional only to m^2, where m is the number of independent nodes in the network at the time the link is added.

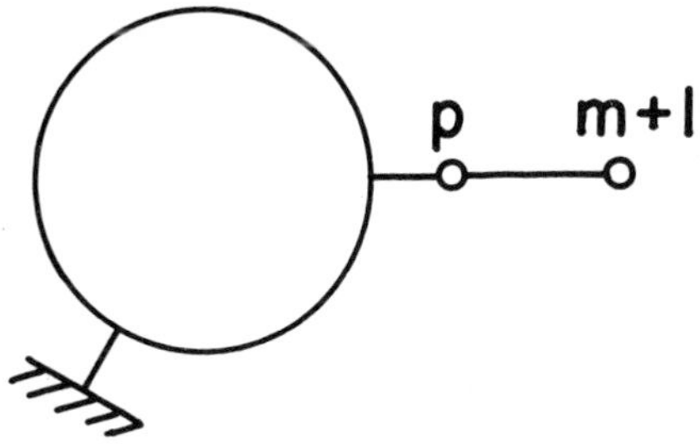

Fig. 1

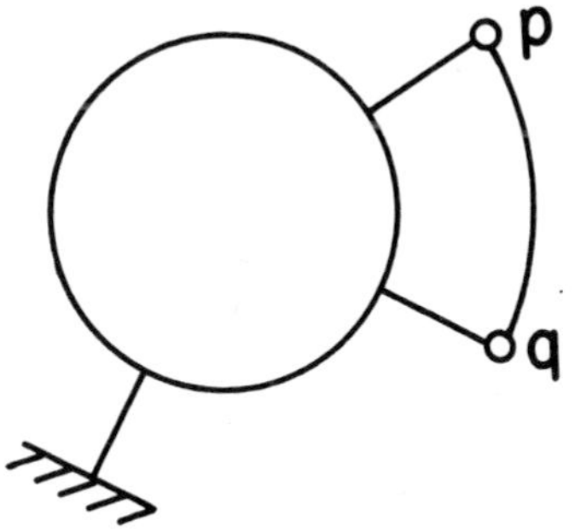

Fig. 2

Often it is not even necessary to update all entries in the intermediate solution matrix when adding a link. Consider, for example, the separable network graph shown in Fig. 3. The essential property of a separable graph is that it can be split into two separate parts by removing only one node; in Fig. 3 subgraphs A and B share only one node as do B and C, so removing the shared nodes divides the graph into three separate parts.

Elementary physical reasoning dictates that adding a link *within* C cannot alter matrix entries in the nodal inverse associated exclusively with A or B; i.e., only those nodes lying exclusively in C need to be updated. If a link is placed *within* A or B, however, the impedance to ground of the node shared by B and C changes and this change must be reflected in the submatrix associated with C. In summary, nodes lying exclusively within a separable part X need not be updated due to a link addition *within* a separable part Y provided node-to-ground paths of X do not pass through Y.

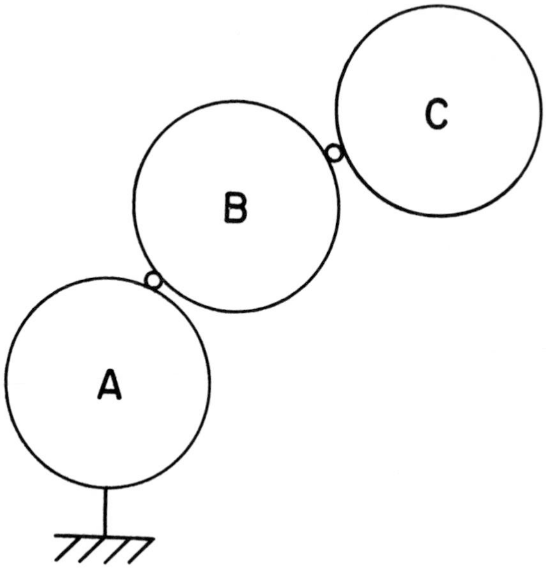

Fig. 3

The branch-at-a-time algorithm shows particular promise in the analysis of sparse networks because graphs of such networks can be decomposed into a large number of separable parts by judiciously removing certain branches, enabling only small portions of many intermediate solution matrices to require updating as the inversion proceeds. On the other hand, significant savings in computational cost can only result if the order in which branches are processed is selected carefully. Consequently, the initial phase of the algorithm must establish an ordering procedure.

(C) *Optimization of the Branch-at-a-time Algorithm*

Since the processing of tree branches is assumed to have zero computational cost, the optimization phase of the solution is in fact a link ordering process. The first step is to select a tree of the network graph and to scan the link list to determine the cost of adding each link individually to the tree. When a link is added to the tree, a minimum nonseparable part containing the link is formed and the cost of adding the link varies as the square of the number of nodes in this subgraph. The cheapest link (or links) is added to the tree and the process is repeated until all links are in place; the corresponding link order is then stepwise optimal. In this way each link defines a minimum nonseparable subgraph, called a link subgraph, containing the nodes which must be updated upon link addition. Table 1 illustrates the results of this procedure for the simple graph in Fig. 4.

In order to realize the stepwise optimal cost, the branch ordering procedure must also take into account that (1) the network graph corresponding to any partial solution must be connected, and (2) links must not be added to separable parts lying in node-to-ground paths of other separable parts (see discussion in Section 2(A)). Both these conditions can be satisfied by processing tree branches in a connected tree order and adding the links as the corresponding link subgraphs are generated. A connected tree order is defined as an ordering of tree branches $a_1, a_2, \ldots, a_n$ such that $a_1, a_2, \ldots, a_k$ is connected for $k = 1, 2, \ldots, n$ and a_1 is incident on the reference.

The final ordering scheme can be displayed succinctly by means of a pointer table or hierarchical tree graph. Fig. 5 illustrates this for the simple network of Fig. 4. Nodes 1 - 8 represent tree branches arranged from left to right in connected tree order. The remaining nodes represent links. The pointer table is constructed as the branches are processed: whenever a link is added, branches of the corresponding link subgraph which do not yet point to another branch are pointed to that link, e.g., branches 12, 5, 3 and 11 point to 13. The branches contained in a link subgraph are simply those encountered when reversing the pointer arrows and tracing back along all paths from the link to the tree branches; e.g., for link 13 they are 8, 7, 6, 5, 4, 3, 2, 10, 12, 11.

TABLE 1

Link Number	No. of Nodes in Link Subgraph	List of Branches in Link Subgraph
9	2	1, 9
10	2	7, 10
11	3	2, 4, 11
12	4	7, 10, 8, 6, 12
13	8	All except 1, 9, 14
14	9	All

The hierarchical tree graph shows in detail the actual order of processing network branches. Since branch 8 is connected to ground, it is processed first. It points to branch 12 which cannot be processed until branches 6 and 10 are added. Hence the pointer returns to the tree and puts in branch 7 followed by branch 10. Once again the pointer returns to the tree and brings in branch 6 which then enables branch 12 to be added. The procedure continues until all network branches are ordered. The final ordering in this example is 8, 7, 10, 6, 12, 5, 4, 3, 2, 11, 13, 1, 9, 14. No optimization is carried out with respect to controlled sources and these are normally added to the network after all two-terminal branches have been added.

The optimization technique is obviously dependent on the selection of the tree used in the analysis. An extreme form of this dependence can be visualized readily

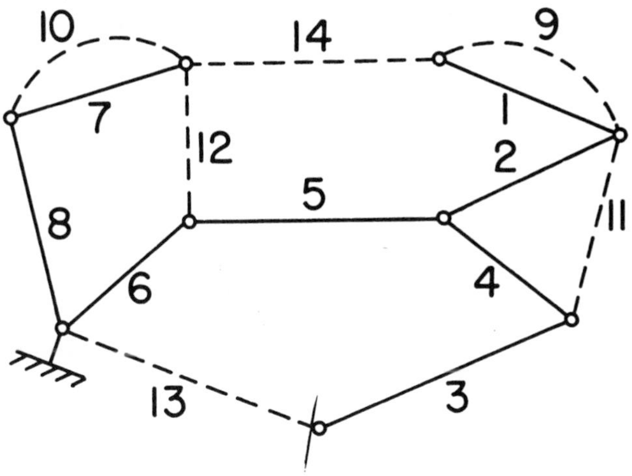

Fig. 4

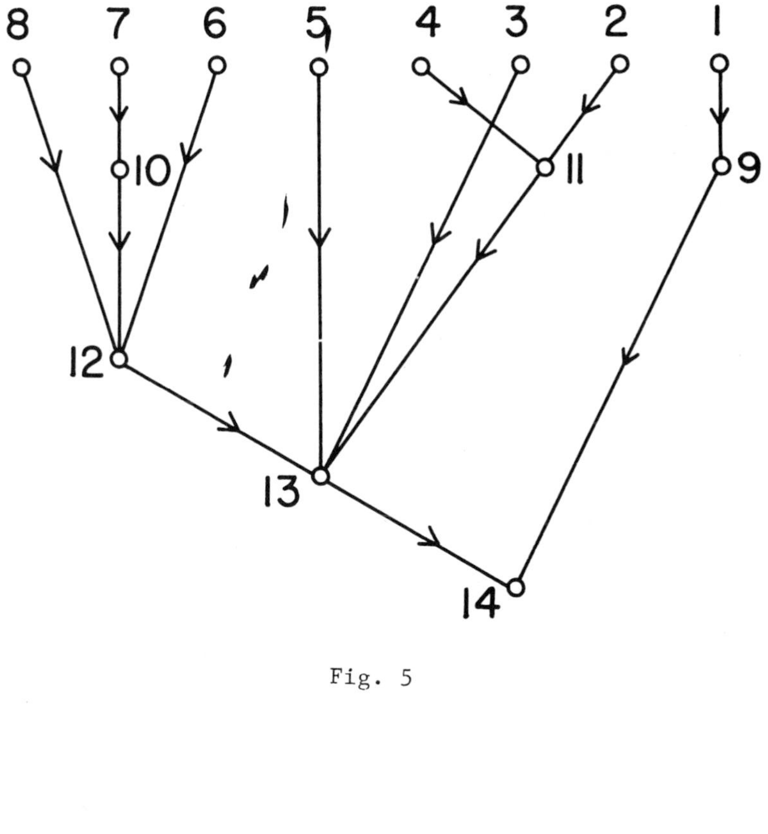

Fig. 5

by considering a simple ladder structure with (n + 1) nodes. It is easy to show that if the shunt branches form the tree, the number of matrix updates required for generating an inverse is approximately n^2, which is entirely equivalent to the various specialized techniques for inverting a symmetrical tridiagonal matrix. On the other hand, if all series branches and one shunt branch form the tree, the corresponding cost is $n^3/6$. Unfortunately, no rigorous method has yet been devised to determine an optimum tree. Two methods have been tried on a number of practical networks and have yielded good results. In the first method, each branch is weighted according to the number of other branches touching it and the tree of maximum weight is selected. In the second method, each branch is weighted according to the number of minimum loops passing through it, where a minimum loop is defined as a loop containing the branch and a minimum number of other branches. Again, the tree of maximum weight is selected. Undoubtedly, further research into this question is necessary.

Although the ordering technique for the algorithm appears complicated, it is quite feasible and requires about 250 FORTRAN IV statements. The optimization time tends to be less than the time to generate an inverse, indicating that the procedure is well-suited for frequency response calculations where many solutions are required for a fixed network topology.

(D) *Node Dropping*

The branch-at-a-time algorithm can be used to generate principle submatrices of the nodal impedance matrix in situations where a full inverse is not needed. Once all branches and controlled sources touching a node have been taken into account, subsequent matrix modifications can ignore the row and column corresponding to the node, unless the node touches the input or output of the circuit. The node is then "dropped" from the working matrix. Since each link cost now depends only on the number of undropped nodes in the link subgraph, a substantial saving in computational effort is effected. For example, if this technique is applied to the ladder problem discussed previously, the cost for finding a transfer function increases as n rather than n^2

Whenever sensitivity coefficients with respect to a set of circuit elements are required, the nodes touching these elements form part of the circuit output and cannot be dropped. A full inverse is then necessary to determine all the sensitivity coefficients for a given transfer function. The discussion in Section 1 implied that, at most, one column and two rows of the inverse are necessary to compute the same data via triangular decomposition; with the branch-at-a-time algorithm, however, the remaining matrix entries are necessary in order to update the pertinent rows and columns. Although the algorithm does not appear advantageous under these circumstances, it retains a definite advantage when the sensitivity elements touch fewer nodes. In practice, the latter represents a more realistic situation.

The nature of the ordering algorithm contributes markedly to the effectiveness of the node dropping technique. Since tree branches are processed in a connected tree order, the network tends to grow out from the ground node as fully connected as possible. This in turn permits nodes to be dropped quickly as branches are processed, so that at any time in the solution mechanism, the working matrix is much smaller than the original nodal admittance matrix.

3. DISCUSSION

An experimental program incorporating the various features of the branch-at-a-time algorithm has been prepared and some preliminary results concerning the effectiveness of the algorithm have been obtained. The program has the facility to compute any set or subset of two-port parameters in a network as well as the corresponding sensitivity coefficients with respect to any circuit element. The program can handle up to sixty-five nodes, several hundred branches and can fit easily into 128,000 bytes of core storage. Single precision arithmetic is used except for the calculation of 1/d in Eq. (1). As yet a study on effective accuracy has not been carried out.

An indication of the remarkable computational advantages of the algorithm can be seen by studying Table 2. The problem under consideration is a three stage amplifier used in telephone applications. The circuit contains 27 nodes, 59 branches and 3 controlled sources. A circuit diagram can be found in Pinel and Blostein [3].

TABLE 2

Analysis of a Three-Stage Amplifier

Solution Technique	Operation Count per Frequency Point
Inversion based on Gaussian elimination (n^3)	17,500
Inversion based on Link-at-time Algorithm	14,000
Inversion based on Branch-at-a-time Algorithm	3,700
Triangular Decomposition	5,900
Branch-at-a-time Algorithm with node dropping	520

The desired output is simply the voltage transfer function of the amplifier *without* sensitivity data so that *full* advantage of node dropping can be taken where applicable. The operation counts do not differentiate between multiply-adds and multiplies and are only approximate - within about 5% of the actual count. The table shows that the branch-at-a-time algorithm can generate the voltage transfer function nearly 12 times faster than triangular decomposition; moreover, the entire nodal inverse can be obtained more than 4 times faster than with the traditional n^3 operations. A comparison has not been made with any of the optimally ordered triangular decomposition techniques suggested by Tinney and Walker [7] and conclusions cannot yet be drawn on the relative effectiveness of these two approaches.

The branch-at-a-time algorithm with the node dropping required about 65 milliseconds per frequency point to generate the voltage transfer function on an IBM 360/75 system using a Fortran IV G compiler. By comparison, CORNAP, the state analysis oriented program prepared by Pottle [10] required before the two methods are competitive (the time to generate a solution via state methods is virtually independent of the number of frequency points).

Clearly, the techniques suggested in this paper can reduce very significantly the time needed to compute transfer functions and sensitivity coefficients in linear circuits. Two aspects, however, require further study: (1) a more rigorous foundation for the tree finding algorithm must be developed and (2) the ordering algorithm must be modified to include controlled sources, since it appears advantageous to process controlled sources earlier in order to make node dropping even more effective.

REFERENCES

1. TEMES, G. and CALAHAN, D. (1967) Computer-Aided Optimization - The State-of-the-Art, *Proc. IEEE* *55*, pp. 1832 - 1864.

2. FORSYTHE, G. and MOLER, C. (1967) *Computer Solution of Linear Algebraic Systems*, Englewood Cliffs, N.J.: Prentice-Hall.

3. PINEL, J. and BLOSTEIN, M.L. (1967) The Frequency Analysis of Linear Electrical Networks, *Proc. IEEE*

55, pp. 1810 - 1819.

4. DIRECTOR, S. (1969) Increased Efficiency of Network Sensitivity Computations by Means of the LU Factorization, *12th Midwest Symposium on Circuit Theory*, Austin, Texas.

5. DIRECTOR, S. and ROHRER, R. (1969) Automated Network Design: The Frequency Domain Case, *IEEE Trans. Circuit Theory CT - 16*, pp. 318 - 322.

6. KLYUYEV, V. and KOKOVKIN-SHEHERBAK, N. (1965) On the Minimization of the Number of Arithmetic Operations for the Solution of Linear Algebraic Systems of Equations, *Computer Science Department, Technical Report CS 24*, Stanford University.

7. TINNEY, W. and WALKER, J. (1967) Direct Solutions of Sparse Network Equations by Optimally Ordered Triangular Factorization, *Proc. IEEE 55*, pp. 1801 - 1810.

8. KINARIWALA, B. (1967) A New Computer Method of System Analysis, *Proc. Fifth Allerton Conference on Circuit and System Theory*, University of Illinois, Urbana, pp. 4 - 7.

9. BRANIN, F.H. (1966) A New Method for Steady State A.C. Analysis of RLC Networks, *IEEE International Convention Record*, Part 7, pp. 218 - 223.

10. POTTLE, C. (to appear) A Textbook Computerized State Space Network Analysis Algorithm, *IEEE Trans. on Circuit Theory*.

11. EL-ABIAD, A. (1963) Digital Computer Analysis of Large Linear Systems, *Proc. First Allerton Conference on Circuit and System Theory*, University of Illinois, Urbana, pp. 205 - 220.

COMPUTER DESIGN OF FILTERS WITH LUMPED-DISTRIBUTED ELEMENTS OR FREQUENCY VARIABLE TERMINATIONS†

HERBERT J. CARLIN and OM P. GUPTA

School of Electrical Engineering,
Cornell University, Ithaca, N.Y.

ABSTRACT

A method for realizing prescribed insertion loss characteristics is presented which is applicable to mixed distributed lumped parameter systems, as well as to transmission line structures which operate into resistive frequency variable terminations. The method utilizes scattering matrix renormalization and is implemented by a computer program. Examples given include capacitor and inductor loaded transmission line sections, as well as a TEM filter terminated in TE_{10} wave impedances. The insertion loss characteristics have equal ripple pass band behavior.

INTRODUCTION

A synthesis technique is presented in this paper to realize filters designed to operate between pure resistances which vary with frequency or in which the filter structure is composed of transmission line sections whose

---

†Part of the doctoral research of Om Gupta under an IBM Fellowship and partially supported under NSF Grant GK-2147.

characteristic impedances are functions of frequency. One important application of the method is to obtain specific design solutions for lump-loaded distributed structures which realize prescribed wide band insertion loss functions. The synthesis technique is based on a cascade realization employing generalized unit element sections.* This is carried out in conjunction with an appropriate renormalization of a prescribed scattering matrix having frequency dependent normalization functions to a new S matrix whose normalization is to real positive constants. The new scattering matrix in effect produces a predistorted amplitude-frequency characteristic which is readily approximated and realized by a cascade of elementary unit elements. When these unit elements are replaced by their complex counterparts and the structure terminated in the prescribed load (which may be frequency dependent), the originally assigned amplitude frequency characteristic results.

Although the technique is quite general, the method is best understood by examining three illustrative examples. The three specific designs considered are: (1) A low pass TEM line filter to operate between frequency variable loads. (2) A distributed parameter filter made up of commensurate transmission lines loaded by lumped capacitors, and (3) A distributed parameter filter made up of commensurate lines loaded by lumped inductors.

TEM LINE FILTER WITH FREQUENCY DEPENDENT RESISTIVE TERMINATIONS

The first example is a filter which is to operate between two terminations whose impedances are real but vary with frequency. The problem chosen could be the model for a TEM mode filter operating into TE_{10} mode wave impedance terminations. Although the design parameters can of course be chosen to fit a given problem, it will be simplest to phrase the discussion here in terms of specific numerical specifications. The cut-off frequency

---

*The final synthesis is based on properties similar to those of commensurate lines. See Appendix.

of the TE_{10} mode, f_c, is 2.077 Ghz. It is required that the pass band insertion loss of the filter be equal ripple and less than 0.4 db between 2.16 Ghz and 3 Ghz, and that the edge of the useful band be 5 Ghz. The impedance Z_0 of the terminations normalized to a unit constant impedance is taken as

$$Z_0(f) = 377/\sqrt{1 - (f_c/f)^2}$$

and this defines the two frequency dependent terminating impedances for the filter.

The range of Z_0 in ohms is $1375 \geq Z_0 \geq 523$ as f varies over $2.16 \leq f \leq 3.0$ (f in Ghz). The specific shape of the filter characteristic below the TE_{10} mode cut-off frequency is not important. We may, therefore, assume arbitrary, but numerically convenient, load variations between d.c. and 2.16 Ghz. In this example, we assume that the load impedance varies linearly from 500 Ω at d.c. to 1375 Ω at 2.16 Ghz, and above this frequency the load is defined by the above equation for $Z_0(f)$. Let it be required that with these load characteristics the coaxial line filter have the insertion loss function

$$|s_{12}(x)|^2 = \frac{1}{1 + 0.1\ T_7^2(x)} \tag{1a}$$

where $T_7(x)$ is a 7th order Tchebyschev polynomial in x, producing a 0.4 db ripple in the pass band.

$$x^2 = \frac{1.528 \tan^2\left(\frac{2\pi\ell}{c}\right) f}{1 + \tan^2\left(\frac{2\pi\ell}{c}\right) f} = \frac{1.528\Omega^2}{1 + \Omega^2} \tag{1b}$$

where

$$\Omega = \tan\left(\frac{2\pi\ell}{c}\right) f = \tan \omega\tau,\ \tau = \frac{\ell}{c} \tag{1c}$$

Ω is a transformed frequency variable, ℓ is the physical length of the unit TEM line elements, c is the velocity of wave propagation in the TEM line, and τ is the delay length of each line.

ℓ is chosen to be 1.5 cm so that the transmission line portion of each unit element is one-quarter wavelength long at 5 Ghz, the edge of the useful band above which periodicity effects set in. The factor 1.528 in (1b) sets the cut-off frequency of the filter at 3.0 Ghz.

Let [S] be the scattering matrix of a network N normalized to the prescribed variable resistive loads. The load resistance functions appear as the two diagonal elements of a diagonal matrix [R]. Then the scattering matrix of the network N, but renormalized to a diagonal matrix of real positive constants $[\hat{R}]$ is [1]

$$[\hat{S}] = E - (\hat{R})^{\frac{1}{2}} [(R)^{\frac{1}{2}} (E-S)^{-1} (R)^{\frac{1}{2}} - \tfrac{1}{2}(R-\hat{R})]^{-1} (\hat{R})^{\frac{1}{2}} \qquad (2)$$

where E is the 2×2 identity matrix. The technique described here aims at finding $[\hat{S}]$, for from it the filter structure can be synthesized by a variant of known techniques.

The diagonal matrices [R] and $[\hat{R}]$ are each 2×2, and the constant elements of $[\hat{R}]$ (denoted as r_0) are both chosen to be 500 Ω in the present example. As in the usual insertion loss problem, only the amplitude $|s_{12}|$ is specified (this off-diagonal scattering coefficient corresponds to the magnitude of the desired transfer function between the frequency variable terminations), and the actual scattering matrix [S] corresponding to this amplitude function is not unique. The crux of the problem is to make the proper choice of the phases (as functions of frequency) of the scattering coefficients in [S]. The amplitudes of these coefficients are readily found, for since the network is reciprocal and lossless, [S] must be unitary and symmetric at real frequencies, which implies that at any frequency

$$|s_{11}|^2 = 1 - |s_{12}|^2 = |s_{22}|^2 \qquad (3)$$

Furthermore, at any frequency, unitarity also requires that

$$\phi_{11} = 2\phi_{12} - \phi_{22} \pm \pi \qquad (4)$$

where ϕ_{ij} are the phases of s_{ij}.

Thus, given $|s_{12}|$, the amplitudes of all the scattering coefficients are determined. Furthermore, if two phases are known, by (4) the third phase characteristic is automatically specified. If, in addition, we restrict the network structure to be symmetric ($s_{11} = s_{22}$), or antimetric ($s_{11} = - s_{22}$), then only one of the phase characteristics is sufficient to specify the phase characteristics of all the elements of [S].

In the present example, we assume the network to be symmetric and the phase $\phi_{11}(\Omega)$ of s_{11} is taken as the unknown phase characteristic which must be determined. [S] can now be specified as a function of $|s_{12}|$ and ϕ_{11} as follows:

$$[S] = \begin{bmatrix} s_{11} & s_{12} \\ s_{21} & s_{11} \end{bmatrix} = \begin{bmatrix} |s_{11}|\, e^{j\phi_{11}} & |s_{12}|\, e^{j\phi_{12}} \\ |s_{12}|\, e^{j\phi_{12}} & |s_{11}|\, e^{j\phi_{11}} \end{bmatrix}$$

$$= \begin{bmatrix} \sqrt{1 - |s_{12}|^2}\, e^{j\phi_{11}} & |s_{12}|\, e^{j(\phi_{11} \pm \frac{\pi}{2})} \\ |s_{12}|\, e^{j(\phi_{11} \pm \frac{\pi}{2})} & \sqrt{1 - |s_{12}|^2}\, e^{j\phi_{11}} \end{bmatrix}$$

Given the desired amplitude response $|s_{12}|$, if we are then able to find ϕ_{11}, we may determine [S], and then, in principle, $[\hat{S}]$ may be calculated by (2). It is more convenient, however, to calculate $[\hat{S}]$ in two steps by the following procedure.

Let s_{11} be the input scattering coefficient of a lossless symmetric reciprocal two-port when both the normalizing numbers are equal to the prescribed frequency variable load impedance $r_L(\Omega)$. Now if the load end termination is changed to the constant r_0, then the input scattering coefficient normalized to $r_L(\Omega)$ is

$$s_{in} = s_{11} + \frac{s_{12}^2\ s_2}{1 - s_2 s_{22}}$$

and applying (3) and (4) with $\phi_{11} = \phi_{22}$

$$s_{in} = \frac{s_{11} - s_2\ e^{2j\phi_{11}}}{1 - s_2 s_{11}} \tag{6}$$

where

$$s_2(j\Omega) = \frac{r_0 - r_L(\Omega)}{r_0 + r_L(\Omega)} \tag{7}$$

Next, the source end normalization is changed to r_0, so as to be equal to the load, to yield an input scattering coefficient

$$\hat{s}_{11} = \frac{k_{12} + s_{in}}{1 + k_{12} s_{in}} \tag{8}$$

where

$$k_{12}(j\Omega) = \frac{r_L(\Omega) - r_0}{r_L(\Omega) + r_0} \tag{9}$$

and $\hat{s}_{11}(j\Omega)$ is the input scattering coefficient with the new normalization.

Noting that k_{12} is the negative of s_2 and substituting s_{in} from (6) into (8) we obtain

$$\hat{s}_{11} = \frac{|s_{11}|(1 + s_2^2) - 2s_2 \cos\phi_{11}}{e^{-j\phi_{11}} + s_2^2\ e^{j\phi_{11}} - 2s_2|s_{11}|} \tag{10}$$

From (10) the amplitude $|\hat{s}_{11}|$ and phase $\hat{\phi}_{11}$ of $\hat{s}_{11}$ can be written explicitly as functions of $|s_{11}|$ and ϕ_{11}

$$|\hat{s}_{11}| = \frac{|s_{11}|(1+s_2^2) - 2s_2 \cos\phi_{11}}{\left\{4s_2^2|s_{11}|^2 + \{1-s_2^2\}^2 + 4s_2\cos\phi_{11}\{s_2\cos\phi_{11} - |s_{11}|(1+s_2^2)\}\right\}^{\frac{1}{2}}} \tag{11}$$

and

$$\hat{\phi}_{11} = \tan^{-1}\left\{\frac{(1-s_2^2)\sin\phi_{11}}{(1+s_2^2)\cos\phi_{11} - 2s_2|s_{11}|}\right\} \tag{12}$$

If ϕ_{11} is known, the $[\hat{S}]$ matrix is completely determined from (11), (12), and if $\hat{s}_{11}$ satisfies the requisite realizability conditions, it may be synthesized by a cascade of commensurate lines (see Appendix) terminated in fixed resistors r_0. When the variable resistors replace r_0, the desired insertion loss results. The problem then is, with the insertion gain function $|s_{12}(j\Omega)|$ prescribed, to determine a compatible $\phi_{11}(\Omega)$ appropriately so that the amplitude and phase of $\hat{s}_{11}$ are related as to be realizable by a cascade of 'n' lines. These conditions for realizability are that $[\hat{S}]$ be rational in $\lambda = j\Omega$ with $|s_{12}(j\Omega)|^2 = \frac{(1+\Omega^2)^n}{G_n(\Omega^2)}$ where $G_n(\Omega^2)$ is an even polynomial of $2n^{th}$ degree in Ω [3]; the s_{ij} satisfy the usual conditions of symmetry and losslessness and, in addition, obey the n-line cascade restraint on phase [2]

$$\hat{\phi}_{11}(\Omega) = \hat{\phi}_{min}(\Omega) - n\tan^{-1}(\Omega) \tag{13}$$

where $\hat{\phi}_{min}(\Omega)$ is the minimum phase calculated from the amplitude characteristic $|\hat{s}_{11}(j\Omega)|$. Substituting for $\hat{\phi}_{min}$ in (13), and using the minimum phase integral relation [4] we obtain

$$\hat{\phi}_{11}(\Omega) = \frac{1}{\pi}\int_0^\infty \frac{d}{dy}(\log|\hat{s}_{11}(y)|)\log\left|\frac{y+\Omega}{y-\Omega}\right| dy - n\tan^{-1}(\Omega) \tag{14}$$

If $\hat{\phi}_{11}$ and $|\hat{s}_{11}|$ are substituted in the above equation from (11) and (12), an integral equation is obtained which must be solved to yield $\phi_{11}(\Omega)$. In general, this integral equation has no solution in closed form. However, if one concentrates on (13), from whence the integral equation arises, the problem can be solved by a computer technique which, as part of its program, employs a subroutine representing the minimum phase in terms of semi-infinite slope approximations for the amplitude [5]. This subroutine stores tables given by Thomas [6] and if a large number of critical frequencies are used, the minimum phase corresponding to a given amplitude characteristic can be calculated with good accuracy.

In effect, then, given $|s_{12}(j\Omega)|$, the computer solution is obtained as follows. Suppose we choose m points $\Omega_1, \Omega_2, \ldots, \Omega_m$. These set the breakpoints for the semi-infinite slope approximation of the amplitude characteristic $|\hat{s}_{11}(j\Omega)|$. If an initial guess is made for the $\phi_{11}(\Omega_K)$ at each of these points, then by (11) $|\hat{s}_{11}(j\Omega_K)|$ may be calculated and the values for the functions $\hat{\phi}_{min}(\Omega_K)$ determined, since we have stored the semi-infinite slope approximation tables and a subroutine gives the minimum phase as a function of amplitude when the breakpoints and slopes are specified. Thus, by (13) trial values for $\hat{\phi}_{11}(\Omega_K)$ are found. At the same time we ask the computer to calculate the $\hat{\phi}_{11}(\Omega_K)$ by (12) and compare these values with the results found from the semi-infinite slope routine. At this point we employ a search technique given by Brown [7] for solving simultaneous nonlinear equations which perturbs the initial choices of $\phi_{11}(\Omega_K)$ until the results by the slope routine and (13) coincide with the results calculated from (12) within a prescribed error. In essence, we are therefore solving, by a numerical method, the set of m simultaneous nonlinear equations in $\phi_{11}(\Omega_j)$

$$\hat{\phi}_{11}\,\{\phi_{11}(\Omega_j)\} =$$

$$\hat{\phi}_{min}\Big\{|\hat{s}_{11}[\phi_{11}(\Omega_1)]|, |\hat{s}_{11}[\phi_{11}(\Omega_2)]|, \ldots, |\hat{s}_{11}[\phi_{11}(\Omega_m)]|\Big\} -$$

$$- n\tan^{-1}(\Omega_j) \qquad j = 1,2,\ldots,m \qquad (15)$$

By choosing m to be a large number, this numerical solution of integral equation (14) can be obtained with reasonable accuracy. In the present example, n, the number of unit elements, was 7 and m was chosen to be 29. After solving the equations and calculating $\phi_{11}(\Omega)$, $|\hat{s}_{11}(j\Omega)|$ and hence $|\hat{s}_{12}(j\Omega)|$ can be easily found numerically by (11). In the final step, $|\hat{s}_{12}(j\Omega)|^2$ is approximated by a rational function in Ω so as to be realizable by a symmetrical cascade of 7 lines. This requires that $|\hat{s}_{12}(j\Omega)|^2$ should be of the form

$$|\hat{s}_{12}(j\Omega)|^2 = \frac{1}{1 + P_7^2\ (x)} \tag{16}$$

Here $P_7(x)$ is to be a 7th order odd polynomial in x (odd to guarantee a symmetric structure); x being defined as in (1b). The rational approximation of $|\hat{s}_{12}(j\Omega)|^2$ in (16) is hence reduced to a polynomial approximation problem. P_7 is calculated to fit the numerical solution for $|\hat{s}_{12}|^2$, via (16), as a truncated series of the orthonormal Tchebyschev polynomials, again using a computer program to determine the coefficients of the series. From (16) we then synthesize a cascade of 7 unit elements [3] which when terminated in fixed loads of $r_0 = 500\ \Omega$ realizes the predistorted transfer function $|\hat{s}_{12}|^2$. However, when the specified frequency variable load is substituted for the fixed terminations, the required transfer function $|s_{12}(j\Omega)|^2$ is obtained. The synthesized filter is shown in Fig. 1, and the actual response calculated with frequency varying resistive terminations is given in Fig. 2. The response is seen to be within the desired 0.4 db. Due to errors in the rational approximation of $|s_{12}(j\Omega)|^2$, particularly near cut-off, the cut-off frequency of the designed filter is slightly less than the desired 3 Ghz. It should be pointed out that in general applications we have not yet determined an a priori condition for convergence of the solution for ϕ_{11}. Experience indicates that provided the loads are not too wildly varying, a convergent solution will be obtained. Of course, once $\phi_{11}(\Omega)$ is found, the realizability of the remainder of the synthesis is guaranteed.

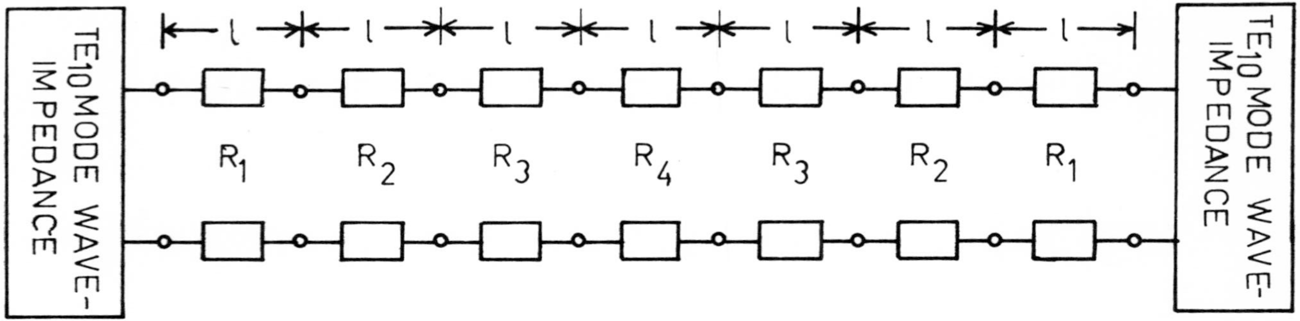

Fig. 1. 0.4 db. TEM mode filter operating between frequency variable (TE_{10} mode wave impedance) terminations:

R_1 = 1476.5 Ohms, R_2 = 733.6 Ohms, R_3 = 1963.6 Ohms, R_4 = 461.8 Ohms. ℓ = 1.5 cms.

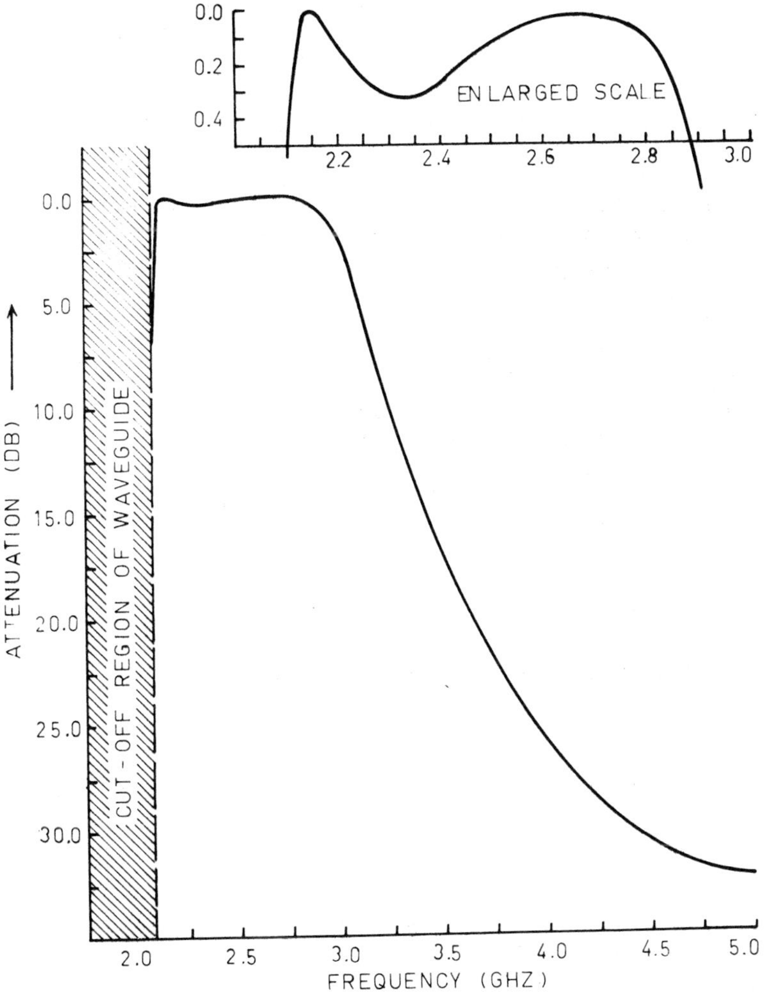

Fig. 2. Response of the filter shown in Fig. 1.

FILTER WITH LUMPED CAPACITOR LOADED TRANSMISSION LINE SECTIONS

The second example, illustrating the application of the method, is the design of a class of mixed lumped-distributed parameter networks. In the preceding example, the characteristic impedance of the unit elements was independent of frequency and the load impedances were frequency dependent. In the present instance, unit elements are employed whose characteristic impedance varies with frequency, but the terminations are frequency independent. Mathematically, this second problem is treated in a manner similar to the first. All the unit element impedances must vary in the same manner as a function of frequency and only differ by multiplicative constants. This case arises in a class of lumped reactance loaded transmission line filters. In the next example the loading is inductive; in the present example the loading is capacitive. The problems can be handled by considering a generalized or complex unit element*, proposed by Rhodes [8], which in the present example is taken as a length of TEM transmission line of characteristic impedance R and two identical lumped capacitors C in shunt, one at each end of the line. The characteristic impedance Z_0 of this generalized unit element is

$$Z_0 = R\{1 + 2RC(p/\lambda) + (RC)^2p^2\}^{-\frac{1}{2}} \tag{17a}$$

p is the actual complex frequency and

$$\lambda = \tanh p\tau \tag{17b}$$

where $\tau = \ell/v$ is the common delay length of each line.

---

*In this paper the term "complex unit element" designates a lossless symmetric section generally containing lumped and distributed elements. All such sections, in a given design problem, have the same propagation function and have characteristic impedances which differ from each other only within multiplicative constants.

The propagation constant α of the composite unit element is defined by

$$\mu = \tanh \alpha = \left\{1 + \frac{\lambda^2 - 1}{(1 + RC\lambda p)^2}\right\}^{\frac{1}{2}} \tag{18}$$

where μ may be taken as a new transformed frequency variable appropriate for the complex capacitor loaded transmission line unit element.

Now if R and C for each complex unit element are constrained so that their product, RC, remains constant, then it can be seen from (17) and (18) that μ varies in the same manner for each unit element, and the characteristic impedance for each section (as a function of frequency) differs only by a multiplicative constant even though Z_0 is frequency dependent. The problem of the design of a filter of a cascade of such generalized unit elements therefore is solved in the same manner as the problem of synthesizing a filter consisting of ordinary cascaded commensurate unit elements provided we work in the transformed frequency μ domain and presume $Z_0(p)$ terminations. The additional step required is to correct for the frequency varying ($Z_0(p)$) terminations so as to synthesize for frequency independent loads. It is clear therefore that exact designs of filters, starting with a prescribed insertion gain, may be obtained by adapting the predistortion method presented in the first example.

It may be remarked that Levy and Rozzi [9] have also used the idea of generalized or complex unit element, but in a manner which only equates the properties of the capacitor loaded section to the unit element parameters at the band adge frequency. The technique presented here can take account of the frequency variation of the composite unit element over the entire band and an example of this is given in the third illustrative problem. In this second example, a simple case is presented of a low-pass Tchebyschev filter of max V.S.W.R. 1.04 with cut-off frequency at 1000 mhz and with the edge of the useful band taken at 3000 mhz as indicated in Fig. 4. In this simpler example, as indicated below, it is sufficient to

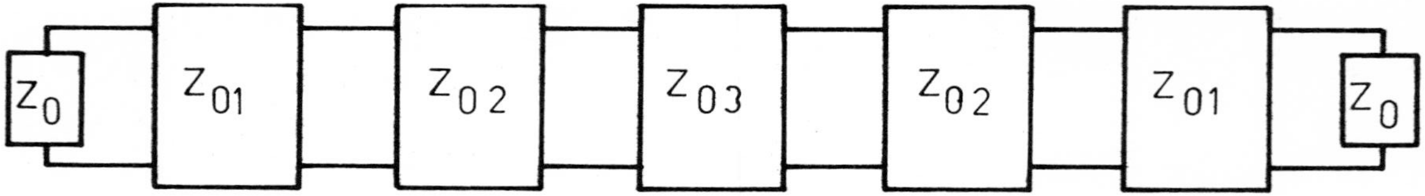

(b)

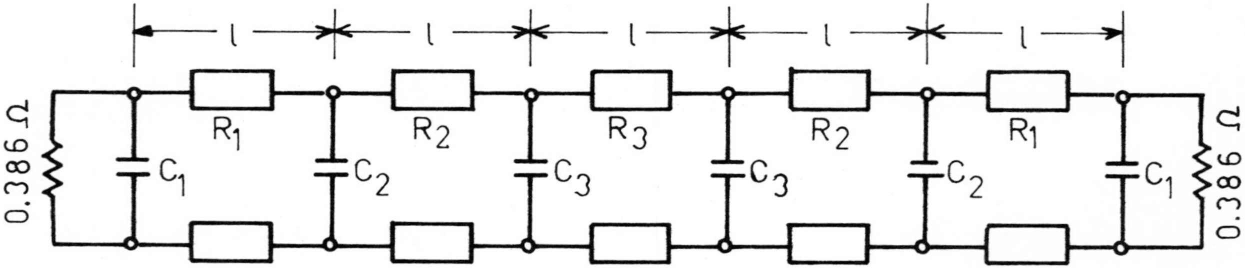

Fig. 3(a). Tchebyschev filter of max. VSWR = 1.04

$Z_{01} = 1.445\ Z_0$, $Z_{02} = 0.488\ Z_0$, $Z_{03} = 2.442\ Z_0$, $Z_{04} = 0.488\ Z_0$, $Z_{05} = 1.445\ Z_0$

Fig. 3(b). Lumped capacitor-transmission line filter of max. VSWR 1.04

$R_1 = 1.445$ Ohms, $R_2 = 0.488$ Ohms, $R_3 = 2.442$ Ohms.

$C_1 = 59.46$ pF, $C_2 = 235.24$ pF, $C_3 = 210.96$ pF

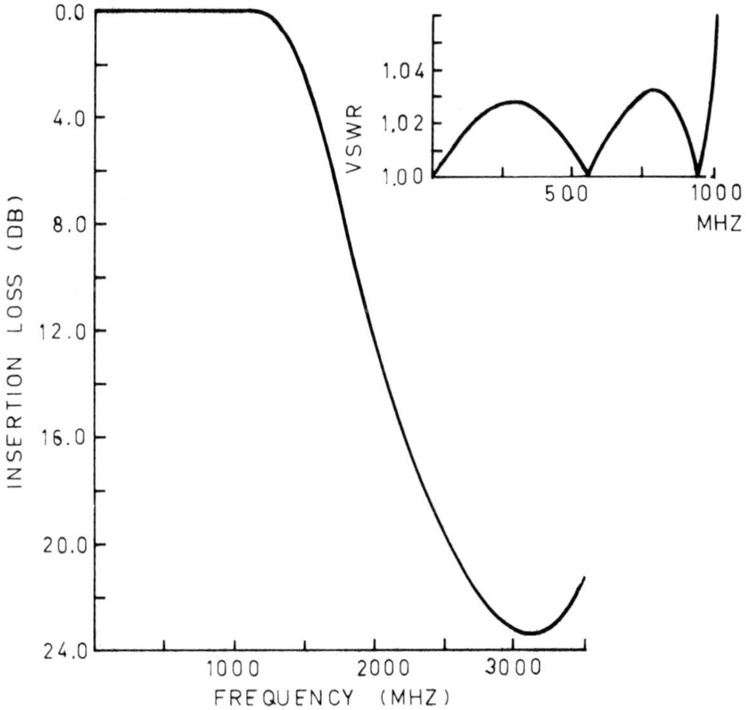

Fig. 4. Response of the capacitor-transmission line filter shown in Fig. 3(b).

choose an average and constant value of characteristic impedance. Thus this second design only illustrates that feature of the synthesis process involving the cascading of complex unit elements.

For the present example, define the impedance Z_{01} according to (17a) to be the characteristic impedance of a complex unit element with R = 1 and the susceptance ωC to be 4.6015 at the quarter wave point for the line (this gives RC = 85.945 × 10^{-12}). Then, in the pass band (0 to 1000 mhz) Z_{01} varies from 0.3818 Ohms to 0.3922 Ohms, although it varies more widely to 0.525 Ohms at the end of useful band. The average value of Z_{01} in the pass band is $Z_{0A} = 0.386$ Ohms and the deviation from this is only $\pm$ 0.15% in the pass band. Thus we may seek a design of a filter consisting of sections of complex unit elements having characteristic impedances which are multiples of Z_{0A}. The terminations are 0.386 Ohms. Note that any other fixed R_0 termination may be obtained by merely scaling the impedance level of the designed structure. The assumption that Z_{01} has the fixed value Z_{0A} introduces some error in the stop band characteristic (1,000-3,000 mhz) but a glance at the response, Fig. 4, shows that this error is not serious.

Having assumed a fixed characteristic impedance unit element, cascade synthesis techniques [10, 11] may be used for the design of the filter structure, although in our problem the unit element is a complex lumped-distributed parameter section. The synthesis is carried out in the frequency variable μ (see (18)) and the amplitude characteristic is specified in a form which guarantees realizability by a cascade of unit elements.

$$|s_{12}(j\hat{\Omega})|^2 = \frac{1}{1 + \{0.02\ T_5(x)\}^2} \tag{19a}$$

where $\hat{\Omega}$ and x are given by

$$\mu = \hat{\Sigma} + j\hat{\Omega}$$

$$x = \frac{\sinh \alpha(\omega)}{\sinh \alpha(\omega_c)} = \frac{\hat{\Omega}\sqrt{1 + \hat{\Omega}_c^{\ 2}}}{\hat{\Omega}_c\sqrt{1 + \hat{\Omega}^2}}$$

α is the propagation function defined in (18) and is purely imaginary; i.e., x is real, well beyond the applicable band of interest for the filter and $\hat{\Omega}_c$ is the cutoff frequency of the filter in the $\hat{\Omega}$ domain ($\hat{\Omega}_c = 0.528$). T_5 is the 5th order Tchebyschev polynomial.

The designed filter consisting of complex unit elements is shown in Fig. 3(a) and actual element values of the characteristic impedances of lines and lumped capacitors are shown in Fig. 3(b). The computed maximum

V.S.W.R. in the pass band is within 1.04 and the cut-off frequency is 1000 mhz. The actual response is shown in Fig. 4. If a wideband filter of maximum V.S.W.R. 1.02 or better is required, then using an average (constant) value for Z_{01} is not sufficiently accurate, and the frequency variation of Z_0 in the pass band (17a) must be explicitly taken into account by calculating a predistorted transfer characteristic as in the first and third examples.

FILTER WITH LUMPED INDUCTANCE LOADED TRANSMISSION LINE SECTIONS

In this final example of a filter made up of inductively loaded transmission line sections, we consider a case where the characteristic impedance of the unit elements varies significantly with frequency and hence cannot be averaged out as in the preceding example of a capacitor loaded transmission line filter. As before, for convenience the technique is illustrated by taking a specific design of one such type of filter which is required to have 0.4 db insertion loss in $0.725 \leq \omega \leq 1.165$, when ω is real (normalized) frequency in radians per second.

The complex unit element is taken to be a length of transmission line of characteristic impedance R loaded by two lumped inductors L, a single inductor placed in shunt at each end of the line section. The characteristic impedance Z_0 and the propagation constant α of this complex unit element can be written as

$$Z_0 = R\{1 + 2(R/L)(1/\lambda p) + (R/L)^2(1/p)^2\}^{-\frac{1}{2}} \tag{20}$$

$$\mu = \tanh \alpha = \left\{1 + \frac{\lambda^2 - 1}{(1 + (R/L)(\lambda/p))^2}\right\}^{\frac{1}{2}} = \hat{\Sigma} + j\hat{\Omega} \tag{21}$$

where λ and p are as defined in (17), and μ of (21) is the transformed frequency variable appropriate to the inductively loaded section.

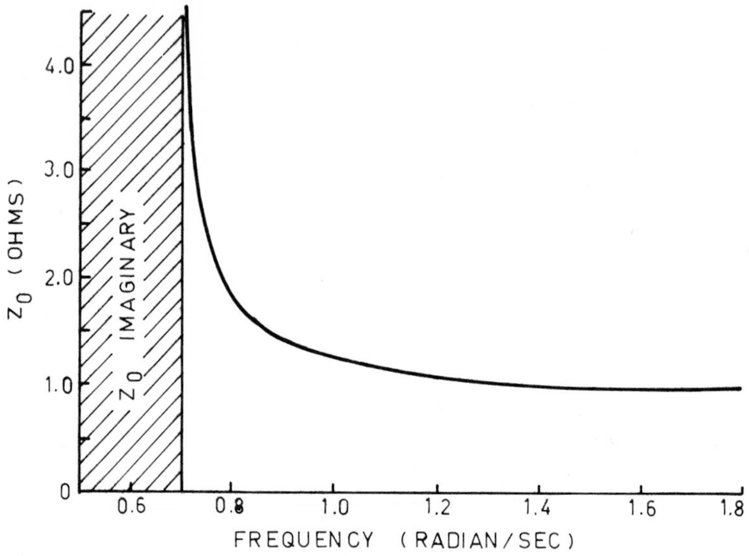

Fig. 5. Normalized characteristic impedance of lumped inductor-transmission line complex unit element.

In this design we choose τ to be unity corresponding to a quarter wave angular frequency of 1.5707 rad/sec. The quantities R and L are chosen so that the ratio (R/L) remains constant for each unit element (in the present numerical example this ratio is taken to be 0.25) analogous to the constant RC product used in the previous example. This forces all unit element propagation constants to be the same at each frequency, and the characteristic impedances of the unit elements track with frequency; that is, they only differ by multiplicative constants. In this case, however, the frequency variation of the unit element characteristic impedance varies significantly (see Fig. 5) even in the pass band (3.25 Ohms to 1.14 Ohms corresponding to R = 1 in (20), over the band $\omega = 0.725$ rad/sec to $\omega = 1.165$ rad/sec), and unacceptable errors result if a constant value of Z_0, equal to the average, is used. Hence we must employ a predistortion technique similar to that presented in the first example. With $p = j\omega$ we can consider the variable μ in (21) to be defined as a function of real frequency ω. For the present numerical example,

$$\lambda = \tanh p \quad (\tau = 1)$$

it is seen from (21) that μ is real for low frequencies up to $\omega = 0.685$. The characteristic impedance Z_0, given by (20) is imaginary for this frequency region and therefore a filter constructed of a cascade of unit elements with characteristic impedances differing by multiplicative constants is automatically cut off in this low-frequency region. This is also physically evident since the inductor loaded transmission line section cannot transmit at d.c. and is essentially a high pass structure At $\omega = 0.685$, $\mu = 0$ and for $0.685 < \omega < 1.72$, $\mu = j\hat{\Omega}$ is purely imaginary ($\hat{\Omega}$ is real), with $\hat{\Omega}$ varying from zero to infinity. Hence any pass band of the filter can only be prescribed in this range of ω. In general, to specify a filter characteristic realizable by unit element inductor loaded transmission line sections we choose τ and (R/L) so that the desired lower cutoff frequency ω_1 corresponds to the value of transformed frequency $\hat{\Omega} = 0$. At the upper cutoff frequency ω_2 of the prescribed filter, the transformed cutoff frequency is $\hat{\Omega} = \hat{\Omega}_c$ and is calculated by (21). We see, therefore, that the inductively loaded filter has a bandpass response in real frequency from ω_1 to ω_2 but performs as a low pass filter in the transformed $\hat{\Omega}$ domain.

An additional precaution has to be observed because $Z_0(j\hat{\Omega})$ becomes infinite as $\hat{\Omega} \to 0$. To take this into account we merely design our filter by arbitrarily assigning finite values to Z_0 over some small region of real frequency corresponding to a positive neighborhood of $\hat{\Omega} = 0$. In our numerical example this region is in the range $0.685 \leq \omega \leq 0.725$. In the final filter, this portion of the real frequency axis merely becomes a slight extension of the low frequency cutoff range above the point where μ starts to take on imaginary values. Above the real angular frequency $\omega = 0.725$, we are of course taking the exact variation of Z_0 into account by using (20).

For the numerical example, as previously indicated, the higher cutoff frequency ω_2 is chosen to be 1.165 rad/sec; corresponding to this by (21) $\hat{\Omega}_c = 1.376$. We now prescribe the desired amplitude characteristic as a low pass response in the domain $0 \leq \hat{\Omega} \leq \infty$, and use precisely the same form as (19) except that with $\mu = \hat{\Sigma} + j\hat{\Omega}$, μ is defined by (21) and

$$|s_{12}(j\hat{\Omega})|^2 = \frac{1}{1 + 0.1\left\{T_7\left[\frac{\hat{\Omega}\sqrt{1+\hat{\Omega}_c^2}}{\hat{\Omega}_c\sqrt{1+\hat{\Omega}^2}}\right]\right\}^2} = \frac{1}{1 + 0.1\,T_7^2(x)} \tag{22}$$

where x is given by (19b). This equation defines a Tchebyschev equal ripple characteristic with a maximum of 0.4 db of insertion loss in the pass band. $|s_{12}|^2$ in (22), of course, corresponds to prescribed fixed terminations and in the renormalization procedure these are chosen to be 1.667 Ohms merely as a matter of convenience. As pointed out earlier, the design for any other fixed termination is obtained by proportionate scaling of the impedance levels of the prototype design.

Having prescribed (22) for fixed terminations we must now calculate $|\hat{s}_{12}(j\hat{\Omega})|^2$, the predistorted response, which is the response when the filter is terminated in the frequency variable unit element characteristic impedance. As discussed earlier, because of the singularity of $Z_0(j\hat{\Omega})$ at $\hat{\Omega} = 0$ we choose the normalization function to be linear in $\hat{\Omega}$ in the region below $\hat{\Omega} = 0.21$ (corresponding to $\omega < 0.725$). The normalization function is real and varies from 1.667 to 3.30 in this low frequency region ($0 \leq \hat{\Omega} \leq 0.21$) and above $\hat{\Omega} = 0.21$ is defined by (20).

Just as in the first problem, $\phi_{11}(j\Omega)$ is now calculated by the computer search technique which provides the numerical solution of integral equation (14). Finally, the predistorted characteristic $|\hat{s}_{12}|^2$ is calculated and approximated as in (16). The resultant predistorted transfer function $|\hat{s}_{12}(j\Omega)|^2$ is realized by a symmetric cascade of complex unit elements terminated at each end in loads taken as the unit element characteristic impedance which in turn is set equal to the frequency variable normalization function. When the frequency variable terminations are replaced by the prescribed frequency independent resistive loads the desired response $|s_{12}(j\hat{\Omega})|^2$ results.

The design of an inductance loaded transmission line filter which realizes (22) with unit constant resistive terminations is shown in Fig. 6 and the computed response of the actual structure is plotted in Fig. 7. The maximum pass band loss is 0.27 db, well within the prescribed 0.4 db. The bandwidth is also slightly less than the desired response. These effects can be corrected by improving the rational approximation to $|\hat{s}_{12}|^2$ (see (16)).

SUMMARY

A technique has been presented in this paper for carrying out the design of transmission line structures from prescribed insertion loss functions where terminating loads or unit element characteristic impedances vary appreciably with respect to frequency. The examples have been chosen mainly to illustrate the method. Thus no particular effort was made to choose unit elements that would lead to especially steep roll-off in the filter stop bands, though this can obviously be done. In illustrating the design approach we considered in Example 1, a system in which the terminating loads varied by a ratio in excess of 2.5 to 1 in the pass band and the method was capable of realizing an insertion loss function which had prescribed equal ripple pass band characteristics. The second and third examples showed how the idea could be extended to lump loaded transmission line circuits. In Example 3, the characteristic impedance of the complex unit element section varied in excess of 2 to 1, in the pass band, with respect to frequency. The method combines the use of a Richards cascade synthesis extended to complex unit elements, together with a scat-

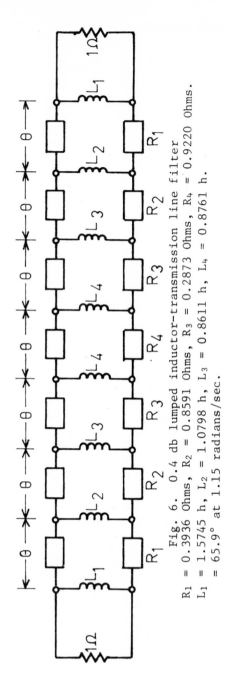

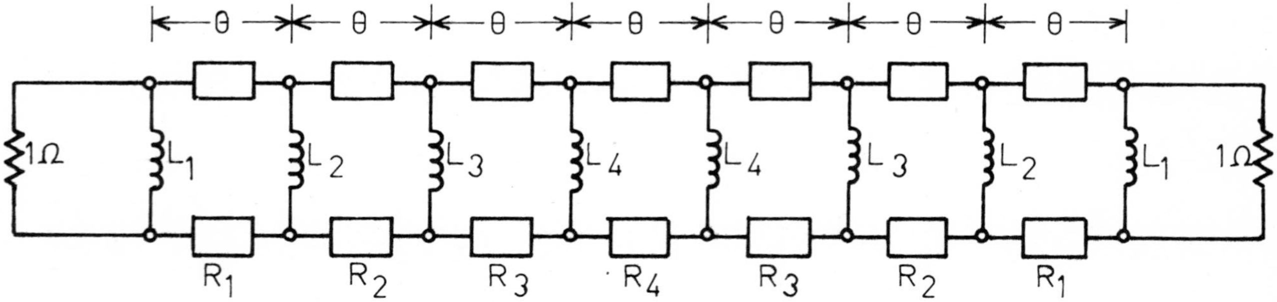

Fig. 6. 0.4 db lumped inductor-transmission line filter
R_1 = 0.3936 Ohms, R_2 = 0.8591 Ohms, R_3 = 0.2873 Ohms, R_4 = 0.9220 Ohms.
L_1 = 1.5745 h, L_2 = 1.0798 h, L_3 = 0.8611 h, L_4 = 0.8761 h.
= 65.9° at 1.15 radians/sec.

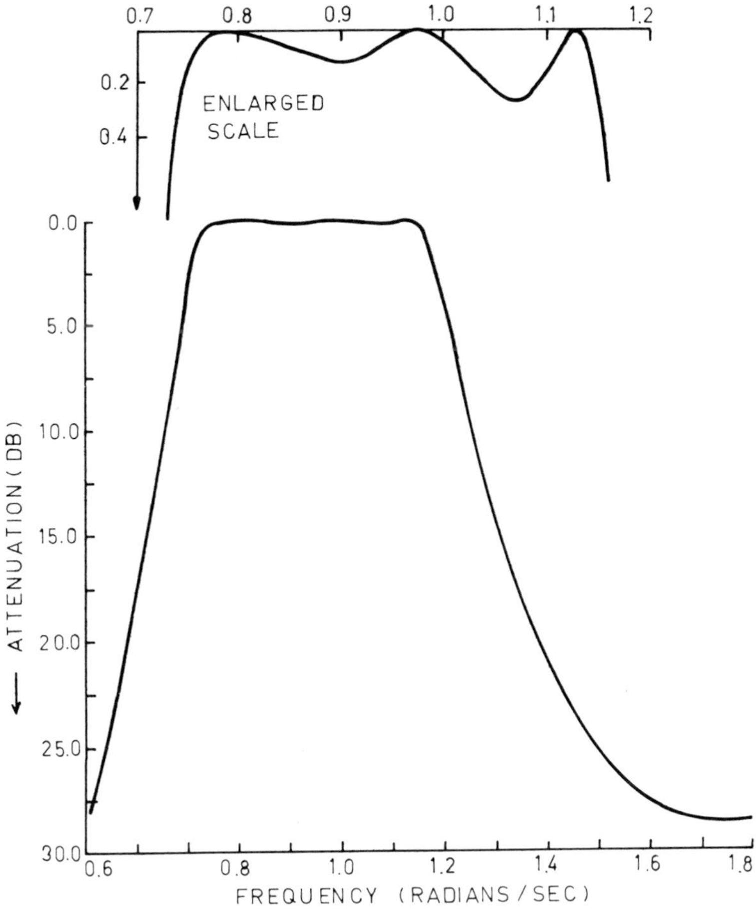

Fig. 7. Response of the filter shown in Fig. 6.

tering matrix renormalization for achieving an amplitude predistortion determination. The predistortion problem is handled by means of a computer solution of an integral equation for an unknown phase function. In every case, the original specification of the problem is taken as a prescribed insertion loss function in the real frequency domain. In the case where the design technique is applied to mixed systems comprising both distributed and lumped elements, the method is a practical means for avoiding the mathematical difficulties of the two frequency variable approach in which one frequency variable is used for the lumped elements, the other for the distributed elements [12]. The two variable method has not been extended to the solution of the insertion loss problem.

It may be remarked that the final synthesis process in all three cases is the same once the predistorted rational function $|\hat{s}_{12}(j\hat{\Omega})|^2$ has been determined. A factorization is performed to obtain the scattering function $\hat{s}_{11}(j\hat{\Omega})$, and the Richards technique for calculating a cascade unit element realization of a prescribed reflection factor is employed [10,11]. The unit elements are of course the complex sections defined in the specific problem. A computer program has been devised to obtain the multiplicative constants for the unit element characteristic impedances in the synthesis. The program is based on a long division algorithm and further information may be obtained on this or other programs used in this paper by writing to the Director's Office, Cornell School of Electrical Engineering, Ithaca, N.Y. 14850.

APPENDIX

Properties of Commensurate Transmission Lines

1. *Richards' Transformation*: The transformation $\lambda = \Sigma + j\Omega = \tanh \gamma\ell = \tanh p$, $p = \sigma + j\omega$, takes the right half of the p plane (via an infinite sequence of p plane strips of width $j\pi$) into the right half of the λ plane provided that at real frequencies $\gamma\ell = j\beta\ell = j\omega = j\hat{\omega}\tau$, $\tau = \ell/v_p$. Here ℓ is the basic line length of a unit element, v_p the phase velocity, τ the delay length of a unit element, and $\hat{\omega}$ the free space angular frequency. Hereafter all such unit elements (or commensurate line sections) are presumed to have common propagation constants, $\gamma = j\beta$ at real frequencies.

2. *Theorem*: The network functions associated with any system of commensurate lines (minimum line length ℓ_0) and resistors are rational under the frequency transformation

$\lambda = \tanh \gamma\ell_0/2 = \tanh p.$

3. *Theorem*: The necessary and sufficient condition for realizing a symmetric scattering matrix (impedance matrix) which is rational in the $\lambda = \tanh \gamma\ell$ domain as a network of interconnected commensurate lines, resistors and ideal transformers is that the scattering matrix be Bounded Real (BR) in λ, or the impedance matrix Positive Real (PR). In the driving point case, by the Bott Doffin theorem, no transformers are necessary.

4. *Theorem on Line Extraction* (Richards' Theorem): Given $s_1(\lambda)$, a BR rational reflection factor in the $\lambda = \tanh \gamma\ell$ domain, and normalized to a real $r_1 > 0$. Corresponding to $s_1(\lambda)$ is a PR $z_1(\lambda) = r_1 \dfrac{1 + s_1(\lambda)}{1 - s_1(\lambda)}$. Then $s_1(\lambda)$ can always be represented as a commensurate line section of length ℓ in cascade with a remainder network whose reflection factor is BR, provided the characteristic impedance of the cascade extracted line is $r_2 = z_1(1)$. The reflection factor function of the remainder network *normalized to* r_2 is

$$s_2(\lambda) = \frac{s_1(\lambda) - s_1(1)}{1 - s_1(1)s_1(\lambda)} \frac{1 + \lambda}{1 - \lambda}$$

The impedance function of the remainder is PR and is given by

$$z_2(\lambda) = z_1(1) \frac{z_1(\lambda) - \lambda z_1(1)}{z_1(1) - \lambda z_1(\lambda)}$$

5. *Foster Theorem*: A lossless rational Foster Function of $\lambda = \tanh \gamma\ell$ can always be realized as a cascade of commensurate lines terminated in an open circuit (if $z(0) = \infty$) or a short circuit (if $z(0) = 0$).

6. *Theorem on Insertion Gain Realizability*: The rational function, even in Ω

$$s_{12}(\lambda)s_{12}(-\lambda)\Big|_{\lambda=j\Omega} = |s_{12}(j\Omega)|^2$$

with $\Omega = \tan \beta\ell$ can be realized as the insertion gain of a lossless cascaded commensurate line 2-port (Insertion Gain $= \frac{\text{Power to Load}}{\text{Power Available}}$) with real resistive terminations if and only if

a) $0 \leq |s_{12}(j\Omega)|^2 \leq 1$

b) $|s_{12}(j\Omega)|^2 = \frac{(1 + \Omega^2)^n}{P_n(\Omega^2)}$

where P_n is an even polynomial in Ω. The number of commensurate or unit element lines, each of length ℓ, required for the realization is n.

REFERENCES

1. CARLIN, H.J. and GIORDANO, A.B. (1964) *Network Theory*, Prentice-Hall, Englewood Cliffs, N.J., pp. 344.

2. CARLIN, H.J. and ZYSMAN, G.I. (April 1966) Linear Phase Microwave Networks, *Polytechnic Institute of Brooklyn Symposium on Generalized Networks*, pp. 193-226.

3. WENZEL, R.J. (January 1964) Exact Design of TEM Microwave Networks Using Quarter Wave Lines, *IEEE Trans. on Microwave Theory and Techniques MTT-12*: (1), pp. 94-111.

4. BODE, H.W. (1945) *Network Analysis and Feedback Amplifier Design*, D. Van Nostrand Co., Inc., Princeton, N.J., Ch. 14, pp. 318.

5. Ibid., Ch. 15, pp. 337-345.

6. THOMAS, D.E. (October 1947) Tables of Phase Associated with a Semi-infinite Unit Slope of Attenuation, *Bell System Technical J. 26*.

7. BROWN, K.M. and CONTE, S.D. (August 1967) The Solution of Simultaneous Nonlinear Equations, *Proc.*

22nd National Conference of ACM, pp. 111-114.

8. RHODES, J.D. (January 1966) General Commensurate Unit Elements, *Electronic Letters 2*: (1) pp. 36-37.

9. LEVY, R. and ROZZI, T.E. (March 1968) Precision Design of Coaxial Low Pass Filters, *IEEE Trans. on Microwave Theory and Techniques MTT-16*: (3), pp. 142-147.

10. CARLIN, H.J. (February 1965) *Network Synthesis with Transmission Line Elements*, Technical Report No. RADC-TDR-64-505, Rome Air Development Center, Air Force Systems Commands, Griffis Air Force Base, N.Y., pp. 11-12. (Polytechnic Institute of Brooklyn Report PIBMRI-1235-64).

11. CARLIN, H.J. (January 1967) Synthèse des circuits à lignes en cascade à partir de la matrice de répartition, *L'Onde Electrique XLVII*: (478), pp. 8-19.

12. SAITO, M. (April 1966) Synthesis of Transmission Line Networks by Multivariable Techniques, *Polytechnic Inst. of Brooklyn Symposium on Generalized Networks*, pp. 353-392.

SOME OPEN QUESTIONS IN THE THEORY OF NONCOMMENSURATE DISTRIBUTED NETWORKS

Y. KAMP

MBLE Research Laboratory

INTRODUCTION

Networks made up of commensurate transmission lines are most easily investigated with the use of Richards' transformation because it allows the problems to be stated in terms of rational single variable functions. The results and techniques developed for lumped networks can thus be used.

For noncommensurate distributed elements, two points of view can be adopted according as the single variable dependance or the rational character of the network functions seems to be more important. The first approach deals with non-rational single variable functions. The second introduces one Richards' variable for each electrical length and leads to the theory of multivariable functions.

The purpose of this note is not to give a complete account of the results obtained by both methods but rather to point out some of their limitations and to list several important unsolved problems in the fields of approximation and synthesis.

1. APPROXIMATION PROBLEMS

The problem consists essentially in approximating

frequency domain requirements by network functions suitable for realization by noncommensurate distributed elements. Up to now, it has been given but little consideration which can be interpreted as a measure of its difficulty.

A few contributions are related to the approximation with non-rational functions and all of them are restricted to cascade structures. Even so, no analytical solution could be obtained except in the simple case of maximally flat criterion and for three cascaded lines [1]. Chebyshev approximations were calculated using optimization techniques [2]. All these results tend to show that noncommensurate cascaded transmission lines cannot achieve a better maximally flat or Chebyshev approximation than equal length transmission lines.

A contribution of Kinariwala [3] gives the type of approximating function which should be used in order to obtain a cascade synthesis. The transmittance must be of the form

$$|s_{12}(\omega)|^2 = \frac{1}{\sum_{k=1}^{m} \alpha_k \cos \omega\beta_k} \qquad (1)$$

Direct approximation of frequency domain specifications by multivariable functions seems almost impossible in the present state of the art and this problem has not even been touched upon.

Unsolved problems

1) The approximation for cascade structures stated above is essentially a nonlinear problem in α_k, β_k and its difficulty is fairly high [4]. In particular, the existence, iniqueness and characterization of a Chebyshev approximation with functions of the type given in (1) have still to be established.

2) Up to now the approximating functions were restricted to cascade structures, apparently because the synthesis of non-rational network functions itself is restricted to these structures. Does it mean that, as long as no general synthesis procedure exists for non-rational functions, the approximation problem will be tied up with the choice of a particular structure?

3) A possible solution for multivariable approximation could be to split the problem in two steps. A first step would consist in obtaining a non-rational approximation in terms of the natural complex frequency variable. The second step would then be to define suitable Richards' variables in order to obtain rational multivariable functions. In this approach, multivariable theory would be restricted to synthesis problems.

4) A synthesis problem is linked to the above suggested approach. It is indeed not quite clear if the choice of the Richards' variables is unique and, if not, how to make the most efficient choice. In particular, since strictly multivariable functions (i.e. functions which are of the first degree in each variable separately) are easier to synthesize, one would like to know what constraints this imposes on the non-rational approximating functions.

2. SYNTHESIS OF NON-RATIONAL NETWORK FUNCTIONS

The behaviour of transmission line networks is usually described in terms of hyperbolic functions. When the latter are replaced by exponentials, any transfer function can be expressed as a ratio of sums of exponentials. The synthesis problem consists then in establishing the necessary and sufficient conditions for a network function of the above type to be realizable with lossless transmission lines. No general solution exists up to now and the problem was solved only for cascade structures loaded by a resistance [3,5,6]. The form of the conditions depends on the particular network function under consideration. The first condition is a passivity constraint (positive - real impedance, bounded - real scattering coefficient) and the second condition results from the fact that a cascade structure is requested. The input impedance must be of the form

$$Z(p) = \frac{N(p)}{D(p)} = \frac{\sum_{k=0}^{m} a_k e^{p i_k}}{\sum_{k=0}^{m} b_k e^{p i_k}}, \tag{2}$$

where i_k are nonegative constants and such that

$$\mathrm{Ev}\ Z(p) = \frac{c}{D(p)\ D(-p)},\ c \geq 0 \tag{3}$$

The analysis of cascaded lines shows in fact that the exponents i_k are partial sums of the electrical line lengths and as such must satisfy a condition of the type

$$\sum_{k=0}^{m} N_k i_k = 0 \tag{4}$$

where N_k are integers.

Equivalently, the transmittance $s_{12}(p)$ must be such that

$$s_{12}(p)\ s_{12}(-p) = \frac{d e^{\gamma p}}{P(e^p)} \tag{5}$$

where $P(e^p)$ is an exponential sum with highest exponent 2γ.

Straightforward synthesis algorithms exist for impedance or reflectance functions satisfying the above conditions. They are either based on the properties of an impedance of the form (2) or on a formulation of the Richards' theorem in terms of exponentials.

Unsolved problems

1) It is seen that any reactance function of the form (2) is synthesizable as an open- or short-circuited cascade of transmission lines. One is therefore tempted to consider that cascade realization is in some sense a canonical synthesis for reactances of the form (2). What this means in the field of non-rational network functions is however not completely clear. Indeed, for rational functions, canonical realization means a synthesis with the minimum number of components. This is certainly not the case here since simple examples can be constructed [5] where the cascade synthesis of a reactance is not minimal.

2) The synthesis algorithms do not guarantee a realization with a finite number of components although each step reduces the largest exponent of exp(p) in the remaining impedance or reflectance. Of course, rational exponents i_k constitute a sufficient condition but it is too restrictive since irrational exponents satisfying (4) do not necessarily lead to a cascade structure with an infinite number of lines. This is easily verified by direct computation of the input impedance of a cascade with noncommensurate line lengths. Thus a more refined criterion is needed to test whether the synthesis ends with a finite number of component lines.

3) No direct synthesis algorithm exists for the transmittance. Thus when a realizable transmittance is given (condition (5)), one is forced to calculate the reflectance by the relation

$$s_{11}(p)\, s_{11}(-p) = 1 - s_{12}(p)\, s_{12}(-p) \qquad (6)$$

and one chooses then for $s_{11}(p)$ the left half-plane poles of (6). This is rather simple to perform for rational functions, but is considerably more difficult when the transmittance is non-rational, since the number of poles and zeros is then generally unknown. A direct synthesis technique starting from the transmittance thus seems desirable.

4) What are the limitations of the theory for what regards its extension to other structures?

3. SYNTHESIS OF RATIONAL MULTIVARIABLE NETWORK FUNCTIONS

Positive real and bounded real multivariable network functions and matrices are defined in the same manner as in the single variable case except that in the definitions the right half-plane of p is replaced by the polydomain Re $\mu_i > 0$ $(i=1,2,...,n)$ of the n variables μ_i.

The importance of two-variable functions stems from their connection with circuits containing lumped and distributed elements. The synthesis of two-variable reactances and reactance matrices has been extensively investigated. Youla's approach [7] seems the most useful as it yields all minimal and non minimal equivalent realizations.

The general solution to the synthesis of positive real multivariable matrices was given by Koga [8] who showed moreover that a symmetric positive real matrix can be synthesized without using gyrators. However, one is frequently interested in obtaining realizations of a prescribed type (especially cascade structures using only transmission lines) and Koga's method seems not to be well adapted for this purpose.

Several contributions [9,10] are devoted to the derivation of necessary and sufficient conditions for which a multivariable network function can be realized by cascaded transmission lines, possibly separated by a certain class of lumped lossless two-ports. These authors restrict themselves to functions $F[\mu_1, \mu_2, \ldots, \mu_k; p]$ which are of the first degree in each variable μ_i and of some degree n in the natural complex frequency variable p. Such functions are strictly multivariable in the set (μ_i) and correspond to transmission line networks where any two line lengths are noncommensurate. These results make use of the multivariable Richards' theorem [9] which gives the conditions for a degree reducing extraction of a transmission line.

A class of strictly multivariable functions $F(\mu_i)$ can be adequately characterized by the behaviour at points $\mu_i = 0$, $\mu_i = 1$ and this method leads to necessary and sufficient conditions for cascade synthesis [11]. Another approach to strictly multivariable functions consists in finding their explicit form for particular structures. In this way, necessary and sufficient conditions can be derived for a realization with cascaded transmission lines separated by series or parallel stubs [12]. The main advantage of this method lies in the fact that synthesis can be performed almost by inspection.

It is noteworthy that all of the above contributions are related to network functions which are of the first degree in each variable except possibly one. This is also true in some sense for the contribution of Koga who, starting from general multivariable functions, introduces new variables so as to obtain strictly multivariable functions.

A first synthesis problem [12], which deals directly with general multivariable functions, considers cascade connections of single-variable blocks consisting of cascaded transmission lines with open-circuited parallel stubs.

Unsolved problems

1) Koga's work [8] gives the solution for the synthesis of general multivariable functions but some related problems are still unsolved like finding all

equivalent realizations or minimal and transformerless realizations. Youla's method [7] seems to be well suited to give an answer to these questions but a first extension [14] to multivariable functions proved unsuccessful because no stable realization can be insured. Does this mean that any multivariable extension of Youla's method is impossible? If so, is there any other promising approach?

2) It is striking that Koga's contribution does not rely on the theory of functions of several complex variables apart from some evident extensions of the single variable theory. This seems to contradict Scanlan's conjecture [15] and raises the question whether the theory of functions of several complex variables is of any use at all for noncommensurate distributed networks.

3) What method should be used to tackle the synthesis of general multivariable functions with prescribed structures?

4) A strictly multivariable reactance which is realizable as an open or short-circuited cascade cannot be realized by a structure using parallel and series stubs. Conversely, many strictly multivariable reactances are not realizable by cascade structures. This is in contrast with the theory of nonrational reactances. Does this mean that a strictly multivariable function contains some information concerning the particular structure to be used for the synthesis? If so, is there any hope to find canonical realizations?

REFERENCES

1. KAMP, Y. and NEIRYNCK, J. (November 1968) Maximally Flat Approximation for Noncommensurate Cascaded Transmission Lines, *Electronics Letters 4 (22)*, pp. 477-478.

2. BANDLER, J.W. (May 1968) Optimum Noncommensurate Stepped Transmission-Line Transformers, *Electronics Letters 4 (11)*, pp. 212-213.

3. KINARIWALA, B.K. (April 1966) Theory of Cascaded Transmission Lines, *Proceedings of the Symposium on Generalized Networks*, Polytechnic Institute of Brooklyn, pp. 345-352.

4. RICE, J.J. (July 1960) The Characterization of Best Nonlinear Tchebycheff Approximations, *Trans. Amer. Math. Soc. 96*, pp.322-340.

5. CARLIN, H.J. (1968) Synthesis of Transmission Line Networks, *Summer School on Circuit Theory*, Prague.

6. KINARIWALA, B.K. (April 1966) Theory of Cascaded Structures: Lossless Tranmission Lines, *Bell Syst. Tech. Jour. 45 (4)*, pp. 631-649.

7. YOULA, D.C. (April 1966) The Synthesis of Networks Containing Lumped and Distributed Elements, *Symposium on Generalized Networks*, Polytechnic Institute of Brooklyn, pp. 289-343.

8. KOGA, T. (March 1968) Synthesis of Finite Passive N-Ports with Prescribed Positive Real Matrices of Several Variables, *IEEE Trans. on Circuit Theory CT-15*, pp. 2-23.

9. SAITO, M. (April 1966) Synthesis of Transmission Line Networks by Multivariable Techniques, *Symposium on Generalized Networks*, Polytechnic Institute of Brooklyn, pp. 353-392.

10. SCANLAN, J.O. and RHODES, J.D. (December 1967) Realizability of a Resistively Terminated Cascade of Lumped Two-Port Networks Seperated by Noncommensurate Transmission Lines, *IEEE Trans. on Circuit Circuit Theory CT-14*, pp. 388-394.

11. PREMOLI, A. (June 1969) Analysis and Synthesis of Cascaded Noncommensurable Transmission Lines, *Electronics Letters 5 (13)*, pp. 286-288.

12. KAMP, Y. (November 1968) Synthesis of a Multivariable 2-Port Using Parallel and Series Stube, *Electronics Letters 4 (22)*, pp. 474-475.

13. SHIRAKAWA, I., TAKAHASHI, M. and OZAKI, H. (June 1968) Synthesis of Some Classes of Multivariable Cascaded Transmission-Line Networks, *IEEE Trans. on Circuit Theory CT-15*, pp. 138-143.

14. NEWCOMB, R.W. (December 1966) On the Realization of Multivariable Transfer Functions, Cornell University, Research Report EERL 58.

15. SCANLAN, J.O. (March 1967) Multivariable Network Functions, *IEEE Int. Conv. Record* Part 5, pp. 37-44.

IMMITTANCE MATRICES OF THREE-LAYER N-PORTS*†

P.P. CIVALLERI

Istituto Elettrotecnico Nazionale "Galileo Ferraris", Torino, Italy

S. RIDELLA

Istituto di Elettrotecnica, Universita di Genova, Genova, Italy

ABSTRACT

The port impedance and admittance matrices of three-layer structures of linear, passive, isotropic materials are defined and their properties briefly investigated; discrete equivalent circuit representations are given, together with an example.

1. INTRODUCTION

The N-ports under investigation are three-layer structures each layer of which has constant thickness and is made of linear, passive, homogeneous, isotropic, time invariant material. The three layers overlap exactly and are laterally bounded by a cylindrical surface of

---

* This work has been partially supported by Consiglio Nazionale delle Ricerche, Italy.

† This paper, in condensed form, will be published in the IEEE Trans. on Circuit Theory.

arbitrary cross-section, parallel to a straight line perpendicular to the plane surfaces of the layers. The system interacts with the outside world through N ports placed on the lateral boundary of the wafer: the latter can be thought of in general as the terminations of N microstrips whose inputs can be connected to the generator and to the loads by various means, e.g. by coaxial cable (Fig. 1).

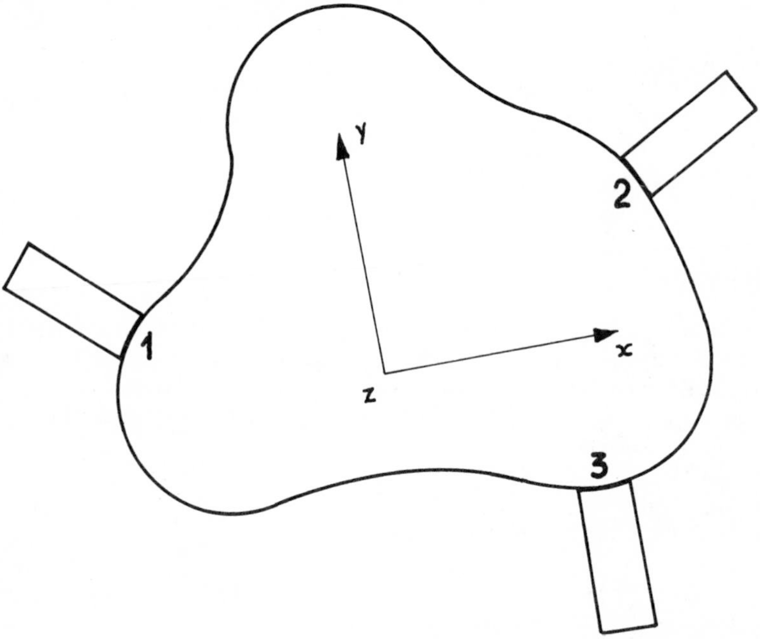

FIG. 1. Cross-section of a three-port structure.

The interest of such structures, which appear to be a broad generalization of various devices commonly used in practice (such as lossless and RC uniform and non-uniform lines, microstrip filters, equalizers and so on), has been pointed out in various recent papers [1, 2, 3] where the basic principles of their electromagnetic and circuit analysis can be found.

In this paper, after a short review of some already established concepts, we set up a partial fraction representation of both the impedance and the admittance matrix of a structure of the said type, together with their equivalent circuit representations and a very elementary example.

2. SYMBOLS AND CONVENTIONS

Layers are distinguished by means of an index i, running from 1 to 3 from the bottom to the top of the structure. In each layer a Cartesian clock-wise oriented reference frame is chosen, such that the lower boundary plane of the layer has equation $z_i = 0$, while the z_i-coordinate of the upper boundary plane is positive.

The following symbols will be currently used:

S : surface of the cross-section of the cylinder.

l : boundary of the cross-section of the cylinder.

l_p: portion of l covered by ports.

l_c: portion of l not covered by ports.

n : unit vector normal to l and directed towards the interior region.

c : unit vector parallel to the z-axis.

Index P will be used to denote either a vector parallel to the (x,y)-plane (or P-plane) or a vector operator acting only on the x, y coordinates.

3. THE STRUCTURE EIGENFUNCTIONS

As it has become evident from broad experience in the theory of resonant cavities since the 1950's [5, 6, 7] the definition of port impedance and admittance matrices is greatly facilitated by the introduction of a complete orthonormal set of eigenfunctions of the partial differential equations (with the proper boundary conditions) describing the field distribution inside the cavity. Here the situation is somewhat different for the following reasons: a) the cylindrical symmetry of the structure suggests the use of eigenfunctions of the wave-guide type rather than the general cavity type; b) the boundary conditions are far more complicated, because the intermediate layer is laterally open (and hence radiation can possibly take place) and its bases are not perfect conductors: the first point will be ignored from here on, as in any practical situation the lateral surface is so small with respect to the whole structure surface that neglecting outflow of energy is fully justified; the second point will be discussed in the following.

We firstly introduce [3] a set of so-called *planar eigenfunctions*, for both the open-circuit and the short-circuit situation, whose complete definition is presented in Table 1.

For each situation, we have two classes of eigenfunctions, the first of which is appropriate to represent a field whose magnetic vector has zero component along the z-axis, while the second refers to the case in which the vertical component of the electric field is absent: the electromagnetic modes pertinent to such two complementary situations will be referred as *plane-magnetic* (or PM) and *plane-electric* (or PE) modes respectively.

For each class (and obviously for both the open-circuit and the short-circuit situation), firstly the scalar quantities ϕ_q, ψ_q, ϕ'_q and ψ'_q are defined as the eigenfunctions of the two-dimensional Helmholtz equation with the boundary conditions suitable for each case; from each set of them, two conditions suitable for each case; from each set of them, two sets of vector functions are then derived, for which the boundary conditions obviously come out as consequences of their definitions.

The functions ϕ_q, $\mathbf{f}_q$, $\mathbf{e}_q$, ϕ'_q, $\mathbf{f}'_q$, $\mathbf{e}'_q$ are called *electric eigenfunctions*, and ψ_q, $\mathbf{h}_q$, $\mathbf{g}_q$, ψ'_q, $\mathbf{h}'_q$, $\mathbf{g}'_q$ *magnetic eigenfunctions*: it could be easily verified that the $\mathbf{f}_q^{\prime}$s and the $\mathbf{g}_q^{\prime}$s are *irrotational* while the $\mathbf{e}_q^{\prime}$s and the $\mathbf{h}_q^{\prime}$s are *solenoidal*: clearly the same thing holds for their short-circuit counterparts. The open-circuit set of eigenfunctions is implemented by one *harmonic function* ϕ_0 = cost corresponding to the eigenvalue $\lambda_0 = 0$, the short-circuit set is implemented by two *harmonic vectors* $f_0^{\prime}$ and $g_0^{\prime}$ corresponding to the eigenvalues $\lambda_0^{\prime} = 0$ and $\mu_0^{\prime} = 0$.

The completeness of these sets of eigenfunctions will not be discussed here but will be taken as granted in all cases of physical interest.*

Owing to the cylindrical shape of the structure, we decompose any vector field in the i-th layer into a component parallel to the P plane and a component parallel to the z-axis: the electric and the magnetic vectors are therefore represented in the form:

$$\mathbf{E}^{(i)} = \mathbf{E}_P^{(i)} + E_z^{(i)}\mathbf{c} \tag{1}$$

$$\mathbf{H}^{(i)} = \mathbf{H}_P^{(i)} + H_z^{(i)}\mathbf{c} \tag{2}$$

Let the open-circuit situation be considered. The plane vectors $\mathbf{E}_P^{(i)}$ and $\mathbf{H}_P^{(i)}$ may be expanded in series of the $\mathbf{f}_q^{\prime}$s, $\mathbf{e}_q^{\prime}$s and the $\mathbf{g}_q^{\prime}$s, $\mathbf{h}_q^{\prime}$s respectively; the $\phi_q^{\prime}$s and the $\psi_q^{\prime}$s are suitable to expand the vertical components $E_z^{(i)}$ and $H_z^{(i)}$. The following representation is therefore obtained:

---

* For dimensional reasons, we choose the scalar eigenfunctions to be normalized to the cross-section surface:

$$\int_S \phi_q^2 dS = S$$

It is easily seen that this implies the same kind of normalization for the vector eigenfunctions too.

TABLE I

Open-circuit eigenfunctions				
	$\nabla_P^2\phi_q + \lambda_q^2\phi_q = 0$	$\dfrac{\partial\phi_q}{\partial n} = 0$	on l	(T1-1)
	$\nabla_P^2\phi_0 = 0$	$\dfrac{\partial\phi_0}{\partial n} = 0$	on l	(T1-2)
PM modes	$\lambda_q\mathbf{f}_q = \mathbf{grad}_P\phi_q$	$\mathbf{n}\cdot\mathbf{f}_q = 0$	on l	(T1-3)
	$\lambda_q\mathbf{h}_q = \mathbf{c}\times\mathbf{grad}_P\phi_q$	$\mathbf{n}\times\mathbf{h}_q = 0$	on l	(T1-4)
	$\nabla_P^2\psi_q + \mu_q^2\psi_q = 0$	$\psi_q = 0$	on l	(T1-5)
PE modes	$\mu_q\mathbf{g}_q = \mathbf{grad}_P\psi_q$	$\mathbf{n}\times\mathbf{g}_q = 0$	on l	(T1-6)
	$\mu_q\mathbf{e}_q = \mathbf{grad}_P\psi_q\times\mathbf{c}$	$\mathbf{n}\cdot\mathbf{e}_q = 0$	on l	(T1-7)

TABLE I (cont.)

Short-circuit eigenfunctions			
$\nabla_P^2\phi'_q + \lambda'^2_q\phi'_q = 0$	$\phi'_q = 0$	on l_p	(T1-8)
	$\frac{\partial\phi'_q}{\partial n} = 0$	on l_c	(T1-9)
$\lambda'_q\mathbf{f}'_q = \mathbf{grad}_P\phi'_q$	$\mathbf{n} \times \mathbf{f}'_q = 0$	on l_p	(T1-10)
	$\mathbf{n} \cdot \mathbf{f}'_q = 0$	on l_c	
$\lambda'_q\mathbf{h}'_q = \mathbf{c} \times \mathbf{grad}_P\phi'_q$	$\mathbf{n} \cdot \mathbf{h}'_q = 0$	on l_p	(T1-11)
	$\mathbf{n} \times \mathbf{h}'_q = 0$	on l_c	
$\mathrm{div}\mathbf{f}'_0 = 0 \ \mathbf{rot}\mathbf{f}'_0 = 0$	$\mathbf{n} \times \mathbf{f}'_0 = 0$	on l_p	(T1-12)
	$\mathbf{n} \cdot \mathbf{f}'_0 = 0$	on l_c	
$\nabla_P^2\psi'_q + \mu'^2_q\psi'_q = 0$	$\frac{\partial\psi'_q}{\partial n} = 0$	on l_p	(T1-13)
	$\psi'_q = 0$	on l_c	
$\mu_q\mathbf{g}'_q = \mathbf{grad}_P\psi'_q$	$n \cdot \mathbf{g}'_q = 0$	on l_p	(T-21)
	$n \times \mathbf{g}'_q = 0$	on l_c	
$\mu_q\mathbf{e}'_q = \mathbf{grad}_P\psi'_q \times \mathbf{c}$	$\mathbf{n} \times \mathbf{e}'_q = 0$	on l_p	(T-22)
	$\mathbf{n} \cdot \mathbf{e}'_q = 0$	on l_c	
$\mathrm{div}\mathbf{g}'_0 = 0 \ \mathbf{rot}\mathbf{g}'_0 = 0$	$\mathbf{n} \cdot \mathbf{g}'_0 = 0$	on l_p	(T-23)
	$\mathbf{n} \times \mathbf{g}'_0 = 0$	on l_c	

$$\mathbf{E}^{(i)} = \sum_q F_q^{(i)} \mathbf{f}_q + \sum_q E_q^{(i)} \mathbf{e}_q + \sum_q E_{zq} \phi_q \mathbf{c} \tag{3}$$

$$\mathbf{H}^{(i)} = \sum_q G_q^{(i)} \mathbf{g}_q + \sum_q H_q^{(i)} \mathbf{h}_q + \sum_q H_{zq}^{(i)} \psi_q \mathbf{c} \tag{4}$$

Similar expressions hold obviously (with adding primes to all quantities) for the short-circuit situation: care must be exercised in order that index q runs for each sum from one or from zero to infinity, so that the harmonic terms are included when necessary.

If one singles out the first and the third sum from (3), and the second sum from (4), one obtains the expansion of a PM field; the remaining sums obviously give the expansion of a PE field. Although it is not strictly necessary we shall often decompose any actual field into a PM and a PE component and analyze them separately.

It is to be pointed out that in Eqs. (3) and (4) the eigenfunctions depend on the coordinates x and y, while the coefficients are functions of the coordinate z_i and of either the time t or the radian frequency ω, according to whether we consider (1) and (2) in the time or the frequency domain. To be precise, only the second interpretation will be used in the following. The dependence of the field components on z can be made explicit by expandin the latter in series of a new set of *depth eigenfunctions* which will be defined soon.

The electric field $\mathbf{E}^{(i)}$ and the magnetic field $\mathbf{H}^{(i)}$ are related by Maxwell's equations:

$$\mathbf{rot}\ \mathbf{H}^{(i)} = j\omega\varepsilon_{ci}\ \mathbf{E}^{(i)} \tag{5}$$

$$\mathbf{rot}\ \mathbf{E}^{(i)} = -j\omega\mu_i\ \mathbf{H}^{(i)} \tag{6}$$

where ε_{ci} is the complex dielectric constant of the material and μ_i its magnetic permeability.

Let now the first of Maxwell's equations be projected upon the $\mathbf{f}'_q$s and the $\phi_q\mathbf{c}$'s, the second upon the $\mathbf{h}'_q$s; through some quite trivial calculations [3] one finds for the i-th layer the following set of equations:

$$
\begin{aligned}
& j\omega\mu_i H_q^{(i)} + \frac{dF_q^{(i)}}{dz_i} - \lambda_q E_{zq}^{(i)} = 0 \\
& \frac{dH_q^{(i)}}{dz} + j\omega\varepsilon_{ci} F_q^{(i)} = 0 \qquad (7) \\
& \lambda_q H_q^{(i)} + j\omega\varepsilon_{ci} E_{zq}^{(i)} = \frac{1}{S}\int_{l_p} H_p^{(i)} \times \mathbf{n} \cdot \phi_q \mathbf{c}\, dl
\end{aligned}
$$

Eqs. (7) clearly describe all open-circuit PM modes or more briefly an open-circuit PM field. Their integration is a typical boundary value problem and is carried out in two steps, the first of which consists in finding the solutions of the homogeneous associate set of differential equations, with the proper boundary conditions at the discontinuity between two layers and between the external layers and the environment; these solutions form in any case a denumerable set and will be called the *depth eigenfunctions* [1, 2]. The second step consists in finding the solutions of the differential system (7) by expanding the $H_q^{(i)}$'s, $F_q^{(i)}$'s, $E_{zq}^{(i)}$'s and the forcing term in series of the latter.

In more precise terms, let $H_q^{(i)}$ and $F_q^{(i)}$ be eliminated from the homogeneous system associated with Eq. (7); the following equation for $E_{zq}^{(i)}$ is then obtained:

$$\frac{d^2E_{zq}^{(i)}}{dz_i^2} + K_c^{(i)2}E_{zq}^{(i)} = 0 \qquad (8)$$

where

$$K_c^{(i)2} = \omega^2\varepsilon_{ci}\mu_i - \lambda_q^2 \qquad (9)$$

Because of the boundary conditions, each of eqs. (8) admits only a denumerable set of properly normalized solutions $f_m^{(i)}(z)$ in correspondence with a denumerable set of values of $K_c^{(i)}$, which will be denoted as $K_{cm}^{(i)}$, the *depth eigenvalues*. Due to the nature of boundary conditions, the latter are in general functions of the frequency [2]. By taking into account Eq. (7) one easily obtains for a general PM field the following representation:

$$\begin{aligned} \mathbf{E}^{(i)} &= \sum_{q,m} F_{qm} \frac{df_m^{(i)}}{dz_i} \mathbf{f}_q + \sum_{q,m} E_{zqm} f_m^{(i)} \phi_q \mathbf{c} \\ \mathbf{H}^{(i)} &= \sum_{q,m} H_{qm} f_m^{(i)} \mathbf{h}_q \frac{\varepsilon_{ci}}{\varepsilon_{c2}} \end{aligned} \tag{10}$$

in terms of which Eqs. (7) become in turn:

$$\begin{cases} \dfrac{\varepsilon_{ci}}{\varepsilon_{c2}} j\omega\mu_i H_{qm} - K_{cm}^{(i)2} F_{qm} - \lambda_q E_{zqm} = 0 \\ \dfrac{\varepsilon_{ci}}{\varepsilon_{c2}} K_{cm}^{(i)2} H_{qm} - j\omega\varepsilon_{ci} K_{cm}^{(i)2} F_{qm} = 0 \\ \dfrac{\varepsilon_{ci}}{\varepsilon_{c2}} \lambda_q H_{qm} + j\omega\varepsilon_{ci} E_{zqm} = \dfrac{1}{S} \dfrac{\varepsilon_{ci}}{\varepsilon_{c2}} \displaystyle\int_{l_p} \mathbf{H}_{Pm} \times \mathbf{n} \cdot \phi_q \mathbf{c} \, dl \end{cases} \tag{11}$$

where $\mathbf{H}_{Pm}$ is defined through the partial expansion:

$$\mathbf{H}^{(i)} = \sum_m \mathbf{H}_{Pm} f_m^{(i)}(z) \frac{\varepsilon_{ci}}{\varepsilon_{c2}} \tag{12}$$

It is not difficult to verify that the three sets of Eqs. (11) corresponding to i = 1, 2, 3 admit the same solution (H_{qm}, F_{qm}, E_{zqm}) independent of i (this must be true, to fulfill the continuity conditions for $\mathbf{E}_P$ and $\mathbf{H}_P$ at the separation surface between two layers); the quantity $K_{cm}^{(i)2} - \omega^2\varepsilon_{ci}\mu_i$ must be the same for all three layers; we shall indicate its value by γ_m^2:

$$\gamma_m^2 = K_{cm}^{(i)2} - \omega^2\varepsilon_{ci}\mu_i \tag{13}$$

Similar procedures can be followed for the short-circuit PM - and for the open - and short-circuit PE modes. It must be pointed out that the depth eigenvalues and eigenfunctions are the same for both the open-circuit and the short-circuit modes of the same kind but they are in general different for PM's and PE's. To take care of this, we shall indicate the depth eigenvalues and eigenfunctions of the PE modes by $h_{cm}^{(i)}$ and $g_m^{(i)}(z)$ respectively. A far more complete approach to the problem of depth eigenmodes can be found in [1, 2].

The whole set of open-circuit and short-circuit equations for both PM and PE modes is found in Table II.

3. VOLTAGE AND CURRENT DEFINITIONS: IMPEDANCE AND ADMITTANCE MATRICES

Since the structures under investigation are not lossless in general, port voltages and currents cannot be defined as simply as in conventional theory of resonant cavities.

A suitable definition for the general case can be given as follows [2].

Let s_k be a linear coordinate on the intersection of port k with plane $z_2 = 0$, ranging from 0 to l_k. The port voltage and current are distinctly defined, according to whether one is considering a PM or a PE field.

a) *PM fields*

We first define a new quantity E_{zm} by the equation:

$$E_{zm} = \sum_q E_{zqm}\phi_q \tag{14}$$

E_{zm} is clearly the component of E_z with respect to the m-th depth mode. Moreover, we indicate by H_{tm} the tangential component of $\mathbf{H}_{Pm}$ on port k:

$$H_{tm} = \left|\mathbf{H}_{Pm} \times \mathbf{n}\right| \tag{15}$$

The voltage and the current at port k are then defined as the coefficients of the Fourier expansion of E_{zm} and H_{zm}

TABLE II

	Open-circuit	Short-circuit
PM MODES	$j\omega\mu\frac{\varepsilon_{ci}}{\varepsilon_{c2}}H_{qm}-K_{cm}^{(i)2}F_{qm}-\lambda_q E_{zqm} = 0$	$j\omega\mu\frac{\varepsilon_{ci}}{\varepsilon_{c2}}H'_{qm}-K_{cim}^{(i)2}F'_{qm}-\lambda'_q E'_{zqm} = \frac{1}{S}\int_{l_P} \mathbf{n}\times E_z\mathbf{c}\cdot\mathbf{h}_q dl$
	$-K_{cm}^{(i)2}\frac{\varepsilon_{ci}}{\varepsilon_{c2}}H_{qm}-j\omega\varepsilon_{ci}K_{cm}^{(i)2}F_{qm} = 0$	$-K_{cm}^{(i)2}\frac{\varepsilon_{ci}}{\varepsilon_{c2}}H'_{qm}-j\omega\varepsilon_{ci}K_{cm}^{(i)2}F'_{qm} = 0$
	$\lambda_q\frac{\varepsilon_{ci}}{\varepsilon_{c2}}H_{qm}+j\omega\varepsilon_{ci}E_{zqm} = \frac{1}{S}\frac{\varepsilon_{ci}}{\varepsilon_{c2}}\int_{l_P} \mathbf{H}_{Pm}\times\mathbf{n}\cdot\phi_q \mathbf{c} dl$	$\lambda'_q H'_{qm}\frac{\varepsilon_{ci}}{\varepsilon_{c2}} + j\omega\varepsilon_{ci}E'_{zqm} = 0$
PE MODES	$-j\omega\mu_i h_{cm}^{(i)2}G_{qm} - h_{cm}^{(i)2}\frac{\mu_i}{\mu_2}E_{qm} = 0$	$-j\omega\mu'_i h_{cm}^{(i)2}G'_{qm} - h_{cm}^{(i)2}\frac{\mu_i}{\mu_2}E'_{qm} = 0$
	$j\omega\mu_i H_{zqm} + \mu_q\frac{\mu_i}{\mu_2}E_{qm} = 0$	$j\omega\mu_2 H'_{zqm}+\mu'_q\frac{\mu_i}{\mu_2}E'_{qm} = \frac{1}{S}\frac{\mu_i}{\mu_2}\int_{l_P} H_z\mathbf{c}\times\mathbf{h}\cdot\mathbf{e}_q\, dl$
	$-h_{cm}^{(i)2}G_{qm}-\mu_q H_{zqm}+j\omega\varepsilon_{ci}\frac{\mu_i}{\mu_2}E_{qm} = \frac{1}{S}\int_{l_P} H_z\mathbf{c}\times\mathbf{n}\cdot l_q dl$	$-h_{cm}^{(i)2}G'_{qm}-\mu'_q H'_{zqm}+j\omega\varepsilon_{ci}\frac{\mu_i}{\mu_2}E'_{qm} = 0$

respectively in the interval $(0, l_k)$. More precisely one has the following definitions:

$$E_{zm}(s_k) = \frac{V_{km0}}{D_{Em}} + \frac{\sqrt{2}}{D_{Em}} \sum_{r=1}^{\infty} \left(V_{kmr}\cos\frac{r\pi s_k}{l_k} + V_{km,-r}\sin\frac{r\pi s_k}{l_k}\right) \tag{16}$$

$$H_{tm}(s_k) = \frac{V_{km0}}{l_k} + \frac{\sqrt{2}}{l_k} \sum_{r=1}^{\infty} \left(I_{kmr}\cos\frac{r\pi s_k}{l_k} + I_{km,-r}\sin\frac{r\pi s_k}{l_k}\right) \tag{17}$$

where

$$D_{Em} = \sum_{i=1}^{3} \int_{d_i} \left(\frac{\varepsilon_{ci}}{\varepsilon_{c2}}\right)^* f_{im}(z)\, f^*_{im}(z)\, dz \tag{18}$$

is a normalization factor introduced to ensure the validity of the conventional definition of power in terms of voltage and current (see, however, below); the same can be said obviously for the factors l_k and $\sqrt{2}$.

b) *PE fields*

In a way perfectly analogous to that of the previous case, one gives the definitions:

$$H_{zm} = \sum_q H_{zmq}\psi_q \tag{19}$$

$$E_{zm} = |\mathbf{E}_{Pm} \times \mathbf{n}| \tag{20}$$

$$H_{zm}(s_k) = \frac{I_{km0}}{D_{Hm}} + \frac{\sqrt{2}}{D_{Hm}} \sum_{r=0}^{\infty} \left(I_{kmr}\cos\frac{r\pi s_k}{l_k} + I_{km,-r}\sin\frac{r\pi s_k}{l_k}\right) \tag{21}$$

$$E_{tm}(s_k) = \frac{V_{km0}}{l_k} + \frac{\sqrt{2}}{l_k} \sum_{r=0}^{\infty} \left(V_{kmr}\cos\frac{r\pi s_k}{l_k} + V_{km,-r}\sin\frac{r\pi s_k}{l_k}\right) \tag{22}$$

with

$$D_{Hm} = \sum_{i=1}^{3} \int_{d_i} \left(\frac{\mu_i}{\mu_2}\right) g_{im}\, g_{im}^{*}\, dz \tag{23}$$

For the sake of convenience, the quantities $\sqrt{2}\cos\frac{r\pi s_k}{l_k}$ and $\sqrt{2}\sin\frac{r\pi s_k}{l_k}$ (1, for r = 0) will be often denoted as e_{Tr} and $e_{T,-r}$, or h_{Tr} and $h_{T,-r}$ according to whether they refer to PM or PE fields: the negative index -r will be systematically ignored by letting r run over the whole set of positive and negative integers.

It can be shown [2] that with the previous definitions the real port of $V_{kmr}\, I_{kmr}^{*}$ just gives the power entering port k associated with the depth mode defined by index m and the *width* mode defined by index r. It must be recognized, however, that if several depth and/or width modes are copresent, the total power is not in general the sum of the terms contributed by each combination of them; in other terms, the port modes (obtained by associating a width and a depth mode) are not orthogonal among themselves in general [2].

With this background, the impedance and admittance matrices for both PM and PE fields are easily derived. To give a short sketch of the procedure to be followed, let the case of the impedance matrix of a PM field be considered. One starts with Eqs. (11) which are solved with respect to E_{zqm}:

$$E_{zqm} = \frac{\gamma_m^2}{j\omega\varepsilon_{c2}(\gamma_m^2+\lambda_q^2)}\,\frac{1}{S}\int_{l} \mathbf{H}_{Pm} \times \mathbf{n} \cdot \phi_q \mathbf{c}\, dl \tag{24}$$

From (14) and (16) it is found:

$$V_{hmr} = D_{Em} \sum_{q} \beta_{hqr}\, E_{zqm} \tag{25}$$

where

$$\beta_{hqr} = \frac{1}{l_h}\int_{l_h} \phi_q\, e_{Tr}\, ds_h \tag{26}$$

On the other hand the following chain of equalities is clearly valid:

$$\frac{1}{S}\int_{l_k} \mathbf{H}_{Pm}\times\mathbf{n}\cdot\phi_q c\, dl = \frac{1}{S}\int_{l_k} H_{tm}\phi_q ds_k = \frac{1}{S l_k}\int_{l_k}\sum_{s=-\infty}^{+\infty} I_{kms} e_s \phi_q ds_k$$

$$= \frac{1}{S}\sum_{s=-\infty}^{+\infty} I_{kms}\frac{1}{l_k}\int_{l_k}\phi_q e_s ds_k = \frac{1}{S}\sum_{s=-\infty}^{+\infty}\beta_{kqs} I_{kms} \tag{27}$$

From (24), (25) and (27):

$$V_{hmr} = D_{Em}\sum_q \beta_{hqr}\frac{\gamma_m^2}{j\omega\varepsilon_{c2}(\gamma^2+\lambda_q^2)}\frac{1}{S}\sum_k\sum_r \beta_{kqs} I_{kms}$$

$$= \sum_k\sum_s\left(\sum_q \beta_{hqr}\beta_{kqs}\frac{\gamma_m^2 D_{Em}}{j\omega\varepsilon_{c2}(\gamma_m^2+\lambda_q^2)}\frac{1}{S}\right) I_{kms} \tag{28}$$

Hence the mutual impedance between port mode rm at port h and port mode sm at port k is given by:

$$Z_{hk}^{rm,sm} = \sum_q \beta_{hqr}\beta_{hqs}\frac{\gamma_m^2 D_{Em}}{j\omega\varepsilon_{c2}(\gamma_m^2+\lambda_m^2)}\frac{1}{S} \tag{29}$$

The other cases are treated in much the same way, by using open-circuit equations to derive PE impedances and short-circuit equations to derive PM and PE admittances; therefore we only report the results in Table III.

Expressions T3-1, T3-2, T3-3, T3-4 show that any self or transmittance function can be developed in partial fractions each of which is a function of *two* complex variables $j\omega$ and γ_m. An explicit expression in terms of $j\omega$ only requires, by the definition (13) of γ_m, the knowledge of $k_{cm}^{(i)}$ and hence the solution of the depth boundary problem.

TABLE III

	Impedance	Admittance
PM modes	$Z_{hk}^{rm,rs} = \sum_q \beta_{hqr}\beta_{hqs} \frac{\gamma_m^2 D_{Em}}{j\omega\varepsilon_{c2}(\gamma_m^2 + \lambda_q^2)} \frac{1}{S}$	$Y_{hk}^{rm,rs} = \sum_q \beta'_{hqr}\beta'_{hqs} \frac{j\omega\varepsilon_{c2}}{\gamma_m^2 + \lambda_q'^2} \frac{1}{D_{Em}}$
	$\beta_{hqr} = \frac{1}{l_h} \int_{l_h} \phi_q\, e_{Tr}\, ds$	$\beta'_{hqr} = \frac{1}{\sqrt{S}} \int_{l_h} h'_q\, e_{Tr}\, dl$
PE modes	$Z_{hk}^{rm,rs} = \sum_q \alpha_{hqr}\alpha_{hqs} \frac{j\omega\mu_2 d_2}{\eta_m^2 + \mu_q^2} \frac{d_2}{D} \frac{1}{S}$	$Y_{hk}^{rm,rs} = \sum_q \alpha'_{hqr}\alpha'_{hqs} \frac{\eta_m^2 D_{hm}}{j\omega\mu_2(\eta_m^2 + \mu_q'^2)S}$
	$\alpha_{hqr} = \frac{1}{d_2} \int_{l_h} e_q\, h_{Tr}\, dl$	$\alpha'_{hqr} = \frac{1}{l_h} \int_{l_h} \psi_q\, h_{Tr}\, dl$

4. APPROXIMATE EXPRESSIONS OF IMPEDANCE AND ADMITTANCE MATRICES. EQUIVALENT CIRCUITS

Various kinds of approximate expressions of the impedance and admittance matrices can be obtained depending on the peculiar physical properties of the structure and some particular performance conditions. We consider in some detail the case in which the operating frequency is high enough, that the skin depth in layers 1 and 3, say δ_1 and δ_3, is very small compared to layer thicknesses. Under such an assumption it can be shown [4] that the following approximations hold:

$$D_{Em} \simeq d_2 \tag{30}$$

$$D_{Hm} \simeq d_2 \tag{31}$$

$$\gamma_m^2 \simeq \frac{\sigma_2 + j\omega\varepsilon_2}{d_2}(\rho + j\omega\mu_2 d_2) + \left(\frac{m\pi}{d_2}\right)^2 \tag{32}$$

where

$$\rho = 2(1 + j)\left(\frac{1}{\sigma_1\delta_1} + \frac{1}{\sigma_3\delta_3}\right) \tag{33}$$

if m is different from zero, and

$$\rho = (1 + j)\left(\frac{1}{\sigma_1\delta_1} + \frac{1}{\sigma_3\delta_3}\right) \tag{34}$$

if m is equal to zero.

$$\eta_m^2 \simeq (\sigma_2 + j\omega\varepsilon_2)j\omega\mu_2 + \rho' + \left(\frac{m\pi}{d_2}\right)^2 \tag{35}$$

where η_m is the analog of γ_m for PE fields:

$$\eta_m^2 = h_{cm}^{(i)2} - \omega^2\varepsilon_{ci}\mu_i \tag{36}$$

and

$$\rho' = -\frac{2(m\pi)^2}{1 + j}\,\frac{\mu_1\delta_1 + \mu_3\delta_3}{\mu_2 d_2^3} \tag{37}$$

The point now reduces to substitute approximate relations (30) to (36) into eqs. T3-1, 2, 3, 4.

The following results are found:

a) *PM-field impedance*

$$Z_{hk}^{rm,sm} \simeq \sum_q \beta_{hqr}\beta_{kqs} \frac{d_2}{Sj\omega\varepsilon_{c2}} \frac{(\sigma_2+j\omega\varepsilon_2)\left(\frac{\rho}{d_2}+j\omega\mu_2\right)+\left(\frac{m\pi}{d_2}\right)^2}{(\sigma_2+j\omega\varepsilon_2)\left(\frac{\rho}{d_2}+j\omega\mu_2\right)+\left(\frac{m\pi}{d_2}\right)^2+\lambda_q^2} \tag{38}$$

b) *PM-field admittance*

$$Y_{hk}^{rm,sm} \simeq \sum_q \beta'_{hqr}\beta'_{kqs} \frac{j\omega\varepsilon_{c2}}{d_2} \frac{1}{(\sigma_2+j\omega\varepsilon_2)\left(\frac{\rho}{d_2}+j\omega\mu_2\right)+\left(\frac{m\pi}{d_2}\right)^2+\lambda'^2_q} \tag{39}$$

c) *PE-field impedance*

$$Z_{hk}^{rm,sm} \simeq \sum_q \alpha_{hqr}\alpha_{kqs} \frac{j\omega\mu_2 d_2}{S} \frac{1}{(\sigma_2+j\omega\varepsilon_2)j\omega\mu_2+\left(\frac{m\pi}{d_2}\right)^2+\rho+\mu_q^2}$$

d) *PE-field admittance*

$$Y_{hk}^{rm,sm} \simeq \sum_q \alpha'_{hqr}\alpha'_{kqs} \frac{d_2}{j\omega\mu_2 S} \frac{(\sigma_2+j\omega\varepsilon_2)j\omega\mu_2+\left(\frac{m\pi}{d_2}\right)^2+\rho'}{(\sigma_2+j\omega\varepsilon_2)j\omega\mu_2+\left(\frac{m\pi}{d_2}\right)^2+\rho'+\mu'^2_q}$$

The equivalent circuits which are easily obtained on both the impedance and the admittance basis are shown in Figs. 2 and 3. The reader can easily draw for himself the particular cases corresponding to a lossless structure ($\sigma_2 = \rho = \rho' = 0$) and to a RC structure ($\mu_2 = 0$); but he must be warned that in the latter case the definition of ρ and ρ' should be considerably modified, since the current distribution in layers 1 and 3 is not confined in a thin depth.

5. EXAMPLE

The previous equations will be applied to a very

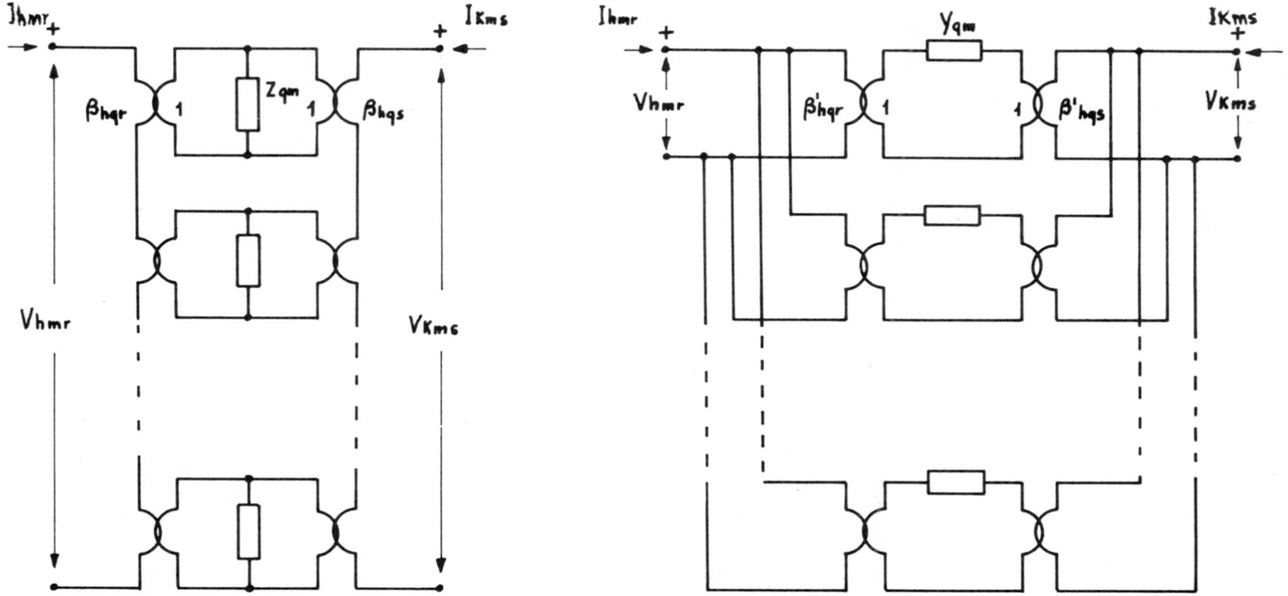

FIG. 2. a) Equivalent circuit for the two-port impedance matrix involving rm-th mode at port h and sm-th mode at port R; b) equivalent circuit for the two-port admittance matrix involving the same index and the same ports.

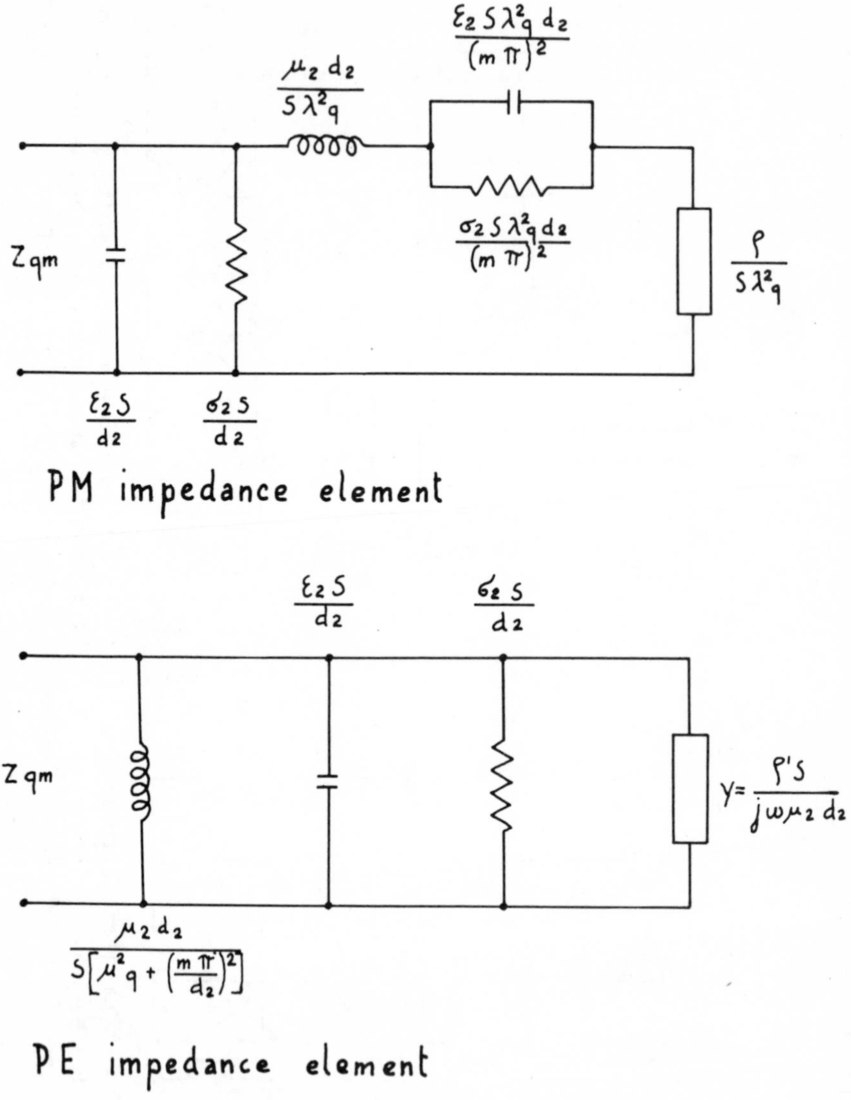

FIG. 3. a) Equivalent circuits for PM and PE impedance elements Z_{qm} (see Fig. 2a) pertinent to the q-th cavity mode and the m-th depth mode.

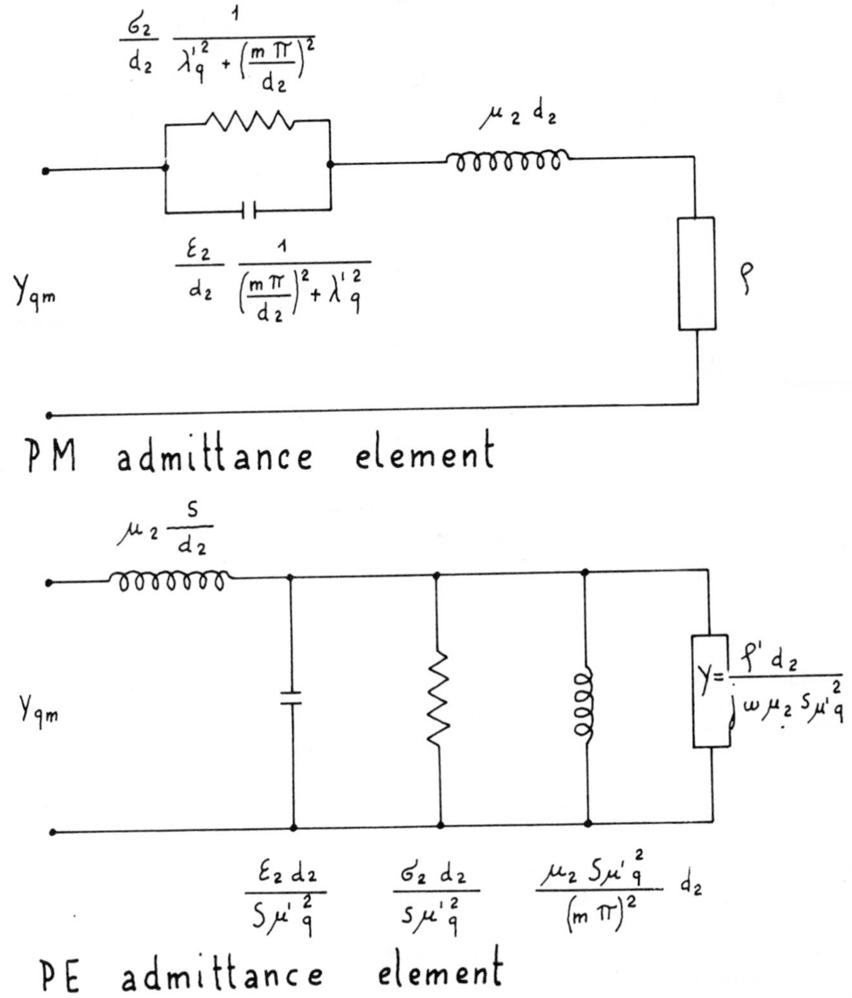

FIG. 3. b) Equivalent circuits for PM and PE admittance elements (see Fig. 2b) pertinent to the same modes.
For all resistors the conductances are given.

simple and well-studied case, that of a uniform microstrip of width a and length l. Only the case in which the line is fed by a TEM mode will be considered: it is immediate that the latter can be identified with the port mode PM_{00}.

We leave to the reader the easy task of carrying out the computations and confine ourselves to giving the results:

a) PM_{00}-impedances:

$$Z_{11} = Z_{22} = \frac{\gamma_0 d_2}{j\omega\varepsilon_{c2} a}\left(\frac{1}{\gamma_0 l} + \sum_{q=1}^{\infty} \frac{2\gamma_0 l}{(\gamma_0 l)^2+(q\pi)^2}\right) = Z_c \mathrm{cth}\gamma_0 l$$

$$Z_{12} = Z_{21} = \frac{\gamma_0 d_2}{j\omega\varepsilon_{c2} a}\cdot\left(\frac{1}{\gamma_0 l} + \sum_{q=1}^{\infty} \frac{2(-1)^n\gamma_0 l}{(\gamma_0 l)^2+(q\pi)^2}\right) = Z_c \frac{1}{\mathrm{sh}\gamma_0 l}$$

b) PM_{00}-admittances:

$$Y_{11} = Y_{22} = \frac{j\omega\varepsilon_{c2} a}{\gamma_0 d_2}\left(\frac{1}{\gamma_0 l} + \sum_{q=1}^{\infty} \frac{2\gamma_0 l}{(\gamma_0 l)^2+(q\pi)^2}\right) = Y_c \mathrm{cth}\gamma_0 l$$

$$Y_{12} = Y_{21} = -\frac{j\omega\varepsilon_{c2} a}{\gamma_0 d_2}\left(\frac{1}{\gamma_0 l} + \sum_{q=1}^{\infty} \frac{2(-1)^n\gamma_0 l}{(\gamma_0 l)^2+(q\pi)^2}\right) = -Y_c \frac{1}{\mathrm{sh}\gamma_0 l}$$

These results coincide with those which are found by conventional uniform line theory.

6. CONCLUSION

In this paper we have given partial fraction expansions for the impedance and admittance matrices of three-layer distributed n-ports: to make them explicit in numerical sense implies solving a two-dimensional and a one-dimensional boundary problem and hence requires the use of a digital computer except for very trivial particular cases; the use of equivalent circuits gives considerable physical insight into the properties of such structures.

7. ACKNOWLEDGEMENTS

The authors warmly thank Prof. G. Biorci, University of Genova and Prof. R. Sartori, Istituto Elettrotecnico Nazionale "Galileo Ferraris," Torino.

REFERENCES

1. RIDELLA, S. (Sept. 1968) Analysis of three-layer distributed structures with n terminal pairs, *Proc. 1968 Int. Symp. on Network Theory*, Belgrade, pp. 687-707.

2. BIORCI, G. and RIDELLA, S. (1969) On the theory of distributed three-layer n-port networks, *Alta Frequenza 38 (8)*: pp. 615-622.

3. BIANCO, B. and CIVALLERI, P.P. (1969) Basic theory of three-layer n-ports, *Alta Frequenza, 38 (8)*: pp. 623-631.

4. RIDELLA, S., Evaluation of depth eigenvalues in three-layer structures (to be presented at Fourth Colloquium on Microwave Communication).

5. KUROKAWA, K. (April 1968) The expansion of electromagnetic fields in cavities, *IRE Trans. on Microwave Theory and Techniques MTT-6*, pp. 178-187.

6. VAN BLADEL, J. (1964) *Electromagnetic Theory*, New York, McGraw-Hill.

7. JONES, D.S. (1964) *The Theory of Electromagnetism*. Oxford, Pergamon Press.

ANALYSIS OF A THREE-LAYER RECTANGULAR STRUCTURE*†

BRUNO BIANCO

Istituto Elettrotecnico Nazionale "Galileo Ferraris", Torino, Italy

SANDRO RIDELLA

Istituto di Elettrotecnica, Università di Genova, Genova, Italy

ABSTRACT

A three-layer rectangular structure endowed with two ports, one of which covers entirely one side while the other covers only partially the opposite side, is considered. The existence of transmission zeros and of related filtering properties is put into evidence.

1. INTRODUCTION

Three-layer tapered lines have been investigated, both in the lossless and in the RC case, by several authors [1-4] with the use of the classical non-uniform

---

* This work has been partially supported by Consiglio Nazionale delle Ricerche, Italy.

† The results of this work will be included in Ref. [7].

line theory. A more general class of distributed three-layer networks has been recently studied by G. Biorci, P.P. Civalleri and us [5-7]; in these works any number of ports and quite a general taper shape have been considered. A correct description of such systems involves the resolution of an elliptic partial differential equation. This is of course very difficult to do and requires, in general, the use of numerical methods [8].

The object of this work is to apply the methods developed in the quoted papers [5-7] to a very simple and nevertheless very interesting structure; these methods are based on an eigenfunction expansion of the solution of a partial differential equation. In spite of the simplicity of the system investigated in this work, the results are not trivial, and they widen the possibilities of applications of a tapered system.

2. TWO-PORTS ANALYSIS

The network under analysis is a three-layer structure of rectangular shape, with two ports, one of which covers a whole side while the other covers part of the opposite side (Fig. 1). The external layers are homogeneous conductors, while the internal one is an homogeneous lossy dielectric.

The methods of analysis of systems of such a type are extensively illustrated in the quoted papers [6-7]; here we report only the essential results.

On the basis of a previous work [5] every component of the electric or of the magnetic field can be expressed as the sum of products of eigenfunctions, a depth eigenfunction and a surface eigenfunction. It is assumed that the field imposed at the ports is the z-component of the electric field, constant over the whole width of the port and corresponding to the first depth eigenfunction. In

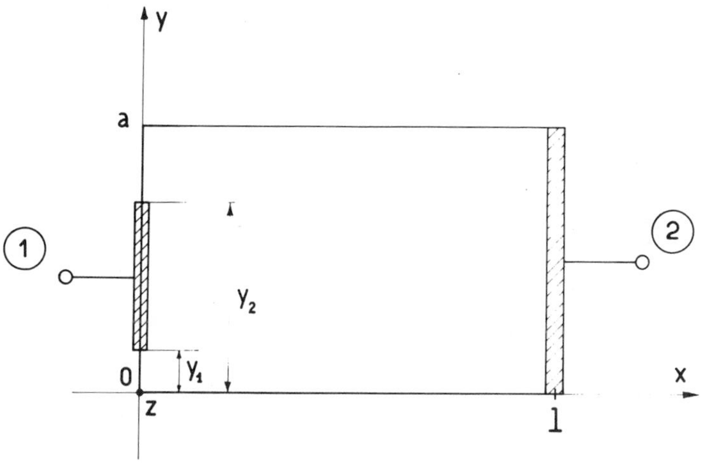

FIG. 1

the lossless case this means that the field imposed at the ports (or at least the only component of the whole field which one wants to consider) is the superposition of at most a TEM forward and a TEM backward wave.

The search for the open-circuit impedance matrix starts from the solution of the partial differential equation [7]:

$$\nabla^2_{xy}\ \phi_{n_1n_2} + \lambda^2_{n_1n_2}\ \phi_{n_1n_2} = 0 \qquad (1)$$

with the boundary condition:

$$\frac{\partial\phi_{n_1n_2}}{\partial n} = 0$$

on the whole perimeter of the rectangle; here $\phi_{n_1n_2}$ is the normalized* eigenfunction associated with the

---

* The normalization is done in such a way that $\int_S \phi^2{}_{n_1n_2} dS = S$ where S is the area of the cross-section of the structure.

eigenvalue $\lambda_{n_1 n_2}$, and **n** is a unit vector perpendicular to the boundary.

Simple computations lead to

$$\phi_{n_1 n_2} = A_{n_1 n_2} \cos \frac{n_1 \pi x}{l} \cos \frac{n_2 \pi y}{a} \quad (n_1, n_2 = 0,1,2,\ldots) \tag{2}$$

where:

$$A_{n_1 n_2} = \begin{cases} 2, & n_1 \neq 0,\ n_2 \neq 0 \\ \sqrt{2}, & n_1 = 0,\ n_2 \neq 0 \quad \text{or} \quad n_1 \neq 0,\ n_2 = 0 \\ 1, & n_1 = 0,\ n_2 = 0 \end{cases} \tag{2'}$$

and

$$\lambda_{n_1 n_2} = \pi[(n_1/l)^2 + (n_2/a)^2]^{\frac{1}{2}}. \tag{3}$$

On the basis of the results of a preceding paper [7] the generic entry of the open-circuit impedance matrix is:

$$Z_{hk} = \sum_{n_1=0}^{\infty} \sum_{n_2=0}^{\infty} \frac{\gamma^2 D_E}{s\varepsilon_{c_2}} \frac{\beta_{h,n_1 n_2} \beta_{k,n_1 n_2}}{(\gamma^2 + \lambda^2_{n_1 n_2})^S} \tag{4}$$

$$(h,k = 1,2)$$

where:

i) γ is the propagation constant of the network, and it is a function of the depth properties of the structure (that is, for each layer, of the complex dielectric constant ε_c, the magnetic permeability μ, and the thickness d) and of the generalized frequency s;

ii) D_E is an equivalent thickness, depending on the same quantities as γ;

iii) $\varepsilon_{c_2} = \varepsilon_2(1 + \sigma_2/s\varepsilon_2)$

is the complex dielectric constant of the internal layer (ε_2 and σ_2 being respectively its dielectric constant and conductivity);

iv) $$\beta_{h,n_1n_2} = \frac{1}{l_h}\int_0^{l_h} \phi_{n_1n_2}(s_h)ds_h \qquad (5)$$

where s_h is a running coordinate on the h-th port, going from 0 to l_h (h-th port width);

v) $S = al$ is the total plane surface of the system.

Substitution of Eqs. (2) (2') into Eq. (5), for port 1, gives:

$$\begin{aligned}
\beta_{1,00} &= 1 \\
\beta_{1,n_10} &= \sqrt{2} && (n_1 \neq 0) \\
\beta_{1,0n_2} &= \frac{\sqrt{2}}{y_2-y_1}\,\frac{a}{n_2\pi}\,c_{n_2} && (n_2 \neq 0) \\
\beta_{1,n_1n_2} &= \frac{2}{y_2-y_1}\,\frac{a}{n_2\pi}\,c_{n_2} && (n_1,n_2 \neq 0)
\end{aligned} \qquad (6)$$

where:

$$c_{n_2} = \sin\frac{n_2\pi y_2}{a} - \sin\frac{n_2\pi y_1}{a}. \qquad (7)$$

Similarly, for port 2:

$$\begin{aligned}
\beta_{2,00} &= 1 \\
\beta_{2,n_10} &= \sqrt{2}(-1)^{n_1} && (n_1 \neq 0) \\
\beta_{2,0n_2} &= 0 && (n_2 \neq 0) \\
\beta_{2,n_1n_2} &= 0 && (n_1,n_2 \neq 0)
\end{aligned} \qquad (8)$$

At this point, straightforward computations give the complete Z-matrix of the two-ports, from Eqs. (4),(6) and (8):

$$Z_{11} = \frac{\gamma D_E}{s\varepsilon_{c_2} a} \text{cth}\, \gamma l + $$

$$+ 2\frac{\gamma^2 D_E}{s\varepsilon_{c_2} a}\left(\frac{a}{y_2-y_1}\right)^2 \sum_{n_2=1}^{\infty}\left(\frac{c_{n_2}}{n_2\pi}\right)^2 \frac{\text{cth}\sqrt{\gamma^2+(n_2\pi/a)^2}\, l}{\sqrt{\gamma^2+(n_2\pi/a)^2}} \quad (9)$$

$$Z_{12} = \frac{\gamma D_E}{s\varepsilon_{c_2} a} \text{csch}\gamma l \quad (10)$$

$$Z_{22} = \frac{\gamma D_E}{s\varepsilon_{c_2} a} \text{cth}\gamma l \quad (11)$$

3. EQUIVALENT CIRCUIT AND PROPERTIES

From Eqs. (10) and (11), it is clear that the terms Z_{12} and Z_{22} of the impedance matrix are those of a uniform line, while Z_{11} is equal to Z_{22} plus a discontinuity impedance

$$Z_{11d} = 2\frac{\gamma^2 D_E}{s\varepsilon_{c_2} a}\left(\frac{a}{y_2-y_1}\right)^2 \sum_{n_2=1}^{\infty}\left(\frac{c_{n_2}}{n_2\pi}\right) \frac{\text{cth}\sqrt{\gamma^2+(n_2\pi/a)^2}\, l}{\sqrt{\gamma^2+(n_2\pi/a)^2}} \quad (12)$$

This suggests a description of the system by the equivalent circuit of Fig. 2. In this circuit the discontinuity impedance Z_{11d} is the series of an infinity of open uniform lines, of length l, propagation constant

$$\gamma_{n_2} = \sqrt{\gamma^2 + (n_2\pi/a)^2} \qquad (n_2 = 1,2,3,\ldots)$$

and characteristic impedance

$$Z_{0n_2} = \frac{2\gamma^2 D_E}{s\varepsilon_{c_2} a\sqrt{\gamma^2+(n_2\pi/a)^2}}\left(\frac{a}{y_2-y_1}\frac{c_{n_2}}{n_2\pi}\right)^2 \quad (n_2 = 1,2,3,\ldots)$$

while the line indicated by $n_2 = 0$ has length l, propagation constant γ, and characteristic impedance

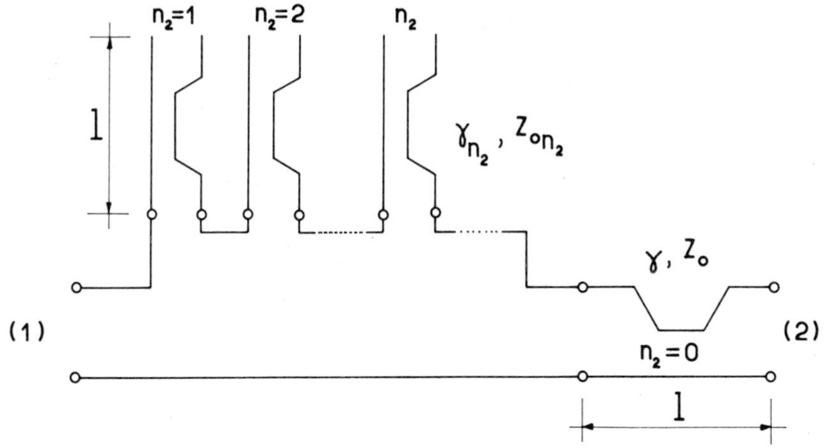

FIG. 2

$$Z_0 = \frac{\gamma D_E}{s\varepsilon_{c_2} a}$$

The discontinuity impedance Z_{11d} has poles in the finite part of the s-plane, which are transmission zeros; this implies that, while the C's and D's terms of the transmission matrix are entire functions, the terms A and B are not (in contrast with the non-uniform line theory [1]).

The poles of Z_{11d} arise from port 1's form. As the width of port 1 does not cover the whole side of the rectangle, the set of the natural frequencies of the impedance Z_{11} has a one-to-one correspondence with the eigenvalues' set of the structure.

Thus the poles of Z_{11} are a double infinity. One can separate these poles in two classes: one corresponding to the length eigenvalues $\lambda_{n_1 0}$, peculiar of a uniform line, and the other to the width eigenvalues $\lambda_{0 n_2}$ and to the "mixed" ones $\lambda_{n_1 n_2}$ $(n_1, n_2 \neq 0)$.

4. NUMERICAL ANALYSIS

The numerical analysis of the two-ports has been carried out in the lossless case, for which [7]:

$$\gamma^2 = -\omega^2 \varepsilon_2 \mu_2$$

$$D_E = d_2$$

The numerical computation of Z_{11d} has been made with the following criterion: for every value of ω we have computed all the terms of Eq. (12) having $\gamma^2 n_2$ less than zero. Such terms are finite in number and for them the hyperbolic cotangent becomes a circular cotangent multiplied by $-j$. For high values of n_2, $\gamma^2 n_2$ becomes positive; the series (12) converges and may be approximately evaluated by truncation.

The following quantities have been computed, for various values of the normalized frequency ($\bar{f} = 2\pi f\sqrt{\varepsilon_2 \mu_2}\, l$):

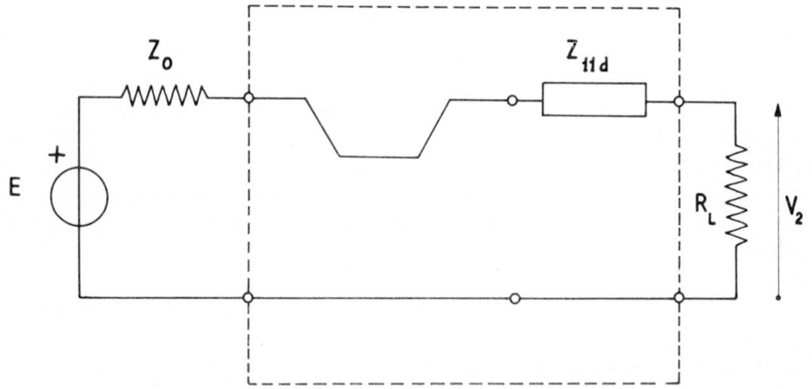

FIG. 3

i) with reference to Fig. 3, amplitude and phase angle of

$$A = \frac{R_L}{Z_0 + Z_{11d} + R_L} \qquad \text{(Figs. 6,7,8)}$$

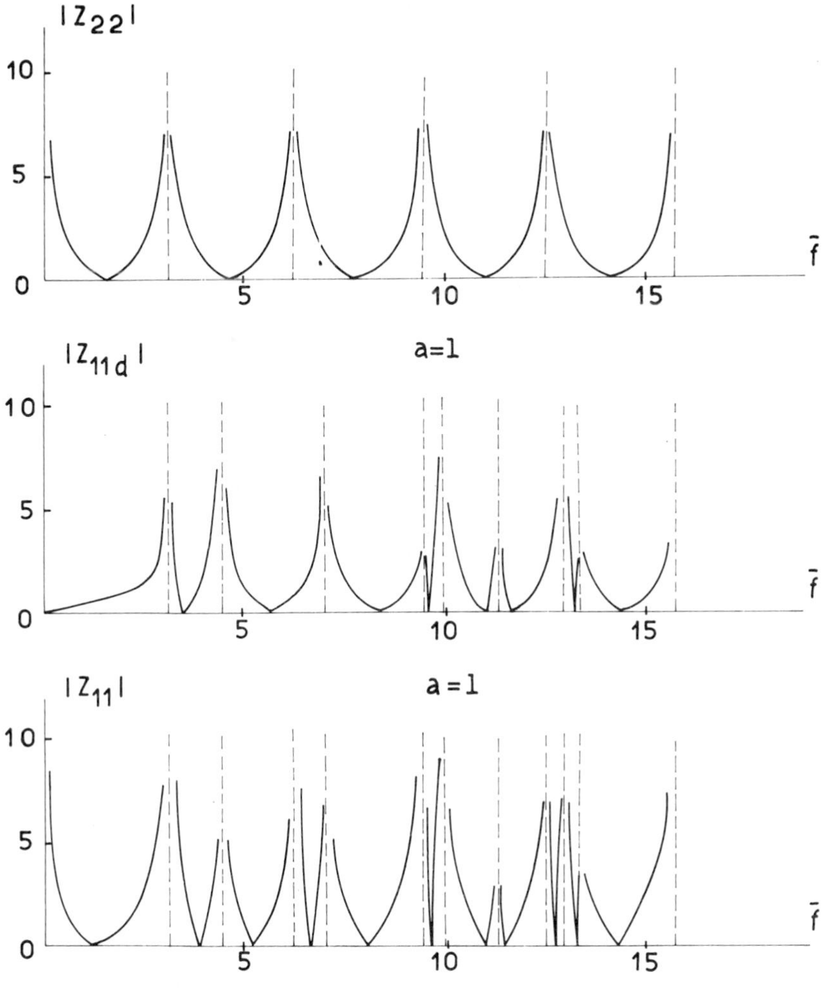

FIG. 4

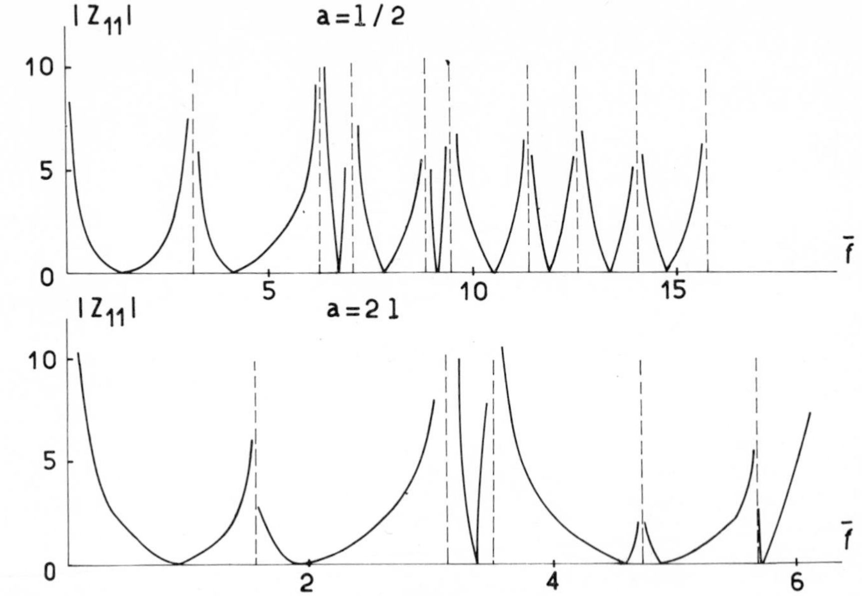
|Z11|
a=1/2
10
5
0
5
10
15
f
|Z11|
a=21
10
5
0
2
4
6
f

FIG. 5

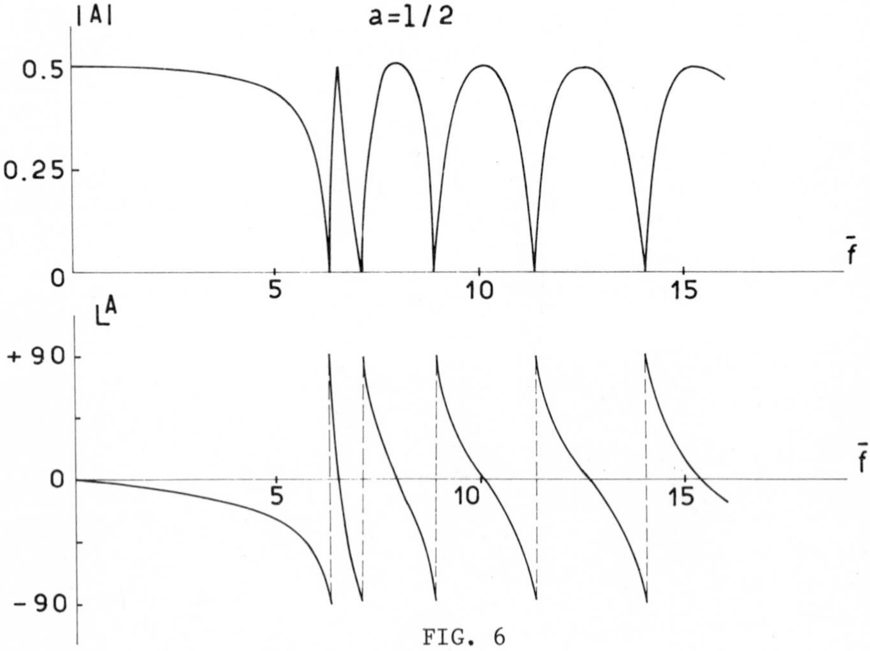
|A|
a=1/2
0.5
0.25
0
5
10
15
f
A
+90
0
-90
5
10
15
f

FIG. 6

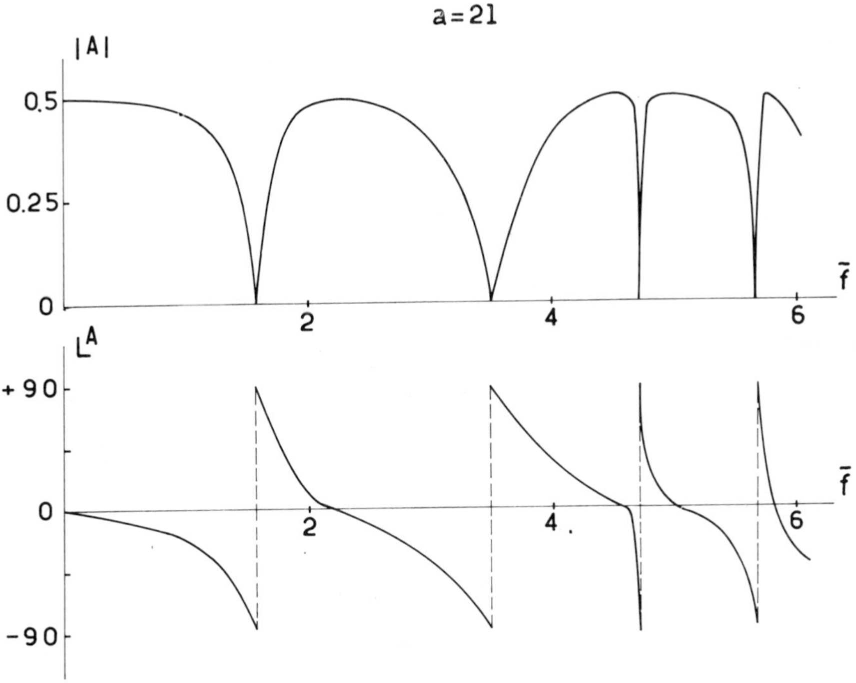

FIG. 7

(notice that $Ae^{-j\omega\sqrt{\varepsilon\mu}l}$ is equal to V_2/E in Fig. 3);

ii) the amplitude of the impedances

$$Z_{22},\ Z_{11},\ Z_{11d} \qquad \text{(Figs. 4,5)}$$

normalized to Z_0.

These data refer to a system with port 1 extending from $y_1 = 0$ to $y_2 = a/2$. The computations have been carried out for various values of a/l, with $R_L = Z_0$.

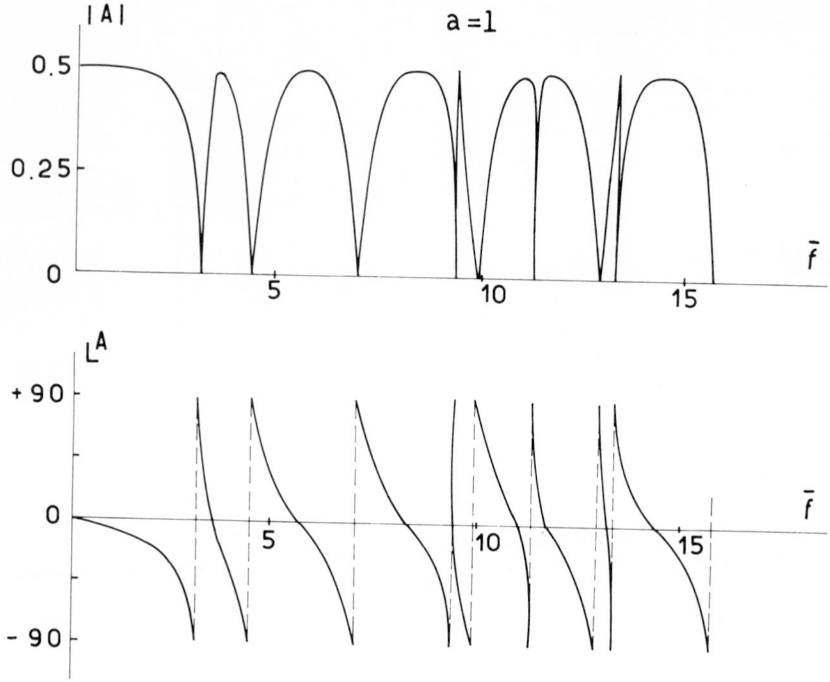

FIG. 8

5. CONCLUSION

We have shown that, by a relatively simple network, one can construct very complicated transfer functions; the most interesting feature seems to be the existence of transmission zeros, in radical contrast with the non-uniform line theory. Figs. 6,7,8 show that this results in low-pass and suppressing-band filtering properties.

The results obtained are significant enough and they encourage the authors to develop this investigation further, especially if one considers their variety and complexity with respect to the simplicity of the structure examined.

6. ACKNOWLEDGEMENT

The authors wish to thank Prof. G. Biorci, Prof. P.P. Civalleri and Prof. R. Sartori for their helpful interest and encouragement in this work.

REFERENCES

1. PROTONOTARIUS, E.N. and WING, O. (March 1967) Theory of nonuniform RC lines. 1: Analytic properties and realizability conditions in the frequency domain. 2: Analytic properties in the time domain, *IEEE Trans. on Circuit Theory CT-14*, pp. 2-20.

2. SU, K.L. (1963) The trigonometric RC transmission lines, *IEEE Internat'l Conv. Rec. 2*, pp. 43-55.

3. WOO, B.B. and BARTLEMAY, J.M. (1963) Characteristics and applications of a tapered, thin film distributed parameter structure, *IEEE Internat'l Conv. Rec. 2*, pp.56-75.

4. WYNDRUM, R.W. (1965) The realization of monomorphic thin film distributed RC networks, *IEEE Internat'l Conv. Rec. 10*, pp. 90-95.

5. BIORCI, G. and RIDELLA, S. (1969) On the theory of distributed three-layer N-port networks, *Alta Frequenza 38* (8), pp. 615-622.

6. BIANCO, B. and CIVALLERI, P.P. (1969) Basic theory of planar distributed N-ports, *Alta Frequenza 38* (8), pp. 623-631.

7. CIVALLERI, P.P. and RIDELLA, S. Impedance and admittance matrices of distributed three-layer N-ports (to be presented at IEEE Symposium on Circuit Theory 1969, and to appear in IEEE Trans. on Circuit Theory).

8. BIANCO, B. and RIDELLA, S. (1969) On the chain matrix of a three-layer RC structure, This volume, pp. 30-3.

ON THE CHAIN MATRIX OF A THREE-LAYER RC STRUCTURE*†

B. BIANCO

Istituto Elettrotecnico Nazionale "Galileo Ferraris", Torino, Italy

S. RIDELLA

Istituto di Elettrotecnica, Università di Genova, Genova, Italy

ABSTRACT

A three-layer tapered distributed *RC* two-port, with a strong discontinuity in the taper, is analyzed. The analysis, which is based on four types of approach with different levels of approximation, leads to the computation of the chain matrix and to the construction of an equivalent network. The aim of the work is to compare the results that are obtained on the basis of the various techniques of analysis now available.

1. INTRODUCTION

The importance of tapered distributed three-layer

---

* This work has been partially supported by CNR (Consiglio Nazionale delle Ricerche), Italy.

† A shorted version of this work will be published in IEEE Transactions on Circuit Theory.

structures has greatly increased in recent years. Several works on analysis and synthesis are available [1 - 5]. A paper by Wyndrum [5] is particularly interesting for our purposes. It gives necessary and sufficient conditions (with the synthesis procedure) for the realizability of a cascade of uniform lines of equal lengths and propagation constants but different widths. This structure can be used to approximate any tapered line.

The methods of analysis of such structures used in the quoted papers [1 - 5] make use of the classical non-uniform line theory, which is valid in some particular cases only [6]. An exact method of analysis for any kind of three-layer structures has been recently developed by G. Biorci, P.P. Civalleri and us [7, 8]. In our approach the linear variable-coefficient second-order ordinary differential equation describing the voltage distribution along the line is substituted by an elliptic partial differential equation. Our method is more general than the classical one, since it is valid for any type of tapered lines, even if their width is strongly discontinuous.

In fact, Wyndrum's synthesis method leads to a structure of the type shown in Fig. 1 where the widths change roughly from one section to the contiguous one.

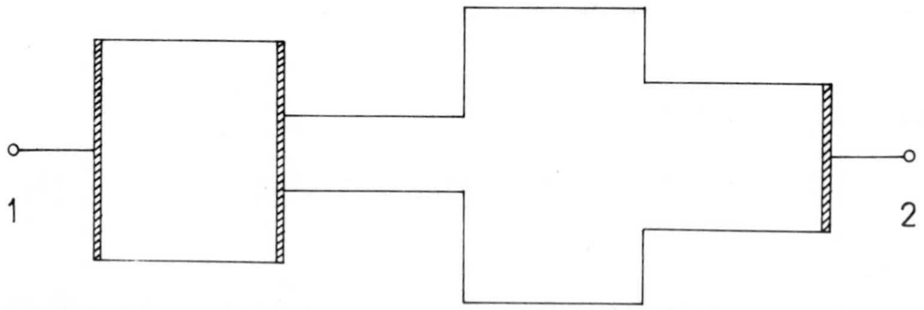

FIG. 1.

It is possible to annihilate the effect of these discontinuities by means of depositions of conducting material in correspondence of the discontinuities, as it is shown in Fig. 1, but this complicates the technological process.

In order to evaluate the approximation of Wyndrum's method, we have started with the analysis of a structure having only one discontinuity (Fig. 2). The methods used in this work are easily extendable to more complicated structures.

Thus, the object of this work is to analyze a strong discontinuity in the taper and to compare the results obtained by various methods of analysis.

2. STRUCTURE DESCRIPTION

The structure that we examine is made of three layers (Fig. 2): the lowest layer is a perfect conductor, the intermediate is a perfect dielectric, and the upper one is a conductor. Each layer is made of isotropic, linear, time-invariant, homogeneous material.

This structure interacts with the environment through two ports. The portion of the boundary of layer 1 which belongs to the ports is covered by perfectly conducting material, in such a way as to make the component of the electric field along the z-axis (over the whole width) sufficiently constant.

3. EQUIVALENT CIRCUIT OF THE SYSTEM

We want to prove that the structure under examination is equivalent to the cascade of two uniform lines with a discontinuity impedance inserted between them.

In the quoted paper [7] it is shown that an exact computation of the chain matrix of the system in Fig. 2 requires the solution of the following partial differential equation:

$$\nabla^2_{xy} E_S - \gamma^2 E_S = 0 \tag{1}$$

The boundary conditions are of the mixed type: on the portion of the boundary which does not belong to the ports, the normal derivative of E_S must be zero; on the ports,

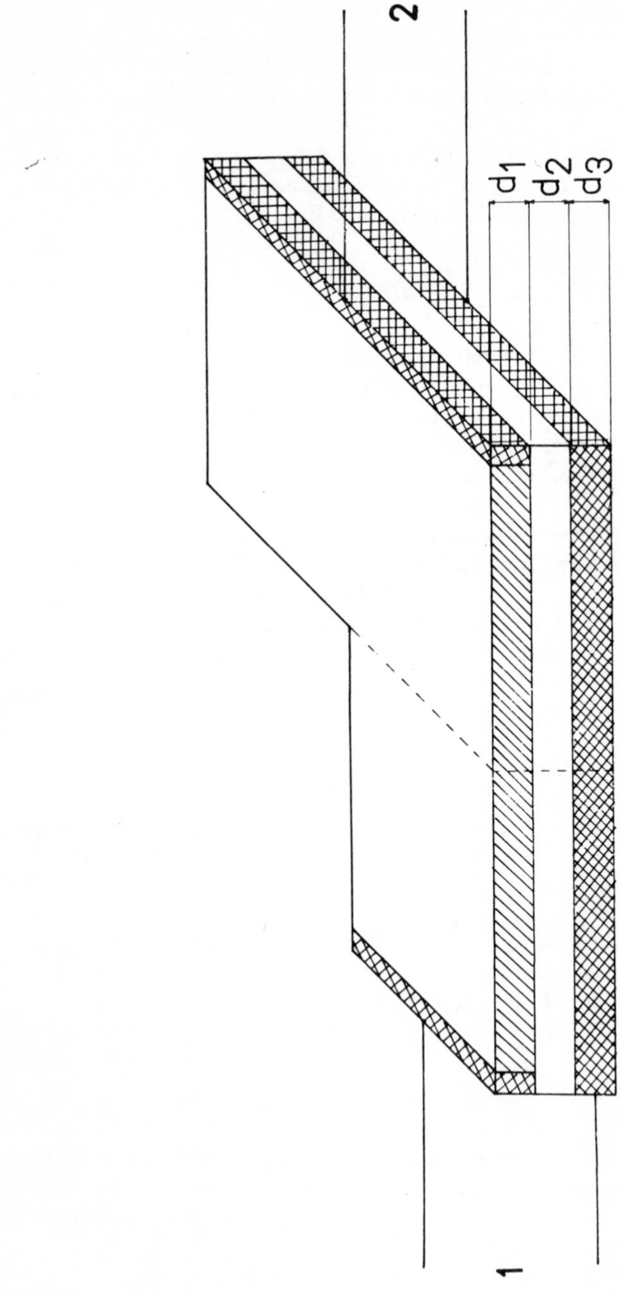

FIG. 2.

different conditions arise for the various elements of the chain matrix.

E_S is a field function whose mean value at a port is proportional to the voltage at the same port [7]. Moreover the current at a port is proportional to the mean value of the normal derivative of E_S.

The constant γ in Eq. (1) is the propagation constant of the system [9], and, for the structure under examination, it is assumed to be

$$\gamma^2 = j\omega R_q C_q \tag{2}$$

where R_q is the resistance per square of layer 1, and C_q the capacitance per square between layers 1 and 3.

From the work previously quoted [7] it is known that the voltage V_k and the current I_k at port k (k = 1, 2) are given by:

$$V_k = \frac{d_2}{l_k}\int_0^{l_k} E_S(s_k)\,ds_k \tag{3}$$

$$I_k = -\frac{d_2}{R_q}\int_0^{l_k} \frac{\partial E_S(s_k)}{\partial n}\,ds_k \tag{4}$$

where s_k is a running coordinate on port k, going from 0 to l_k (k-th port width).

Eq. (1) must be solved by numerical methods. First, Eq. (1) is replaced by a finite-difference equation. A linear system of equations is then derived. To have a good convergence of Seidel's iteration process, used to solve this system, it is convenient to start with the parameter B; the boundary conditions imposed at the ports are then

$$E_S = 1 \quad \text{on port 1,}$$

$$E_S = 0 \quad \text{on port 2.}$$

When $E_S(x,y)$ is obtained, the computer program gives the mean value of the normal derivative of E_S on port 2:

$$\left(\frac{\partial E_S}{\partial n}\right)_2 = \frac{1}{b}\int_0^b \frac{\partial E_S(s_2)}{\partial n}\, ds_2 \;. \tag{5}$$

And then, one obtains B:

$$B = \frac{R_q}{b\left(\dfrac{\partial E_S}{\partial n}\right)_2} \;. \tag{6}$$

Now we will prove that the other elements of the chain matrix can be calculated from the value of B, without solving Eq. (1).

In order to find the chain parameters of the network, the expressions (3) and (4) must be calculated at port 1 and at port 2: that is, the mean values of $E_S(x,y)$ and of $\partial E_S(x,y)/\partial n$ in the y-direction at $x = 0$ and $x = l$ ($l = l_a + l_b$) must be known.

Thus it is convenient to define $E_a(x)$ as the mean value of $E_S(x,y)$ (in the y direction) for $0 \le x \le l_a$:

$$E_a(x) = \frac{1}{a}\int_0^a E_S(x,y)\,dy \tag{7'}$$

and $E_b(x)$, the same quantity for $l_a \le x \le l$:

$$E_b(x) = \frac{1}{b}\int_0^b E_S(x,y)\,dy \tag{7''}$$

Explicit expressions for $E_a(x)$ and $E_b(x)$ can be readily found. Integration of Eq. (1) over the closed area ABCD (Fig. 3) gives:

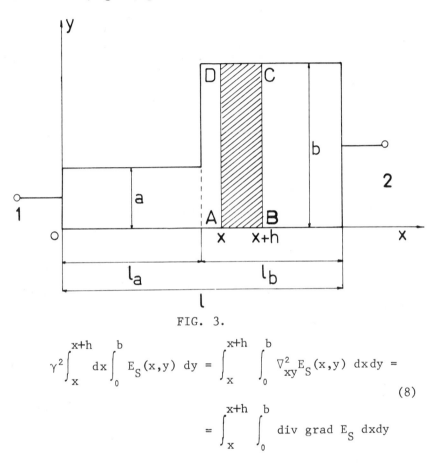

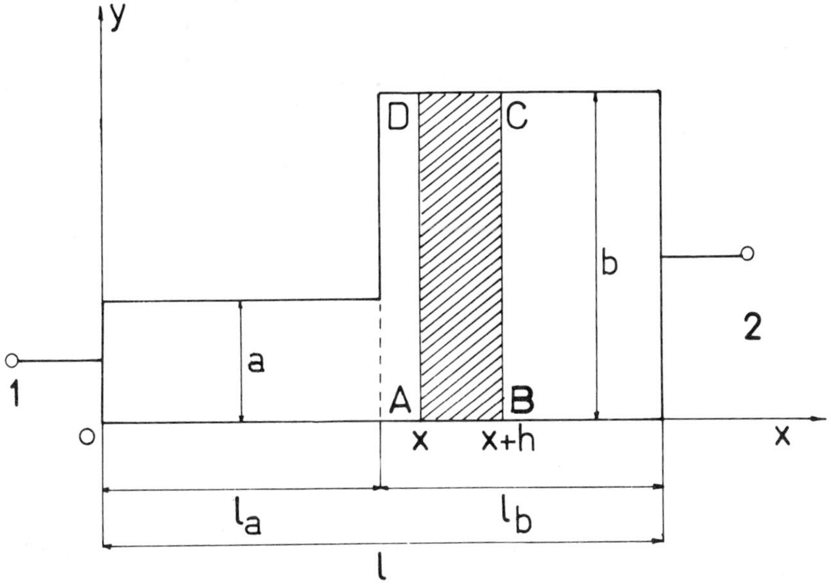

FIG. 3.

$$\gamma^2 \int_x^{x+h} dx \int_0^b E_S(x,y)\, dy = \int_x^{x+h} \int_0^b \nabla_{xy}^2 E_S(x,y)\, dx dy = \tag{8}$$

$$= \int_x^{x+h} \int_0^b \text{div grad } E_S \, dxdy$$

Straightforward use of the divergence theorem simplifies the preceding expression; one obtains:

$$\gamma^2 \int_x^{x+h} dx \int_0^b E_S(x,y)\,dy = \int_0^b \left(\frac{\partial E_S}{\partial x}\right)_{x+h} dy - \int_0^b \left(\frac{\partial E_S}{\partial x}\right)_x dy \tag{9}$$

as the boundary conditions impose that the normal derivative of $E_S(x,y)$ is zero on the edges of the structure not belonging to the ports.

By substitution of the expression (7") into Eq. (9) one obtains:

$$\frac{d^2E_b}{dx^2} - \gamma^2 E_b = 0, \tag{10"}$$

when the parameter h goes to zero. This equation is valid only for $l_a \leq x \leq l$. A similar equation can be obtained for $E_a(x)$ and it is valid for $0 \leq x \leq l_a$:

$$\frac{d^2E_a}{dx^2} - \gamma^2 E_a = 0 . \tag{10'}$$

From Eqs. (10') and (10") explicit expressions for $E_a(x)$ and $E_b(x)$ are obtained:

$$E_a(x) = m \operatorname{ch}\gamma x + n \operatorname{sh}\gamma x \tag{11'}$$

$$E_b(x) = p \operatorname{ch}\gamma x + q \operatorname{sh}\gamma x \tag{11"}$$

It is worth noticing that the coefficients m, n, p, q must be determined. Two equations are given by the boundary conditions at the ports. These conditions, for the parameter B, are:

$$\left.\begin{aligned} E_a(0) &= 1 \text{ (port 1)} , \\ E_b(l) &= 0 \text{ (port 2)} . \end{aligned}\right\} \tag{12}$$

Thus, Eqs. (11') and (11") become:

$$E_a(x) = \mathrm{ch}\gamma x + n\ \mathrm{sh}\gamma x\ , \tag{13'}$$

$$E_b(x) = t\ \mathrm{sh}\gamma(x - l)\ , \tag{13''}$$

where t is a new constant which substitutes p and q. Moreover, one relation among the constants n and t is given by integration of Eq. (1) over the whole xy area of the structure:

$$\int_S \gamma^2 E_S\, dxdy = \int_S \nabla^2_{xy} E_S\, dxdy\ . \tag{14}$$

The use of the divergence theorem and of the above-stated boundary conditions gives:

$$\gamma^2 \int_0^{l_a} dx \int_0^a E_S\, dy + \gamma^2 \int_{l_b}^{l} dx \int_0^b E_S\, dy =$$

$$= \int_0^b \left(\frac{\partial E_S}{\partial x}\right)_l dy - \int_0^a \left(\frac{\partial E_S}{\partial x}\right)_0 dy\ . \tag{15}$$

Straightforward manipulations of this equation, after substitution of Eqs. (11') and (11"), give:

$$a\left(\frac{dE_a}{dx}\right)_{l_a} = b\left(\frac{dE_b}{dx}\right)_{l_a}\ . \tag{16}$$

With this equation one can find the constant t as a function of the constant n:

$$t = \frac{a}{b}\ \frac{\mathrm{sh}\gamma l_a + n\ \mathrm{ch}\gamma l_a}{\mathrm{ch}\gamma l_b}\ . \tag{17}$$

That is, Eqs. (13') and (13") become:

$$E_a = \text{ch}\gamma x + n\ \text{sh}\gamma x\ , \tag{18'}$$

$$E_b = \frac{a}{b}\ \frac{\text{sh}\gamma l_a + n\ \text{ch}\gamma l_a}{\text{ch}\gamma l_b}\ \text{sh}\gamma(x - l). \tag{18"}$$

From the knowledge of the parameter B one can obtain the constant n. It is well known that

$$B = \left(\frac{V_1}{-I_2}\right)_{V_2=0} \quad ;$$

from this and from the definition (4) of the current one can obtain:

$$n = -\frac{\text{ch}\gamma l_b}{\text{ch}\gamma l_a}\left(\frac{R_q}{B\gamma a} + \frac{\text{sh}\gamma l_a}{\text{ch}\gamma l_b}\right) . \tag{19}$$

Now it is known that

$$D = \left(\frac{I_1}{-I_2}\right)_{V_2=0} \quad ;$$

thus one obtains:

$$D = \left(1 + \frac{a\gamma B}{R_q}\ \frac{\text{sh}\gamma l_a}{\text{ch}\gamma l_b}\right)\frac{\text{ch}\gamma l_b}{\text{ch}\gamma l_a} \ . \tag{20'}$$

Thus, D is obtained as a function of B. A similar procedure can be used to obtain A, changing port 1 with port 2 (since the system is reciprocal):

$$A = \left(1 + \frac{b\gamma B}{R_q}\ \frac{\text{sh}\gamma l_b}{\text{ch}\gamma l_a}\right)\frac{\text{ch}\gamma l_a}{\text{ch}\gamma l_b} \ . \tag{20"}$$

Finally, from the knowledge of A, B, D, one has C through the reciprocity:

$$C = \frac{AD - 1}{B} \quad . \qquad (20''')$$

In conclusion, it is necessary to solve Eq. (1), by numerical methods, only once to find the B parameter of the chain matrix; the other elements are easily obtained from Eqs.(20'), (20"), (20"').

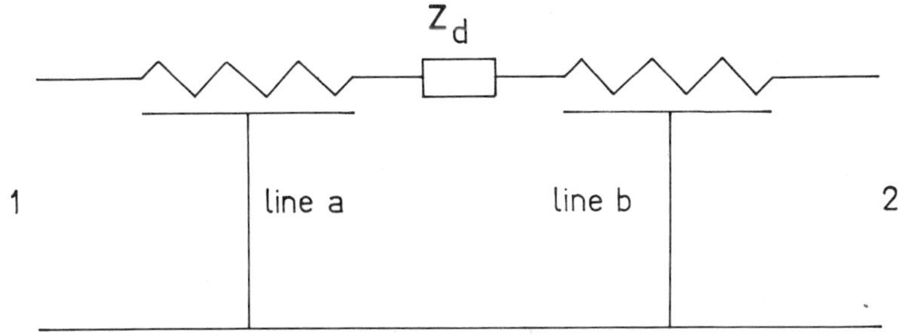

FIG. 4.

One can consider the network of Fig. 4: it is composed of a discontinuity impedance Z_d inserted between two uniform RC lines with the same propagation constant γ, different lengths (l_a and l_b), and different characteristic impedances:

$$\frac{1}{a}\sqrt{R_q/j\omega C_q} \text{ (line a) and } \frac{1}{b}\sqrt{R_q/j\omega C_q} \text{ (line b).}$$

Straightforward calculations give the chain matrix of this network:

$$A = \mathrm{ch}\gamma l + \left(\frac{b}{a} - 1\right) \mathrm{sh}\gamma l_a \ \mathrm{sh}\gamma l_b +$$

$$+ Z_d \sqrt{\frac{j\omega C_q}{R_q}}\, b \ \mathrm{ch}\gamma l_a \ \mathrm{sh}\gamma l_b \tag{21'}$$

$$B = \sqrt{\frac{R_q}{j\omega C_q}}\, \frac{1}{b} \left\{ \mathrm{sh}\gamma l + \left(\frac{b}{a} - 1\right) \mathrm{sh}\gamma l_a \ \mathrm{ch}\gamma l_b \right\} +$$

$$+ Z_d \ \mathrm{ch}\gamma l_a \ \mathrm{ch}\gamma l_b \tag{21''}$$

$$C = \sqrt{\frac{j\omega C_q}{R_q}}\, a \left\{ \mathrm{sh}\gamma l + \left(\frac{b}{a} - 1\right) \mathrm{ch}\gamma l_a \ \mathrm{sh}\gamma l_b \right\} +$$

$$+ Z_d \frac{j\omega C_q \ ab}{R_q} \mathrm{sh}\gamma l_a \ \mathrm{sh}\gamma l_b \tag{21'''}$$

$$D = \mathrm{ch}\gamma l + \left(\frac{a}{b} - 1\right) \mathrm{sh}\gamma l_a \ \mathrm{sh}\gamma l_b +$$

$$+ Z_d \sqrt{\frac{j\omega C_q}{R_q}}\, a \ \mathrm{sh}\gamma l_a \ \mathrm{ch}\gamma l_b \tag{21^{iv}}$$

Now, if one chooses the discontinuity impedance Z_d in such a way that the B parameters for the network in Fig. 4 and for the structure in Fig. 2 are equal, that is:

$$Z_d = \left\{ B - \sqrt{\frac{R_q}{j\omega C_q}}\, \frac{1}{b}\left(sh\gamma l + \left(\frac{b}{a} - 1\right) sh\gamma l_a \; ch\gamma l_b \right) \right\} \times$$

$$\times \frac{1}{ch\gamma l_a \; ch\gamma l_b}\,, \tag{22}$$

then the chain matrixes of the two networks are equal; for, notice that Eqs. (21) satisfy Eqs. (20); on the other hand, the B parameters for the two networks that we consider are equal and therefore the parameters A, D and C obtained from Eqs. (20) are the same.

Thus, the structure in Fig. 2 can be represented by the equivalent network of Fig. 4.

In conclusion, for every value of the frequency, Eq. (1) must be solved only once to determine B; while the other elements of the chain matrix and the discontinuity impedance Z_d are easily calculated from Eqs. (20'), (20"), (20"') and (22).

We remark that Eq. (22) does not give simple information on the nature of Z_d; further investigations are needed on this subject.

4. NUMERICAL ANALYSIS

A tapered structure has been analysed, whose dimensions are:

$$l_a = l_b = \frac{a}{2} = \frac{b}{10} = 0.5 \text{ cm.}$$

These dimensions are not unusual: from Wyndrum's papers one can see that the ratio b/a can assume values in a very large range (1 to 250). The length $l = l_a + l_b$ cannot be made as large as desired because it is the frequency scale factor, while the amplitude scale factor imposes the width a, once R_q and C_q are chosen. Thus, they are determined from the properties of the impedance to be realized.

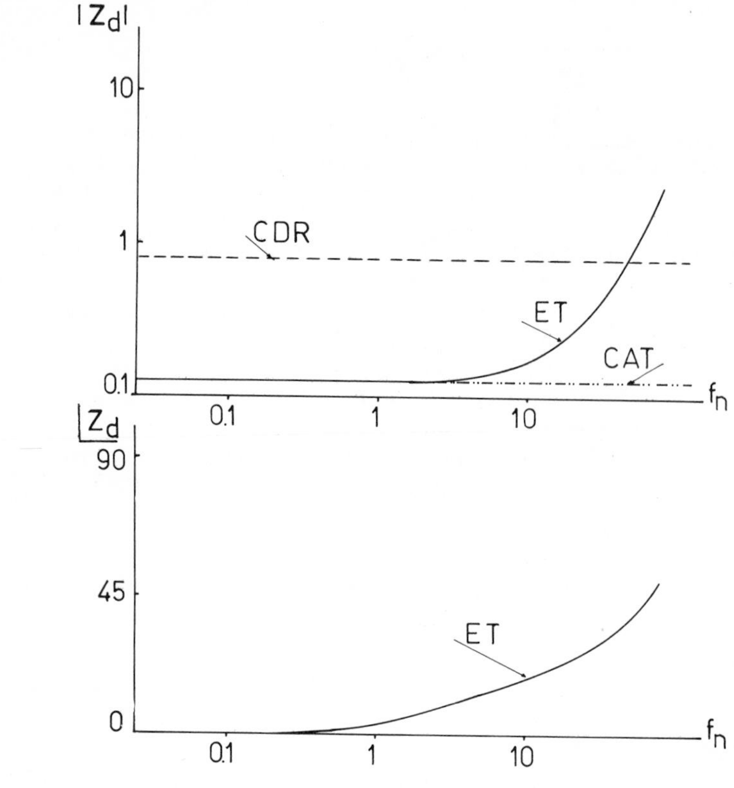

FIG. 5.

In FIG. 5.one can find the discontinuity impedance obtained by Eq. (22), and in Figs. 6, 7, 8, 9 one can find the elements of the chain matrix of the network, obtained with different methods. In these figures the impedances are normalized to R_q, the admittances to $j\omega C_q\left(a l_a + b l_b\right)$; and the normalized frequency is given by

$$f_n = \omega R_q C_q l^2 .$$

In Figs. 5 to 9 the plots obtained with numerical solution of Eq. (1) are indicated with ET (Exact Theory).

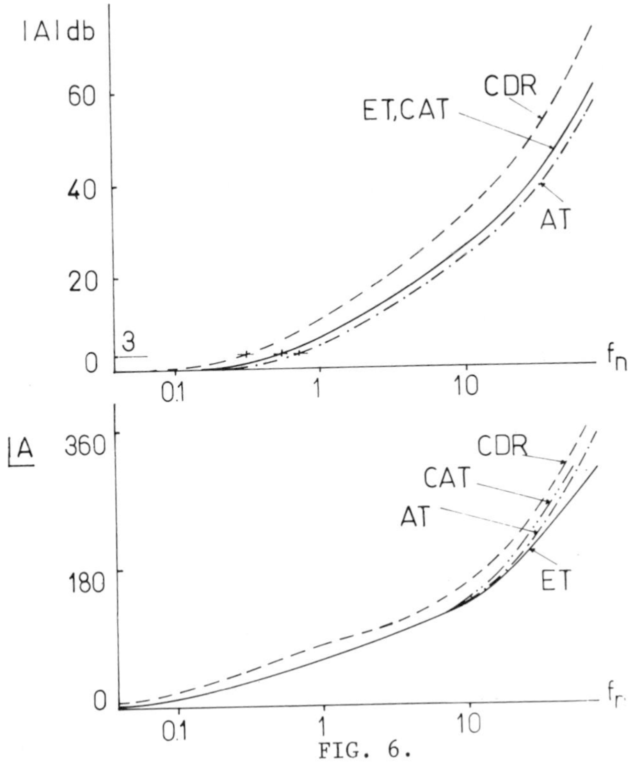

FIG. 6.

One can obtain, in a first approximation, the chain matrix neglecting the discontinuity impedance in the equivalent circuit of Fig. 4. With this approximation one has the usual equivalent circuit, used in the quoted papers [1 - 5]. The plots obtained with this approximation are indicated in the figures with AT (Approximated Theory).

In the simple AT model the d.c. measured total capacitance at port 1 (with port 2 open) is the real capacitance of the structure, between layers and 3. On the contrary the d.c. resistance at port 1 (with port 2 shorted) is less than the real resistance of the structure.

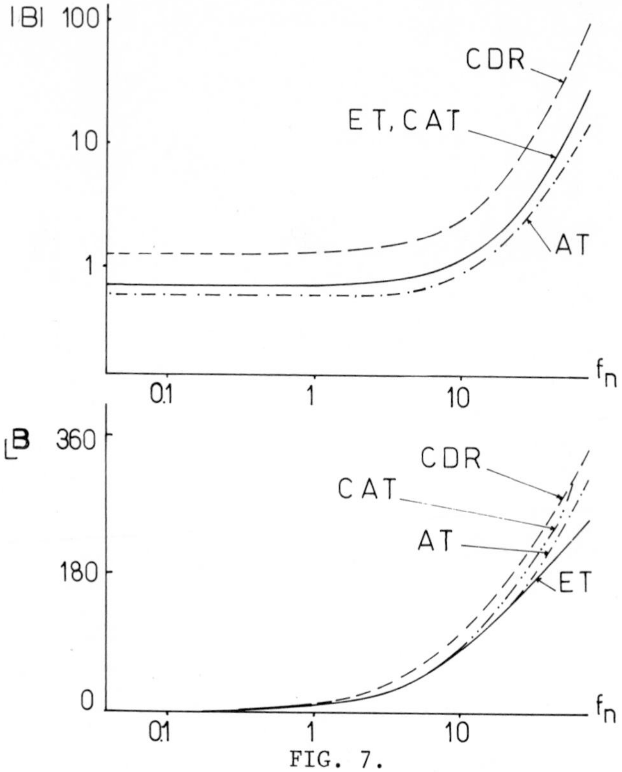

FIG. 7.

Thus one can think that a better approximation can be obtained inserting a discontinuity resistance between the two lines. It can be calculated exactly as the d.c. difference between the B parameter of the Exact Theory and of the Approximated Theory. The results obtained with this method are indicated in the figures with CAT (Corrected Approximated Theory).

It is worth noticing that the knowledge of this discontinuity resistance is possible only if a Laplace equation is solved. In many cases an expression of the discontinuity resistance can be interesting; by a conformal-mapping technique [10] one has:

$$R_d = \frac{R_q}{\pi}\left\{\left(\frac{1}{\sqrt{\alpha}} + \sqrt{\alpha}\right)\ln\frac{1 + \alpha + 2\sqrt{\alpha}}{\alpha - 1} - 2\ln\frac{4}{\alpha - 1} - \ln\alpha\right\} \qquad (23)$$

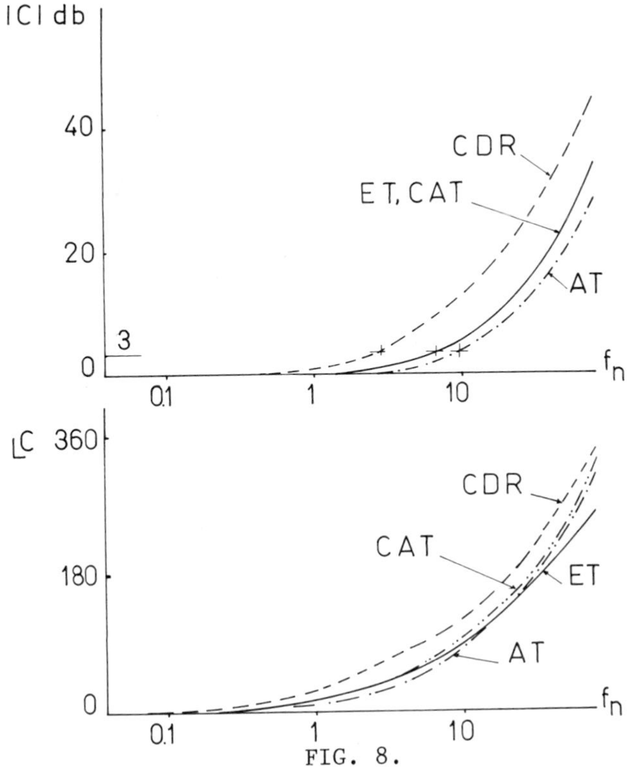

FIG. 8.

where $\alpha = (b/a)^2$. The method used to derive this expression is very simple because it does not take into account the distances l_a and l_b between the ports and the discontinuity; thus, Eq. (23) is valid only if l_a and l_b are much greater than a and b [10]. This condition is satisfied in many practical cases [5].

The plots obtained when R_d is calculated by Eq. (23) are indicated in the figures with CDR (Conformal Discontinuity Resistance).

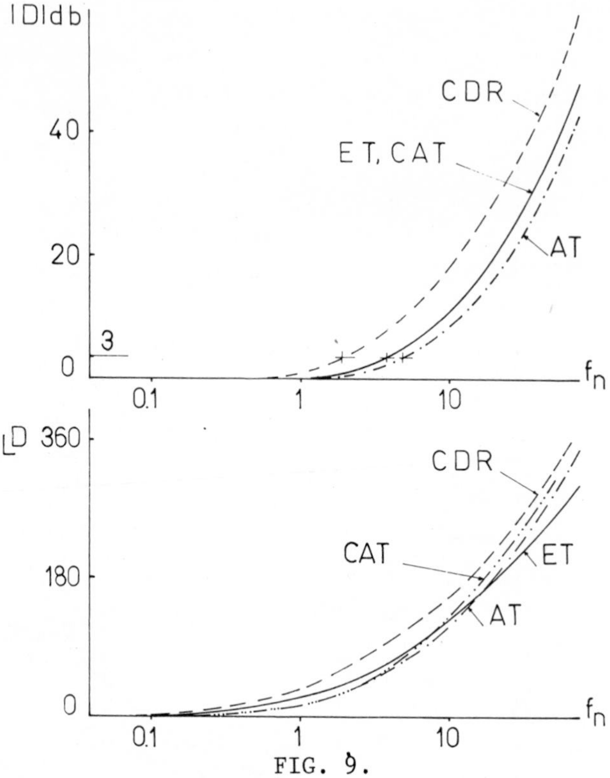

FIG. 9.

5. COMMENTS ON NUMERICAL RESULTS

Meaningful results have been obtained. In fact, there are appreciable differences among the values of the network parameters calculated with the different methods.

First of all, the exact discontinuity impedance is not purely resistive: for the structure that we have examined numerically, Z_d has an inductive part. Obviously the difference between two RC network functions can be inductive.

The CDR method has given results strongly different from those of ET: this is caused by the presence of the ports very near to the discontinuity. On the contrary the CAT has given results very similar to ET: the amplitudes of A, B, C, D in the range of frequency considered are equal, while there is a difference for the phases. This difference is caused by the presence in the discontinuity impedance of an inductive part.

It is worth noticing the differences in the 3 dB points of the parameters calculated by ET and AT. For the parameter A this difference is almost 35%. Moreover there is a difference of almost 20% between the B parameter of ET and AT.

6. CONCLUSION

The chain matrix of a three-layer distributed network with an abrupt taper has been obtained with different methods: the exact theory, the approximated theory and two approximated intermediate methods.

The results obtained suggest:

(i) the structure is equivalent to two uniform RC lines with a discontinuity impedance inserted between them; this impedance can be found, for every value of the frequency, by solving an elliptical partial differential equation;

(ii) the chain matrix of the structure can be obtained with great precision approximating Z_d with the d.c. increased resistance of the system;

(iii) if one ignores Z_d, there are meaningful differences with respect to the exact values.

7. ACKNOWLEDGEMENTS

The authors warmly thank Prof. R. Sartori and

Prof. P.P. Civalleri, Istituto Elettrotecnico Nazionale "Galileo Ferraris", Torino, and Prof. G. Biorci, University of Genova.

REFERENCES

1. PROTONOTARIUS, E.N. and WING, O. (March 1967) Theory of non-uniform RC lines. Part 1: Analytic properties and realizability conditions in the frequency domain. Part 2: Analytic properties in the time domain, *IEEE Trans. on Circuit Theory CT-14*, pp. 2-20.

2. SU, K.L. (1963) The trigonometric RC transmission lines, *IEEE International Convention Record*, Part 2, pp. 43-55.

3. WOO, B.B. and BARTLEMAY, J.M. (1963) Characteristic and applications of a tapered, thin film distributed parameter structure, *IEEE International Convention Record*, Part 2, pp. 56-75.

4. WYNDRUM, R.W. (1965) The realization of monomorphic thin film distributed RC networks, *IEEE International Convention Record*, Part 10, pp. 90-95.

5. WYNDRUM, R.W. (May 1963) *The exact synthesis of distributed RC networks*, Tech. Rept. 400-76, New York University.

6. RIDELLA, S. (Sept. 1968) Analysis of a three-layer distributed structure with N terminal pairs, *Proc. 1968 Int. Symp. on Network Theory*, Belgrade, pp. 687-707.

7. BIORCI, G. and RIDELLA, S. (1969) On the theory of distributed three-layer N-port networks, *Alta Frequenza 38*, 8, pp. 615-622.

8. BIANCO, B. and CIVALLERI, P.P. (1969) Basic Theory of three-layer N-ports, *Alta Frequenza 38*, 8, pp. 623-631.

9. CIVALLERI, P.P. and RIDELLA, S. Impedance and admittance matrices of distributed three-layer N-ports, (to be presented at IEEE Symposium on Circuit Theory 1969, and to appear in *IEEE Transactions on Circuit Theory*).

10. RIDELLA, S. *Microstrisce profilate* (Thesis).

BANACH SYSTEMS, HILBERT PORTS, AND ∞-PORTS*

A.H. ZEMANIAN

State University of New York at Stony Brook

TABLE OF CONTENTS

* The research culminating in this report was supported partly by the National Science Foundation of the United States under Grant No. GP-7577 and partly by the Science Research Council of Great Britain during the 1968-1969 academic year when the author was a Research Fellow at the Mathematical Institute, University of Edinburgh.

1. INTRODUCTION

This work is a development in gradual stages of several concepts that may become of some value in the analysis and perhaps eventually in the synthesis of physical systems. A typical concept is the idea of an n-port where $n = \infty$. This idea arises not only as a natural mathematical extension of the n-port but, more importantly for the engineer, as a representation of certain physical systems. For example, consider a modal analysis [1; pp. 21-27] of a microwave transmission system. Each mode can be taken to be the excitation at a port of a black box with a separate port for each mode, and therefore the black box has in general an infinity of ports. In network theory it is common, indeed almost the rule, to assume that all but a finite number of modes can be neglected (see, for example, [2; p. 3]) so that it is sufficient to represent the system by an n-port.

But is it? To do so in every situation makes as much sense as would the replacement of every Fourier series occurring in network theory by a finite sum. Network theorists do not resort to the latter simplification since Fourier series are well understood and quite usable. The situation is very different for the systems considered here. The subject is in its infancy and possesses from the engineering (i.e., synthesis) point of view very few results. Moreover, from the analysis viewpoint, a variety of mathematical difficulties arise. Indeed, the systems considered herein possess input and output signals which take their instantaneous values in Banach spaces, whereas for an n-port these values occur in n-dimensional euclidean space E_n. It happens disconcertingly often that a readily established fact concerning operators on E_n-valued functions is very difficult if not impossible to extend to operators on Banach-space-valued functions.

Nevertheless, we propose to explore this subject. For the mathematician no justification for doing so is needed. (The mountain is there; let's climb it.) The engineer asks, and indeed should ask, "Is it worth it?" We offer no reply to this question other than the hope that perhaps someday it may be. The fact that the electromagnetic waves within microwave systems, solid-state devices, integrated networks, etc. are better represented by Banach-space-valued functions rather than by E_n-valued functions offers some basis for this hope.

To put all this another way, this paper is motivated by physics, but its content is mathematics.

2. SOME DEFINITIONS AND A SUMMARY

The first in the order of business is to define the phrases appearing in the title. Throughout this work A and B will denote complex Banach spaces and H a complex Hilbert space. A Banach-space-valued function f is a mapping of some domain, which in this work will always be the real line R, into a Banach space, say, A. Thus, for each fixed $t \in R$, $f(t)$ is a member of A, and, as t varies in R, $f(t)$ may vary in A. Such a function is a typical signal in a variety of physical systems, as for example a microwave-transmission system. Indeed, we can conceive of a physical system (more precisely, a model of a physical system) whose signals are Banach-space-valued functions or even Banach-space-valued distributions. It seems natural to call such a system a "Banach system", and this we shall do.

A Banach system may have many different Banach spaces associated with it. Thus, at one location x in the system the signal representing a particular physical variable may be an A-valued distribution and at another location y the signal for another physical variable may be a B-valued distribution. Moreover, the system defines an operator that maps the signal at x into the signal at y. In general, it defines many different operators depending on the choices of the locations x and y and of the variables of interest (i.e., electric-field intensity, magnetic-field intensity, etc.) We shall always make this distinction between the model of the physical system and the operators that it generates. The term "Banach system" refers to the model and not to any particular operator.

By a "Hilbert port" we mean the following. Assume that in a given Banach system we have singled out two physical variables u and v that are complementary in the following sense: Both u and v take their values in a Hilbert space H and the real part of their inner product (u, v) represents the instantaneous power entering the Banach system. Then, the Banach system with these two variables so singled out is called a "Hilbert port". When discussing a Hilbert port we in general pay no attention to the other variables within the system. We become interested exclusively in the variables u and v and the three operators $\mathfrak{N}: v \to u$, $\mathfrak{B}: u \to v$, and $\mathfrak{W}: v + u \to v - u$. (In microwave transmission systems, it is customary to choose v as the electric-field intensity on a closed surface containing the system and cutting all the wave guides to the system on transverse planes. Also, u is taken to be the magnetic-field intensity on the same closed surface. Then, $\mathfrak{N}$ is called the admittance operator, $\mathfrak{B}$ the impedance operator, and $\mathfrak{W}$ the scattering operator. Furthermore $v_+ \overset{\Delta}{=} v + u$ can be related to the incident electric-field wave and $v_- \overset{\Delta}{=} v - u$ to the reflected electric-field wave.)

How about "∞-ports"? Assume that the space H for a given Hilbert port is separable; it will be for all practical systems. Choose an orthonormal basis for H. The separability of H implies that the basis will be countable. (Moreover, a modal analysis of v and u suggests a natural orthonormal basis for H.) Then, v and u can be represented by their sequences of Fourier coefficients $\{v_n\}$ and $\{u_n\}$ respectively. Thus, for each n, we have a pair v_n, u_n of complex-valued functions or distributions on R. We can for the sake of analysis assume that each pair v_n, u_n occurs on a separate port, and thus we are lead to a system having a countable infinity of ports. We call such a paper network [3] corresponding to the given Hilbert port an "∞-port". In this case the operators $\mathfrak{N}$, $\mathfrak{B}$, and $\mathfrak{W}$ mentioned above can be represented by $\infty \times \infty$ matrices, as will be indicated in Sec. 8.

These are the kinds of systems with which we will be concerned in this work. Our primary objective is to develop characterizations and representations for various operators generated by such systems when they satisfy various idealized physical properties such as linearity, time-invariance, and passivity. Our theory depends crucially on the concept of a Banach-space-valued distribution, and therefore we present an introduction to this concept in the next section. Time-varying Banach systems are taken up in Sec. 4 and both a kernel representation and a composition representation for their continuous linear operators are established. To the author's knowledge, this theory does not appear elsewhere in the literature and so we also present proofs. Time-invariant Banach systems are discussed in Sec. 5; in this case the kernel and composition representations become convolution representations. The results of Secs. 4 and 5 apply to both active and passive systems. Next, we restrict our attention to passive Hilbert ports and observe that such systems have frequency-domain descriptions. The frequency-domain characterizations of their admittance and scattering formulisms are given in Secs. 6 and 7 respectively. The proofs of the results in Secs. 5, 6 and 7 appear elsewhere [4,5], and so in these sections we merely survey some pertinent results but omit all proofs. In sec. 8 and its tail we return to detailed arguments. There the $\infty \times \infty$ matrix representation for the admittance operator of an ∞-port is developed, and the problem of synthessizing an ∞-port is considered but not resolved.

3. BANACH-SPACE VALUED DISTRIBUTIONS

The natural language for the theory of continuous linear systems is that provided by distribution theory.

It not only simplifies proofs but permits one to establish a number of theorems that simply could not be formulated in terms of conventional functions. Moreover, it allows the consideration of many idealized systems, such as those that differentiate an arbitrary number of times, without requiring that the domain of inputs be excessively restricted. The signals we will be concerned with are Banach-space-valued distributions on the real-time line $-\infty < t < \infty$, and so our first objective is to present a brief introduction to the theory of such distributions. This we do in the present section.

Throughout this work we use the following symbolism. If U and V are two topological linear spaces, the symbol [U; V] denotes the linear space of all continuous linear mappings of U into V. Moreover, if $\phi \in U$ and $f \in [U; V]$, then $f\phi$, $f(\phi)$, and $\langle f, \phi \rangle$ all denote that element of V assigned by f to ϕ. H will always denote a complex Hilbert space, and $(\cdot,\cdot)$ will denote its inner product. On the other hand, both A and B will always represent complex Banach spaces. Furthermore, R^n is n-dimensional euclidean space, $R = R^1$ is the real line, and C the complex plane. C_+ denotes the open right-half of C; i.e., $C_+ = \{\zeta: \zeta \in C, \text{Re } \zeta > 0\}$

A function ϕ mapping an open set $I \subset R$ into A is said to be strongly continuous or strongly differentiable at a point $t \in I$ if the standard difference expression that defines continuity or the derivative at t converges in the norm topology of A. ϕ is said to be smooth on I if it possesses strong derivatives of all orders at all points of I. Similarly, if ϕ maps an open domain $J \subset C$ into A, ϕ is said to be analytic on J if, for every point $\zeta \in J$, the standard difference expression that defines the derivative at ζ converges in the norm of A independently of the path in which the increment $\Delta\zeta$ is taken to zero. Derivatives are denoted alternatively by

$$\frac{d}{dx}\phi = D\phi = D_x\phi(x) = \phi^{(1)}.$$

The support of a continuous function ϕ mapping R into A is the smallest closed set outside of which ϕ is the zero function; it is denoted by supp ϕ. $||\cdot||_A$ or simply $||\cdot||$ denotes the norm of a Banach space A. If nothing else is explicitly stated, it will be understood that [A; B] carries the usual operator-norm topology (i.e., the uniform topology).

If T and X are either members of R, $+\infty$, or $-\infty$ with $T < X$, then (T, X) and [T, X] represent open and respectively closed intervals in R, and similarly for the semi-open and semi-closed intervals (T, X] and [T, X).

Let K be a compact (i.e., closed bounded) interval in R. $\mathcal{D}_K(A)$ denotes the linear space of all smooth functions ϕ on R into A whose supports are contained in K. We assign to $\mathcal{D}_K(A)$ the topology generated by the sequence $\{\gamma_k\}_{k=0}^{\infty}$ of seminorms defined by

$$(3\text{-}1) \qquad \gamma_k(\phi) \triangleq \sup_{t\in R} ||D^k \phi(t)||_A \qquad \phi \in \mathcal{D}_K(A) .$$

Thus, a sequence $\{\phi_\nu\}_{\nu=1}^{\infty}$ is convergent in $\mathcal{D}_K(A)$ if and only if every $\phi_\nu \in \mathcal{D}_K(A)$ and there exists a $\phi \in \mathcal{D}_K(A)$ such that, for every $k=0,1,\ldots,\gamma_k(\phi_\nu - \phi) \to 0$ as $\nu \to \infty$.

For any nonnegative integer m, $\mathcal{D}_K^m(A)$ is defined similarly except that we impose the conditions on the derivatives $D^k\phi$ for only $k = 0, 1, \ldots, m$.

Next, let K_n denote the compact interval $-n \leq t \leq n$. Set

$$\mathcal{D}(A) = \bigcup_{n=1}^{\infty} \mathcal{D}_{K_n}(A) ,$$

and equip $\mathcal{D}(A)$ with the inductive-limit topology [6; vol. 15, pp. 61-62]. This implies that a sequence $\{\phi_\nu\}_{\nu=1}^{\infty}$ converges in $\mathcal{D}(A)$ if and only if the entire sequence is contained in some fixed space $\mathcal{D}_{K_n}(A)$ and converges in that space. It is also a fact that a linear mapping f of $\mathcal{D}(A)$ into some locally convex space V (every topological linear space considered in this paper is locally convex) is continuous if and only if its restriction to each $\mathcal{D}_{K_n}(A)$ is sequentially continuous [6; vol. 15, p. 62]. The latter means that the convergence of $\{\phi_\nu\}$ to zero in $\mathcal{D}_{K_n}(A)$ implies the convergence of $\{<f, \phi_\nu>\}$ to zero in V. Moreover, a set Ω is bounded in $\mathcal{D}(A)$ if and only if there exists an n such that $\Omega \subset \mathcal{D}_{K_n}$ and, for every k, γ_k remains bounded on Ω. When A is C, we denote $\mathcal{D}(C)$ simply by $\mathcal{D}$.

$\mathcal{D}^m(A)$ is defined as the inductive-limit space $\bigcup_{n=1} \mathcal{D}^m_{K_n}(A)$ and has similar properties.

The linear space of all continuous linear mappings of $\mathcal{D}(A)$ into B is denoted by $[\mathcal{D}(A); B]$. It is a space of vector-valued distributions. We assign to it the so-called "topology of bounded convergence". This is the topology generated by the collection of seminorms $\{\sigma_\Omega\}_\Omega$ where Ω varies through the bounded sets of $\mathcal{D}(A)$ and

$$\sigma_\Omega(f) \triangleq \sup_{\phi \in \Omega} \| <f, \phi> \|_B \qquad f \in [\mathcal{D}(A); B]. \tag{3-2}$$

When A and B are both C, we get the customary space $\mathcal{D}' = [\mathcal{D}(C); C]$ of scalar distributions on the real line, and the topology of $\mathcal{D}'$ is then the so-called strong topology.

$[\mathcal{D}^m(A); B]$ and its topology of bounded convergence are defined in just the same way.

We will on occasion use a weaker topology for $[\mathcal{D}(A); B]$, namely, the "topology of pointwise convergence"; this is generated by the collection of seminorms (3-2) where now Ω is restricted to the finite sets in $\mathcal{D}(A)$. It corresponds to the weak topology of $\mathcal{D}'$. The symbol $[\mathcal{D}(A); B]^w$ will denote the space $[\mathcal{D}(A); B]$ equipped with this weaker topology.

On just one occasion (when we discuss a kernel theorem in Sec. 4) we will need functions and distributions defined on the two-dimensional euclidean space R^2. In this case, the definitions of $\mathcal{D}_K(A)$, $\mathcal{D}(A)$, and $[\mathcal{D}(A); B]$, as well as $\mathcal{D}_K^m(A)$, $\mathcal{D}^m(A)$, and $[\mathcal{D}^m(A); B]$, are the same as above except that now K_n is a compact interval in R^2 (i.e., $K_n = \{z: z \in R^2, |z| \leq n\}$), k is a nonnegative integer in R^2, and D^k is a partial differentiation corresponding to k. In the definition of $\mathcal{D}_K^m(A)$, $k = \{k_1, k_2\}$ is restricted to those values for which $0 \leq k_1 + k_2 \leq m$. In this two-dimensional case, we will always denote $\mathcal{D}(A)$ by $\mathcal{D}_{t,x}(A)$ and $\mathcal{D}^m(A)$ by $\mathcal{D}_{t,x}^m(A)$; moreover, we will on occasion replace $\langle\cdot,\cdot\rangle$ by $\langle\cdot,\cdot\rangle_{t,x}$ to emphasize that we are dealing with the two-dimensional case. Henceforth, the symbols $\mathcal{D}$, $\mathcal{D}(A)$, and $\mathcal{D}^m(A)$ will always signify that their members are defined on $R = R^1$.

We return now to the one-dimensional case. Of importance to us is the space $[\mathcal{D}; [A; B]]$ of continuous linear mappings of $\mathcal{D} = \mathcal{D}(C)$ into the space $[A; B]$. The topology of $[\mathcal{D}; [A; B]]$ is that generated by $\{\sigma_\Omega\}_\Omega$ with σ_Ω defined by

$$\sigma_\Omega(f) \overset{\Delta}{=} \sup_{\phi\in\Omega} ||\langle f, \phi\rangle||_{[A;\, B]}$$

where now $f \in [\mathcal{D}; [A; B]]$ and Ω traverses the bounded sets in $\mathcal{D}$. A crucial result for our theory is the following [4; Theorem 3-1].

Theorem 3-1

There is a bijection (i.e., a one-to-one onto mapping) from $[\mathcal{D}(A); B]$ *onto* $[\mathcal{D}; [A; B]]$ *defined by*

$$\langle g, \phi a \rangle = \langle f, \phi \rangle\, a \tag{3-3}$$

where $\phi \in \mathcal{D}$, $a \in A$, $g \in [\mathcal{D}(A);\, B]$, *and* $f \in [\mathcal{D};\, [A;\, B]]$.

Because of this theorem, we can identify the members of $[\mathcal{D}(A);\, B]$ with those of $[\mathcal{D};\, [A;\, B]]$, and in place of (3-3) we will write $\langle f, \phi a \rangle = \langle f, \phi \rangle a$.

Here are some examples.

Example 3-1

Let F denote a fixed number of $[A;\, B]$ and let δ denote the customary delta functional. $F\delta$ is defined on any $\theta \in \mathcal{D}(A)$ by

$$\langle F\delta, \theta \rangle \overset{\Delta}{=} F\theta(0) \in B\ . \tag{3-4}$$

Clearly, $F\delta \in [\mathcal{D}(A);\, B]$. On the other hand, we define $F\delta$ on any $\phi \in \mathcal{D}$ by

$$\langle F\delta, \phi \rangle \overset{\Delta}{=} F\phi(0) \in [A;\, B] \tag{3-5}$$

and obtain thereby a member of $[\mathcal{D};\, [A;\, B]]$. That these two definitions accord with (3-3) follows from the fact that, for any $a \in A$, we have $\phi a \in \mathcal{D}(A)$ so that by (3-4) and (3-5)

$$\langle F\delta, \phi a \rangle = F\phi(0)a = \langle F\delta, \phi \rangle\, a\ .$$

Thus, $F\delta$ is simultaneously a member of $[\mathcal{D}(A);\, B]$ and $[\mathcal{D};\, [A;\, B]]$.

Example 3-2

Let h be an $[A;\, B]$-valued function on R that is continuous in the norm topology of $[A;\, B]$. Define a mapping (also denoted by h) on any $\theta \in \mathcal{D}(A)$ by

$$\langle h, \theta \rangle = \int_R h(t)\, \theta(t)\, dt \in B\ .$$

The integral on the right-hand side is the strong limit in B of the corresponding Riemann sums because h(t) $\theta(t)$ is strongly continuous from R into B. That h is a linear mapping follows from the fact that integration is a linear process. The continuity of h from $\mathcal{D}(A)$ into B is implied by the estimate:

$$||\langle h,\theta\rangle||_B \le \int_R ||h(t)||_{[A;B]}||\theta(t)||_A \, dt$$

$$\le \sup_{t\varepsilon R}||\theta(t)||_A \int_K ||h(t)||_{[A;B]} dt$$

where K is a compact interval that contains supp ϕ. Thus, h ε $[\mathcal{D}(A); B]$.

On the other hand, we can define h as a mapping of $\mathcal{D}$ into [A; B] by

$$\langle h, \phi\rangle = \int_R h(t)\, \phi(t)\, dt \ \varepsilon\ [A; B] \qquad \phi \ \varepsilon\ \mathcal{D},$$

and considerations similar to those above show that h ε $[\mathcal{D}; [A; B]]$. Thus, the originally given [A; B]-valued function h on R generates both a member of $[\mathcal{D}(A); B]$ and a member of $[\mathcal{D}; [A; B]]$.

We can generate still other members of $[\mathcal{D}(A); B]$ or $[\mathcal{D}; [A; B]]$ by differentiating in a generalized sense the distributions of the preceding examples. Such a differentiation D^p of order p is defined on any f in $[\mathcal{D}(A); B]$ (or in $[\mathcal{D}; [A; B]]$) by

$$\text{(3-6)} \qquad \langle D^p f, \phi\rangle = (-1)^p \langle f; D^p \phi\rangle$$

where ϕ is an arbitrary member of $\mathcal{D}(A)$ (or respectively of $\mathcal{D}$). Since D^p is a continuous linear mapping of $\mathcal{D}(A)$ (or $\mathcal{D}$) into itself and hence maps bounded sets into bounded sets, it follows that (generalized) differentiation D^p is a continuous linear mapping of $[\mathcal{D}(A); B]$ (or $[\mathcal{D}; [A; B]]$) into itself.

Example 3-3

We can now define a member of $[\mathcal{D}(A); B]$ or of $[\mathcal{D}; [A; B]]$ by applying D^p to the $F\delta$ of Example 3-1. Upon applying (3-6), we obtain

$$\langle D^p(F\delta), \phi\rangle = (-1)^p F\phi^{(p)}(0) \tag{3-7}$$

where ϕ is either in $\mathcal{D}(A)$ or respectively in $\mathcal{D}$. Since $\langle D^p\delta, \phi\rangle = (-1)^p \phi^{(p)}(0)$, we have that $D^p(F\delta) = FD^p\delta$, when $FD^p\delta$ is defined by $\langle FD^p\delta, \phi\rangle = F\langle D^p\delta, \phi\rangle$.

Example 3-4

Similarly, by applying D^p to the results of Example 3-2, we get the member D^ph of $[\mathcal{D}(A); B]$ or of $[\mathcal{D}; [A; B]]$ defined by

$$\langle D^ph, \phi\rangle = (-1)^p \int_R h(t)\, \phi^{(p)}(t)\, dt \tag{3-8}$$

where again $\phi \in \mathcal{D}(A)$ or respectively $\phi \in \mathcal{D}$.

Let us also state the definition of the shifting operator σ_τ; we shall have need of it later on. For any given $\tau \in R$, σ_τ is defined on any $\phi \in \mathcal{D}(A)$ by $\sigma_\tau \phi(t) \overset{\Delta}{=} \phi(t - \tau)$. σ_τ is a continuous linear mapping of $\mathcal{D}(A)$ into itself. Next, σ_τ is defined on any $f \in [\mathcal{D}(A); B]$ by

$$\langle \sigma_\tau f, \phi\rangle = \langle f, \sigma_{-\tau} \phi\rangle ,$$

and consequently σ_τ is a continuous linear mapping of $[\mathcal{D}(A); B]$ into itself.

There are several other spaces of Banach-space-valued functions and mappings on them that we shall need. One of these is $\mathcal{E}(A)$, the space of all smooth A-valued functions on R equipped with the topology generated by the collection of seminorms $\{\gamma_{K,k}\}_{K,k}$ where K traverses the compact sets in R, k traverses the nonnegative integers, and

$$\gamma_{K,k}(\phi) \overset{\Delta}{=} \sup_{t \varepsilon K} ||D^k\phi(t)||_A \qquad \phi \;\varepsilon\; E(A) \;. \tag{3-9}$$

Here again, when A is C, we denote $E(C)$ by simply E. A sequence $\{\phi_\nu\}$ converges in $E(A)$ if and only if there exists a $\phi \;\varepsilon\; E(A)$ such that, for every K and k, $\gamma_{K,k}(\phi_\nu - \phi) \to 0$ as $\nu \to \infty$. A set in $E(A)$ is bounded if and only if, for every K and k, $\gamma_{K,k}$ remains bounded on Ω. A linear mapping f of $E(A)$ into any topological linear space V is continuous if and only if the convergence of $\{\phi_\nu\}$ to zero in $E(A)$ implies the convergence of $\{\langle f, \phi_\nu\rangle\}$ to zero in V.

Let m be a nonnegative integer. We define $E^m(A)$ as was $E(A)$ except that conditions on $D^k\phi$ are imposed only for $k = 0, 1, \ldots, m$.

$[E(A); B]$ denotes the linear space of all continuous linear mappings of $E(A)$ into B. The distribution $DP(F\delta)$, when extended onto $E(A)$ in accordance with (3-7), is a member of $[E(A); B]$. So too is DPh, as defined by (3-8) with $\phi \;\varepsilon\; E(A)$, so long as supp h is a compact set. In fact, it can be shown that every member f of $[E(A); B]$ has a compact support (i.e., there exists some compact set K such that $\langle f, \phi\rangle = 0$ for every $\phi \;\varepsilon\; E(A)$ with supp ϕ contained in the complement of K). We assign to $[E(A); B]$ the topology of bounded convergence, that is, the topology generated by the collection $\{\sigma_\Omega\}_\Omega$ of seminorms defined on any $f \;\varepsilon\; [E(A), B]$ by (3-2) where now Ω traverses the bounded sets in $E(A)$.

We obtain the definition of the space $[E^m(A); B]$ by replacing $E(A)$ by $E^m(A)$ in the preceding paragraph. $[E^m(A); B]$ is also a space of distributions of bounded support.

Here's another space of importance to us. Let $p \in R$ be such that $1 \leq p < \infty$. $\mathcal{D}_{L_p}(A)$ is the set of all smooth functions ϕ on R such that, for every nonnegative integer k,

$$\alpha_k(\phi) \stackrel{\Delta}{=} [\int_R ||D^k \phi(t)||_A^p \, dt]^{1/p} < \infty . \tag{3-10}$$

We assign to $\mathcal{D}_{L_p}(A)$ the topology generated by $\{\alpha_k\}_{k=0}^{\infty}$. As before, we use the notation $\mathcal{D}_{L_p}(C) = \mathcal{D}_{L_p}$. We can make the same comments concerning convergent sequences, bounded sets, and continuous linear mappings on $\mathcal{D}_{L_p}(A)$ as those made for $\mathcal{E}(A)$ except now $\gamma_{K,k}$ is replaced by α_k.

$[\mathcal{D}_{L_p}(A); B]$ is the topological linear space of all continuous linear mappings of $\mathcal{D}_{L_p}(A)$ into B equipped with the topology of bounded convergence.

We now turn our attention to the spaces $L(0,\infty; A)$ and $[L(0,\infty; A); B]$. A detailed discussion of the scalar version of these spaces, where A and B are both C, is given in [7; Chapter 3]. First of all, for any $c \in R$ and $d \in R$ we define $L_{c,d}(A)$ as the space of all smooth functions on R into A such that, for every nonnegative integer k,

$$\gamma_{c,d,k}(\phi) \stackrel{\Delta}{=} \sup_{t \in R} ||\kappa_{c,d}(t) \, D^k \phi(t)||_A < \infty \tag{3-11}$$

where

$$\kappa_{c,d}(t) \stackrel{\Delta}{=} \begin{cases} e^{ct} & t \geq 0 \\ e^{dt} & t < 0 . \end{cases}$$

The topology of $L_{c,d}(A)$ is that generated by the collection $\{\gamma_{c,d,k}\}_{k=0}^{\infty}$ of seminorms on $L_{c,d}(A)$. Once again, the comments concerning convergent sequences, bounded sets, and continuous linear mappings of $L_{c,d}(A)$ are quite the same as those made for $E(A)$.

Next, let w be either a member of R or $-\infty$ and let z be either a member of R or $+\infty$. Also, let $\{a_n\}_{n=1}^{\infty}$ and $\{b_n\}_{n=1}^{\infty}$ be two monotonic sequences in R such that $a_\nu \to w+$ and $b_\nu \to z-$. It follows that, for $m > n$, $L_{a_n,b_n}(A) \subset L_{a_m,b_m}(A)$ and the topology of $L_{a_n,b_n}(A)$ is stronger than that induced on it by $L_{a_m,b_m}(A)$. We set

$$L(w,z;\ A) = \bigcup_{n=1}^{\infty} L_{a_n,b_n}(A)$$

and assign to $L(w,z;\ A)$ the inductive-limit topology. This means among other things that a linear mapping f on $L(w,z;\ A)$ into a locally convex space V is continuous if and only if the restriction of f to each $L_{a_n,b_n}(A)$ is continuous. Here again, we set $L(w,z;\ C) = L(w,z)$

$[L(w,z;\ A);\ B]$ is the linear space of all continuous linear mappings of $L(w,z;\ A)$ into B. We assign to it the following topology. Let $\mathfrak{S}_n$ denote the collection of bounded sets in $L_{a_n,b_n}(A)$ (i.e., $\Omega \in \mathfrak{S}_n$ if and only if $\Omega \subset L_{a_n,b_n}(A)$ and, for each k, the seminorm $\gamma_{a_n,b_n,k}$ remains bounded on $\mathfrak{S}_n$). Let $\mathfrak{S} = \bigcup_{n=1}^{\infty} \mathfrak{S}_n$. Then, the topology of $[L(w,z;\ A);\ B]$ is that generated by the collection $\{\sigma_\Omega\}_{\Omega\in\mathfrak{S}}$ of seminorms where each σ_Ω is defined by

$$\sigma_\Omega(f) \triangleq \sup_{\phi\in\Omega} \|\langle f,\phi\rangle\|_B \qquad f \in [L(w,z;\ A);\ B]$$

If $w < z$ and if $y \in [L(w,z;\ A);\ B]$, we define the Laplace transform $Y = \mathfrak{L}y$ of y by

$$\text{(3-12)} \qquad Y(\zeta)\,a \triangleq \langle y(t),\ e^{-\zeta t}a\rangle \qquad w < \operatorname{Re}\zeta < z$$

where a is any member of A. The right-hand side of (3-12) has a sense because $e^{-\zeta t}a \in L(w,z;\ A)$ for every ζ such that $w < \operatorname{Re}\zeta < z$. It turns out that Y is an $[A;\ B]$-valued analytic function on the strip

$$\Omega_y = \{\zeta : w < \mathrm{Re}\zeta < z\} \quad .$$

(See [4; Sec. 5].) Moreover, the Laplace transformation is unique in the following sense: If $Y(\zeta)$ is the zero member of $[A;B]$ on any open subset of Ω_y, then y is the zero member of $[L(w,z; A); B]$. (See [4; Corollary 5-2a].)

A fact that we shall subsequently employ is the following: If $f \in [\mathcal{D}_{L_p}(A); B]$ where $1 \le p < \infty$ and if $\operatorname{supp} f \subset [T,\infty]$ for some $T \in R$, then $f \in [L(0,\infty; A); B]$ (see [4; Lemma 7-2] for the case $p = 1$ and [5; Lemma 2-1] for the case $p = 2$) so that $(\mathfrak{L}\, y)(\zeta)$ exists on at least the right-half plane $C_+ = \{\zeta : 0 < \mathrm{Re}\zeta < \infty\}$.

We now take note of certain relations between some of the above spaces. We have

(3-13) $$\mathcal{D}(A) \subset \mathcal{D}_{L_p}(A) \subset \mathcal{E}(A)$$

and

(3-14) $$\mathcal{D}(A) \subset L(w,z; A) \subset \mathcal{E}(A).$$

In both (3-13) and (3-14), any space therein is a dense subspace of any space occurring to the right of it and the topology of the former space is stronger than the topology induced on it by the latter space. By virtue of these facts, we have

(3-15) $$[\mathcal{E}(A); B] \subset [\mathcal{D}_{L_p}(A); B] \subset [\mathcal{D}(A); B]$$

and

(3-16) $$[\mathcal{E}(A); B] \subset [L(w,z; A); B] \subset [\mathcal{D}(A); B].$$

4. TIME-VARYING BANACH SYSTEMS, A KERNEL THEOREM, AND COMPOSITION

We wish to consider a model of a physical system which determines an operator that maps a class of input signals into a class of output signals. Our purpose is to investigate the relationship between various analytic properties of the operator and certain (idealized) physical properties. In this section the only physical properties we shall impose are single-valuedness, linearity, and continuity. (Throughout this paper, whenever we specify that an operator is linear on some domain Ω, it will also be understood that it is single-valued on Ω. On occasion we will allow multivalued operators but these will not be called linear.) Thus, time-varying systems are allowed at this point. Active systems are also allowed. The adjective "active" signifies merely that no requirement of passivity, as defined in Secs. 6 and 7 below, is being imposed but may nevertheless be satisfied. Thus, we view passive systems as being special cases of active ones.

The kind of operator we have in mind is one that is generated by a Banach system and therefore maps, say, A-valued distributions into B-valued distributions. We can obtain an analytic representation for the operator by extending Schwartz's kernel theorem [8; p. 531] to Banach-space-valued distributions. This extension is given by Theorem 4-1 below. (A proof of Theorem 4-1 is provided in a paper by Bogdanowicz [9]. Actually, Bogdanowicz's work restricts the range space B to the complex plane C, but it is not difficult to extend his argument to the more general case considered here.)

To say that $\mathfrak{M}$ is a separately continuous bilinear mapping of $\mathcal{D} \times \mathcal{D}(A)$ into B means the following: $\mathfrak{M}$ maps any ordered pair ϕ, v of a $\phi \in \mathcal{D}$ and a $v \in \mathcal{D}(A)$ into a member $\mathfrak{M}(\phi,v)$ of B, and, in addition, if v (respectively ϕ) is kept fixed, then $\mathfrak{M}$ is linear and continuous with respect to ϕ (respectively v).

Theorem 4-1

If $\mathfrak{M}$ *is a separately continuous bilinear mapping of* $\mathcal{D} \times \mathcal{D}(A)$ *into* B, *then there exists a unique continuous linear mapping* $f = f(t,x)$ *of* $\mathcal{D}_{t,x}(A)$ *into* B *such that*

$$\mathfrak{M}(\phi,v) = \langle f(t,x),\ \phi(t)\ v(x)\rangle \tag{4-1}$$

for every $\phi \in \mathcal{D}$ *and every* $v \in \mathcal{D}(A)$.

Next step: We define a composition product $f\cdot v$ on any $f \in [\mathcal{D}_{t,x}(A);\ B]$ and any $v \in \mathcal{D}(A)$ by

$$\langle f\cdot v, \phi\rangle \overset{\Delta}{=} \langle f(t,x),\ \phi(t)\ v(x)\rangle \qquad \phi \in \mathcal{D}. \tag{4-2}$$

We shall refer to the process of forming this product $f\cdot v$ as "composition•" to distinguish it from another such process, called "composition∘," which will be defined subsequently. The right-hand side of (4-2) has a sense and determines a member of B because $\phi(t)v(x) \in \mathcal{D}_{t,x}(A)$. Thus, we can consider the composition operator $f\cdot : v \mapsto f\cdot v$ as a mapping of $\mathcal{D}(A)$ into the space of mappings of $\mathcal{D}$ into B.

Theorem 4-2

For any given $f \in [\mathcal{D}_{t,x}(A);\ B]$, *the composition operator* $f\cdot$ *is a continuous linear mapping of* $\mathcal{D}(A)$ *into* $[\mathcal{D};\ B]$.

Proof

We first observe that $f\cdot v \in [\mathcal{D};\ B]$. The linearity of $f\cdot v$ on $\mathcal{D}$ is obvious. Its continuity follows from the fact that, if $\phi_\nu \to 0$ in $\mathcal{D}$, then $\phi_\nu(t)\ v(x) \to 0$ in $\mathcal{D}_{t,x}$ so that $\langle f\cdot v, \phi_\nu\rangle \to 0$ in B.

Now consider $f\cdot$. Its linearity on $\mathcal{D}(A)$ is again clear. To show its continuity, let Ω be any bounded set in $\mathcal{D}$ and let $v_\nu \to 0$ in $\mathcal{D}(A)$. Consequently, $\phi(t)\ v_\nu(x) \to 0$ in $\mathcal{D}_{t,x}(A)$ uniformly for all $\phi \in \Omega$. Moreover, there exists a compact set $K \subset R^2$, a nonnegative integer $m \in R$, and a constant $Q > 0$ such that $\operatorname{supp} \phi(t)\ v_\nu(x) \subset K$ for all ν and all $\phi \in \Omega$, and, in addition,

$$\begin{aligned} \phi_\Omega(f\cdot v_\nu) &= \sup_{\phi\in\Omega} \|\langle f\cdot v_\nu, \phi\rangle\|_B \\ &= \sup_{\phi\in\Omega} \|\langle f(t,x), \phi(t)v_\nu(x)\rangle\|_B \\ &\le \sup_{\phi\in\Omega} Q \max_{0\le|k|\le m} \sup_{t,x} \|D^k\phi(t)v_\nu(x)\|_A \end{aligned} \tag{4-3}$$

where $k = \{k_1, k_2\}$ is a nonnegative integer in R^2 and $|k| = k_1 + k_2$. In view of the uniformity of the convergence of $\phi(t)\, v_\nu(x)$ with respect to all $\phi \in \Omega$, the right-hand side of (4-3) tends to zero as $\nu \to \infty$. Since Ω was arbitrary, we conclude that $f \cdot v_\nu \to 0$ in $[\mathcal{D};\, B]$.

Theorem 4-2 possesses a converse. In order to obtain it, we will need

Lemma 4-1

Let $\mathfrak{N}$ *be a continuous linear mapping of* $\mathcal{D}(A)$ *into* $[\mathcal{D};\, B]^w$. *Define* $\mathfrak{M}$ *from* $\mathfrak{N}$ *by*

$$\text{(4-4)} \qquad \mathfrak{M}(\phi, v) \triangleq \langle \mathfrak{N} v, \phi \rangle \qquad v \in \mathcal{D}(A),\ \phi \in \mathcal{D}.$$

Then, $\mathfrak{M}$ *is a uniquely defined separately continuous bilinear mapping of* $\mathcal{D} \times \mathcal{D}(A)$ *into* B.

Proof

Since $\mathfrak{N} v \in [\mathcal{D};\, B]$, the right-hand side of (4-4) is a member of B. Thus, $\mathfrak{M}$ maps $\mathcal{D} \times \mathcal{D}(A)$ into B.

Next, fix ϕ. Let $\alpha \in C$, $\beta \in C$, $v_1 \in \mathcal{D}(A)$, and $v_2 \in \mathcal{D}(A)$. Then, the linearity of $\mathfrak{M}$ with respect to v is established by

$$\begin{aligned} \mathfrak{M}(\phi, \alpha v_1 + \beta v_2) &= \langle \mathfrak{N}(\alpha v_1 + \beta v_2), \phi \rangle = \langle \alpha \mathfrak{N} v_1 + \beta \mathfrak{N} v_2, \phi \rangle \\ &= \alpha \langle \mathfrak{N} v_1, \phi \rangle + \beta \langle \mathfrak{N} v_2, \phi \rangle \\ &= \alpha \mathfrak{M}(\phi, v_1) + \beta \mathfrak{M}(\phi, v_2). \end{aligned}$$

To show the continuity of $\mathfrak{M}$ with respect to v, let $v_\nu \to 0$ in $\mathcal{D}(A)$. We have that $\mathfrak{M}(v_\nu, \phi) = \langle \mathfrak{N} v_\nu, \phi \rangle$, and this tends to zero in B because $\mathfrak{N}$ is continuous from $\mathcal{D}(A)$ into $[\mathcal{D};\, B]^w$.

Similar arguments establish the linearity and continuity of $\mathfrak{M}$ with respect to ϕ when v is held fixed.

We may now combine Lemma 4-1, Theorem 4-1, and the definition (4-2) to get the aforementioned converse to theorem 4-2.

Theorem 4-3

For every continuous linear mapping $\mathfrak{N}$ *of* $\mathcal{D}(A)$ *into* $[\mathcal{D};\, B]^W$, *there exists a unique* $f \in [\mathcal{D}_{t,x}(A);\, B]$ *such that, for all* $v \in \mathcal{D}(A)$, $\mathfrak{N}v = f\bullet v$ *in the sense of equality in* $[\mathcal{D};\, B]$.

Theorem 4-3 provides an explicit analytic representation for a sufficiently well-behaved operator $\mathfrak{N}$ of a time-varying active Banach system, but it does so for only a very restricted domain for the representation, namely, $\mathcal{D}(A)$. We can construct a composition representation for certain such $\mathfrak{N}$ with wider domains for the representation by appropriately extending to Banach-space-valued distributions the concept of the composition of distributions as developed by Cristescu [10], Cristescu and Marinescu [11], Sabac [12], Wexler [13], Cioranescu [14], Pondelicek [15], and Dolezal [16]. But, before doing so, let us present some examples of composition$\bullet$ operators.

Example 4-1

We first present the composition$\bullet$ representation of an nth order differential operator hD^n with a variable coefficient $h \in \mathcal{E}^0([A;\, B])$. That is, h is a $[A;\, B]$-valued function on R which is continuous in the norm topology of $[A;\, B]$. Define the operator $I(t,x) \in [D^0_{t,x}(B), B]$ by

$$\langle I(t,x),\, \theta(t,x)\rangle \triangleq \int_R \theta(t,t)\, dt \in B \qquad \theta \in \mathcal{D}^0_{t,x}(B).$$

Consider $f(t,x) = h(t)(-D_x)^n\, I(t,x)$; it is a member of $[\mathcal{D}^n_{t,x}(A);\, B] \subset [\mathcal{D}_{t,x}(A);\, B]$, as can be seen from the equation:

$$\langle f(t,x),\, \psi(t,x)\rangle = \int_R h(t)\, [D^n_x\, \psi(t,x)]_{x=t} dt \qquad \psi \in \mathcal{D}^n_{t,x}(A)$$

With this choice of f, we have for any $v \in \mathcal{D}(A)$ and $\phi \in \mathcal{D}$

$$\langle f\bullet v,\phi\rangle = \langle f(t,x),\, \phi(t)\, v(x)\rangle = \int_R h(t)\, \phi(t)\, D^n v(t) dt$$
$$= \langle hD^n v,\phi\rangle\ .$$

It follows from this expression that the composition• operator f• is a continuous linear mapping of $\mathcal{D}(A)$ into $[\mathcal{D}; B]$. Thus, we have arrived at the desired representation for hD^n, namely, in the sense of equality in $[\mathcal{D}; B]$,

$$hD^n v = f \bullet v \tag{4-5}$$

$$v \in \mathcal{D}(A),\ h \in E^0([A;B]),$$

$$f(t,x) = h(t)\ (-D_x)^n\ I(t,x) \in [\mathcal{D}^n_{t,x}(A); B].$$

Example 4-2

Here's the composition• representation for the operator $h\sigma_\tau$ where $h \in E^0([A; B])$ again, $\tau \in R$, and $\sigma_\tau(x)$ is the shifting operator defined on any $\theta \in \mathcal{D}^0_{t,x}(A)$ by

$$\sigma_\tau(x)\ \theta(t,x) \triangleq \theta(t,x-\tau)$$

and on any $f \in [\mathcal{D}^0_{t,x}(A); B]$ by

$$\langle \sigma_\tau(x)\ f(t,x),\ \theta(t,x)\rangle \triangleq \langle f(t,x),\ \theta(t,x+\tau)\rangle\ .$$

Then, $h\sigma_\tau \triangleq h(t)\ \sigma_\tau(t)$ denotes the operation of first shifting and then multiplying-by-h quantities defined on, say, the t axis.

Let $f(t,x) = h(t)\ \sigma_{-\tau}(x)\ I(t,x)$. It is easily seen that $f(t,x) \in [\mathcal{D}^0_{t,x}(A); B] \subset [\mathcal{D}_{t,x}(A); B]$. Then, for all $v \in \mathcal{D}(A)$ and $\phi \in \mathcal{D}$,

$$\begin{aligned}\langle f\bullet v,\phi\rangle &= \langle h(t)\sigma_{-\tau}(x)\ I(t,x),\ \phi(t)\ v(x)\rangle \\ &= \langle I(t,x),\ h(t)\ \phi(t)\ v(x-\tau)\rangle \\ &= \int_R h(t)\ \phi(t)\ v(t-\tau)dt = \langle h\sigma_\tau v,\phi\rangle \in B\ .\end{aligned}$$

This equation shows that f• is a continuous linear mapping of $\mathcal{D}(A)$ into $[\mathcal{D}; B]$. Thus, in the sense of equality in $[\mathcal{D}; B]$, we have

$$h\sigma_\tau v = f\bullet v \tag{4-6}$$

$$v \in \mathcal{D}(A),\ h \in E^0([A; B]),$$

$$f(t,x) = h(t)\ \sigma_{-\tau}(x)\ I(t,x) \in [\mathcal{D}^0_{t,x}(A); B]\ .$$

Example 4-3

We now show that convolution is a special case of composition. Let $y \in [\mathcal{D}(A); B]$. Define $y(t - x)$ as a member of $[\mathcal{D}_{t,x}(A); B]$ by

$$\langle y(t-x), \psi(t,x)\rangle_{t,x} \triangleq \langle y(t), \textstyle\int_R \psi(t+x,x)\,dx\rangle \ . \tag{4-7}$$

Let $f(t,x) = y(t - x)$. Then, for all $v \in \mathcal{D}(A)$ and $\phi \in \mathcal{D}$,

$$\begin{aligned}\langle f\cdot v,\phi\rangle &= \langle y(t - x), \phi(t)\, v(x)\rangle_{t,x} \\ &= \langle y(t), \textstyle\int_R \phi(t + x)\, v(x)dx\rangle \\ &= \langle y(t), \langle v(x), \phi(t + x)\rangle\rangle \ .\end{aligned}$$

The last expression is the definition of the convolution product $y * v$ applied to ϕ [4; Sec.4]. Thus, in the sense of equality in $[\mathcal{D}; B]$, we have

$$y * v = f\cdot v \tag{4-8}$$

$v \in \mathcal{D}(A)$, $y \in [\mathcal{D}(A); B]$, $f(t,x) = y(t - x) \in [\mathcal{D}_{t,x}(A); B]$.

We now attack the problem of finding a composition procedure that is explicitly defined for a pair of Banach-space-valued distributions neither of which are in $\mathcal{D}(A)$. (In fact, they may both be singular distributions.) We shall refer to this latter procedure as "composition∘" to distinguish it from the previously discussed composition• procedure. We first develop some properties of Banach-space-valued distributions that we shall need.

In the following N is a compact interval in R. Also, $L_{n,N}$ with $n = 1$ or 2 is the customary Banach space of complex Lebesgue-measurable functions f on N with $|f|^n$ Lebesgue integrable on N. The norm for $L_{n,N}$ is $||\cdot||_{n,N}$ where

$$||f||_{n,N} = \left(\textstyle\int_N |f(t)|^n\, dt\right)^{1/n}$$

By the Schwarz inequality,

$$||f||_{1,N} \le c||f||_{2,N}$$

where c denotes the square root of the length of N.

Theorem 4-4

Let the sequence $\{v_j\}_{j=1}^{\infty}$ *tend to zero in* $[\mathcal{D}; A]$, *and let* N *and* K *be compact intervals in* R *such that* K *contains a neighborhood of* N. *Then, there exists an integer* $p \geq 0$ *such that the following two conditions are satisfied.*

(*i*) *For each* j *there exists a strongly continuous* A-*valued function* g_j *on* R *such that* $\operatorname{supp} g_j \subset K$ *and* $v_j = D^p g_j$ *in the sense of equality in* $[\mathcal{D}_N; A]$.

(*ii*) *The sequence* $\{g_j(x)\}$ *tends to zero in* A *uniformly for all* $x \in R$.

Proof

By hypothesis, $v_j \to 0$ in $[\mathcal{D}_N; A]$. By [4; Lemma 3-1], there exists a constant $M > 0$ and an integer $r \geq 0$ such that, for all $\phi \in \mathcal{D}_N$,

$$(4\text{-}9) \qquad \sup_j \|\langle v_j, \phi\rangle\|_A \leq M \max_{0\leq k\leq r} \sup_{t\in R} |D^k\phi(t)|.$$

Next, for each derivative $D^k\phi$, we may write

$$D^k\phi = \int_{-\infty}^{t} dt_1 \int_{-\infty}^{t_1} dt_2 \cdots \int_{-\infty}^{t_{r-k}} D^{r+1} \phi(t_{r+1-k})\, dt_{r+1-k} .$$

Let $T = \min N$ and $\tau = \max N$. We get

$$\sup_{t\in R} |D^k\phi(t)| \leq (\tau - T)^{r-k} \sup_{t\in R} \int_T^t |D^{r+1}\phi(x)|\, dx$$
$$= (\tau - T)^{r-k} \int_N |D^{r+1}\phi(x)|\, dx .$$

We see from (4-9) that there exists a constant $M_1 > 0$ such that

$$(4\text{-}10) \quad \sup_j \|\langle v_j, \phi\rangle\|_A \leq M_1 \int_N |D^{r+1}\phi(x)|\,dx = M_1 \|D^{r+1}\phi\|_{1,N}.$$

Next, consider the following linear subspace of $\mathcal{D}_N$:

$$\Delta = \{\psi : \psi = D^{r+1}\phi, \ \phi \in \mathcal{D}_N\}$$

Let U_j be the linear mapping of Δ into A defined by

$$(4\text{-}11) \qquad \langle U_j, \psi\rangle \overset{\Delta}{=} \langle v_j, \phi\rangle \quad .$$

This defines U_j uniquely because, if $\psi \equiv 0$ and $\phi \in \mathcal{D}_N$, then $\phi \equiv 0$ so that $\langle U_j, 0\rangle = \langle v_j, 0\rangle = 0$. Moreover, U_j is continuous when Δ is supplied the topology induced by $L_{2,N}$ because by (4-10)

$$(4\text{-}12) \quad ||\langle U_j,\psi\rangle||_A = ||\langle v_j,\phi\rangle||_A \le M_1||\psi||_{1,N} \le M_1 c||\psi||_{2,N}.$$

We extend U_j onto the closure $\bar{\Delta}$ of Δ in $L_{2,N}$ by continuity. The extended mapping, which we also denote by U_j, is continuous and linear on $\bar{\Delta}$ and satisfies

$$||\langle U_j,\psi\rangle||_A \le M_1 c||\psi||_{2,N}$$

for all $\psi \in \bar{\Delta}$. Furthermore, let $C\Delta$ be the orthogonal complement of Δ in $L_{2,N}$, and define $\langle U_j,\psi\rangle = 0$ on every $\psi \in C\Delta$. Finally, define U_j on all of $L_{2,N}$ by linearity. Thus we have arrived at a uniquely defined continuous linear mapping U_j of $L_{2,N}$ into A.

We now invoke a result [17; p. 259, Theorem 1] which asserts that there exists an A-valued measure m_j defined on the Borel subsets of N such that

$$(4\text{-}13) \qquad \langle U_j,\psi\rangle = \int_N \psi \, dm_j$$

for each $\psi \in L_{2,N}$. In addition, (4-13) tends to zero as $j \to \infty$. Indeed, $\psi = \psi_1 + \psi_2$ where $\psi_1 \in \bar{\Delta}$ and $\psi_2 \in C\Delta$. Thus,

$$\int_N \psi dm_j = \langle U_j,\psi\rangle = \langle U_j,\psi_1 + \psi_2\rangle = \langle U_j,\psi_1\rangle$$
$$= \langle U_j,\psi_1 - \theta\rangle + \langle U_j,\theta\rangle$$

for any $\theta \in \Delta$. Hence,

$$||\int_N \psi dm_j||_A \le M_1 c||\psi_1 - \theta||_{2,N} + ||\langle U_j,\theta\rangle||_A$$

Given any $\varepsilon > 0$, we can choose $\theta \in \Delta$ such that

$$M_1 c||\psi_1 - \theta||_{2,N} < \frac{\varepsilon}{2}.$$

By the hypothesis on v_j and (4-11), $||\langle U_j, \theta\rangle||_A$ is also less than $\varepsilon/2$ for all sufficiently large j. Thus, (4-13) truly tends to zero as $j \to \infty$.

Moreover, we also have that, for every $\psi \in L_{2,N}$,

$$(4\text{-}14) \qquad ||\int_N \psi dm_j||_A = ||\langle U_j, \psi_1\rangle|| \le M_1 c||\psi_1||_{2,N} \le M_1 c||\psi||_{2,N}.$$

Next consider the function

$$K_x(t) = \begin{cases} 1 \text{ for } t \le x,\ t \in N \\ 0 \text{ otherwise,} \end{cases}$$

where x is any member of R. As before, we set T = min N and define an A-valued function f_j on R by

$$f_j(x) \overset{\Delta}{=} \int_T^x dm_j = \int_N K_x dm_j \qquad x \in R$$

Since the restriction of $K_x(t)$ to N is a member of $L_{2,N}$ for each $x \in R$, $f_j(x) \to 0$ as $j \to \infty$. Moreover, $\{f_j\}_j$ in a strongly equicontinuous set because by (4-14)

$$||f_j(x) - f_j(y)||_A = ||\int_N (K_x - K_y)dm_j||_A \le M_1 c||K_x - K_y||_{2,N} = M_1 c\sqrt{|y - x|}$$

Consequently, $\{f_j(x)\}_j$ tends to 0 in A uniformly for all $x \in R$.

Finally, we have from (4-11), (4-13), and an integration by parts that, for every $\phi \in \mathcal{D}_N$,

$$\langle v_j, \phi\rangle = \langle U_j, D^{r+1}\phi\rangle = \int_N D^{r+1}\phi dm_j = -\int_N f_j D^{r+2}\phi dt .$$

Let $g_j = (-1)^{r+1}\alpha f_j$ where $\alpha \in \mathcal{D}$ is such that $\alpha \equiv 1$ on a neighborhood of N, $\alpha \equiv 0$ outside K. Then, for all $\phi \in \mathcal{D}_N$,

$$\langle v_j,\phi\rangle = \langle -f_j,\alpha D^{r+2}\phi\rangle = \langle (-1)^{r+1}D^{r+2}(\alpha f_j),\phi\rangle$$
$$= \langle D^{r+2}g_j,\phi\rangle \ .$$

This completes the proof of theorem 4-4.

We note in passing that the foregoing proof can be modified to eliminate the integral representation in the right-hand side of (4-13). One need merely work with the left-hand side of (4-13) and define $f_j(x)$ as $\langle U_j,K_x\rangle$. But then, this will require the use of the concepts of the primitive of U_j and its differentiation.

Theorem 4-5

If the sequence $\{v_j\}_{j=1}^{\infty}$ *converges in* $[\mathcal{E};\ A]$ *to zero, then there exist a compact interval* N, *an integer* $p \geq 0$, *and strongly continuous* A*-valued functions* $h_{k,j}$ *on* R *such that, in the sense of equality in* $[\mathcal{E};\ A]$,

$$v_j = \sum_{k=0}^{p} D^k h_{k,j}$$

where $\text{supp } h_{k,j} \subset N$ *for all* k *and* j *and, for each fixed* k, $h_{k,j} \to 0$ *in* A *uniformly for all* $x \in R$.

Proof

That $\{v_j\}$ converges in $[\mathcal{E};\ A]$ implies that there exists a compact interval $G \subset R$ such that $\text{supp } v_j \subset G$ for all j [18; Chapter 1, pp. 62-63]. Let N be a compact interval in R containing a neighborhood of G. Choose $\lambda \in \mathcal{D}$ such that $\lambda \equiv 1$ on a neighborhood of G and $\lambda \equiv 0$ outside N. Then, for any $\phi \in \mathcal{E}$, we have that $\lambda\phi \in \mathcal{D}_N$ and $\langle v_j,\phi\rangle = \langle v_j,\lambda\phi\rangle$. By Theorem 4-4 and the fact that convergence in $[\mathcal{E};\ A]$ implies convergence in $[\mathcal{D};\ A]$,

$$\langle v_j,\phi\rangle = \langle v_j,\lambda\phi\rangle = \langle D^p g_j,\lambda\phi\rangle = (-1)^p\langle g_j,D^p(\lambda\phi)\rangle$$
$$= (-1)^p\langle g_j, \sum_{k=0}^{p}\binom{p}{k}(D^{p-k}\lambda)(D^k\phi)\rangle = \langle \sum_{k=0}^{p} D^k h_{k,j},\phi\rangle$$

where

$$h_{k,j} = (-1)^{p+k}\binom{p}{k}(D^{p-k}\lambda)g_j \ .$$

The functions $h_{k,j}$ possess the properties stated in the theorem.

Lemma 4-2

Any given v ε [*E*; A] *generates a unique* $\hat{v}$ ε [*E*([A; B]); B] *by means of the definition:*

(4-15) $$\langle \hat{v}, J\theta\rangle \triangleq J\langle v,\theta\rangle \qquad J \in [A;\, B],\ \theta \in E \ .$$

Moreover, the mapping v → $\hat{v}$ *is a sequentially continuous linear injection of* [*E*; A] *into* [*E*([A; B]); B].

Proof

Let P denote the set of all elements in *E*([A; B]) of the form Jθ. Equation (4-15) uniquely defines from the given v ε [*E*; A] a mapping g_v of P into B.

Next, as a consequence of Theorem 4-5, we have that v is a member of [*E*; A] if and only if there exist an integer $p \geq 0$, a compact interval N, and a finite set $\{h_k\}_{k=0}^{p}$ of strongly continuous A-valued functions h_k on R with supp h_k N such that

$$v = \sum_{k=0}^{p} D^k h_k$$

in the sense of equality in [*E*; A]. We define a linear mapping $\hat{v}$ of *E*([A; B]) into B by

(4-16) $$\langle \hat{v},\psi\rangle = \langle \sum_{k=0}^{p} D^k h_k, \psi\rangle = \sum_{k=0}^{p} (-1)^k \int_N h_k D^k \psi\, dt \in B$$

$$\psi \in E([A;\, B]) \ .$$

(For any J ε [A; B] and any a ε A, we denote the application of J to a by either Ja or aJ. Thus, $h_k D^k \psi$ denotes a B-valued function on R, which also happens to be strongly continuous.) $\hat{v}$ is continuous because

$$||<\hat{v},\psi>||_B \le \sum_{k=0}^{p} \int_N ||h_k||_A ||D^k\psi||_{[A;B]} dt$$

$$\le \sum_{k=0}^{p} \int_N ||h_k||_A dt \sup_{t\in N} ||D^k\psi(t)||_{[A;B]}$$

Thus, $\hat{v} \in [E([A;\ B]);\ B]$.

We now observe that the substitution of $J\theta$ for ψ in (4-16) yields (4-15). Thus, $\hat{v}$ coincides with g_V on P. Moreover, P is total in $E([A;\ B])$; that is, the span of P is dense in $E([A;\ B])$. Indeed, $Q \triangleq \{J\chi : J \in [A;\ B],\ \chi \in \mathcal{D}\}$ is total in $\mathcal{D}([A;\ B])$ according to [4; Lemma 6-1]. But, $\mathcal{D}([A;\ B])$ is dense in $E([A;\ B])$ and the topology of $\mathcal{D}([A;\ B])$ is stronger than that induced on it by $E([A;\ B])$. Hence, Q is total in $E([A;\ B])$. Since $Q \subset P$, so too is P. This implies that $\hat{v}$ is uniquely determined by its restriction to P. This restriction is g_V, which is uniquely determined by v, as was noted above. This proves the first sentence of the lemma.

Turning to the second sentence, we first show that $v \to \hat{v}$ is an injection (i.e., a one-to-one mapping). Assume that v_1 and v_2 are both members of $[E;\ A]$ and that $<v_1,\theta> \ne <v_2,\theta>$ for some $\theta \in E$. There exists at least one $J \in [A;\ B]$ such that $J<v_1,\theta> \ne J<v_2,\theta>$. (Indeed, by the Hahn-Banach theorem [8; p. 187, Corollary 2], there exists a continuous linear functional F on A such that $F<v_1,\theta> \ne F<v_2,\theta>$. Now, set $J = bF$ where b is any member of B other than the zero member.) Next, let $\hat{v}_1$ and $\hat{v}_2$ be the members of $[E([A;\ B]);\ B]$ defined by v_1 and v_2 respectively in accordance with (4-15). Therefore, $<\hat{v}_1,J\theta> \ne <\hat{v}_2,J\theta>$. So truly, $v \to \hat{v}$ is an injection.

That $v \to \hat{v}$ is a linear mapping is clear. Its sequential continuity follows directly from (4-16) and Theorem 4-5. Lemma 4-2 is now established.

The concept of "composition∘" employs the idea of a distribution y_x depending on a parameter $x \in R$. For our purpose, we will impose upon y_x the

Conditions G

G1. *For each fixed* $x \in R$, y_x *is a member of* $[\mathcal{D};[A;B]]$. *Thus, for any given* $\phi \in \mathcal{D}$, *the equation:*

$$\psi(x) \triangleq \langle y_x,\phi\rangle \tag{4-17}$$

defines an $[A; B]$*-valued function* ψ *on* R.

G2. $\phi \mapsto \psi$ *is a mapping of* $\mathcal{D}$ *into* $\mathcal{E}([A; B])$.

The next lemma is due to Dolezal [16].

Lemma 4-3

Assume that y_x *satisfies contitions* G. *Then, the mapping* $\phi \mapsto \psi$ *of* $\mathcal{D}$ *into* $\mathcal{E}([A; B])$ *is linear and continuous.*

We now define the composition∘ product $v \circ y_x$ of any $v \in [\mathcal{E}; A]$ with y_x by

$$\langle v \circ y_x,\phi\rangle \triangleq \langle \hat{v},\psi\rangle \triangleq \langle \hat{v}(x),\langle y_x(t),\phi(t)\rangle\rangle \qquad \phi \in \mathcal{D} \tag{4-18}$$

where $\hat{v} \in [\mathcal{E}([A; B]); B]$ is defined by (4-15) or equivalently by (4-16). At times, we will denote the operator $v \to v \circ y_x$ by $\circ y_x$.

Theorem 4-6

Let $v \in [\mathcal{E}; A]$ *and let* y *satisfy conditions* G. *Then,* $v \mapsto v \circ y_x$ *is a sequentially continuous linear mapping of* $[\mathcal{E}; A]$ *into* $[\mathcal{D}; B]$.

Proof

That $v \circ y_x \in [\mathcal{D}; B]$ follows immediately from the definition (4-18) and the following two facts: $\phi \mapsto \psi$ is a continuous linear mapping of $\mathcal{D}$ into $\mathcal{E}([A; B])$ according to Lemma 4-3. $\psi \mapsto \langle \hat{v},\psi\rangle$ is a continuous linear mapping of $\mathcal{E}([A; B])$ into B according to Lemma 4-2.

It is clear that $v \mapsto v \circ y_x$ is a linear mapping. To verify its sequential continuity, let Ω be an arbitrarily chosen bounded set in $\mathcal{D}$, let $\{v_j\}_{j=1}^{\infty}$ tend to zero in $[\mathcal{E}; A]$, and define $\hat{v}_j$ from v_j as in Lemma 4-2. By Lemma 4-3, ψ traverses a bounded set Λ in $\mathcal{E}([A; B])$ when ϕ traverses Ω. So, for the corresponding seminorm σ_Ω on $[\mathcal{D}; B]$, we have

$$\sigma_\Omega(v_j \circ y_x) \triangleq \sup_{\phi\varepsilon\Omega} ||<v_j \circ y_x, \phi>||_B = \sup_{\psi\varepsilon\Lambda} ||<\hat{v}, \psi>||_B$$

By Theorem 4-5 and (4-16),

$$\sigma_\Omega(v_j \circ y_x) = \sup_{\psi\varepsilon\Omega} ||\sum_{k=0}^{p} (-1)^k \int_N \psi^{(k)}(t) h_{k,j}(t)dt||_B$$

$$\leq \sum_{k=0}^{p} \sup_{t\varepsilon N} ||h_{k,j}(t)||_A \sup_{\psi\varepsilon\Lambda} \int_N ||\psi^{(k)}(t)_{[A;B]} dt ,$$

and the right-hand side tends to zero as $j \to \infty$. This shows that $v_j \circ y_x$ tends to zero in $[\mathcal{D}; B]$ and completes the proof.

In much the same way, a variation of Theorem 4-6 can be established. In this case we assume that $v \;\varepsilon\; [\mathcal{D}; A]$ and that y_x satisfies the following two conditions:

Condition G'

G1'. y_x *satisfies condition A1.*

G2'. $\phi \to \psi$ *is a mapping of* $\mathcal{D}$ *into* $\mathcal{D}([A;B])$.

Dolezal [16] has shown that conditions G' imply that the mapping $\phi \mapsto \psi$ of $\mathcal{D}$ into $\mathcal{D}([A; B])$ is linear and continuous. Furthermore, by modifying the proof of Lemma 4-2 (we now use Theorem 4-4 instead of 4-5), we can also show that any given $v \;\varepsilon\; [\mathcal{D}; A]$ defines a unique $\hat{v} \;\varepsilon\; [\mathcal{D}([A;B]);B]$ via the equation (4-15), where now $\theta \;\varepsilon\; \mathcal{D}$ and, in addition, $v \mapsto \hat{v}$ is a sequentially continuous linear injection of $[\mathcal{D}; A]$ into $[\mathcal{D}([A; B]); B]$. Once again, we define $v \circ y_x$ by (4-18). Then, an argument almost identical to the proof of Theorem 4-6 establishes

Theorem 4-7

Let $v \in [\mathcal{D};\ A]$ *and let* y_x *satisfy conditions* G'. *Then* $v \mapsto v \circ y_x$ *is a sequentially continuous linear mapping of* $[\mathcal{D};\ A]$ *into* $[\mathcal{D};\ B]$.

We can relate composition∘ mappings to composition• mappings in the following way. Let there be given a composition∘ mapping $\mathfrak{M}: v \to v \circ y_x$ where $v \in [\mathcal{E};\ A]$ and y_x satisfies conditions G. Then, the restriction of $\mathfrak{M}$ to $\mathcal{D}(A)$ is a continuous linear mapping $\mathfrak{N}$ of $\mathcal{D}(A)$ into $[\mathcal{D};B]^w$ given by $v \mapsto f \cdot v$, $v \in \mathcal{D}(A)$, where $f \in [\mathcal{D}_{t,x}(A);\ B]$ is uniquely determined. Indeed, $\mathcal{D}(A) \subset [\mathcal{E};\ A]$, and the topology of $\mathcal{D}(A)$ is stronger than that induced on it by $[\mathcal{E};A]$. On the other hand, the topology of $[\mathcal{D};\ B]$ is stronger than that of $[\mathcal{D};\ B]^w$. Hence, by Theorem 4-6, the restriction $\mathfrak{N}$ of $\mathfrak{M}$ to $\mathcal{D}(A)$ is a sequentially continuous linear mapping of $\mathcal{D}(A)$ into $[\mathcal{D};\ B]^w$. It is even continuous since $\mathcal{D}(A)$ is the inductive limit of Fréchet spaces. Thus, we may invoke Theorem 4-3 to conclude that $\mathfrak{N} = f\cdot$ on $\mathcal{D}(A)$, where $f \in [\mathcal{D}_{t,x}(A);\ B]$ is uniquely determined.

It is also true that the mapping $\mathfrak{M} \mapsto \mathfrak{N}$ is injective because $\mathcal{D}(A)$ is dense in $[\mathcal{E};\ A]$. We shall see later on (Theorem 4-8) that any given continuous linear mapping of $[\mathcal{E};\ A]$ into $[\mathcal{D};\ B]$ uniquely determines a y_x satisfying conditions G such that $\mathfrak{M}v = v \circ y_x$ for every $v \in [\mathcal{E};\ A]$. It follows from these two facts that, if $f \in [\mathcal{D}_{t,x}(A);\ B]$ is given and if $f\cdot$ can be extended into a continuous linear mapping $\mathfrak{M}$ of $[\mathcal{E};\ A]$ into $[\mathcal{D};\ B]$, then there exists a unique y_x satisfying conditions G such that $f \cdot v = v \circ y_x$ for all $v \in \mathcal{D}(A)$.

As examples we now develop the composition∘ operators corresponding to the composition• operators presented in Examples 4-1, 4-2 and 4-3.

Example 4-1a

For $x \in R$ and n a positive integer, we set

$$y_x(t) = h(t) D_t^n \delta_x(t) \tag{4-19}$$

where now we assume that $h \in \mathcal{E}([A; B])$ in contrast to the less restrictive assumption made in Example 4-1. (The symbol δ_x denotes the shifted delta functional; that is, $\langle\delta_x,\phi\rangle = \phi(x)$.) For any $\phi \in \mathcal{D}$

$$\langle y_x,\phi\rangle = \langle hD^n\delta_x,\phi\rangle = (-D_x)^n[h(x)\phi(x)] \in \mathcal{D}([A; B]) .$$

Thus, y_x satisfies both conditions G and G′. So, by Theorem 4-7, $v\circ y_x \in [\mathcal{D}; B]$ for any $v \in [\mathcal{D}; A]$. In particular, for $\phi \in \mathcal{D}$, we have

$$\langle v\circ y_x,\phi\rangle = \langle\hat{v}(x),\ \langle h(t)D_t^n\delta_x(t),\ \phi(t)\rangle\rangle$$

$$= \langle\hat{v}(x),\ (-D_x)^n[h(x)\phi(x)]\rangle \in B.$$

By virtue of the representation $v = D^p g$, $g \in \mathcal{D}_k^0(A)$, from which $\hat{v}$ is defined via the analogue to (4-16), the right-hand side is equal to

$$\int_N g(x)(-D_x)^{n+p}[h(x)\phi(x)]dx = \langle hD^n v,\phi\rangle .$$

Thus, in the sense of equality in $[\mathcal{D}; B]$,

$$v\circ y_x = hD^n v . \qquad (4\text{-}20)$$

Compare this to (4-5).

Example 4-2a

Let $\tau \in R$, let σ_τ be the shifting operator as before, let $h \in \mathcal{E}([A; B])$, and set

$$y_x = h\sigma_\tau\delta_x = h\delta_{x+\tau} . \qquad (4\text{-}21)$$

Here too, y_x satisfies conditions G and G′. For any $v \in [\mathcal{D}; A]$ and $\phi \in \mathcal{D}$,

$$\langle v\circ y_x,\phi\rangle = \langle\hat{v}(x),\ \langle h(t)\delta_{x+\tau}(t),\ \phi(t)\rangle\rangle$$

$$= \langle\hat{v}(x),\ h(x+\tau)\phi(x+\tau)\rangle \in B .$$

As in the previous example, the right-hand side can be shown to be equal to

$$\langle h(x)v(x - \tau),\ \phi(x)\rangle = \langle h\sigma_\tau v, \phi\rangle \ .$$

Thus, in the sense of equality in $[\mathcal{D};\ B]$,

$$v \circ y_x = h\sigma_\tau v \ . \tag{4-22}$$

Example 4-3a

We now turn to convolution once again. Let $y \in [\mathcal{D}(A);\ B]$ and $v \in [\mathcal{E};\ A]$. By Theorem 3-1, y is also a member of $[\mathcal{D};[A;\ B]]$. For each $x \in R$, we define y_x as either a member of $[\mathcal{D}(A);\ B]$ or $[\mathcal{D}[A;\ B]]$ by $y_x = \sigma_x y$. So, for any $\phi \in \mathcal{D}$,

$$\begin{aligned} \langle v \circ y_x, \phi\rangle &= \langle \hat{v}(x),\ \langle y_x(t),\ \phi(t)\rangle\rangle \\ &= \langle \hat{v}(x),\ \langle y(t),\ \phi(t + x)\rangle\rangle. \end{aligned} \tag{4-23}$$

Here, the right-hand side has the sense of the application of $\hat{v} \in [\mathcal{E}([A;\ B]);\ B]$ to $\langle y(t),\ \phi(t + x)\rangle \in \mathcal{E}([A;\ B])$. See [4; Theorems 3-1 and 4-3].) We now employ the representation of v as a sum of derivatives of strongly continuous A-valued functions on R of compact support (see Theorem 4-5) and the representation of y on $\mathcal{D}_K$ for any given compact interval K as the derivative of an $[A;\ B]$-valued function on R that is continuous in the norm topology of $[A;\ B]$ (see [4; Theorem 3-1 and equations (3-8) and (3-9)]). This allows us to invoke Fubini's theorem and then to rewrite (4-23) as follows, where now $y \in [\mathcal{D}(A);\ B]$:

$$\langle v \circ y_x, \phi\rangle = \langle y(t),\ \langle v(x),\ \phi(t + x)\rangle\rangle = \langle y * v,\ \phi\rangle$$

Thus, in the sense of equality in $[\mathcal{D};\ B]$, $v \circ y_x = y * v$.

Theorem 4-3 states that every continuous linear mapping of $\mathcal{D}(A)$ into $[\mathcal{D};\ B]^w$ has a composition$\bullet$ representation. On the other hand, the examples of this section show that at least in three particular cases a composition$\bullet$ operator has a corresponding composition$\circ$ operator. A natural conjecture therefore is that every continuous linear mapping of $[\mathcal{E};\ A]$ into $[\mathcal{D};\ B]$ has a composition$\circ$ representation. Theorem 4-8 below, which is a partial converse to Theorem 4-6, states that this is indeed the case. Its proof makes use of

Lemma 4-4

Every $v \in [E; A]$ *is the limit in* $[E; A]$ *of a sequence* $\{v_\nu\}_{\nu=1}^{\infty}$ *such that each* v_ν *is a finite sum of the form:*

$$v_\nu(t) = \sum_{\mu=1}^{m_\nu} a_{\nu,\mu} D_x^r \delta_x(t)\big|_{x=\tau_{\nu,\mu}}$$

where $a_{\nu,\mu} \in A$; $\tau_{\nu,\mu} \in R$, *and* r *is a nonnegative integer not depending on* μ *or* ν.

The proof of this lemma mimics that of Lemma 2 in [19; Sec. 5.8]. In this case, we use Theorem 4-5 to write $v = (-D)^{r+2}h$ where $h \in E^0(A)$. (Here, h need not have a compact support.) As in [19; p. 145], we then set up the function $h_\nu \in E^0(A)$, all of which coincide with h outside some compact interval I containing supp v. This allows us to write, for any $\phi \in E$,

$$\langle v - (-D)^{r+2}h_\nu, \phi\rangle = \int_I (h - h_\nu)D^{r+2}\phi dt.$$

Lemma 4-4 is established by estimating a bound on this integral.

Theorem 4-8

Let $\mathfrak{N}$ *be a continuous linear mapping of* $[E;A]$ *into* $[\mathcal{D}; B]$. *Then, there exists a unique* y_x *satisfying conditions* G *such that for every* $v \in [E; A]$, $\mathfrak{N} v = v \circ y_x$ *in the sense of equality in* $[\mathcal{D}; B]$.

Proof

Define an operator $\mathfrak{M}$ from $\mathfrak{N}$ by

$$\langle \mathfrak{M}_g, \phi\rangle a \triangleq \langle \mathfrak{N}(ga), \phi\rangle \tag{4-24}$$

where $g \in [E; C]$, $a \in A$, and $\phi \in \mathcal{D}$. By [4; Theorem 3-2],

$\mathfrak{M}$ is a continuous linear mapping of $[\mathcal{E};\ C]$ into $[\mathcal{D};[A;B]]$. Moreover, the equation $\langle a\mathfrak{M}g,\phi\rangle = \langle \mathfrak{M}g,\phi\rangle a$ defines $a\mathfrak{M}g$ as a member of $[\mathcal{D};\ B]$ (see [4; Sec. 3]). Therefore, in the sense of equality in $[\mathcal{D};\ B]$,

$$a\mathfrak{M}g = \mathfrak{N}(ga) \ . \tag{4-25}$$

For each $x \in R$, set $y_x = \mathfrak{M}\,\delta_x \in [\mathcal{D};\ [A;\ B]]$. We now establish two facts: (i) For any given $\phi \in \mathcal{D}$, $\langle y_x,\phi\rangle$ as a function of x is a member of $\mathcal{E}([A;\ B])$ so that y_x satisfies conditions G. (ii) Moreover, for any nonnegative integer k,

$$D_x^k y_x = \mathfrak{M}\, D_x^k \delta_x \tag{4-26}$$

in the sense of equality in $[\mathcal{D};\ [A;\ B]]$. ($D_x^k y_x$ denotes the kth-order *parametric* derivative of y_x defined by $\langle D_x^k y_x,\ \phi\rangle \triangleq D_x^k\langle y_x,\phi\rangle$, $\phi \in \mathcal{D}$.) We use an inductive argument. First note that (4-26) is true for k = 0 by definition. Next, fix x and choose any $\Delta x \in R$, $\Delta x \neq 0$. Assuming that (4-26) is true for some k, we may write

$$\frac{1}{\Delta x}(D_{x+\Delta x}^k y_{x+\Delta x} - D_x^k y_x) = \mathfrak{M}\frac{1}{\Delta x}[D_{x+\Delta x}^k \delta_{x+\Delta x} - D_x^k \delta_x] \tag{4-27}$$

The quantity on the right-hand side upon which $\mathfrak{M}$ operates converges in $[\mathcal{E};\ C]$ to $D_x^{k+1}\delta_x$. Therefore, (4-27) converges in $[\mathcal{D};\ [A;\ B]]$ to

$$D_x^{k+1} y_x = \mathfrak{M}\, D_x^{k+1} \delta_x,$$

and, in addition, $\langle y_x,\phi\rangle$ is a smooth $[A;\ B]$-valued function on R by the definition of parametric differentiation. The statements (i) and (ii) are hereby established.

We now employ the sequence $\{v_\nu\}$ indicated in Lemma 4-4. By the linearity of $\mathfrak{N}$ and equations (4-25) and (4-26), we may write

$$\mathfrak{N} v_\nu = \sum_{\mu=1}^{m_\nu} a_{\nu,\mu} D_x^r y_x \Big|_{x=\tau_{\nu,\mu}} \tag{4-29}$$

By Lemma 4-4 and the continuity of $\mathfrak{N}$, the left-hand side of (4-29) converges in $[\mathcal{D}; B]$ to $\mathfrak{N}v$. On the other hand, the application of the right hand side or (4-29) to any $\phi \in \mathcal{D}$ yields

$$\sum_{\mu=1}^{m_\nu} a_{\nu,\mu} D_x^r \langle y_x, \phi\rangle |_{x=\tau_{\nu,\mu}}$$

where $\langle y_x, \phi\rangle \in \mathcal{E}([A; B])$ as was noted above. The last quantity is equal to

$$\sum_{\mu=1}^{m_\nu} a_{\nu,\mu} \langle D^r_{\tau_{\nu,\mu}} \delta_{\tau_{\nu,\mu}}(x), \langle y_x(t), \phi(t)\rangle\rangle = \langle v_\nu \circ y_x, \phi\rangle .$$

But, this tends to $v \circ y_x$ since $f \mapsto f \circ y_x$ is continuous on $[\mathcal{E}; A]$. Thus Theorem 4-8 is proven.

Finally, we define the concept of causality and point out how it affects the composition$\bullet$ and composition$\circ$ representations of operators generated by time-varying Banach systems.

Definition 4-1

Let $\mathfrak{N}$ *be an operator mapping a set* $X \subset [\mathcal{D}; A]$ *into* $[\mathcal{D}; B]$. $\mathfrak{N}$ *is said to be causal on* X *if, for every* $t_0 \in R$, *we have that* $(\mathfrak{N}v_1)(t) = (\mathfrak{N}v_2)(t)$ *on* $-\infty < t < t_0$ *whenever* $v_1 \in X$, $v_2 \in X$, *and* $v_1(t) = v_2(t)$ *on* $-\infty < t < t_0$.

The following theorems are established in the same way as in the scalar case [20].

Theorem 4-9

Let $f \in [\mathcal{D}_{t,x}(A); B]$. *The operator* $f\bullet$ *is causal on* $\mathcal{D}(A)$ *if and only if* supp f *is contained in the half plane* $\{t,x\colon t \geq x\}$

Theorem

Let y_x *satisfy conditions* G. *The operator* $\circ y_x$ *is causal on* $[\mathcal{E}; A]$ *if and only if, for each* $x \in R$, supp y_x

contained in the semi-infinite line $[x,\infty)$.

5. TIME-INVARIANT BANACH SYSTEMS AND CONVOLUTION

A time-invariant Banach system is of course one whose components do not vary with time. One can define such a system in a mathematical way by saying that every operator generated by the system commutes with the shifting operator σ_τ whatever be the choice of $\tau \in R$. (We shall also call every such operator time-invariant.) In this case the composition• and composition∘ operators become convolution operators. The latter is defined as follows [4; Sec. 4].

Let y be a fixed member of $[\mathcal{D}(A); B]$. Then, the convolution product $y * v$ of y and any $v \in [\mathcal{E}; A]$ is defined by

$$\langle y * v, \phi\rangle \triangleq \langle y(t), \langle v(x), \phi(t + x)\rangle\rangle \qquad \phi \in \mathcal{D} \tag{5-1}$$

The right-hand side has a sense because $\psi(t) \triangleq \langle v(x), \phi(t + x)\rangle \in \mathcal{D}(A)$. This also defines the convolution operator $v \mapsto y * v$, which we denote by $y*$. This operator is a continuous linear mapping of $[\mathcal{E}; A]$ into $[\mathcal{D}; B]$ (see [4; Theorem 4-1]). Moreover, it is time-invariant, which means, as was noted above, that it commutes with the shifting operator σ_τ for every $\tau \in R$ [4; Proposition 4-1]. In addition, $y*$ is a continuous linear mapping of $\mathcal{D}(A)$ into $\mathcal{E}(B)$, and in this case we have that

$$(y * v)(t) = \langle y(x), v(t - x)\rangle \qquad v \in \mathcal{D}(A) \tag{5-2}$$

in the sense of equality in $[\mathcal{D}; B]$ (see [4; Theorem 4-3]). Similar results hold for several other spaces of Banach-space-valued distributions.

Conversely, every continuous linear mapping of $\mathcal{D}(A)$ into $[\mathcal{D}; B]$ that commutes with the shifting operator σ_τ for every $\tau \in R$ has a convolution representation [4; Theorem 6-1]. In particular, we have

Theorem 5-1

$\mathfrak{N}$ *is a continuous linear time-invariant mapping of* $\mathcal{D}(A)$ *into* $[\mathcal{D}; B]$ *if and only if there exists a* $y \in [\mathcal{D}(A); B]$ *such that* $\mathfrak{N} = y*$ *on* $\mathcal{D}(A)$ *(i.e., in the sense of equality in* $[\mathcal{D}; B]$, $\mathfrak{N} v = y * v$ *for all* $v \in \mathcal{D}(A)$*).* y *is uniquely determined by* $\mathfrak{N}$*, and conversely.*

(This theorem can be refined by replacing $\mathcal{D}(A)$ by the space $\mathcal{D} \otimes A$ which is the span of all elements of the form ϕa where $\phi \in \mathcal{D}$ and $a \in A$ [4].)

In view of Theorem 4-3, 4-8, and 5-1, we see that every convolution operator is a special case of a composition• operator as well as of a composition∘ operator. This was also observed in Examples 4-3 and 4-3a. By theorem 3-1 and the analysis of Example 4-3a, we see that the composition∘ operator $\circ y_x$ corresponding to any given convolution operator $y *$ is obtained simply by setting $y_x = \sigma_x y$. Thus, Theorem 4-10 immediately yields

Theorem 5-2

Let $y \in [\mathcal{D}(A); B]$*. Then, the convolution operator* $y*$ *is causal on* $[\mathcal{E}; A]$ *if and only if* $\text{supp}\, y \subset [0,\infty)$.

A causality criterion for $y*$ can be stated in terms of the Laplace transform $\mathfrak{L}y$ of y if $\mathfrak{L}y$ happens to exist in the sense stated at the end of Sec. 3 [4; Theorem 6-2 and Proposition 6-3].

Theorem 5-3

Assume that $y \in [\mathfrak{L}(w,z; A); B]$ *for some* w *and* z*. Necessary and sufficient conditions for* $y*$ *to be causal on* $[\mathcal{E}; A]$ *(and, in fact, on* $[\mathfrak{L}(w,z); A]$*) are that* $z = \infty$ *and, on some half plane* $\{\zeta : \text{Re}\,\zeta \geq a,\ a \in R\}$*, we have*

$$||(\mathfrak{L}y)(\zeta)||_{[A;B]} \leq P(|\zeta|)$$

where P *is a polynomial.*

6. HILBERT PORTS AND PASSIVITY: AN ADMITTANCE FORMULISM

As was explained in Sec. 2, the concept of a Hilbert port arises when two physical variables v and u in a system take their values in a complex Hilbert space H and, in addition, are complementary in the sense that their inner product (u,v) = (u(t), v(t)) represents the instantaneous complex power entering the system. If this power is Lebesgue integrable in the interval $(-\infty,x)$, then the integral

$$\text{(6-1)} \qquad \operatorname{Re} \int_{-\infty}^{x} (u,v)\,dt$$

represents the total energy entering the system during the time interval $-\infty < t < x$. This allows us to define the passivity of the admittance operator $\mathfrak{N} : v \mapsto u$.

Definition 6-1

Let $\mathcal{W}(H)$ *be a set of* H*-valued functions on* R *contained in the domain of an operator* $\mathfrak{N}$. $\mathfrak{N}$ *is said to be a passive mapping on* $\mathcal{W}(H)$ *if, for every* $v \in \mathcal{W}(H)$, *for* $u = \mathfrak{N}v$, *and for every finite real number* x, *we have that* (u(t), v(t)) *is Lebesgue integrable on* $-\infty < t < x$ *and the integral* (6-1) *is nonnegative.*

When $\mathfrak{N}$ is the admittance operator of a Hilbert port and is passive, we shall also call the Hilbert port passive.

If $\mathfrak{N}$ is a convolution operator y*, $y \in [\mathcal{D}(H); H]$, then it turns out that, for every $v \in \mathcal{D}(H)$, $u = y * v$ is a member of $\mathcal{E}(H)$ and that $(u,v) \in \mathcal{D}$ [4; Theorem 4-3 and Lemma 7-1]. Thus, (6-1) certainly exists for every $x \in R$, and we may establish the passivity of $\mathfrak{N}$ on $\mathcal{D}(H)$ merely by checking the nonnegativity of (6-1).

Every passive convolution operator y* has a frequency-domain description; namely, its impulse response y possesses a Laplace transform Y which is positive*. The last word is defined as follows.

Definition 6-2

Given a complex Hilbert space H, *a function* Y *of the complex variable* ζ *is called a positive* mapping of* H *into* H *(or simply positive*) if, on the half plane* $C_+ \triangleq \{\zeta : \text{Re}\zeta > 0\}$, Y *is an* [H; H]*-valued analytic function such that* $\text{Re}(Y(\zeta)a,a) \geq 0$ *for every* $a \in H$.

The principal theorem in the admittance formulism of passive Hilbert ports is the following [4].

Theorem 6-1

Assume that $\mathfrak{N}$ *is a continuous linear time-invariant passive mapping of* $\mathcal{D}(H)$ *into* $[\mathcal{D}; H]$. $\mathfrak{N}$ *has a convolution representation* $\mathfrak{N} = y*$ *where* $y \in [\mathcal{D}_{L_1}(H); H]$ *and* $\text{supp}\, y \subset [0,\infty)$. *Moreover,* y *possesses a Laplace transform* Y *on at least* C_+ *which is positive*.*

Conversely, assume that Y *is positive*. Then, there exists a unique convolution operator* $\mathfrak{N} = y*$ *such that* $y \in [\mathcal{D}_{L_1}(H); H]$, $\text{supp}\, y \subset [0,\infty)$, *and* $\mathfrak{L} y = Y$ *on* C_+. *Moreover,* $\mathfrak{N}$ *is a continuous linear time-invariant passive mapping of* $\mathcal{D}(H)$ *into* $[\mathcal{D}; H]$.

Note that the fact that $\text{supp}\, y \subset [0,\infty)$ implies that $y*$ is also causal on $[\mathcal{E}; H]$ as was indicated in *Then,* 5-2.

To introduce the concept of positive*-reality, we must first consider real operators in [H; H], and this in turn requires that we assign to H a somewhat more complicated structure. In particular, we shall now assume that the complex Hilbert space H is generated from a real Hilbert space H_r through complexification [21; Sec. 2.1]. This implies among other things that $H_r \subset H$. Then, an $F \in [H; H]$ is called real if $F \in [H_r; H_r]$.

Definition 6-3

Given H *and* H_r *as stated, a function* Y *of the complex variable* ζ *is called positive *-real if it is a positive* mapping of* H *into* H *and, for each real positive number* σ, $Y(\sigma)$ *is real.*

Corollary 6-1a

Theorem 6-1 remains valid when H *is replaced by* H_r, *"positive*" by "positive*-real", and* $\mathcal{D}$ *by the space* $\mathcal{D}(R)$ *of real-valued functions in* $\mathcal{D}$.

(Note: The spaces $\mathcal{D}(H_r)$, $\mathcal{D}_{L_1}(H_r)$, and $[\mathcal{D}(R); H_r]$, their topologies, and the properties of passivity, time-invariance, etc. are defined just as they are in the complex case.)

7. HILBERT PORTS AND PASSIVITY: A SCATTERING FORMULISM

Let us consider once again a Hilbert port and its variables v, u which determine its admittance operator $\mathfrak{N} : v \to u$. A scattering formulism for the Hilbert port is generated by working with the variables

(7-1) $$v_+ \triangleq v + u, \qquad v_- \triangleq v - u .$$

We call v_+ the incident wave and v_- the reflected wave. The mapping $\mathfrak{W} : v_+ \to v_-$ is the scattering operator of the Hilbert port. In terms of v_+ and v_-, the energy integral (6-1) becomes

(7-2) $$\int_{-\infty}^{x} [(v_+,v_+) - (v_-,v_-)]dt .$$

In the following definition, $L_2(H)$ denotes the space of all (equivalence classes of) H-valued locally Lebesgue-integrable functions f on R such that $\int_{-\infty}^{\infty} ||f(t)||^2 dt < \infty$; moreover, 1_+ denotes the Heaviside unit step function.

Definition 7-1

Let $\mathcal{W}(H)$ *be a set of* H*-valued functions on* R *contained in the domain of an operator* $\mathfrak{W}$. $\mathfrak{W}$ *is said to be scatter-passive on* $\mathcal{W}(H)$ *if, for every* $x \in R$, *for every* $v_+ \in \mathcal{W}(H)$, *and for* $v_- = \mathfrak{W} v_+$, *we have that* $v_+(t)1_+(t-x) \in L_2(H)$, $v_-(t)1_+(t-x) \in L_2(H)$, *and* (7-2) *is nonnegative.* $\mathfrak{W}$ *is said to be scatter-passive-at-infinity on* $\mathcal{W}(H)$ *if, for every* $v_+ \in \mathcal{W}(H)$ *and for* $v_- = \mathfrak{W}v_+$, *we have that* $v_+ \in L_2(H)$, $v_- \in L_2(H)$, *and*

(7-3) $$\int_{-\infty}^{\infty} [(v_+,v_+) - (v_-,v_-)]dt$$

is nonnegative.

If $\mathfrak{W}$ is a linear mapping of $\mathcal{D}(H)$ into $[\mathcal{D}; H]$ and is scatter-passive-at-infinity on $\mathcal{D}(H)$, then $\mathfrak{W}$ is also continuous from $\mathcal{D}(H)$ into $[\mathcal{D}; H]$ (see [5; Sec. 3]).

There is an interesting relationship between scatter-passivity and the two properties of causality and scatter-passivity-at-infinity. It was discovered by Wohlers and Beltrami for the scalar case [22]. Its extension to Hilbert ports, which is stated in the next theorem, is established in [5; Sec. 3].

Theorem 7-1

Assume that $\mathfrak{W}$ *is a linear time-invariant mapping of* $\mathcal{D}(H)$ *into* $[\mathcal{D}; H]$. *Then,* $\mathfrak{W}$ *is a scatter-passive on* $\mathcal{D}(H)$ *if and only if* $\mathfrak{W}$ *is causal and scatter-passive-at-infinity on* $\mathcal{D}(H)$.

For the frequency-domain description of our scattering formulism, we will need

Definition 7-2

Given a complex Hilbert space H, *a function* S *of the complex variable* ζ *is said to be a bounded* mapping of* H *into* H *(or simply bounded*) if, on the half plane* $C_+ \triangleq \{\zeta : \mathrm{Re}\zeta > 0\}$, S *is an* $[H; H]$*-valued analytic function such that* $||S(\zeta)||_{[H;H]} \leq 1$.

A description for the scattering formulism of a passive Hilbert port [5; Theorems 4-2 and 5-1] is given by

Theorem 7-2

Assume that $\mathfrak{W}$ *is a linear time-invariant causal scatter-passive-at-infinity mapping of* $\mathcal{D}(H)$ *into* $[\mathcal{D}; H]$. *Then,* $\mathfrak{W}$ *has a convolution representation* $\mathfrak{W} = s*$, *where* $s \in [\mathcal{D}_{L_2}(H); H]$ *and* $\mathrm{supp}\ s \subset [0,\infty)$. *Moreover,* s *possesses a Laplace transform* S *on at least* C_+ *which is* *bounded*.*

Conversely, assume that S *is bounded*. Then, there exists a unique convolution operator* $\mathfrak{W} = s*$ *such that* $s \in [\mathcal{D}_{L_2}(H); H]$, supp $s \subset [0,\infty)$, *and* $\mathfrak{L}\, s = S$ *on* C_+. *Moreover,* $\mathfrak{W}$ *is a continuous linear time-invariant causal scatter-passive-at-infinity mapping of* $\mathcal{D}(H)$ *into* $[\mathcal{D}; H]$.

Now, for bounded*-reality [5; Corollaries 4-2a and 5-1a]. Assume once again that H is generated from a real Hilbert space H_r through complexification.

Definition 7-3

Given H *and* H_r *as stated, a function* S *of the complex variable* ζ *is called bounded*-real if it is a bounded* mapping of* H *into* H *and, for each real positive number* σ, $S(\sigma)$ *is real.*

Corollary 7-2a

Theorem 7-1 remains valid when H *is replaced by* H_r, *"bounded*" by "bounded*-real", and* $\mathcal{D}$ *by* $\mathcal{D}(R)$.

We end this section by stating the connection between the admittance and scattering formulisms. Given any Hilbert port, whose operators need not satisfy any assumptions of linearity, continuity, etc., we see immediately from (7-1) that the admittance operator $\mathfrak{N} : v \mapsto u$ uniquely determines and is uniquely determined by the scattering operator $\mathfrak{W} : v_+ \mapsto v_-$. In this case, either one or both of these operators may be multivalued. However, when the aforementioned assumptions are imposed, we get the following theorem (see 5; Theorem 6-1 and 6-2]).

Theorem 7-3

If the admittance operator $\mathfrak{N} : v \to u$ *of a Hilbert port is a continuous linear time-invariant passive mapping of* $\mathcal{D}(H)$ *into* $[\mathcal{D}; H]$, *then its scattering operator* $\mathfrak{W} : v_+ \to v_-$ *is a linear time-invariant causal scatter-passive-at-infinity mapping of* $\mathcal{D}(H)$ *into* $[\mathcal{D}; H]$.

The converse statement is also true if the scattering transform $S \triangleq \mathfrak{L} s$ *corresponding to* $\mathfrak{W} = s*$ *is such that, for every* $\zeta \in C_+$, $(I + S)^{-1}$ *exists. Here,* I *denotes the identity operator on* H.

8. ∞-PORTS

Henceforth, we assume that the complex Hilbert space H is separable. This allows us to exploit the isomorphism between any such Hilbert space and the space ℓ_2 in order to introduce an infinite-dimensional extension to the concept of an n-port. We also assume throughout the rest of this paper that an orthonormal basis $\{\theta_k\}_{k=1}^{\infty}$ has been chosen in H. (When analysing systems such as microwave-transmission networks, it is natural to fix upon the orthonormal basis generated by a modal analysis.)

Lemma 8-1

Let $v \in \mathcal{D}(H)$. *Then, in the sense of convergence in* $\mathcal{D}(H)$,

$$v = \sum_{k=1}^{\infty} (v, e_k) e_k \tag{8-1}$$

Proof

Since, for any nonnegative integer p and any fixed t, $D^p v(t) \in H$, we have that

$$D^p v = \sum_{k=1}^{\infty} (D^p v, e_k) e_k \ .$$

By Parseval's equation,

$$\left\| \sum_{k=m+1}^{\infty} (D^p v, e_k) e_k \right\|_H^2 = \sum_{k=m+1}^{\infty} |(D^p v, e_k)|^2$$

As $m \to \infty$, the right-hand side tends to zero monotonically at every point of R. By Dini's theorem [23, p. 117], it therefore tends to zero uniformly on every compact interval in R. But, $D^p v$ has a compact support. Therefore,

$$\sum_{k=1}^{m} (v,e_k)e_k$$

tends to v in $\mathcal{D}(H)$ since it and every one of its derivatives have their supports contained in a fixed compact interval.

Lemma 8-2

Let $u \in [\mathcal{D}; H]$. *Then, in the sense of convergence in* $[\mathcal{D}; H]$,

$$u = \sum_{k=1}^{\infty} (u,e_k)e_k \quad . \tag{8-2}$$

Moreover, the series is unique; that is, if two such series converge in $[\mathcal{D}; H]$ *to the same limit, they must have the same coefficients.*

Proof

We first note that, for each k, $(u,e_k) \in [\mathcal{D}; C]$ according to [4; Sec. 3] so that $(u,e_k)e_k \in [\mathcal{D}; H]$. Moreover, for each $\phi \in \mathcal{D}$, we have that $\langle u,\phi\rangle \in H$ and

$$(\langle u,\phi\rangle, e_k)e_k = \langle (u,e_k), \phi\rangle e_k = \langle (u,e_k)e_k, \phi\rangle$$

again according to [4; Sec. 3]. Hence, we may set up the orthonormal series expansion:

$$\langle u,\phi\rangle = \sum_{k=1}^{\infty} (\langle u,\phi\rangle, e_k)e_k = \sum_{k=1}^{\infty} \langle (u,e_k)e_k, \phi\rangle \tag{8-3}$$

We wish to show that this series converges uniformly with respect to all ϕ in any given bounded set Ω in $\mathcal{D}$. Since $\mathcal{D}$ is a Montel space [8; p. 357], the closure $\overline{\Omega}$ of Ω is a compact set in $\mathcal{D}$. So, we need merely establish the uniformity of the convergence on any compact set Λ in $\mathcal{D}$.

Consider the function F_m on $\mathcal{D}$ into R defined by

$$F_m(\phi) \triangleq ||\langle u - \sum_{k=1}^{m-1} (u,e_k)e_k, \phi\rangle||^2 = \sum_{k=m}^{\infty} (\langle u,\phi\rangle, e_k)e_k||^2$$

F_m is continuous on $\mathcal{D}$. Moreover, by Parseval's equation,

$$F_m(\phi) = \sum_{k=m}^{\infty} |(\langle u,\phi\rangle, e_k)|^2 ,$$

and therefore, for each $\phi \in \mathcal{D}$, $F_m(\phi)$ tends monotonically to zero as $m \to \infty$. By the standard argument, we can conclude that $F_m(\phi)$ tends to zero uniformly on any compact set Λ in $\mathcal{D}$. This turn implies that the series in (8-3) converges uniformly on Λ.

The uniqueness of the expansion follows from the fact that, for any $a \in H$, the mapping $u \mapsto (u,a)$ is a continuous linear mapping of $[\mathcal{D}; H]$ into $[\mathcal{D}; C]$ (see [4; Sec. 3] again). Indeed, we need merely set $a = e_p$ and then apply this mapping to both sides of

$$u = \sum_{k=1}^{\infty} b_k e_k ,$$

where $b_k \in [\mathcal{D}; C]$, doing this term by term on the right-hand side, to get $b_p = (u, e_p)$.

Lemma 8-3

Let $\mathfrak{N}$ *be a continuous linear mapping of* $\mathcal{D}(H)$ *into* $[\mathcal{D}; H]$. *Then, for every choice of the positive integers* j *and* k, *there exists an* $f_{j,k} \in [\mathcal{D}_{t,x}; C]$ *such that, for all* $\phi \in \mathcal{D}$,

$$\mathfrak{N}\phi e_k = \sum_{j=1}^{\infty} (f_{j,k} \cdot \phi) e_j \tag{8-4}$$

where the series converges in $[\mathcal{D}; H]$.

Proof

Since $\mathfrak{N}\phi e_k \in [\mathcal{D}; H]$, we can expand it according to

Lemma 8-2 to obtain

$$\mathfrak{N}\phi e_k = \sum_{j=1}^{\infty} (\mathfrak{N}_{j,k}\phi)e_j \tag{8-5}$$

where

$$\mathfrak{N}_{j,k}\phi \triangleq (\mathfrak{N}\phi e_k, e_j) .$$

$\mathfrak{N}_{j,k}$ maps $\mathcal{D}$ into $[\mathcal{D}; C]$ by [4; Sec. 3].

In addition, $\mathfrak{N}_{j,k}$ is both linear and continuous. Indeed, its linearity being clear, consider its continuity. Assume that $\phi_\nu \to 0$ in $\mathcal{D}$ as $\nu \to \infty$. Then, $\phi_\nu e_k \to 0$ in $\mathcal{D}(H)$, and therefore $\mathfrak{N}(\phi_\nu e_k) \to 0$ in $[\mathcal{D}; H]$. We may also write for any $\psi \in \mathcal{D}$

$$\begin{aligned} |<\mathfrak{N}_{j,k}\phi_\nu, \psi>| &= |<(\mathfrak{N}\phi_\nu e_k, e_j), \psi>| \\ &= |(<\mathfrak{N}\phi_\nu e_k, \psi>, e_j) \le ||<\mathfrak{N}\phi_\nu e_k, \psi>|| \end{aligned} \tag{8-6}$$

because $||e_j|| = 1$. Since $\phi_\nu e_k \to 0$ in $[\mathcal{D}; H]$ as $\nu \to \infty$, the right-hand side of (8-6) tends to zero uniformly for all ψ in any bounded set in $\mathcal{D}$. This proves the asserted continuity of $\mathfrak{N}_{j,k}$.

We may now invoke Theorem 4-3 with A = B = C to conclude that $\mathfrak{N}_{j,k}\phi = f_{j,k} \cdot \phi$. (Here, $f_{j,k}$ is defined as in (8-7) below.) Inserting this result into (8-5), we complete the proof.

We are at last ready to establish for the mapping $\mathfrak{N}$ a composition• representation that is amenable to an ∞-port interpretation.

Theorem 8-1

If $\mathfrak{N}$ *is a continuous linear mapping of* $\mathcal{D}(H)$ *into* $[\mathcal{D}; H]$, *then there exists a collection* $\{f_{j,k}\}_{j,k}$ *of distributions in* $[\mathcal{D}_{t,x}, C]$ *defined by*

$$<f_{j,k}(t,x), \psi(t)\phi(x)> \triangleq <\mathfrak{N}_{j,k}\phi, \psi> \triangleq <(\mathfrak{N}\phi e_k, e_j), \psi>$$

$$\phi \in \mathcal{D};\ \psi \in \mathcal{D};\ j = 1, 2, \ldots;\ k = 1, 2, \ldots \tag{8-7}$$

such that, for any $v \in \mathcal{D}(H)$,

(8-8)
$$\mathfrak{N}v = \sum_{k=1}^{\infty} \sum_{j=1}^{\infty} [f_{j,k} \cdot (v, e_k)] e_j = \sum_{j=1}^{\infty} \sum_{k=1}^{\infty} [f_{j,k} \cdot (v, e_k)] e_j$$

where the series converge in $[\mathcal{D}; H]$.

Proof

We may apply $\mathfrak{N}$ term by term to the series indicated in Lemma 8-1 to get

$$\mathfrak{N}\, v = \sum_{k=1}^{\infty} \mathfrak{N}(v, e_k) e_k$$

Upon replacing ϕ by (v, e_k) in (8-4) and invoking Lemma 8-3, we can rewrite the last equation as

(8-9)
$$\mathfrak{N}\, v = \sum_{k=1}^{\infty} \sum_{j=1}^{\infty} [f_{j,k} \cdot (v, e_k)] e_j \ ,$$

which is the first series in (8-8). Here, it is understood that we sum first on j and then on k to obtain in both cases a limit in $[\mathcal{D}; H]$. To show that this order of summation can be reversed, we expand $\mathfrak{N}\, v \in [\mathcal{D}; H]$ into a series according to Lemma 8-2 to get

(8-10)
$$\mathfrak{N}\, v = \sum_{j=1}^{\infty} (\mathfrak{N}\, v, e_j) e_j \ .$$

Upon applying the operator $u \to (u, e_p)$ to (8-9) as in the proof of Lemma 8-2, we obtain

$$(\mathfrak{N}\, v, e_p) = \sum_{k=1}^{\infty} [f_{p,k} \cdot (v, e_k)].$$

Then, substituting this result into (8-10) with p replaced by j, we arrive at the second series in (8-8).

We can interpret the representation (8-8) in terms of an ∞-port. Think of a black box to whose interior we have access only through a collection of electrical ports which are countably infinite in number. Number these ports 1, 2, 3, ... Then, given the $\mathfrak{N}$ of Theorem 8-1 and any $v \in \mathcal{D}(H)$, set $u = \mathfrak{N} v$. Also, for each positive integer j, assume that the voltage impressed on the jth port is $v_j = (v, e_j) \in \mathcal{D}$ so that the corresponding current is

$$u_j = \sum_{k=1}^{\infty} f_{j,k} \cdot v_k \in [\mathcal{D};\ C] .$$

The operator that maps the vector $\{v_k\}_{k=1}^{\infty}$ into the vector $\{u_j\}_{j=1}^{\infty}$ can be represented by an $\infty \times \infty$ matrix $[f_{j,k}]$. In applying this matrix to any $\{v_k\}$ to get $\{u_j\}$, we follow the customary rule for the multiplication of matrices. Thus, the matrix equation corresponding to the composition representation (8-8) is

$$\{u_j\} = [f_{j,k}] \cdot \{v_k\} . \tag{8-11}$$

It represents the behavior at the ports of a time-varying ∞-port corresponding to the Hilbert port whose admittance operator is the $\mathfrak{N}$ of Theorem 8-1 and to the given choice of the orthonormal basis $\{e_k\}$.

We now take up the case where $\mathfrak{N}$ commutes with the shifting operator and develops the matrix representation of a time-invariant ∞-port. We first note that each $\mathfrak{N}_{j,k}$ in (8-7) also commutes with the shifting operator. Indeed, for any $\tau \in R$, any $\psi \in \mathcal{D}$ and any $f \in [\mathcal{D};\ C]$, we have that

$$\langle \sigma_\tau (f, e_j), \psi \rangle = \langle (f, e_j), \sigma_{-\tau}\psi \rangle = (\langle f, \sigma_{-\tau}\psi \rangle, e_j)$$

$$= (\langle \sigma_\tau f, \psi \rangle, e_j) = \langle (\sigma_\tau f, e_j), \psi \rangle .$$

Hence, in the sense of equality in $[\mathcal{D};\ C]$, we have for any $\phi \in \mathcal{D}$,

$$\sigma_\tau \mathfrak{N}_{j,k}\phi = \sigma_\tau (\mathfrak{N} \phi e_k, e_j) = (\sigma_\tau \mathfrak{N}\phi e_k, e_j)$$

$$= (\mathfrak{N} \sigma_\tau \phi e_k, e_j) = \mathfrak{N}_{j,k}\sigma_\tau\phi ,$$

as was asserted.

By virtue of Theorem 5-1 with A = B = C, the composition• representations in Theorem 8-1 become convolution representations, and we have

Corollary 8-1a

If $\mathfrak{N}$ *is a continuous linear time-invariant mapping of* $\mathcal{D}(H)$ *into* $[\mathcal{D}; H]$, *then there exists a collection* $\{y_{j,k}\}$ *of distributions in* $[\mathcal{D}; C]$, *defined by*

(8-12)

$$y_{j,k} * \phi \triangleq \mathfrak{N}_{j,k}\phi \triangleq (\mathfrak{N}\phi e_k, e_j) = (y * \phi e_k, e_j)$$

$$\phi \in \mathcal{D};\ j = 1, 2, \ldots;\ k = 1, 2, \ldots$$

such that, for any $v \in \mathcal{D}(H)$,

(8-13) $$\mathfrak{N} v = \sum_{j=1}^{\infty} \sum_{k=1}^{\infty} [y_{j,k} * (v, e_k)] e_j$$

where the series converges in $[\mathcal{D}; H]$ *and the order of summation can be reversed.*

The matrix representation for the admittance equation of a time-invariant ∞-port corresponding to the chosen $\{e_k\}$ and the Hilbert port whose admittance operator is the $\mathfrak{N}$ of this corollary, is

(8-14) $$\{u_j\} = [y_{j,k}] * \{v_k\} .$$

Here again, we use the customary rules for matrix multiplication.

Finally, we turn to the case where $\mathfrak{N}$ is not only continuous, linear, and time-invariant as a mapping of $\mathcal{D}(H)$ into $[\mathcal{D}; H]$ but also passive on $\mathcal{D}(H)$. As was stated in Theorem 6-1, this implies that $y \in [\mathcal{D}_{L_1}(H); H]$ and that $\operatorname{supp} y \subset [0,\infty)$. Let us show that these conditions on y imply that, for each j and k, $y_{j,k} \in [\mathcal{D}_{L_1}; C]$ and $\operatorname{supp} y_{j,k} \subset [0,\infty)$.

Indeed, first consider $y * (e_k\delta)$. As before, we drop the parentheses in this expression and simply write $y * e_k\delta$. By the standard definition of convolution, we have for any $\phi \in \mathcal{D}_{L_1}$ the expression

$$\begin{aligned} \langle y * e_k\delta, \phi\rangle &\triangleq \langle y(t), \langle e_k\delta(x), \phi(t+x)\rangle\rangle \\ &= \langle y(t), e_k\phi(t)\rangle = \langle y,\phi\rangle e_k \end{aligned} \tag{8-15}$$

In the right-hand side, y now denotes a member of $[\mathcal{D}_{L_1}; [H; H]]$. Thus, $y * e_k\delta$ is a mapping of $\mathcal{D}_{L_1}$ into H. Moreover, this mapping is both linear and continuous, as is easily seen from the right-hand side of (8-15). Hence, $y * e_k\delta \in [\mathcal{D}_{L_1}; H]$.

Next, consider $y_{j,k} = (y * e_k\delta, e_j)$. We get this expression from (8-12) by letting ϕ converge in $[\mathcal{E}; C]$ to δ and invoking the continuity of the operators $y_{j,k}*$ and $y*$ on $[\mathcal{E}; C]$ and $[\mathcal{E}; H]$ respectively. So, again for any $\phi \in \mathcal{D}_{L_1}$, we have from [4; Sec. 3] that

$$\langle y_{j,k}, \phi\rangle = \langle (y * e_k\delta, e_j), \phi\rangle = (\langle y * e_k\delta, \phi\rangle, e_j) . \tag{8-16}$$

This shows that $y_{j,k}$ maps $\mathcal{D}_{L_1}$ into C in a continuous linear fashion. That is, $y_{j,k} \in [\mathcal{D}_{L_1}; C]$.

As for the support of $y_{j,k}$, let $\phi \in \mathcal{D}$ and rewrite (8-12) as a regularization process [4; Theorem 4-3].

$$\langle y_{j,k}(x), \phi(t-x)\rangle = (\langle y(x), \phi(t-x)e_k\rangle, e_j) \tag{8-17}$$

Given any $\psi \in \mathcal{D}$ with supp $\psi \subset (-\infty,0)$, we can choose $t \in R$ and $\phi \in \mathcal{D}$ such that $\phi(t-x) = \psi(x)$ for all $x \in R$. Since supp $y \subset [0,\infty)$, it follows immediately that the right-hand side of (8-17) is equal to zero. Thus, supp $y_{j,k} \subset [0,\infty)$.

We noted in Sec. 3 that these properties of y and $y_{j,k}$ imply that they have Laplace transforms Y and $Y_{j,k}$ respectively on at least the half-plane $C_+ \triangleq \{\zeta : \mathrm{Re}\zeta > 0\}$. We can relate $Y_{j,k}$ to Y as follows. For any $\zeta \in C_+$,

$$Y_{j,k}(\zeta) = \langle y_{j,k}, e^{-\zeta t}\rangle = \langle (y * e_k\delta, e_j), e^{-\zeta t}\rangle$$
$$= (\langle y * e_k\delta, e^{-\zeta t}\rangle, e_j) \ .$$

The last equality holds because $y * e_k\delta \in [L_{c,d}; H]$ for any $c \in R$ with $c > 0$, and any $d \in R$ and moreover $\text{supp } y * e_k\delta \subset [0,\infty)$. (See again [4; Sec. 3].) Thus,

$$Y_{j,k}(\zeta) = (\langle y(t), \langle e_k\delta(x), e^{-\zeta(t + x)}\rangle\rangle, e_j)$$
$$= (\langle y(t), e_k e^{-\zeta t}\rangle, e_j)$$

or

$$(8\text{-}18) \qquad Y_{j,k}(\zeta) = (Y(\zeta)\, e_k, e_j) \qquad \zeta \in C_+ \ .$$

In the following ℓ_2 represents the standard Hilbert space of all sequences $\alpha = \{\alpha_k\}$ of complex numbers for which the norm

$$||\alpha|| = \left(\sum_{k=1}^{\infty} |\alpha_k|^2\right)^{\frac{1}{2}}$$

exists. We know from Theorem 6-1 that, under our stated assumptions on $\mathfrak{N}$, $Y(\zeta) \in [H; H]$ for each fixed $\zeta \in C_+$. Having fixed upon the orthonormal basis $\{e_k\}$ in H, we construct an $\infty \times \infty$ matrix $[Y_{j,k}(\zeta)]$ that represents the operator in $[\ell_2; \ell_2]$ corresponding to $Y(\zeta) \in [H; H]$ under the isomorphism existing between H and ℓ_2. The elements of this matrix are given precisely by (8-18). (See [24; Sec. 3.1].) Moreover, if $v \in [\mathcal{D}; H]$ has a Laplace transform V whose strip of definition contains the chosen $\zeta \in C_+$, then the Laplace transform of the equation (8-14) yields

$$(8\text{-}19) \qquad \{U_j(\zeta)\} = [Y_{j,k}(\zeta)]\, \{V_k(\zeta)\}$$

where $\{U_j\}$ and $\{V_j\}$ denotes the componentwise transformations of $\{u_j\}$ and $\{v_k\}$ respectively and, for the given ζ, $\{U_j(\zeta)\} \in \ell_2$ and $\{V_k(\zeta)\} \in \ell_2$.

We conclude this section by relating the positivity*

of Y to the positivity* of $[Y_{j,k}]$.

Theorem 8-2

Y *is a positive* mapping of* H *into* H *if and only if* $[Y_{j,k}]$, *as defined by* (8-18), *is a positive* mapping of* ℓ_2 *into* ℓ_2. *Thus,* Y *and* $[Y_{j,k}]$ *are positive* if and only if* $\mathfrak{N}$ *is a continuous linear time-invariant passive mapping of* $\mathcal{D}(H)$ *into* $[\mathcal{D}; H]$.

Proof

We have already noted that, for any fixed $\zeta \in C_+$, $Y(\zeta) \in [H; H]$ if and only if $[Y_{j,k}(\zeta)] \in [\ell_2; \ell_2]$.

Now let a and b be arbitrary members of H and let $\{a_k\}$ and $\{b_k\}$ be the corresponding members of ℓ_2 (i.e., $\{a_k\}$ is the sequence of Fourier coefficients of a with respect to $\{e_k\}$). Then, we have that

$$(Ya,b) = ([Y_{j,k}]\ \{a_k\},\ \{b_j\}) \ . \tag{8-20}$$

Thus, Y is weakly analytic on C_+ if and only if $[Y_{j,k}]$ is weakly analytic on C_+. But, weak analyticity is equivalent to analyticity in the norm topology [25; p. 93].

We check the nonnegativity condition and thereby complete the proof of the first sentence of this theorem by setting a = b in (8-20) and then taking real parts. The second sentence follows immediately from Theorem 6-1.

Finally, assume once again that H is obtained from a real separable Hilbert space H_r through complexification and assume that the orthonormal basis $\{e_k\}$ is contained in H_r. Through the isomorphism between H and ℓ_2 we have that H_r corresponds to the subspace $\ell_{2,r}$ of ℓ_2, where $\ell_{2,r}$ consists of all sequences of real numbers in ℓ_2. Thus, for any $\sigma > 0$, $[Y_{j,k}(\sigma)]$ maps $\ell_{2,r}$ into $\ell_{2,r}$ if and only if $Y(\sigma)$ maps H_r into H_r. This allows us to state

Corollary 8-2a

The first sentence of Theorem 8-2 *remains true if*

"positive" is replaced by "positive*-real." In this case, for every* $\sigma > 0$ *and every* j *and* k, $Y_{j,k}(\sigma)$ *is a real number. In addition, the second sentence of Theorem* 8-2 *remains true if "positive*" is replaced by "positive*-real,"* H *by* H_r *and* $\mathcal{D}$ *by* $\mathcal{D}(R)$.

We have treated only the admittance formulism of the ∞-port. An analysis of the scattering formulism can also be made, but, since it is quite similar to the foregoing, we omit it.

8½. A FIRST THRUST AT THE SYNTHESIS OF AN ∞-PORT

Given a positive*-real mapping $[Y_{j,k}]$ of ℓ_2 into ℓ_2, can one synthesize an ∞-port to realize it? Here are a few thoughts on the subject.

First of all, let it be said that we are trying to synthesize a "paper network" [3], which is a perfectly legitimate mathematical idea that can only be approximated by a physical system. (So too is the ideal one-ohm resistor.)

Assume once again that H is the complexification of a real separable Hilbert space H_r and that $\{e_k\} \subset H_r$. Let $[Y_{j,k}]_n$ be the $\infty \times \infty$ matrix obtained from $[Y_{j,k}]$ by replacing each element $Y_{j,k}$ in $[Y_{j,k}]$ by 0 if either $j > n$ or $k > n$. This corresponds to the following alteration of the ∞-port whose admittance matrix is $[Y_{j,k}]$: For every port beyond the nth, disconnect the wires to the port terminals and short them together. The resulting system which is in fact an n-port, possesses $[Y_{j,k}]_n$ as its admittance matrix. $[Y_{j,k}]_n$ can be identified with the $n \times n$ matrix obtained by dropping all rows and columns in $[Y_{j,k}]$ beyond the nth row and n^{th} column, and the latter is a positive-real $n \times n$ matrix in the usual sense. Indeed, let $a = \{a_k\}$ be any member of ℓ_2 such that $a_k = 0$ if $k > n$. Then, for $\zeta \in C_+$,

$$\text{Re} \sum_{j=1}^{n} \sum_{k=1}^{n} Y_{j,k}(\zeta)\bar{a}_j a_k = \text{Re}([Y_{j,k}]\{a_k\}, \{a_j\}) \geq 0$$

by virtue of the positivity* of $[Y_{j,k}]$. The analyticity of each $Y_{j,k}(\zeta)$ on C_+ and the reality of $Y_{j,k}(\sigma)$ for $\sigma > 0$ is equally clear.

If we now assume in addition that every $Y_{j,k}$ is a rational function of ζ, then we can apply known synthesis procudures [26] to realize $[Y_{j,k}]_n$ as an ∞-port whose first n ports connect to a lumped passive network and whose ports beyond the nth all have broken terminal wires. Thus, we have synthesized in this way a sequence $\{[Y_{j,k}]_n\}_{n=1}^{\infty}$ of ∞-ports.

$\{[Y_{j,k}]_n\}_{n=1}^{\infty}$ is an approximating sequence for the originally given $[Y_{j,k}]$ in the sense that, for each fixed $\zeta \in C_+$, $[Y_{j,k}(\zeta)]_n \to [Y_{j,k}(\zeta)]$ in the weak topology of $[\ell_2; \ell_2]$. To show this, let $a = \{a_k\}$ and $b = \{b_k\}$ be arbitrary members of ℓ_2. Then,

$$|([Y_{j,k}(\zeta)]a, b) - ([Y_{j,k}(\zeta)]_n a, b)|$$

$$= \left| \left(\sum_{j=n+1}^{\infty} \sum_{k=1}^{n} + \sum_{j=1}^{\infty} \sum_{k=n+1}^{\infty} \right) Y_{j,k}(\zeta) a_k \bar{b}_j \right|$$

By applying the Schwarz inequality to the summations on j, we bound the last expression by

$$\left(\sum_{j=n+1}^{\infty} |b_j|^2 \sum_{j=n+1}^{\infty} \left| \sum_{k=1}^{n} Y_{j,k}(\zeta) a_k \right|^2 \right)^{\frac{1}{2}} +$$

$$\left(\sum_{j=1}^{\infty} |b_j|^2 \sum_{j=1}^{\infty} \left| \sum_{k=n+1}^{\infty} Y_{j,k}(\zeta) a_k \right|^2 \right)^{\frac{1}{2}}$$

We now invoke a result [24; Sec. 3.1], which states that the $\infty \times \infty$ matrix $[W_{j,k}]$ represents a member of $[\ell_2; \ell_2]$ if and only if there exists a constant $M > 0$ such that, for every pair p,q of positive integers and every choice of the complex numbers $a_1, \ldots, a_p$,

$$\sum_{j=1}^{q} \left| \sum_{k=1}^{p} W_{j,k}(\zeta) a_k \right|^2 \le M^2 \sum_{k=1}^{p} |a_k|^2$$

Under the additional assumption that $\{a_k\} \in \ell_2$, we may take $p \to \infty$ and/or $q \to \infty$ and still obtain a valid inequality. In view of this fact, we have that

$$\left|([Y_{j,k}(\zeta)]a,\ b) - ([Y_{j,k}(\zeta)]_n a,\ b)\right| \le$$

$$\left(M^2 \sum_{j=n+1}^{\infty} |b_j|^2 \sum_{k=1}^{n} |a_k|^2 \right)^{\frac{1}{2}} + \left(M^2 \sum_{j=1}^{\infty} |b_j|^2 \sum_{k=n+1}^{\infty} |a_k|^2 \right)^{\frac{1}{2}}$$

The right-hand side tends to zero as $n \to \infty$, which establishes our assertion.

If it happens that, for the chosen $\zeta \in C_+$,

$$\sum_{j=1}^{\infty} \sum_{k=1}^{\infty} |Y_{j,k}(\zeta)|^2 < \infty ,$$

then, the convergence of $[Y_{j,k}(\zeta)]_n$ to $[Y_{j,k}(\zeta)]$ occurs in the norm topology of $[\ell_2;\ \ell_2]$. Indeed, if $b = [Y_{j,k}(\zeta)]a$, $a \in \ell_2$, then by the Schwarz inequality,

$$||b||^2 = \sum_{j=1}^{\infty} \left| \sum_{k=1}^{\infty} Y_{j,k}(\zeta)\ a_k \right|^2 \le \sum_{j=1}^{\infty} \sum_{k=1}^{\infty} |Y_{j,k}(\zeta)|^2 \sum_{k=1}^{\infty} |a_k|^2$$

$$= ||a||^2 \sum_{j=1}^{\infty} \sum_{k=1}^{\infty} |Y_{j,k}(\zeta)|^2$$

Therefore,

$$||[Y_{j,k}(\zeta)] - [Y_{j,k}(\zeta)]_n\ || \triangleq$$

$$\sup_{||a||=1} ||\{[Y_{j,k}(\zeta)] - [Y_{j,k}(\zeta)]_n\}\ a\ || \le$$

$$\left(\sum_{j=n+1}^{\infty} \sum_{k=1}^{n} |Y_{j,k}(\zeta)|^2 + \sum_{j=1}^{\infty} \sum_{k=n+1}^{\infty} |Y_{j,k}(\zeta)|^2 \right)^{\frac{1}{2}}$$

The right-hand side tends to zero as $n \to \infty$.

In summary, we have not obtained a synthesis of the given positive*-real $[Y_{j,k}]$ but instead have constructed an approximating sequence $\{[Y_{j,k}]_n\}$ whose members have n-port realizations whenever the $[Y_{j,k}]_n$ are rational.

REFERENCES

1. WOHLERS, M. R., (1969) *Lumped and Distributed Passive Networks*, Academic Press, New York.

2. YOULA, D. C., CASTRIOTA, L. J. and CARLIN, H. J., (September 10, 1957) *Scattering Matrices and the Foundations of Linear Passive Network Theory*, Report R-594-57, PIB-522, Polytechnic Institute of Brooklyn.

3. ZEMANIAN, A. H., (September, 1965 and March, 1966) Paper Networks, I and II, *IEEE Trans. on Circuit Theory CT-12*, 425-426 and *CT-13*, 110-111.

4. ZEMANIAN, A. H., The Hilbert Port, *SIAM J. App. Math.* (to appear).

5. ZEMANIAN, A. H., A Scattering Formalism for the Hilbert Port (to appear).

6. BOURBAKI, N., (1953) *Eléments de Mathématique*, Vol. 15, Herman, Paris.

7. ZEMANIAN, A. H., (1968) *Generalized Integral Transformations*, Interscience Publishers, John Wiley, New York.

8. TREVES, F., (1967) *Topological Vector Spaces, Distributions and Kernels*, Academic Press, New York.

9. BOGDANOWICZ, W., (1961) A Proof of Schwartz's Theorem on Kernels, *Studia Mathematica 20*, 77-85.

10. CRISTESCU, R., (1964) Familles Composables de Distributions et Systèmes Physiques Linèaires, *Rev. Roum. Math. Pures et Appl. 9*, 703-711.

11. CRISTESCU, R. and MARINESCU, G., (1966) *Unele Aplicattii ale Teoriei Distributiilor*, Editura Academiei Republicii Socialiste România, Bucuresti.

12. ŞABAC, M., (1965) Familii Compozabile de Distribuţii şi Transformata Fourier, *St. Cerc. Mat. 17*, 607-613.

13. WEXLER, D., (1966) Solutions Périodiques et Presque-périodiques des Systémes d'équations Différentielles Linéaires en Distributions, *J. Differential Equations 2*, 12-32.

14. CIORĂNESCU, I., (1967) Familii Compozabile de Operatori, *St. Cerc. Mat. 19*, 449-454.

15. PONDELICEK, B., A Contribution to the Foundation of Network Theory Using Distribution Theory, *Aplikace Matematiky* (to appear).

16. DOLEZAL, V., On Convergence of Testing Functions (to appear).

17. DINCULEANU, N., (1967) *Vector Measures*, Pergamon Press, New York.

18. SCHWARTZ, L., (1957 and 1959) Théorie des Distributions à Valeurs Vectorielles, *Annales de l'Institut Fourier 7*, 1-139 and *8*, 1-206.

19. ZEMANIAN, A. H., (1965) *Distribution Theory and Transform Analysis*, McGraw-Hill Book Co., New York.

20. ZEMANIAN, A. H., (1968) The Postulational Foundations of Linear Systems, *J. Math. Anal. Appl. 24*, 409-429.

21. BACHMAN, G. and NARICI, L., (1966) *Functional Analysis*, Academic Press, New York.

22. WOHLERS, M. R. and BELTRAMI, E. J. (1965) Distribution Theory as the Basis of Generalized Passive-network Analysis, *IEEE Trans. on Circuit Theory CT-12*, 164-170.

23. HOBSON, E. W., (1957) *The Theory of Functions of a Real Variable*, Vol. 2, Dover Publications, New York.

24. KANTOROVICH, L. V. and AKILOV, G. P., (1964) *Functional Analysis in Normed Spaces*, Pergamon Press, New York.

25. HILLE, E. and PHILLIPS, R. S., (1957) *Functional Analysis and Semi-groups*, American Mathematical Society Colloquium Publications, Vol. 31, Providence, Rhode Island.

26. NEWCOMB, R. W., (1966) *Linear Multiport Synthesis*, McGraw-Hill Book Co., New York.

[illegible] (1957) [illegible] Lubrication[illegible] New York.

[illegible] Pergamon Press, [illegible]

[illegible]

[illegible] Rhode Island.

[illegible]

SMALL SIGNAL THEORY OF NONLINEAR LUMPED NETWORKS

C. A. DESOER

University of California, Berkeley

Under very general conditions ([1], [2], [3], [7], [8], [10]), the equations of nonlinear *lumped* networks can be written in the form

$$\dot{x} = f(x, \tilde{u}, t) \qquad t \geq 0 \tag{1}$$

with

$$f(0, 0, t) = 0 \qquad \forall\, t \geq 0 \tag{2}$$

For simplicity, we assume throughout that inputs are applied at t = 0 and that

$$x(0) = 0 \tag{3}$$

We assume that

(F1) for every fixed $x \in R^n$, $u \in f(x, u, \cdot): R_+ \to R^n$ is regulated, for every fixed $u \in R^m$ and $t \in R_+$, f is locally Lipschitz in x.

To the network described by (1) we apply a "bias" U: $R_+ \to R^m$ assumed to be regulated. Corresponding to this bias is the *operating point* X(t), which for each $t \geq 0$ is the solution of

$$\dot{X}(t) = f(X(t), U(t), t) \tag{4}$$

$$X(0) = 0 \tag{5}$$

We think of $t \mapsto X(t)$ as of the *nominal trajectory*.

Suppose the input $\tilde{u}$ in (1) consists of the bias U and a "small signal" u:

$$\tilde{u} = U + u \tag{6}$$

The exact solution corresponding to the input $\tilde{u}$ given by (6) is

$$\dot{x} = f(x, U + u, t), \quad x(0) = 0 \tag{7}$$

We think of x as the sum

$$x(t) = X(t) + \xi(t) \tag{8}$$

By Taylor expansion of the right-hand side of (7),

$$f(X(t) + \xi(t), U(t) + u(t), t) = f(X(t), U(t), t)$$

$$+ D_1 f(X(t), U(t), t) \cdot \xi + D_2 f(X(t), U(t), t) \cdot u + g(\xi, u, t)$$

where $D_i f$ denotes the derivative of f with respect to its ith argument, g represents "second-order terms".

Let ξ_m, u_m be (finite) real numbers, define $C(\xi_m, u_m)$ to be the cylinder

$$C(\xi_m, u_m) = \left\{ (\xi, u, t) \;\middle|\; |\xi| \le \xi_m, \ |u| \le u_m, \ t \ge 0 \right\}$$

where $|\cdot|$ denotes an appropriate norm (chosen once and for all) in R^n or R^m. If the mapping f is C^2 in the cylinder C, then there is some finite S such that for all $(\xi, u, t) \in C$,

$$|g(\xi, u, t)| \le (|\xi|^2 + |u|^2)\, S \tag{9}$$

Using obvious notations, the *exact* equation for ξ, defined by (8), is

$$\left.\begin{aligned} \dot{\xi}(t) &= A(t)\,\xi(t) + B(t)\,u(t) + g(\xi(t), u(t), t) \\ \xi(0) &= 0 \end{aligned}\right\} \tag{10}$$

The *linearized* equation is

$$\left.\begin{aligned} \dot{\xi}_0(t) &= A(t)\ \xi_0(t) + B(t)\ u(t) \\ \xi_0(0) &= 0 \end{aligned}\right\} \tag{11}$$

Note that

$$\xi_0(t) = \int_0^t \Phi(t,\tau)\ B(\tau)\ u(\tau)\ d\tau \tag{12}$$

where $\Phi(\cdot,\cdot)$ denotes the state transition matrix of the first equation (11).

The question we consider below is: under what conditions can we guarantee that for all $t \geq 0$ $|\xi(t) - \xi_0(t)|$ is small? The precise answer is given by the theorem below. We use sup norms throughout:

$$||x|| = \sup_{t \geq 0} |x(t)|, \qquad ||u|| = \sup_{t \geq 0} |u(t)|$$

Theorem

Consider a nonlinear network described by Eqs. (1), (2) and (3). Define the operating point by (4), (5); let $\xi(\cdot)$ and $\xi_0(\cdot)$ be defined by (10) and (11). Assume that (F1) holds and (A1) there is an $M > 0$ such that $\forall\ t \geq 0$

$$\int_0^t |\Phi(t,r)|\ d\tau \leq M \tag{13}$$

(A2)

$$\sup_{t \geq 0} |B(t)| = B_m < \infty \tag{14}$$

(A3) for any $\varepsilon > 0$ there is a $\delta(\varepsilon) > 0$ such that $\forall\ t \geq 0$

$$(|\xi| + |u|) < 2\delta \Longrightarrow |g(\xi, u, t)| < (|\xi| + |u|)\,\varepsilon \tag{15}$$

Under these conditions, if

$$||u|| \leq \frac{1 - \varepsilon M}{M(B_m + \varepsilon)}\delta \leq \delta \tag{16}$$

then

(a) $||\xi_0|| \leq M\, B_m\, ||u|| \leq (1 - \varepsilon M)\,\delta \leq \delta$

(b) $||\xi|| \leq \delta$

(c) $\dfrac{||\xi - \xi_0||}{||u||} \leq \dfrac{\varepsilon M(1 + M\, B_m)}{1 - \varepsilon M}$

Remarks

I. If (1) were to read $x = f(x,t) + B(t)\,u$ then the conclusion (c) can be sharpened to

$$\frac{||\xi - \xi_0||}{||\xi_0||} \leq \frac{\varepsilon M}{1 - \varepsilon M}$$

II. To test assumptions (F1), (A2) and (A3) is straightforward; it requires only checking on the source waveforms and the branch characteristics of the network. (In this connection, note that if (9) holds, so does (A3)). To test (A1) requires in principle the solution of differential equations. A limiting case is worth mentioning: if for all $t \geq 0$ the eigenvalues of $A(t)$ (i.e., the natural frequencies of the "frozen" small signal network about the operating point $X(t)$) have real parts smaller than some fixed negative number and if $\dot{A}(t)$ is sufficiently small then (A1) is automatically satisfied. Rosenbrock [9] established that fact, and Desoer [4] gave explicit bounds on $|\dot{A}(t)|$ for that to be true.

III. Another approach would be to calculate the "*frozen operating point*"

$$f(X_0(t), U(t), t) = 0 \quad \forall\, t \geq 0$$

and linearize the equations about that point [5]. It is interesting to note that the linear equations require then a correction term, a sort of penalty for performing the expansion about the wrong operating point. In many instances, the frozen operating point can be obtained by inspection.

Proof

Pick ε small so that $\varepsilon M < 1$. If necessary increase B_m so that $(1 - \varepsilon M)/[M(B_m + \varepsilon)] < 1$. Assume *temporarily* that with the ε just chosen, it is true that for all $(\xi, u, t) \in R^n \times R^m \times R_+$

$$|g(\xi, u, t)| < (|\xi| + |u|)\varepsilon \tag{18}$$

Now

$$\xi(t) = \xi_0(t) + \int_0^t \Phi(t,\tau)\, g(\xi, u, \tau)\, d\tau \tag{19}$$

hence

$$|\xi(t)| \le |\xi_0(t)| + \varepsilon \int_0^t |\Phi(t,\tau)|(|\xi(\tau)| + |u(\tau)|)\, d\tau \tag{20}$$

Note that from (12), we obtain directly conclusion (a)

$$||\xi_0|| \le M\, B_m ||u||$$

We assert that for all $T > 0$, $||\xi_T|| < \infty$ where $||\xi_T|| \triangleq \sup_{0 \le t \le T} |\xi(t)|$: indeed, $A(\cdot)$ is regulated (hence bounded on any $[0,T]$), also by (18) and the Gronwall lemma, for any $T > 0$, $||\xi_T|| < \infty$. Hence from (20), for all $T > 0$,

$$||\xi_T|| \le M\, B_m\, ||u|| + \varepsilon M\, ||\xi_T|| + \varepsilon M\, ||u||$$

Thus, with $\varepsilon M < 1$,

$$||\xi_T|| < \frac{M(B_m + \varepsilon)}{1 - \varepsilon M} ||u|| \qquad \forall T \in R_+$$

Since the right-hand side is independent of T, we may let $T \to \infty$ and

$$||\xi|| < \frac{M(B_m + \varepsilon)}{1 - \varepsilon M} ||u||.$$

Thus (b) has been proved under the temporary assumption. Looking back over the derivation, we note that over R_+, $|\xi(t)| < \delta$ and $|u(t)| < \delta$, hence we need only require that (18) hold for all $(\xi, u, t) \in C(\delta, \delta) = \{(\xi,u,t) | \; |\xi| \leq \delta, |u| \leq \delta$ and $t \geq 0\}$. Hence (b) holds under the assumption of the theorem.

Using again (19),

$$|\xi(t) - \xi_0(t)| \leq \varepsilon \int_0^t |\Phi(t,\tau)| \; [|\xi(\tau)| + |u(\tau)|] \, d\tau$$

$$\leq \varepsilon M(||\xi|| + ||u||)$$

$$\leq \varepsilon M(||\xi-\xi_0|| + ||\xi_0|| + ||u||)$$

Hence

$$||\xi-\xi_0|| \leq \frac{1}{1 - \varepsilon M} \varepsilon M (||\xi_0|| + ||u||)$$

$$\leq \frac{\varepsilon M(1 + MB_m)}{1 - \varepsilon M} ||u||$$

This completes the derivation.

It is interesting to note that under the assumptions (13) and (14), the *relative* error between ξ, the *exact* solution of the *nonlinear* equations, and ξ_0, the solution of the linearized equation can be made arbitrarily small by taking u sufficiently small. Computer verification of this fact can be found in [6].

REFERENCES

1. BRAYTON, R.K. and MOSER, J.K. (April and July, 1964) A Theory of Nonlinear Networks, *Q. Appl. Math. 22*, pp. 1-33 and 81-104.

2. CHUA, L.O. and ROHRER, R.A. (December, 1965) On the Dynamic Equations of a Class of Nonlinear RLC Networks, *IEEE Trans. Circuit Theory CT-12*, pp. 475-489.

3. DESOER, C.A. and KATZENELSON, J.K. (January 1965) Nonlinear RLC Networks, *Bell Syst. Tech. J. 44*, pp. 161-198.

4. DESOER, C.A. Slowly Varying System $\dot{x} = A(t)x$, to appear in *Trans. IEEE AC-16*, 6, December 1969).

5. DESOER, C.A. and PEIKARI, B. The Frozen Operating Point Method of Small Signal Analysis (to appear in *IEEE Trans. on Circuit Theory*).

6. DESOER, C.A. and WONG, K.K. (January 1968) Small Signal Behavior of Nonlinear Lumped Networks, *Proc. IEEE 56 (1)*, pp. 14-22.

7. KUH, E.S. and ROHRER, R.A. (July 1965) The State-Variable Approach to Network Analysis, *Proc. IEEE 53 (7)*, pp. 672-686.

8. OHTSUKI, T. and WATANABE, H. (February 1969) State-Variable Analysis of RLC Networks Containing Nonlinear Coupling Elements, *IEEE Trans. CT-16, (1)*, pp. 26-39.

9. ROSENBROCK, H.H. (1963) The Stability of Linear Time-Dependent Control Systems, *Jour. Electr. and Contr. 15, (1)*, pp. 73-80.

10. STERN, T.E. (March 1966) On the Equations of Non-linear Networks, *Trans. IEEE CT-13*, pp. 74-81.

STABILITY OF LINEAR TIME-INVARIANT NETWORKS

C.A. DESOER

University of California, Berkeley

By well known flow graph techniques (Mason, 1956), any interconnection of multiple-input multiple-output linear time-invariant n-ports can be reduced to a single-loop feedback system shown in Fig. 1. Concerning this

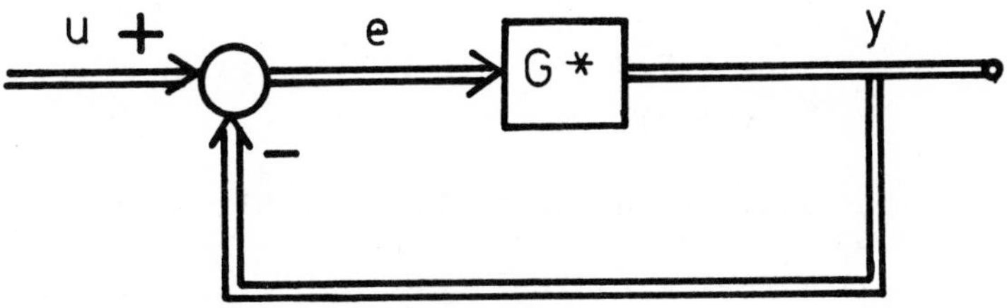

FIG. 1.

system we shall discuss the existence, uniqueness and continuous dependence of solutions; input-output properties, and applications.

This lecture is based on work carried out over the last few years with the purpose of extending Nyquist's result (1932) and elucidating the input-output properties of feedback systems. The generalization of the Nyquist criterion to the input-output point of view as well as to the classes of linear systems considered is due to Desoer (1965) and Desoer and Wu (1968). The implications of these results for nonlinear time-varying systems follow immediately from the work of Sandberg (1965) and Zames (1966); see for example Desoer and Wu (1969).

NOTATION

R: Set of real numbers; R_+: set of non-negative real numbers; R^2: set of ordered real n-tuples; $R^{n\times n}$: set of $n \times n$ matrices with real elements. $|\cdot|$ denotes the absolute value of a complex number, the norm of a vector in R^n and the induced norm of a matrix $\in R^{n\times m}$. $||\cdot||$ denotes the norm of a function mapping R into R^n, or the induced norm of a linear operator. $|| \, ||_p$ denotes the L^p norm.

We say that A is an *algebra over the field* F iff
(i) A is linear space over F,
(ii) Multiplication is defined: $(x,y) \mapsto x \cdot y$, from $A \times A$ into A
(iii) Addition and multiplication are distributive

$$x \cdot (y + z) = x \cdot y + x \cdot z,$$
$$(y + z) \cdot x = y \cdot x + z \cdot x$$

(iv) Multiplication and scalar multiplication commute

$$(\alpha x) \cdot (\beta y) = (\alpha\beta)\, x \cdot y \;\; \forall \alpha, \beta \in F, \forall x, y \in A$$

If multiplication is commutative, the Algebra is said to be commutative. Consider the class $\mathcal{A}$ of elements of the form: $g \in \mathcal{A}$ iff

$$g(t) = 0 \text{ for } t < 0, \; g(t) = g_a(t) + \sum_{i=0}^{\infty} g_i \, \delta(t - t_i) \quad \text{for } t \geq 0$$

where

$$g_a(\cdot) \in L^1, \; \sum_{i=0}^{\infty} |g_i| < \infty, \quad 0 = t_0 < t_1 < \ldots$$

If $g, u \in A$, the product is defined by convolution

$$(g * u)(t) = (g_a * u_a)(t) + \sum_i g_i\, u_a(t-t_i) + \sum_k u_k\, g(t-t_k') + \sum_{i,k} g_i\, u_k\, \delta(t-t_i-t_k').$$

With these operations, A is a commutative algebra with a unit element. If $g \in A$, then

$$||g||_A \triangleq \int_{\infty}^{0} |g_a(t)|dt + \sum_0^{\infty} |g_i|.$$

If $f \in L^p(R_+)$, $g \in A$,

$$||f * g||_p \leq ||g||_A\, ||f||_p.$$

If $G \in A^{n\times n}$,

$$||G||_A \triangleq \int_0^{\infty} |G_a(t)|\; dt + \sum_{\gamma=0}^{\infty} |G_\nu|$$

We denote Laplace transforms by ^, viz. if $g \in A$, $\hat{g}(s) = L(g)$ and we write $\hat{g} \in \hat{A}$. We use the subscript T to denote the result of the truncation:
if $f:R_+ \to R^n$,

$$f_T(t) = \begin{cases} f(t) & 0 \leq t \leq T \\ 0 & t > T \end{cases}$$

If $f:R_+ \to R^n$ is such that, for all $T > 0$, $f_T \in L_n^p$ then we say that $f \in L_{ne}^p$.

I. EXISTENCE, UNIQUENESS AND CONTINUOUS DEPENDENCE

Theorem I. (Desoer, 1968)

Let $u, e, y:R_+ \to R^n$, locally measurable. Let

$$G(t) = \begin{cases} 0 & \text{for } t < 0 \\ G_a(t) + \sum_0^\infty G_i\,\delta(t-t_i) & \text{for } t \geq 0 \end{cases} \tag{1}$$

where $G_a(\cdot)$ is locally integrable,
$G_i \in R^{n\times n}$ $(i = 0, 1,\ldots)$, $0 = t_0 < t_1 < t_2 \ldots$, and $t_k \to \infty$. Let p be fixed and $p \in [1,\infty]$. If $u(t) = 0$ for $t < 0$ and $u \in L^p_{ne}$, then, provided

$$\det(I + G_0) \neq 0 \tag{2}$$

the equation

$$u = e + G * e \tag{3}$$

has a *unique* solution which is in L^p_{ne} and the solution depends continuously on u in the sense of (8) below.

Proof

Without loss of generality (and this may require a little preliminary manipulation) we may assume that

$$|G_0| < 1 \tag{4}$$

Inequality (4) guarantees that we can choose some $T > 0$ $T \neq t_i$ $(i = 1, 2, \ldots)$, such that

$$\int_0^T |G(t)|\, dt \triangleq r < 1 \tag{5}$$

Restrict our attention to $[0,T]$ and construct the solution by iteration on $[0,T]$

$$x_T = u_T - G_T * x_T \triangleq f(x_T)$$

Now, $f: L^p[0,T] \to L^p[0,T]$, furthermore f is a contraction: indeed, for all $x_1, x_2 \in L^p[0,T]$,

$$\begin{aligned} ||f(x_1) - f(x_2)||_p &= ||G_T * x_1 - G_T * x_2||_p \\ &\le ||G_T||_A \, ||x_1 - x_2||_p \\ &\le r \, ||x_1 - x_2||_p \end{aligned}$$

Hence (Dieudonné, 1960) for any $x_0 \in L^p[0,T]$, the iterations

$$x_{i+1} = f(x_i) \qquad i = 0, 1, 2, \ldots$$

converge to *the unique* solution x_T and, because $L^p[0,T]$ is complete, $x_T \in L^p[0,T]$.

The next step is to construct the solution on $[T,2T]$, $[2T,3T]$, ... The procedure is obvious.

The continuous dependence of the truncated solutions is obvious: for the T above, consider any two inputs u, u' and *the* corresponding solutions, e, e'; then by (3)

$$(u-u')_T = (e-e')_T + G_T * (e-e')_T$$

and

$$||(e-e')_T||_p \le \frac{1}{1-r} \, ||(u-u')_T||_p$$

Remarks

1. If $u, e \in L^p_{ne}$ then $y \in L^p_{ne}$, since $y = u-e$.

2. If $u \in A^n$, then, provided $\det(I + G_0) \neq 0$, e and $y \in A^n$.

3. Not all convolution equations have a solution: Let $I \in \mathcal{D}'_+$, $\phi \in \mathcal{D}_+$ and $1(\cdot)$ be the unit step; try to solve

$$I * \phi = 1$$

This is impossible because $\phi \in \mathcal{D}_+$, $I \in \mathcal{D}'_+ \Longrightarrow I * \phi \in C^\infty$ (Schwartz, 1966).

System Theoretic Point of View

Consider an arbitrary interconnection of multiple-input multiple-output linear time-invariant systems characterized by convolutions with kernels in $A^{n\times n}$. Provided that, after reduction to a single loop system, inequality (2) holds, the interconnection is *determinate* (Zadeh, 1963): i.e., any set of inputs produces a unique set of outputs. Furthermore, the output depends continuously on the input.

II. INPUT OUTPUT PROPERTIES

Given a commutative algebra A, the subset I is said to be an *ideal of A* iff

(a) I is a linear space over A,

(b) $i \in I \Longrightarrow ix \in I$ for all $x \in A$.

The ideal I is said to be *proper* iff I is neither ϕ nor A itself.

Ex. 1.

Let A be the set of all *diagonal* matrices $\in R^{n\times n}$ Pick for I, the subset of all those whose (1,1) element zero.

Ex. 2.

Let

$$A = \{f:R_+ \to R;\ f \in L^1(R);\ \text{convolution product}\}.$$

Pick for I,

$$I = \{f \in A \mid \hat{f}(j\omega_0) = 0 \text{ for some fixed } \omega_0\}.$$

Ex. 3.

The algebra A and for I,

$$I = \{f \in A \mid \hat{f}(s_0) = 0 \text{ for some fixed } s_0 \text{ with } \operatorname{Re} s_0 \geq 0\}.$$

Useful Fact: Let A be a *commutative* algebra with unit element e; let I be a *proper ideal* of A; then $i \in I \Longrightarrow i$ has no inverse in A (that is, $ix \neq e, \forall x \in A$).

This is immediate by contradiction. Suppose i has an inverse in A; call it i^{-1}. Then

$$i \cdot i^{-1} = e$$

By associativity, $\forall x \in A$,

$$i \cdot (i^{-1} \cdot x) = (i \cdot i^{-1}) \cdot x = e \cdot x = x \in I$$

Hence $I = A$, a contradiction with the requirement of being proper.

Useful Theorem. (Hille-Phillips)

Let $g \in \mathcal{A}$.

$$\inf_{\text{Re } s \geq 0} |\hat{g}(s)| > 0 \Longleftrightarrow g \text{ has an inverse in } \mathcal{A}.$$

Proof. Call x the inverse of g. Since $x \in \mathcal{A}$.

$$\hat{g}(s)\, \hat{x}(s) = 1 \quad \forall s \text{ with Re } s \geq 0.$$

Now $x \in \mathcal{A}$ implies

$$\sup_{\text{Re } s \geq 0} |\hat{x}(s)| \leq ||x||_{\mathcal{A}} < \infty$$

Hence

$$|\hat{g}(s)| = \frac{1}{|\hat{x}(s)|} \geq \frac{1}{||x||_{\mathcal{A}}} > 0 \quad \forall s \text{ with Re } s \geq 0.$$

Hence $\hat{g}$ is bounded below in the closed right half plane.

$\Longrightarrow$ (Plausibility argument)

If, for some s_0 with Re $s_0 \geq 0$, $\hat{g}(s_0) = 0$, then g would be in a *proper ideal* of $\mathcal{A}$, hence could not have an inverse in $\mathcal{A}$. So to guarantee the existence of the inverse requires at least $|\hat{g}(s)| > 0$, $\forall s$ in Re $s \geq 0$.

To show that $\inf |\hat{g}(s)| > 0$ suffices requires delicate limiting arguments.

III. INPUT-OUTPUT PROPERTIES

Consider the system shown in Fig. 2:

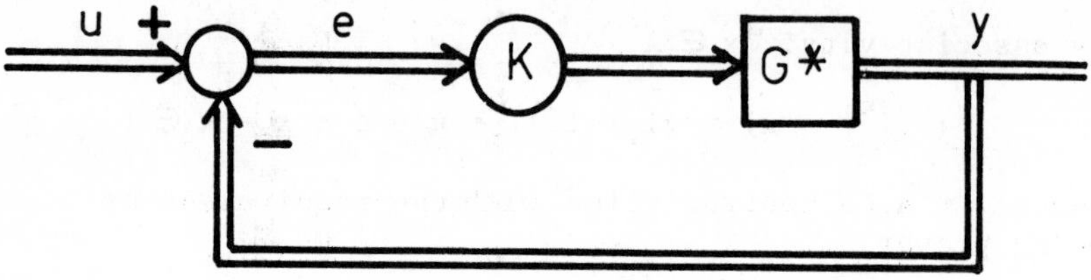

FIG. 2.

where $u, e, y: R_+ \to R^n$, $K \in R^{n\times n}$ and

$$G(t) = \begin{cases} R + G_\ell(t) & \text{for } t \geq 0 \\ 0 & \text{for } t < 0 \end{cases} \tag{1}$$

with R a constant matrix and

$$G_\ell(t) = G_a(t) + \sum_{\nu=0}^{\infty} G_\nu \, \delta(t-t_\nu) \tag{2}$$

We now require that

$$G_a \in L^1_{n\times n}, \quad \sum_{\nu=0}^{\infty} |G_\nu| < \infty \tag{3}$$

Under these conditions, if

$$\inf_{\operatorname{Re} s \geq 0} |\det(I + \hat{G}(s)\, K)| > 0 \tag{4}$$

and

$$\text{if} \begin{cases} \text{either} & R = 0 \quad \text{(no integration)} \\ \text{or} & -RK \text{ has all its eigenvalues in the open left half plane} \end{cases} \tag{5}$$

then

(a) $H \in A$, where H is the impulse response matrix of the closed loop system ($||H||_A < \infty$).

(b) $u \in A^n \Longrightarrow y \in A^n$

(c) for any $p \in [1,\infty]$, $u \in L_n^p \Longrightarrow y \in L_n^p$

(d) provided $R \neq 0$, $u \in L_n^\infty$ and $u(t) \to u_\infty$ as $t \to \infty$ imply that $y \in L_n^\infty$ and $y(t) \to u_\infty$.

Proof

We have

$$y = G * K(u-y)$$

or

$$G * Ku = y + G * Ky$$

By definition of the impulse response,

$$GK = H + G * KH$$

Now with s real and $s \to \infty$, $\hat{G}(s) \to G_0$, hence (4) implies that $\det [I + G_0 K] \neq 0$; hence by the result above, H is uniquely defined and $H \in L_{ne}^p \quad \forall p \in [1,\infty]$.

Case I: $R = 0$ (no integrators).

Hence $G = G_\ell$ and $G\ K \in A^{n\times n}$. Using Cramer

$$[I + \hat{G}_\ell(s)\ K]^{-1} = \hat{A}(s)/\det[I + \hat{G}(s)\ K]$$

where $\hat{A}(s)$ is the matrix of cofactors. Each element of $\hat{A}(s)$ is the L-transform of elements in A. Consider

$$\frac{\hat{a}_{ij}(s)}{\det\,[I + \hat{G}(s)\,K]} \tag{9}$$

From (1), (2), (3) and (4), and the theorem above, (9) is the L-transform of an element in A. Consequently the matrix-valued function H_c defined by

$$H_c(\cdot) \triangleq L^{-1}\,\{[I + \hat{G}(s)\,K]^{-1}\,\hat{G}(s)\,K\} \in A^{n\times n} \tag{10}$$

By Laplace transforms,

$$H_c + G * K\,H_c = GK$$

Hence, by uniqueness, $H_c = H$, the impulse response matrix. Therefore we have $H \in A^{n\times n}$.

Case II: $R \neq 0$ and all e.v. of $-RK$ have negative real parts.

Now

$$\hat{G}(s) = \frac{R}{s} + \hat{G}_\ell(s) \quad \text{Re } s \geq 0, \quad s \neq 0 \tag{11}$$

From (10),

$$\begin{aligned}\hat{H}_c(s) &= [I + s^{-1}RK + \hat{G}_\ell(s)K]^{-1}[s^{-1}RK + \hat{G}_\ell(s)K] \\ &= \underbrace{[I+s(sI+RK)^{-1}\hat{G}_\ell(s)K]^{-1}}_{[\hat{M}_1(s)]^{-1}}\underbrace{[s(sI+RK)^{-1}(s^{-1}RK+\hat{G}_\ell(s)K)]}_{\hat{M}_2(s)}\end{aligned} \tag{12}$$

Now $L^{-1}[s(sI + RK)^{-1}] \in A^{n\times n}$ hence $\hat{M}_1(s) \in \hat{A}^{n\times n}$. Since

$$\hat{M}_2(s) = (sI + RK)^{-1}RK + s(sI + RK)^{-1}\hat{G}_\ell(s)RK$$

we have $M_2(t) \in A^{n\times n}$

Now

$$\begin{aligned}\det[\hat{M}_1(s)] &= \det[(sI + RK)^{-1}(sI + RK + s\hat{G}_\ell(s)K)] \\ &= \det[(I + s^{-1}RK)^{-1}\;(I + s^{-1}RK + \hat{G}_\ell(s)K)]\end{aligned}$$

$$= \frac{1}{\det[I + s^{-1}RK]} \det(I + s^{-1}RK + \hat{G}_\ell(s)K)$$

The first determinant is zero only at the e.v. of -RK, which all have negative real parts: hence this determinant is bounded away from zero over Re s ≥ 0. The second determinant is bounded away from zero by assumption (4). Hence

$$\inf_{\text{Re } s \geq 0} \det[\hat{M}_1(s)] > 0,$$

and since

$$\hat{M}_1(s) = I + s(sI + RK)^{-1}\hat{G}_\ell(s)K \in \hat{A}^{n\times n}$$

it follows, by the previous theorem, that $[\hat{M}_1(s)]^{-1} \in \hat{A}^{n\times n}$. Hence, by (12), $H_c(s) \in A^{n\times n}$. As before, by direct calculation, $H_c = H \in A^{n\times n}$. This completes the proof of (a).

Conclusions (b) and (c) follow immediately from the properties of the algebra A.

(d) This can be proved directly [Desoer, 1968]. However borrowing an idea from [Vakharia], let us prove that $e(t) \to 0$ if $u(t) = 1(t)u_\infty$, where u_∞ is a constant vector. We have

$$s\,\hat{e}(s) = [I + \hat{G}(s)K]^{-1}u_\infty \qquad (14)$$

We have shown that $[\hat{M}_1(s)]^{-1} = [I + \hat{G}(s)K]^{-1} \in \hat{A}^{n\times n}$, hence its inverse transform, H_e, is in $A^{n\times n}$. From (14) we have

$$\frac{de}{dt} = -H_e(t)u_\infty \in A^n.$$

Consequently,

$$e(T) - e(\infty) = \int_T^\infty \frac{de}{dt}\,dt \to 0 \text{ as } T \to \infty$$

Thus $e(t) \to e_\infty$ a constant. Furthermore this constant must be zero because, if it were not, $y(t) \sim RKe_\infty t$ which contradicts, in the limit, the feedback equation

$$u_\infty = e(t) + y(t) \tag{15}$$

Thus $e(t) \to 0$ as $t \to \infty$. This implies, by (15), that $y(t) \to u_\infty$. Q.E.D.

Extension. Using a similar technique, Vakharia (1969) has verified for the single-input single-output case that these results still hold when $\hat{g}(s)$ has *multiple* poles on the $j\omega$-axis and in the open right half plane.

REFERENCES

DESOER, C.A. (1965) A General Formulation of the Nyquist Criterion, *IEEE Trans. Circuit Theory CT-12*, pp. 230-234.

DESOER, C.A. and WU, M.Y. (1968) Stability of Linear Time-Invariant Systems, *Trans. IEEE CT-15*, pp. 245-250.

DESOER, C.A. and WU, M.Y. (July 1968) Stability of Multiple Loop Feedback Linear Time-Invariant Systems, *Jour. Math. Anal. and Appl. 23: (1)* pp. 121-130.

DESOER, C.A. and WU, M.Y. (1969) L^p-Stability of Nonlinear Time-Varying Feedback Systems, *SIAM Jour. Control 7: (2)* (Also Allerton Confer. Proc., Oct. 1968).

DESOER, C.A. and WU, M.Y. (March, 1969) Stability of Nonlinear Time-Invariant Feedback Systems Under Almost Constant Inputs, *Automatica 5: (2)* (Also Asilomar Confer. Proc., Oct. 1968).

DESOER, C.A., *Lecture Notes of EECS 290D*, University of California, Berkeley.

DIEUDONNÉ, J. (1960) *Foundations of Modern Analysis*,

Academic Press, New York.

MASON, S.J. (July 1956) Feedback Theory - Further Properties of Signal Flow Graphs, *Proc. IRE 44*, pp. 920-926.

SANDBERG, I.W. (1965) Some Results in the Theory of Physical Systems Governed by Nonlinear Functional Equation, *Bell Syst. Tech. Jour. 44*, pp. 871-898.

VAKHARIA, D.J. (June 1969) *Stability of Linear Systems*, M.S. Thesis, Washington State University, Pullman, Washington.

SCHWARTZ, L. (1966) *Théorie des Distributions*, Rev. Ed., Hermann, Paris.

ZAMES, G. (1966) On the Input-Output Stability of Nonlinear Time-Varying Feedback Systems, Part I, *IEEE Trans. AC-11*, pp. 228-238; Part II, pp. 465-477.

ZADEH, L.A. and DESOER, C.A. (1963) *Linear System Theory*, McGraw-Hill, New York.

Added in proof:

Recent sharpenings of the results reported above are in process of publication:

DESOER, C.A. and VIDYASAGAR, M. "General Necessary Conditions for Input Output Stability", Letter to the Editor, Proc. IEEE, 1971.

DESOER, C.A. and LAM, F.L. "On Input-Output Properties of Linear Time-Invariant Systems", Proc. Princeton Conference, March 1971; also Letter to the Editor, Trans. IEEE, C.T., 1971.

THEOREMS ON THE ANALYSIS OF NONLINEAR TRANSISTOR NETWORKS*

I.W. SANDBERG

Bell Telephone Laboratories, Incorporated,
Murray Hill, New Jersey

ABSTRACT

The purpose of this paper is to report on further results concerning nonlinear equations of the form $F(x) + Ax = B$, in which $F(\cdot)$ is a "diagonal nonlinear mapping" of real Euclidean n-space E^n into itself, A is a real $n \times n$ matrix, and B is an element of E^n. Such equations play a central role in the dc analysis of transistor networks, the computation of the transient response of transistor networks, and the numerical solution of certain nonlinear partial-differential equations.

Here a nonuniqueness result, which focuses attention on a simple special property of transistor-type nonlinearities, is proved; it shows that under certain conditions the equation $F(x) + Ax = B$ has at least two solutions for some $B \in E^n$. The result proves that some earlier conditions for the existence of a unique solution cannot be improved by taking into account more information concerning the nonlinearities, and therefore makes more clear that the set

---

* The material of Dr. Sandberg's three lectures, "Some recent results on the theory of nonlinear networks," was drawn from three papers, of which this is the third. The previous two papers were published as [8] and [1].

of matrices denoted in earlier work by P_0 plays a very basic role in the theory of nonlinear transistor networks. In addition, some material concerned with the convergence of algorithms for computing the solution of the equation $F(x) + Ax = B$ is presented, and some theorems are proved which provide more of a theoretical basis for the efficient computation of the transient response of transistor networks. In particular, the following proposition is proved. If the dc equations of a certain general type of transistor network possess at most one solution for all $B \in E^n$ for "the original set of α's as well as for an arbitrary set of not-larger α's", then the nonlinear equations encountered at each time step in the use of certain implicit numerical integration algorithms possess a unique solution for all values of the step size, and hence then for all step-size values it is possible to carry out the algorithms.

I. INTRODUCTION AND DISCUSSION OF RESULTS

In [1] and [2], some results are presented concerning the equation

$$F(x) + Ax = B, \tag{1}$$

in which, with n an arbitrary positive integer, A is a real $n \times n$ matrix, B is an element of real Euclidean n-space E^n, and $F(\cdot)$ is a mapping of E^n into E^n defined by the condition that* for all $x = (x_1, x_2, \ldots, x_n)^{tr} \in E^n$,

$$F(x) = (f_1(x_1), f_2(x_2), \ldots, f_n(x_n))^{tr} \tag{2}$$

with each $f_i(\cdot)$ a strictly monotone increasing mapping of E^1 into itself. Eq. (1) plays a central role in the dc analysis of transistor networks,† the transient analysis

* Throughout the paper the superscript "tr" denotes transpose.

† See [1] for a derivation of the equation within the context of the transistor dc-analysis problem.

of transistor networks,* and the numerical solution of certain nonlinear partial differential equations.

In [1] it is proved that there exists a unique solution x of equation (1) for each strictly monotone increasing mapping $F(\cdot)$ of E^n onto E^n (i.e., for each set of strictly monotone increasing mappings $f_i(\cdot)$ of E^1 onto itself) and each $B \varepsilon E^n$ if and only if A is a member of the set P_0 of real $n \times n$ matrices with all principal minors nonnegative. It is also proved in [1] that Eq. (1) possesses a unique solution x for each continuous monotone nondecreasing mapping $F(\cdot)$ of E^n into E^n (i.e., for each set of continuous monotone nondecreasing mappings of E^1 into E^1) and each $B \varepsilon E^n$ if A belongs to the set P of all real $n \times n$ matrices with all principal minors positive†, and a direct modification of the existence proof given in [1], as indicated in [2], shows that (1) possesses a unique solution for each strictly monotone increasing mapping $F(\cdot)$ of E^n onto $(\alpha_1, \beta_1) \times (\alpha_2, \beta_2) \times \ldots \times (\alpha_n, \beta_n)$ with each α_i and β_i elements of the extended real line†† (real line) such that $\alpha_i < \beta_i$ and each $B \varepsilon E^n$ if (and only if) $A \varepsilon P_0$ and $\det A \neq 0$. Some network theoretic implications of these and related results are discussed in [1] and [2] where the matter of determining whether or not $A \varepsilon P_0$ or $A \varepsilon P$ is considered in some detail.

In this paper a nonuniqueness result, whose proof focuses attention on a simple special property of transistor-type nonlinearities, is proved; it shows that under certain conditions Eq. (1) has at least two solutions for some $B \varepsilon E^n$. In addition, some material con-

---

* See Section 1.4.

† There are some interesting applications of this result, in the study of numerical methods for solving certain nonlinear partial-differential equations, in which A has nonpositive off-diagonal terms and is irreducibly diagonally dominant [3].

†† The numbers α_i and β_i are members of the *extended real line* if $-\infty \leq \alpha_i \leq \infty$ and $-\infty \leq \beta_i \leq \infty$.

cerned with the convergence of algorithms for computing the solution of (1) is presented, and some theorems are proved which provide more of a theoretical basis for the efficient computation of the transient response of transistor networks. The remaining portion of Section I is concerned with a detailed discussion of the results and their significance.

1.1 *An Application of the Nonuniqueness Theorem*

The standard Ebers-Moll transistor model, which is widely used, gives rise to functions $f_i(\cdot)$ which, while continuous and strictly monotone increasing, are mappings of E^1 onto open semi-infinite intervals. For such $f_i(\cdot)$, the results stated above assert that the equation (1) possesses at most one solution x for each $B \in E^n$ if $A \in P_0$, and if $A \in P_0$ and $\det A \neq 0$, then (1) possesses a solution for each $B \in E^n$. Since, as indicated in [1], $A = T^{-1}G$ with T a nonsingular matrix which takes into account the forward and reverse transistor α's, and G is the short circuit conductance matrix of the linear portion of the network, the condition that det A not vanish is equivalent to the rather weak assumption that the linear portion of the network possesses an open-circuit resistance matrix.

It is natural to ask whether the use of more detailed information concerning the nonlinearities of the transistor model would enable us to make assertions concerning the existence of a unique solution of (1) under weaker assumptions on A. In particular, can the condition that A belong to P_0 be relaxed? The first result proved in this paper, Theorem 1 of Section II, shows that if the $f_i(\cdot)$ are exponential nonlinearities of the type associated with the Ebers-Moll model, then the condition that A belong to P_0 cannot be replaced by a weaker condition. More explicitly, in Section II a set $\mathcal{F}_0^n$ of mappings of E^n into E^n is defined, and $\mathcal{F}_0^n$ contains all of the mappings $F(\cdot)$ that correspond to Ebers-Moll type $f_i(\cdot)$'s. It is proved there that if $A \notin P_0$, then for *any* $F(\cdot) \in \mathcal{F}_0^n$, there is a $B \in E^n$ such that (1) possesses at least two solutions. In fact, it is proved that if $A \notin P_0$ and if δ is an arbitrary positive number, then for any $F(\cdot) \varepsilon F_0^n$, there is a $B \varepsilon E^n$ such that (1) possesses two solutions such that the distance in E^n between the two solutions is δ.

Thus Theorem 1 together with the earlier results mentioned above concerning existence of solutions show that the set of matrices P_0 plays a quite fundamental role in the theory of nonlinear transistor networks.

1.2 *An Algorithm for Computing the Solution of* (1)

Several results which assert that $A \varepsilon P_0$ under certain conditions on the transistor α's and the short-circuit conductance matrix of the linear portion of the network are proved in [1] and [2]. In particular, it is proved in [1] that $A \varepsilon P$, and hence that $A \varepsilon P_0$, if $A = P^{-1}Q$ with P and Q real $n \times n$ matrices such that for all $j = 1, 2, \ldots, n$

$$p_{jj} > \sum_{i \neq j} |p_{ij}| \text{ and } q_{jj} > \sum_{i \neq j} |q_{ij}|.^*$$

Theorem 2 of Section II shows that a relatively simple and entirely constructive algorithm can be used to generate a sequence $x^{(0)}, x^{(1)}, \ldots$ of elements of E^n that converges to the unique solution of (1) if $A = P^{-1}Q$ with P and Q as defined above and each $f_i(\cdot)$ is a continuous (but not necessarily differentiable) monotone nondecreasing mapping of E^1 into E^1.†

1.3 *Palais' Theorem, Existence of Solutions of* (1), *and Algorithms for Computing the Solution of* (1)

There are two existence proofs concerning (1) given in [1]. One proof, the more basic of the two, is based

---

* It is proved also that $A \varepsilon P_0$ if $A = P^{-1}Q$ with

$$p_{jj} > \sum_{i \neq j} |p_{ij}| \text{ and } q_{jj} \geq \sum_{i \neq j} |q_{ij}| \text{ for all } j.$$

† A related result given in [4] is not directly applicable here because of assumptions made in [4] concerning the existence and boundedness of a certain Jacobian matrix.

on first principles and employs an inductive argument in which, with k an arbitrary positive integer less than n, the existence proposition is assumed to be true with n replaced by k and it is proved that then the proposition is true with n replaced by (k + 1). The second proof employs a theorem of R.S. Palais and requires that the $f_i(\cdot)$ be continuously differentiable throughout E^1. More explicity, Palais' theorem* asserts that if $R(\cdot)$ is a continuously differentiable mapping of E^n into itself with values $R(q)$ for $q \in E^n$, then $R(\cdot)$ is a diffeomorphism† of E^n onto itself if and only if

(i) $\det J_q \neq 0$ for all $q \in E^n$, in which J_q is the Jacobian matrix of $R(\cdot)$ with respect to q, and

(ii) $||R(q)|| \to \infty$ as $||q|| \to \infty$,††

and the second proof of Ref. [1] shows that, with

$$R(q) = F(q) + Aq$$

for all $q \in E^n$, the two conditions (i) and (ii) are met when $A \in P_0$ and each $f_i(\cdot)$ is a continuously differentiable strictly-monotone-increasing function which maps E^1 onto E^1 and whose slope is positive throughout E^1.¶

---

* See Ref. [5] and the appendix of Ref. [6].

† A diffeomorphism of E^n onto itself is a continuously differentiable mapping of E^n into E^n which possesses a continuously differentiable inverse.

†† Here $||\cdot||$ denotes any norm on E^n.

¶The reasons that two proofs were presented in Ref. [1], with the second proof a proof of a somewhat weaker result, are that the arguments needed for the application of Palais' theorem had already been developed in Ref. [1] and used for other purposes there, and it was felt desirable to indicate an alternative approach to essentially the same problem.

There are some problems which arise in connection with, for example, the numerical solution of certain nonlinear partial differential equations* in which one encounters an equation of the form (1) with $A \in P_0$ and $\det A \neq 0$, but with functions $f_i(\cdot)$ which, while continuously differentiable, are monotone nondecreasing (rather than strictly monotone increasing) mappings of E^1 into E^1. We can prove that even in such cases (1) possesses a unique solution for each $B \in E^n$ as follows. Here the Jacobian matrix of $F(q) + Aq$ exists and is of the form $D(q) + A$ in which $D(q)$ is a diagonal matrix with nonnegative diagonal elements. Since $A \in P_0$ and $\det A \neq 0$, we have [2] $\det [D(q) + A] \neq 0$ for all $q \in E^n$. An immediate application of Theorem 3 of Section II shows that $||F(q) + Aq|| \to \infty$ as $||q|| \to \infty$.† Therefore, by Palais' theorem, $F(x) + Ax = B$ possesses a unique solution for each B.

Theorem 3 is of use not only in connection with the proof given in the preceding paragraph; it also plays a key role in showing that there is an algorithm which generates a sequence of elements of E^n: $x^{(0)}, x^{(1)}, \ldots$ that converges to the unique solution of $F(x) + Ax = B$ whenever each $f_i(\cdot)$ is twice continuously differentiable on E^1 and the conditions on A and $F(\cdot)$ of the preceding paragraph are satisfied.††

More generally, if $R(\cdot)$ is any twice-differentiable mapping of E^n into itself such that conditions (i) and (ii) of Palais' theorem are satisfied, then E^n contains a

---

* The writer is indebted to J. McKenna and E. Wasserstrom for bringing this fact to his attention.

† More explicitly, Theorem 3 shows that there is a vector $C \in E^n$ such that $||F(q) + Aq + C|| \to \infty$ as $||q|| \to \infty$, which is equivalent to the statement concerning $||F(q) + Aq||$ made above.

†† The differentiability assumption here is introduced as a matter of convenience, and is certainly satisfied when the $f_i(\cdot)$ are Ebers-Moll exponential-type nonlinearities.

unique element x such that $R(x) = \theta$ in which θ is the zero element of E^n, and there are steepest decent as well as Newton-type algorithms each of which generates a sequence in E^n that converges to x. To show this, let [7] $f(y) = ||R(y)||^2$ for all $y \in E^n$ in which $||\cdot||$ denotes the usual Euclidean norm (i.e., the square-root of the sum of squares). Since condition (i) of Palais' theorem is satisfied, the gradient ∇f of $f(\cdot)$ satisfies $(\nabla f)(y) \neq \theta$ unless $f(y) = 0$,* and since condition (ii) of Palais' theorem is satisfied the set $S = \{y \in E^n: f(y) \leq f(x^{(0)})\}$ is bounded for any $x^{(0)} \in E^n$. Therefore we may appeal to, for example, the theorem of p. 43 of [7] according to which for any $x^{(0)} \in E^n$, for any member of a certain class of mappings $\phi(\cdot)$ of S into E^n, and for suitably chosen constants $\gamma_0, \gamma_1, \ldots$, the sequence $x^{(0)}, x^{(1)}, \ldots$ defined by

$$x^{(k+1)} = x^{(k)} + \gamma_k \phi(x^{(k)}) \text{ for all } k \geq 0$$

is such that $||R(x^{(k)})|| \to 0$ as $k \to \infty$. However, since $R^{-1}(\cdot)$ exists and is continuous,† it follows from

$$x^{(k)} = R^{-1}[R(x^{(k)})] \text{ for all } k \geq 0$$

and the fact that $R(x^{(k)}) \to \theta$ as $k \to \infty$, that $\lim_{k\to\infty} x^{(k)}$ exists and

$$\lim_{k\to\infty} x^{(k)} = R^{-1}(\theta),$$

which means that $\lim_{k\to\infty} x^{(k)}$ is the unique solution x of $R(y) = \theta$.††

---

* Here we have used the fact that $(\nabla f)(y) = 2J_y^{tr}R(y)$ for all $y \in E^n$ [7].

† By Palais' theorem $R(\cdot)$ is a diffeomorphism of E^n onto itself.

†† The material of the second part of Section 1.3 was motivated by previous recent work of the writer's

1.4 *Transient Response of Transistor-Diode Networks and Implicit Numerical-Integration Formulas*

At this point we briefly consider some aspects of the manner in which the previous material bears on the important problem of providing more of a theoretical basis for numerically integrating the ordinary differential equations which govern the transient response of nonlinear transistor networks. Although we consider explicitly only networks containing transistors, diodes, and resistors, the material to be presented can be extended to take into account other types of elements as well. In addition, we shall focus attention on the use of linear multipoint integration formulas of closed (i.e., of implicit) type, since such formulas are of considerable use in connection with the typically "stiff systems" of differential equations encountered.

A very large class of networks containing resistors, transistors, and diodes modeled in a standard manner is governed by the equation [8]

$$\frac{du}{dt} + TF[C^{-1}(u)] + (I + GR)^{-1}GC^{-1}(u) = B(t), \; t \geq 0 \qquad (3)$$

where, assuming that there are q diodes and p transistors,

(i) $T = I_q \oplus T_1 \oplus T_2 \oplus \ldots \oplus T_p$, the direct sum of the identity matrix of order q and p 2 × 2 matrices T_k in which

$$T_k = \begin{pmatrix} 1 & -\alpha_r^{(k)} \\ -\alpha_f^{(k)} & 1 \end{pmatrix}$$

---

colleague A. Gersho who made the observation that the convergence of an algorithm for the solution of (1) could be shown by combining results of [1] with the approaches described by Goldstein [7]. Gersho showed also that an alternative existence proof concerning (1) can be obtained in this manner.

with $0 < \alpha_r^{(k)} < 1$ and $0 < \alpha_f^{(k)} < 1$ for $k = 1, 2, \ldots, p$.

(ii) $R = R_0 \oplus R_1 \oplus R_2 \oplus \ldots \oplus R_p$, the direct sum of a diagonal matrix $R_0 = \text{diag}(r_1, r_2, \ldots, r_q)$ with $r_k \geq 0$ for $k = 1, 2, \ldots, q$ and p 2×2 matrices R_k in which for all $k = 1, 2, \ldots, p$

$$R_k = \begin{pmatrix} r_e^{(k)} + r_b^{(k)} & r_b^{(k)} \\ r_b^{(k)} & r_c^{(k)} + r_b^{(k)} \end{pmatrix}$$

with $r_e^{(k)} \geq 0$, $r_b^{(k)} \geq 0$, and $r_c^{(k)} \geq 0$. (The matrix R takes into account the presence of bulk resistance in series with the diodes and the emitter, base, and collector leads of the transistors.)

(iii) G is the short-circuit conductance matrix associated with the resistors of the network. (It does not take into account the bulk resistances of the semiconductor devices.)

(iv) $F(\cdot)$ is a mapping of $E^{(2p+q)}$ into $E^{(2p+q)}$ defined by the condition that

$$F(x) = [f_1(x_1), f_2(x_2), \ldots, f_{2p+q}(x_{2p+q})]^{tr}$$

for all $x \in E^{(2p+q)}$ with each $f_i(\cdot)$ a continuously differentiable strictly-monotone increasing mapping of E^1 into E^1.

(v) $C^{-1}(\cdot)$ is the inverse of the mapping $C(\cdot)$, of $E^{(2p+q)}$ into itself, defined by

$$C(x) = \text{diag}(c_1, c_2, \ldots, c_{2p+q})x$$
$$+ \text{diag}(\tau_1, \tau_2, \ldots, \tau_{2p+q})F(x)$$

for all $x \in E^{(2p+q)}$ with each c_i and each τ_i a positive constant.

(vi) $B(t)$ is a $(2p + q)$-vector which takes into account the voltage and current generators present in the network, and

(vii) u is related to v the vector of junction voltages of the semiconductor devices through $C(v) = u$ for all $v \varepsilon E^{(2p+q)}$.

Eq. (3) is equivalent to*

$$\dot{u} + f(u,t) = \theta_{(2p+q)}, \quad t \geq 0 \tag{4}$$

in which of course

$$f(u,t) = TF[C^{-1}(u)] + (I + GR)^{-1}G\dot{C}^{-1}(u) - B(t) \tag{5}$$

and $\theta_{(2p+q)}$ is the zero vector of order (2p+q).

It is well known that certain specializations of the general multipoint formula [9, 10]

$$y_{n+1} = \sum_{k=0}^{r} a_k y_{n-k} + h \sum_{k=-1}^{r} b_k \tilde{y}_{n-k} \tag{6}$$

in which

$$\tilde{y}_{n-k} = - f(y_{n-k}, (n-k)h) \tag{7}$$

can be used as a basis for computing the solution of (4). Here h, a positive number, is the step size, the a_k and the b_k are real numbers, and of course y_n is the approximation to u(nh) for $n \geq 1$.

In the literature dealing with formulas of the type (6) in connection with systems of equations of the type (4), information concerning the location of the eigenvalues of the Jacobian matrix J_u of f(u,t) with respect to u plays an important role in determining whether or not a given formula will be (in some suitable sense) stable. In particular, an assumption often made is that all of the eigenvalues of J_u lie in the strict right-half

---

* In [8] it is shown that if B(•) is a continuous mapping of $[0,\infty)$ into $E^{(2p+q)}$, then for any initial condition $u^{(0)} \varepsilon E^{(2p+q)}$ there exists a unique continuous (2p+q)-vector-valued function u(•) such that $u(0) = u^{(0)}$ and (3) is satisfied for all $t > 0$.

plane for all $t \geq 0$ and all u. For f(u,t) given by (5), we have

$$J_u = T \operatorname{diag}\left\{\frac{f_j'[g_j(u_j)]}{c_j + \tau_j f_j'[g_j(u_j)]}\right\}$$

$$+ (I+GR)^{-1}G \operatorname{diag}\left\{\frac{1}{c_j + \tau_j f_j'[g_j(u_j)]}\right\} \tag{8}$$

in which for $j = 1, 2, \ldots, (2p+q)$ $g_j(u_j)$ is the jth component of $C^{-1}(u)$. Thus here J_u is a matrix of the form

$$TD_1 + (I+GR)^{-1}GD_2 \tag{9}$$

where D_1 and D_2 are diagonal matrices with positive diagonal elements. A simple result concerning (9), Theorem 4 of Section II, asserts that if there exists a diagonal matrix D with positive diagonal elements such that*

(i) DT is strongly column-sum dominant, and

(ii) $D(I+GR)^{-1}G$ is weakly column-sum dominant, then for all diagonal matrices D_1 and D_2 with positive diagonal elements, all eigenvalues of (9) lie in the strict right-half plane. This condition on T, G, and R is often satisfied.†

The subclass of numerical integration formulas (6) defined by the condition that $b_{-1} > 0$ are of considerable use [11, 12, 13] in applications involving the typically "stiff systems" of differential equations encountered in the analysis of nonlinear transistor networks. With $b_{-1} > 0$, y_{n+1} is defined *implicitly* through

---

* The terms "strongly-column sum dominant" and "weakly column-sum dominant" are reasonably standard. However they are defined in Section II.

† See [8] for examples.

$$y_{n+1} + hb_{-1}f(y_{n+1},(n+1)h) = \sum_{k=0}^{r} a_k y_{n-k} + h \sum_{k=0}^{r} b_k \tilde{y}_{n-k}$$

in which the right side depends on y_{n-k} only for $k \in \{0, 1, 2, \ldots, r\}$, and for $f(u,t)$ given by (5), we have

$$y_{n+1} + hb_{-1}\{TF[C^{-1}(y_{n+1})] + (I+GR)^{-1}GC^{-1}(y_{n+1})\} = q_n \quad (10)$$

in which

$$q_n = \sum_{k=0}^{r} a_k y_{n-k} + h \sum_{k=0}^{r} b_k \tilde{y}_{n-k} + hb_{-1}B[(n+1)h].$$

Obviously, the numerical integration formula (10) makes sense only if there exists for each n a $y_{n+1} \in E^{(2p+q)}$ such that (1) is satisfied.

Let $x_{n+1} = C^{-1}(y_{n+1})$ for each n. Then (10) possesses a unique solution y_{n+1} if and only if there exists a unique $x_{n+1} \in E^{(2p+q)}$ such that

$$C(x_{n+1}) + hb_{-1}[TF(x_{n+1}) + (I+GR)^{-1}Gx_{n+1}] = q_n. \quad (11)$$

Since $C(x_{n+1}) = cx_{n+1} + \tau F(x_{n+1})$, in which

$$c = \mathrm{diag}(c_1, c_2, \ldots, c_{2p+q})$$

and

$$\tau = \mathrm{diag}(\tau_1, \tau_2, \ldots, \tau_{2p+q}),$$

(11) is equivalent to

$$[\tau + hb_{-1}T]F(x_{n+1}) + [c + hb_{-1}G]x_{n+1} = q_n. \quad (12)$$

The matrices τ and c are both diagonal with positive diagonal elements. Thus it is clear that for all positive h

$$\det[\tau + hb_{-1}T] \neq 0$$

and

$$\det[c + hb_{-1}(I+GR)^{-1}G] \neq 0.^{*}$$

For all sufficiently small positive h

$$[\tau+hb_{-1}T]^{-1}[c + hb_{-1}(I+GR)^{-1}G] \in P_0.^{\dagger}$$

Consequently†† for all sufficiently small $h > 0$, (12) possesses a unique solution for each q_n. ¶ However, our interest in (12) is primarily in connection with "large-h" algorithms.

Suppose that $\det G \neq 0$ and that $T^{-1}G \in P_0$ for all possible combinations of α_r and α_f $(0 < \alpha_r < 1, 0 < \alpha_f < 1)$ for each transistor.¶¶ Then, according to Theorem 6 of Section II, for any particular T and R

$$[\tau+hb_{-1}T]^{-1}[c + hb_{-1}(I+GR)^{-1}G] \in P_0$$

for all $h > 0$, and hence (10) possesses a unique solution y_{n+1} for all positive values of h.

An important and general proposition concerning (10) is as follows. Suppose that

$$T^{-1}[(I+GR)^{-1}G] \in P_0 \tag{13}$$

and that (13) is satisfied whenever $\alpha_r^{(k)}$ and $\alpha_f^{(k)}$ are replaced with positive constants $\delta_r^{(k)}$ and $\delta_f^{(k)}$ such that

---

* Here we have used the fact that $(I+GR)^{-1}G$ is positive semidefinite.

† See Section 1.2.

†† See Section 1.1.

¶ Alternatively, this conclusion could have been obtained by applying the contraction-mapping fixed-point principle to (10), in view of the fact that each of the elements of J_u is bounded on $u \in E^{(2p+q)}$ and $t \in [0,\infty)$.

¶¶ See [1] for examples.

$\delta_r^{(k)} \le \alpha_r^{(k)}$ and $\delta_f^{(k)} \le \alpha_f^{(k)}$ for $k = 1, 2, \ldots, p$. In other words, assuming that $F(\cdot)$ is as defined in this section and that* $F(\cdot) \in \mathcal{F}_0^{(2p+q)}$, suppose that the dc equation

$$F(x) + T^{-1}[(I+GR)^{-1}G]x = B$$

possesses at most one solution x for each $B \in E^{(2p+q)}$ for "the original set of α's as well as for an arbitrary set of *not-larger* α's." Then an immediate application of Theorem 5 of Section II shows that

$$[\tau+hb_{-1}T]^{-1}[c + hb_{-1}(I+GR)^{-1}G] \in P_0$$

for all $h > 0$, and hence that (10) possesses a unique solution y_{n+1} for all $h > 0$ and all $q_n \in E^{(2p+q)}$.

II. THEOREMS, PROOFS, AND SOME DISCUSSION

Throughout this section,

(i) n is an arbitrary positive integer

(ii) P_0 denotes the set of all real $n \times n$ matrices M such that all principal minors of M are nonnegative.

(iii) real Euclidean n-space is denoted by E^n, and θ is the zero element of E^n

(iv) v^{tr} denoted the transpose of the row vector $v = (v_1, v_2, \ldots, v_n)$

(v) $||v||$ denotes $\left(\sum_{j=1}^{n} v_j^2\right)^{\frac{1}{2}}$ for all $v \in E^n$

(vi) if D is a real diagonal matrix, then $D > 0$ ($D \geq 0$) means that the diagonal elements of D are positive (nonnegative)

---

* See Definition 1 of Section 2.1.

(vii) I_q denotes the identity matrix of order q, and I denotes the identity matrix of order determined by the context in which the symbol is used, and

(viii) we shall say that a real $n \times n$ matrix M is strongly (weakly) column-sum dominant if and only if for $j = 1, 2, \ldots, n$

$$m_{jj} > (\geq) \sum_{i \neq j} |m_{ij}|.$$

2.1 *Definition 1*

For each positive integer n, let F_0^n denote that collection of mappings of E^n into itself defined by: $F \in F_0^n$ if and only if there exist for $j = 1, 2, \ldots, n$, continuous functions $f_j(\cdot)$ mapping E^1 into E^1 such that for each $x = (x_1, x_2, \ldots, x_n) \in E^n$, $F(x) = [f_1(x_1), f_2(x_2), \ldots, f_n(x_n)]^{tr}$, and

(i) $$\inf_{\alpha \in (-\infty,\infty)} [f_j(\alpha+\beta) - f_j(\alpha)] = 0$$

(ii) $$\sup_{\alpha \in (-\infty,\infty)} [f_j(\alpha+\beta) - f_j(\alpha)] = +\infty$$

for all $\beta > 0$ and all $j = 1, 2, \ldots, n$.

2.2 *Theorem 1*

Let $F \in F_0^n$, let A be a real $n \times n$ matrix such that $A \notin P_0$, and let δ be a positive constant. Then there exist $B \in E^n$, $x \in E^n$, and $y \in E^n$ such that

(i) $$F(x) + Ax = B$$

(ii) $$F(y) + Ay = B,$$

and

(iii) $$||x-y|| = \delta.$$

2.3 *Proof of Theorem 1*

Since $A \notin P_0$, there exists [2] a real diagonal matrix $D > 0$ such that $\det(D+A) = 0$. Thus there exists a $x^* \varepsilon E^n$ such that $||x^*|| = \delta$ and $(D+A)x^* = \theta$.

Since $F \varepsilon \mathcal{F}_0^n$, there exists a $x \varepsilon E^n$ such that

$$f_j(x_j) - f_j(x_j - x_j^*) = x_j^* d_j$$

for all $j = 1, 2, \ldots, n$ in which d_j is the jth diagonal element of D. Let

$$B = F(x) + Ax,$$

and let $y = x - x^*$. Then $A(x-y) = Ax^* = -Dx^*$, and

$$F(x) - F(y) + A(x-y) = \theta.$$

2.4 *Remarks Concerning Theorem 1*

If, as in the case of standard transistor models,

$$f_j(x_j) = e^{\lambda_j x_j} - 1$$

or

$$f_j(x_j) = 1 - e^{-\lambda_j x_j}$$

with $\lambda_j > 0$, we have, respectively,

$$f_j(\alpha+\beta) - f_j(\alpha) = e^{\lambda_j \alpha}(e^{\lambda_j \beta} - 1)$$

or

$$f_j(\alpha+\beta) - f_j(\alpha) = e^{-\lambda_j \alpha}(1 - e^{-\lambda_j \beta})$$

and it is clear that for either type of function conditions (i) and (ii) of Definition 1 are satisfied.

In [1] it is proved that if $F(\cdot) \varepsilon M$, the set of all $F(\cdot)$ of the form (2) with each $f_i(\cdot)$ a strictly monotone increasing mapping of E^1 into E^1, and if $A \varepsilon P_0$, then (1) possesses at most one solution. Thus, using Theorem 1, we see that for each $F(\cdot) \in \mathcal{M} \cap \mathcal{F}_0^n$ there exists at most one solution of $F(x) + Ax = B$ for each $B \in E^n$ if and only if $A \in P_0$. Similarly, with $\mathcal{M}_0$ the set of all $F(\cdot)$ of the form (2) with each $f_i(\cdot)$ a strictly-monotone increasing mapping of E^1 onto (α_i, β_i) with each α_i and β_i such that $-\infty \leq \alpha_i < \beta_i \leq \infty$, if $F(\cdot) \in \mathcal{M}_0 \cap \mathcal{F}_0^n$ and $\det A \neq 0$, then there exists a unique $x \in E^n$ such that $F(x) + Ax = B$ for each $B \in E^n$ if and only if $A \in P_0$.*

There may be a temptation to conjecture that whenever $F(\cdot) \in \mathcal{M} \cap \mathcal{F}_0^n$ and $A \notin P_0$ then the equation $F(x) + Ax = B$ does not possess a solution for some $B \in E^n$. The conjecture is false. In fact, with $n = 2$, $f_1(x_1) = e^{x_1}$, $f_2(x_2) = e^{x_2}$, and

$$A = \begin{pmatrix} 0 & 1 \\ 1 & 0 \end{pmatrix},$$

we have a situation in which (it is easy to show that) there exists a solution for all $B \in E^2$. Of course here for some choices of B the solution is not unique.

2.5 *Theorem 2*

Let P and Q denote real $n \times n$ matrices such that

$$p_{jj} > \sum_{i \neq j} |p_{ij}| \text{ and } q_{jj} > \sum_{i \neq j} |q_{ij}|$$

for all $j = 1, 2, \ldots, n$. For $j = 1, 2, \ldots, n$ let $f_j(\cdot)$ denote a continuous monotone nondecreasing (but not necessarily differentiable) mapping of E^1 into itself, and let $F(x) = [f_1(x_1), f_2(x_2), \ldots, f_n(x_n)]^{tr}$ for all $x \in E^n$. Then for each $R \in E^n$, there exists a unique $x \in E^n$ such that

$$PF(x) + Qx = R,$$

---

* The "if" part of this statement is proved in [2].

and, for any $y_0 \in E^n$, x is the limit of the sequence $x^{(0)}, x^{(1)}, \ldots$ defined by

$$y^{(n)} = D_P F(x^{(n)}) + D_Q x^{(n)}$$

$$y^{(n+1)} + (P-D_P)F(x^{(n)}) + (Q-D_Q)x^{(n)} = R$$

for $n \geq 0$, in which D_P and D_Q are diagonal matrices whose diagonal elements coincide with those of P and Q, respectively.

2.6 *Proof of Theorem 2*

Since the continuous mapping $[D_P F(\cdot) + D_Q]$ of E^n into E^n possesses an inverse $[D_P F(\cdot) + D_Q]^{-1}$, the equation

$$PF(x) + Qx = R$$

possesses a unique solution x if and only if $y = D_P F(x) + D_Q x$ is the unique solution of

$$y + \tilde{P}F[(D_P F(\cdot) + D_Q)^{-1}y] + \tilde{Q}[(D_P F(\cdot) + D_Q)^{-1}y] = R$$

in which $\tilde{P} = (P-D_P)$ and $\tilde{Q} = (Q-D_Q)$.

Therefore, by Banach's contraction-mapping fixed-point theorem, it suffices to show that with the metric $\rho(y,z) = \sum_{i=1}^{n} |y_i - z_i|$, the operator H defined by

$$H(y) = \tilde{P}F[(D_P F(\cdot) + D_Q)^{-1}y] + \tilde{Q}[(D_P F(\cdot) + D_Q)^{-1}y]$$

for all $y \in E^n$, is a contraction mapping of E^n into itself. We show this as follows. Let $y \in E^n$ and $z \in E^n$. Using the fact that

$$\alpha = d_{Qj}[(d_{Pj}f_j(\cdot) + d_{Qj})^{-1}\alpha] + d_{Pj}f_j[(d_{Pj}f_j(\cdot) + d_{Qj})^{-1}\alpha]$$

for all real α and all $j = 1, 2, \ldots, n$, in which d_{Pj} and d_{Qj} is the jth diagonal element of D_P and D_Q, respectively, it is a simple matter to verify that for all j:

$$f_j[(d_{Pj}f_j(\cdot) + d_{Qj})^{-1}y_j] - f_j[(d_{Pj}f_j(\cdot) + d_{Qj})^{-1}z_j]$$

$$= \frac{r_j}{d_{Qj}+d_{Pj}r_j}(y_j-z_j),$$

and

$$(d_{Pj}f_j(\cdot) + d_{Qj})^{-1}y_j - (d_{Pj}f_j(\cdot) + d_{Qj})^{-1}z_j$$

$$= \frac{1}{d_{Qj}+d_{Pj}r_j}(y_j-z_j)$$

in which $r_j = 1$ if $y_j = z_j$, and if $y_j \neq z_j$,

$$r_j = \frac{f_j[(d_{Pj}f_j(\cdot) + d_{Qj})^{-1}y_j] - f_j[(d_{Pj}f_j(\cdot) + d_{Qj})^{-1}z_j]}{(d_{Pj}f_j(\cdot) + d_{Qj})^{-1}y_j - (d_{Pj}f_j(\cdot) + d_{Qj})^{-1}z_j}$$

Thus

$$H(y) - H(z)$$

$$= \tilde{P}\,\mathrm{diag}\left\{\frac{r_j}{d_{Qj}+d_{Pj}r_j}\right\}(y-z) + \tilde{Q}\,\mathrm{diag}\left\{\frac{1}{d_{Qj}+d_{Pj}r_j}\right\}(y-z)$$

in which $r_j \geq 0$. Therefore

$$\rho(H(y),H(z)) \leq \max_j\left(\frac{\sigma_{Qj}+\sigma_{Pj}r_j}{d_{Qj}+d_{Pj}r_j}\right)\rho(y,z)$$

in which $\sigma_{Qj} = \sum_{i\neq j}|q_{ij}|$ and $\sigma_{Pj} = \sum_{i\neq j}|p_{ij}|$. Since $\sigma_{Qj} < d_{Qj}$ and $\sigma_{Pj} < d_{Pj}$ for all j, there exists a positive constant $\beta < 1$ such that

$$\max_j\left(\frac{\sigma_{Qj}+\sigma_{Pj}r_j}{d_{Qj}+d_{Pj}r_j}\right) \leq \beta$$

for all $r_j \geq 0$.

2.7 *Theorem 3*

If $A \varepsilon P_0$ and det $A \neq 0$, if for each $j = 1, 2, \ldots, n$: $f_j(\cdot)$ is a continuous mapping of E^1 into itself such that

$$f_j(x_j) = 0 \text{ for all } x_j$$

or

$$f_j(x_j) > 0 \text{ for all } x_j > c$$

and

$$f_j(x_j) < 0 \text{ for all } x_j < -c$$

for some $c \geq 0$, then, with $F(x) = [f_1(x_1), f_2(x_2), \ldots, f_n(x_n)]^{tr}$ for all $x \varepsilon E^n$,

$$||F(x) + Ax|| \to \infty \text{ as } ||x|| \to \infty.$$

2.8 *Proof of Theorem 3*

We note that

$$||F(x) + Ax|| \to \infty \text{ as } ||x|| \to \infty$$

if and only if

$$||A^{-1}F(x) + x|| \to \infty \text{ as } ||x|| \to \infty.$$

With $M = A^{-1}$, let

$$MF(x) + x = q. \tag{14}$$

Since $A \varepsilon P_0$, we have $M \varepsilon P_0$ [1]. Since $M \varepsilon P_0$, we have [1] for any $y \varepsilon E^n$ and $y \neq \theta$

$$y_k(My)_k \geq 0$$

for some index k such that $y_k \neq 0$.

Suppose that $F(x) \neq \theta$. Then there exists an index k_1 such that

$$f_{k_1}(x_{k_1})[MF(x)]_{k_1} \geq 0$$

with $f_{k_1}(x_{k_1}) \neq 0$. Thus, using (14),

$$f_{k_1}(x_{k_1})[MF(x)]_{k_1} + f_{k_1}(x_{k_1})x_{k_1} = f_{k_1}(x_{k_1})q_{k_1}$$

and

$$f_{k_1}(x_{k_1})x_{k_1} \leq f_{k_1}(x_{k_1})q_{k_1}.$$

Either $x_{k_1} \in [-c,c]$ or not. If not, then $f_{k_1}(x_{k_1})x_{k_1} > 0$ and $|x_{k_1}| \leq |q_{k_1}|$. Therefore for some index k_1, $|x_{k_1}| \leq \delta_1 \stackrel{\Delta}{=} \max(c, |q_{k_1}|)$, whether or not $F(x) = \theta$.

Let $M^{(k_1)}$ denote the matrix obtained from M by deleting the k_1 row and column, and let $M_{(k_1)}$ denote the k_1 column of M with the k_1 entry removed. Similarly, let $x_{(k_1)}$, $q_{(k_1)}$ and $F_{(k_1)}(x_{(k_1)})$ denote the (n-1)-vectors obtained from x, q, and F(x), respectively, by removing the k_1 entry. Then

$$M^{(k_1)}F_{(k_1)}(x_{(k_1)}) + x_{(k_1)} = q_{(k_1)} - M_{(k_1)}f_{k_1}(x_{k_1}).$$

Since $M^{(k_1)} \in P_0$, we can repeat the argument given above. Thus there exists an index k_2, different from k_1, such that

$$|x_{k_2}| \leq \delta_2 \stackrel{\Delta}{=} \max(c, |q_{(k_1,k_2)}|)$$

in which

$$|q_{(k_1,k_2)}| = \max_{|x_{k_1}| \leq \delta_1} \left| [q_{(k_1)} - M^{(k_1)}f_{k_1}(x_{k_1})]_{\ell_2} \right|$$

and ℓ_2 is the index of the component of $x_{(k_1)}$ that corresponds to the k_2 component of x. By continuing in this manner we can determine positive constants $\delta_1, \delta_2, \ldots, \delta_n$ depending only on q, F, M, and c such that, with $\delta = \max_j \{\delta_j\}$,

$$|x_j| \leq \delta \text{ for all } j = 1, 2, \ldots, n$$

and each δ_i depends on q such that for any positive constant α, there exists a constant $\beta_i(\alpha)$ with the property that $\delta_i \leq \beta_i(\alpha)$ for all $||q|| \leq \alpha$. Therefore for any $\alpha > 0$ there is a $\beta(\alpha)$ such that $||x|| \leq \beta(\alpha)$ whenever $||q|| \leq \alpha$, which implies that $||q|| \to \infty$ as $||x|| \to \infty$.

2.9 *Theorem 4*

Let P and Q denote real $n \times n$ matrices with P strongly column-sum dominant. Suppose that there exists a real diagonal matrix $D > 0$ such that DP is strongly column-sum dominant and DQ is weakly column-sum dominant. Then for all real diagonal matrices $D_1 > 0$ and $D_2 > 0$, all eigenvalues of $(PD_1 + QD_2)$ lie in the strict (i.e., open) right-half plane.

2.10 *Proof of Theorem 4*

Since the strict right-half plane contains all of the eigenvalues of P, there exist choices of $D_1 > 0$ and $D_2 > 0$ such that every eigenvalue of $(PD_1 + QD_2)$ lies in the strict right-half plane. Thus it suffices to show that $(PD_1 + QD_2)$ does not possess an eigenvalue on the boundary of the complex plane for all $D_1 > 0$ and all $D_2 > 0$. In other words, it suffices to show that (with $i = \sqrt{-1}$)

$$PD_1 + QD_2 + i\omega I \tag{15}$$

is nonsingular for all $D_1 > 0$, all $D_2 > 0$, and all real constants ω.

Suppose that (15) is singular for some ω and some $D_1 > 0$ and some $D_2 > 0$. Then $(DPD_1 + DQD_2 + i\omega D)$ singular. But DPD_1 is strongly column-sum dominant and DQD_2 is weakly column-sum dominant. Thus $M = (DPD_1 + DQD_2)$ is strongly column-sum dominant, and, since

$$|m_{jj} + i\omega d_j| > \sum_{i \neq j} |m_{ij}|$$

for all j, in which d_j is the jth diagonal element of j D, it follows that $\det(M+i\omega D) \neq 0$, which is a contradiction.

2.11 *Definition 2*

With q and p nonnegative integers such that $(p+q) > 0$, let J denote the set of all matrices M such that $M = I_q \oplus M_1 \oplus M_2 \oplus \ldots \oplus M_p$ with

$$M_k = \begin{pmatrix} 1 & -\delta_r^{(k)} \\ -\delta_f^{(k)} & 1 \end{pmatrix}$$

and

$$0 < \delta_r^{(k)} < 1$$

$$0 < \delta_f^{(k)} < 1$$

for all k = 1, 2, ..., p.*

2.12 *Definition 3*

With q and p nonnegative integers such that $(p+q) > 0$, let $J(\alpha)$ denote the set of all matrices M such that $M = I_q \oplus M_1 \oplus M_2 \oplus \ldots \oplus M_p$ with

$$M_k = \begin{pmatrix} 1 & -\delta_r^{(k)} \\ -\delta_f^{(k)} & 1 \end{pmatrix}$$

and

$$0 < \delta_r^{(k)} \leq \alpha_r^{(k)}$$

$$0 < \delta_f^{(k)} \leq \alpha_f^{(k)}$$

for all k = 1, 2, ..., p.*

* As suggested, if q = 0, then $M = M_1 \oplus M_2 \oplus \ldots \oplus M_p$, while if p = 0, then $M = I_q$.

2.13 *Theorem 5*

Let $T \in \mathcal{J}$, let H be a real matrix of order (2p+q), and suppose that $M^{-1}H \in P_0$ for all $M \in \mathcal{J}(\alpha)$. Then

$$(T+D_1)^{-1}(H+D_2) \in P_0$$

for all diagonal matrices $D_1 \geq 0$ and $D_2 \geq 0$.

2.14 *Proof of Theorem 5*

Suppose that for some $D_1 \geq 0$ and $D_2 \geq 0$

$$(T+D_1)^{-1}(H+D_2) \notin P_0.$$

Then there exists [2] a diagonal matrix $D > 0$ such that

$$(T+D_1)^{-1}(H+D_2) + D$$

is singular. It follows that

$$H + \Delta + TD$$

is singular, in which $\Delta = D_2 + D_1D$. Since

$$\Delta + TD = M(\Delta + D)$$

in which $M \in \mathcal{J}(\alpha)$, it follows that

$$H + M(\Delta + D)$$

is singular, and therefore that

$$M^{-1}H + (\Delta + D)$$

is singular, which is a contradiction since [1] $M^{-1}H \in P_0$ and $(\Delta+D)$ is a diagonal matrix with positive diagonal elements.

2.15 *Theorem 6*

Let $M^{-1}G \in P_0$ for all $M \in \mathcal{J}$, and let $\det G \neq 0$. Let R be as defined in Section 1.4. Then for any $T \in \mathcal{J}$

$$(T+D_1)^{-1}[(I+GR)^{-1}G + D_2] \in P_0$$

for all diagonal matrices $D_1 \geq 0$ and $D_2 \geq 0$.

2.16 *Proof of Theorem 6*

Since $\det G \neq 0$ and $M^{-1}G \in P_0$ for all $M \in J$, it follows (see proof of Theorem 7 of [2]) that

$$M^{-1}(I+GR)^{-1}G \in P_0$$

for all $M \in J$.

Suppose that for some $T \in J$ and some $D_1 \geq 0$ and $D_2 \geq 0$

$$(T+D_1)^{-1}[(I+GR)^{-1}G + D_2] \notin P_0.$$

Then, following the proof of Theorem 5, we would have

$$\det\{M^{-1}(I+GR)^{-1}G + (\Delta+D)\} = 0$$

for some $M \in J$ and some matrix $(\Delta+D)$ with positive diagonal elements, which is a contradiction.

ACKNOWLEDGEMENT

The writer is indebted to A.N. Willson, Jr. for carefully reading the draft.

REFERENCES

1. SANDBERG, I.W. and WILLSON, A.N. JR., (January 1969) Some Theorems on Properties of DC Equations of Nonlinear Networks, *B.S.T.J.*, *48*: (1), pp. 1-34.

2. SANDBERG, I.W. and WILLSON, A.N. JR., (May-June 1969) Some Network-Theoretic Properties of Non-Linear DC Transistor Networks, *B.S.T.J.*, *48*: (5), pp. 1293-1312.

3. VARGA, R.S. (1962) *Matrix Iterative Analysis*, Prentice-Hall, Englewood Cliffs, New Jersey, p. 23.

4. STERN, T.E. (1965) *Theory of Nonlinear Networks and Systems*, Addison-Wesley, Reading, Mass., pp. 42-43.

5. PALAIS, R.S., Natural Operations on Differential Forms, *Trans. Amer. Math. Soc., 92*: (1), pp. 125-141.

6. HOLZMANN, C.A. and LIU, R. (1965) On the Dynamical Equations of Nonlinear Networks with n-Coupled Elements, *Proc. Third Ann. Allerton Conf. on Circuit and System Theory*, U. of Illinois, pp. 536-545.

7. GOLDSTEIN, A.A. (1967) *Constructive Real Analysis*, Harper & Row, New York, pp. 41-45.

8. SANDBERG, I.W. (January 1969) Some Theorems on the Dynamic Response of Nonlinear Transistor Networks, *B.S.T.J., 48*: (1), pp. 35-54.

9. HAMMING, R.W. (1962) *Numerical Methods for Scientists and Engineers*, McGraw-Hill Book Co., New York.

10. RALSTON, A.A. (1965) *A First Course in Numerical Analysis*, McGraw-Hill Book Co., New York.

11. HACHTEL, G.D. and ROHRER, R.A. (November 1967) Technique for the Optimal Design and Synthesis of Switching Circuits, *Proc. of the IEEE, 55*: (11) pp. 1864-1876.

12. SANDBERG, I.W. and SCHIMAN, H. (April 1968) Numerical Integration of Systems of Stiff Nonlinear Differential Equations, *B.S.T.J., 47*: (4), pp. 511-527.

13. CALAHAN, D.A. (1968) Efficient Numerical Analysis of Non-Linear Circuits, *Proc. Sixth Ann. Allerton Conf. on Circuit and System Theory*, U. of Illinois, pp. 321-331.

FUNDAMENTALS OF NONLINEAR NETWORK ANALYSIS

THOMAS E. STERN

Department of Electrical Engineering
Columbia University, New York

The purpose of these notes is to survey the techniques of analysis of nonlinear networks by digital computer. In part I we consider the problem of reducing information of network topology and element terminal characteristics to a canonical form suitable for computer use. In part II we develop a class of numerical methods especially suited to integration of the differential equations of these networks. Emphasis is placed on some of the more recent developments in the field.

PART I. FORMULATION OF THE NETWORK EQUATIONS

In dealing with nonlinear networks there is no limit to the amount of generality one may accept. Therefore, to keep the problem within bounds we shall postulate a certain limited class of lumped networks which, nevertheless, includes almost all of the practical problems one would like to deal with at present. We shall show that for this class of networks it is always possible by a systematic procedure, to formulate a set of constrained differential equations in the form

$$\dot{x} = f(t, x, u) \tag{1a}$$

$$0 = g(t, x, u) \tag{1b}$$

where x represents a set of n state variables, u a set of k auxiliary variables, (1a) a set of n first order differential equations and (1b) a set of k implicit constraint relations. In general it will *not* be possible to eliminate the auxiliary variables u by solving (1b) for

$$u = h(t, x)$$

and substituting into (1a). Therefore we shall develop methods of dealing with these equations in their constrained form.

1.1 *Equation Formulation*

Let the network be described by a connected graph G in which each branch is classified as either independent voltage source (V), capacitive (C), resistive (R), inductive (L) or independent current source (J). It will be assumed that:

1) The terminal characteristics of all capacitive branches are given in the "charge-controlled" form:

$$e_C = f_C(t, q_C) \qquad \dot{q}_C = i_C \tag{2}$$

where e_C, q_C, i_C are vectors representing respectively the capacitor voltages, charges and currents. Any capacitor whose terminal relation can be written in the form

$$e_{C_i} = S_i(t)\, q_{C_i} \qquad \text{where } S_i(t) > 0 \; \forall \; t$$

will be called a linear positive capacitor (LPC). Coupling is permitted among all capacitive branches except LPC's.

2) The terminal characteristics of all inductive branches are given in the "flux-controlled" form:

$$i_L = f_L(t, \lambda_L) \qquad \dot{\lambda}_L = e_L \tag{3}$$

where i_L, λ_L, and e_L represent respectively the inductor currents, flux-linkages and voltages. Linear positive inductors (LPL) are defined in the same way as LPC's. Coupling is permitted among all inductive branches except LPL's.

3) The terminal characteristics of the resistive branches are given in the form

$$w = f_R(t, y) \tag{4}$$

where each element w_i of the vector w and each element y_i of y is a resistive branch voltage or current, and y_i is a current (voltage) whenever w_i is a voltage (current). We partition the resistive branches into three disjoint classes:

a) Linear positive resistance (LPR), defined as is the LPC.

b) Voltage controlled nonlinear resistance (VCNR) whenever the branch is not LPR and w_i is a *current*.

c) Current controlled nonlinear resistance (CCNR) whenever the branch is not LPR and w_i is a *voltage*.

We shall call w the controlled variables and y the controlling variables. (All elements which are not LPC, LPL or LPR will be termed *nonlinear* even though their terminal relations may be linear.) Coupling is permitted among all resistive branches except LPR's. (Note that the classifications VCNR and CCNR are based on the way the terminal relations are *given*, even though it may be possible to express them in other forms.)

4) There are no loops consisting of voltage sources only or of voltage sources and capacitors only, nor cut-sets consisting of current sources only or current sources and inductors.

(Certain of these assumptions were made merely to simplify the exposition. For example, assumption (4) could be partially relaxed without any radical change in the results which follow.)

The network equations are obtained by appropriately combining the terminal relations (2,3,4) with the Kirchhoff law constraints defined by the graph G. We begin by selecting a "proper tree" defined to be one whose elements are chosen in the following order of priority: LPC's, nonlinear capacitors, independent voltage sources, VCNR's, LPR's, CCNR's, independent current sources, nonlinear inductors, LPL's. Following [1] the state variables are chosen as cutset charges q and loop flux linkages λ. When there are no all-capacitor loops and no all inductor cutsets, each capacitor charge and inductor flux-linkage constitutes a state variable. Otherwise the state variables are linear combinations of these charges and flux linkages. (See Appendix.)

With respect to any proper tree T, we define certain "excess" elements,* associating with each excess element an "auxiliary variable" as follows:

EXCESS ELEMENTS	AUXILIARY VARIABLES
All nonlinear capacitive chords of T	charges q_c
All nonlinear inductive branches of T	flux linkages λ_ℓ
All VCNR branches of T	voltages e_y
All CCNR chords of T	currents i_z

Note that for an element to be excess it must be nonlinear and it must appear in a certain topological configuration. (For example, a nonlinear capacitor would be excess if and only if it occurred in an all-capacitor loop.)

It is now possible to reduce all of the network data by *linear operations* only to the form

---

*Our usage of the term "excess" is somewhat different from that commonly appearing in the literature.

$$\dot{q} = f_q(t, q, \lambda, q_c, \lambda_\ell, e_y, i_z) \tag{5a}$$

$$\dot{\lambda} = f_\lambda(t, q, \lambda, q_c, \lambda_\ell, e_y, i_z) \tag{5b}$$

$$0 = g_c(t, q, \lambda, q_c, \lambda_\ell, e_y, i_z) \tag{5c}$$

$$0 = g_\ell(\quad '' \quad) \tag{5d}$$

$$0 = g_y(\quad '' \quad) \tag{5e}$$

$$0 = g_z(\quad '' \quad) \tag{5f}$$

Note that Eqs. (5) are in the general canonical form (1). The state variables are $x = (q, \lambda)$, the auxiliary variables, $u = (q_c, \lambda_\ell, e_y, i_z)$, and there are exactly as many constraint equations as there are auxiliary variables. In fact, there is a one-one correspondence between constraint equations and excess elements. A derivation and explicit representation of the functions f and g in (5) is presented in the Appendix. It is worth noting that the selection of the proper tree and the formulation of these functions is a straightforward procedure for a digital computer.

Example

Consider the network of Fig. (1a) (ignoring the dashed elements). A proper tree chosen according to the aforementioned priority of branch selection is shown in Fig. (1b). Following (A-7a) and (A-7b) the state variables are chosen as

$$Q_1 = q_7 - q_1 \qquad \Lambda_1 = \lambda_{13} - \lambda_5$$

$$Q_2 = q_8 + q_1 \qquad \Lambda_2 = \lambda_{13} - \lambda_6$$

$$Q_3 = q_9$$

The excess elements are branches 1,13,11,3, and the

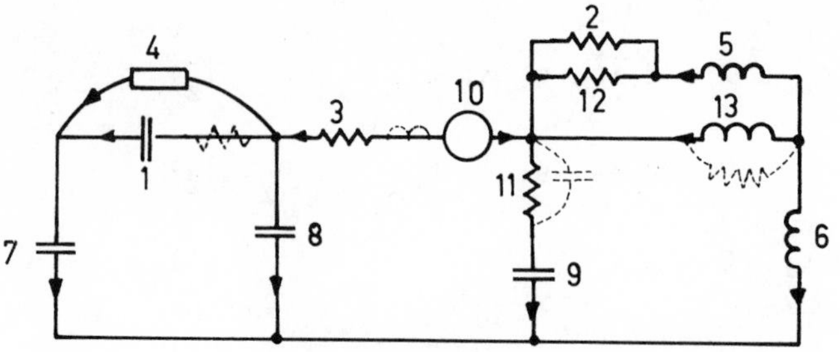

NETWORK

(Links)	Terminal Relations (Tree branches)
$e_1 = f_1(q_1)$	$q_7 = C_7 e_7$
$e_2 = R_2 i_2$	$q_8 = C_8 e_8$
$e_3 = f_3(e_{11}, i_3)$	$e_9 = f_9(q_9)$
$i_4 = J$	$e_{10} = v$
$\lambda_5 = L_5 i_5$	$i_{11} = f_{11}(e_{11}, i_3)$
$\lambda_6 = L_6 i_6$	$e_{12} = R_{12} i_{12}$
	$i_{13} = f_{13}(\lambda_{13})$

1a

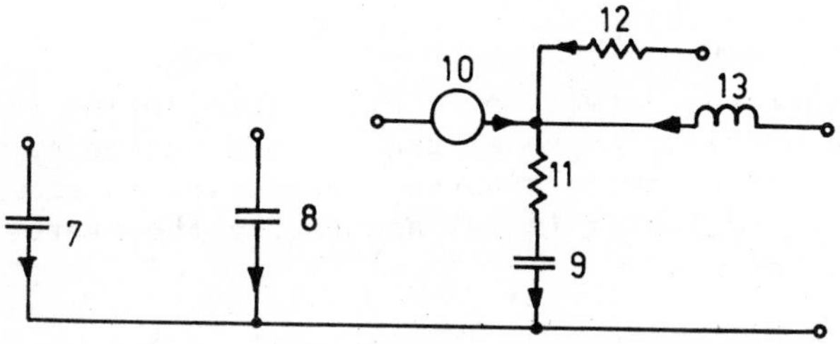

TREES

Fig. 1.

auxiliary variables are q_1, λ_{13}, e_{11}, i_3. The differential equations become

$$\dot{Q}_1 = J$$

$$\dot{Q}_2 = -J + i_3$$

$$\dot{Q}_3 = -i_3 - (\lambda_{13} - \Lambda_2)/L_6 \qquad (6a)$$

$$\dot{\Lambda}_1 = R(\lambda_{13} - \Lambda_1)/L_5$$

$$\dot{\Lambda}_2 = -e_{11} - f_9(Q_3)$$

where $R = R_{12}\, R_2/(R_{12} + R_2)$

and the constraint relations are

$$0 = -f_1(q_1) - (Q_1 + q_1)/C_7 + (Q_2 - q_1)/C_8$$

$$0 = -f_{13}(\lambda_{13}) - (\lambda_{13} - \Lambda_1)/L_5 - (\lambda_{13} - \Lambda_2)/L_6$$

$$0 = -f_{11}(e_{11}, i_3) - i_3 - (\lambda_{13} - \Lambda_2)/L_6$$

$$0 = -f_3(e_{11}, i_3) - (Q_2 - q_1)/C_8 - f_9(Q_3) - e_{11} - V$$

Eq. (6) is in the canonical form (5) where the variables in (5) are identified with those in (6) as follows:

$$q = (Q_1, Q_2, Q_3) \qquad q_c = q_1, \qquad \lambda_\ell = \lambda_{13}$$

$$\lambda = (\Lambda_1, \Lambda_2) \qquad e_y = e_{11}, \qquad i_z = i_3$$

1.2 *Augmentation*

The constrained Eqs. (5) are not in a form suitable for numerical integration. Furthermore, in the general case, there will be no systematic way of eliminating the constraints and the auxiliary variables. Therefore as an artifice for handling the constraints we shall *augment* the original network by inserting a number of "stray" network elements of *infinitesimally small values*. The augmentation will be performed in such a way as to convert each implicit constraint equation to a differential equation. The procedure is illustrated in Fig. 2. We insert exactly as many augmenting elements as there are "excess" elements in the original network. Each of the

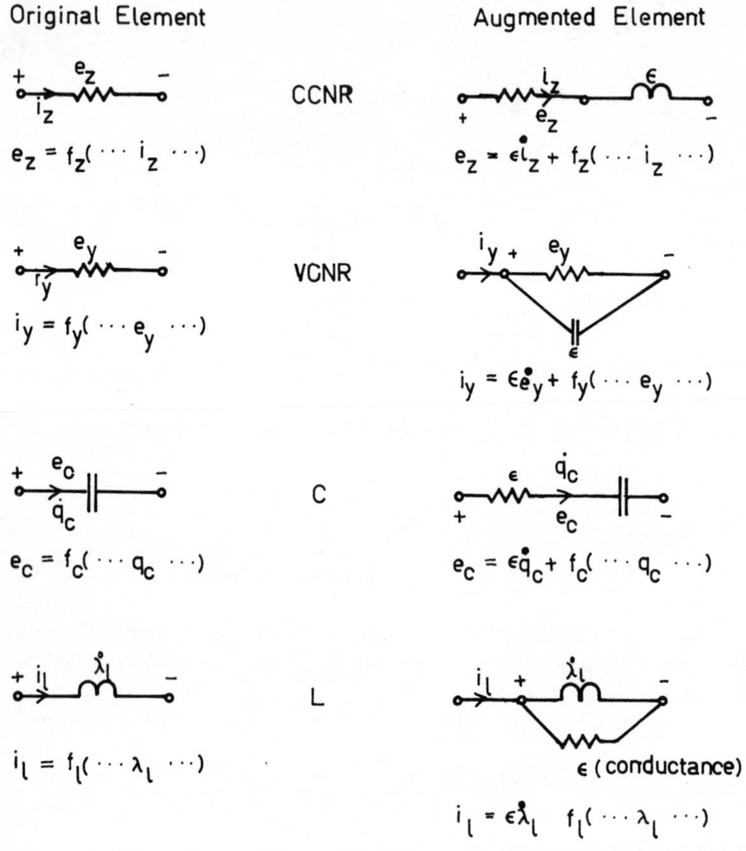

Fig. 2. Augmentation

four classes of excess elements requires a different type of augmentation. The case of the excess VCNR's will be taken as an illustrative example. As shown in Fig. 2, each excess VCNR is augmented by a stray capacitance of arbitrarily small positive value ε. Before augmentation the terminal characteristic of the element is of the form $i_y = f_y(\ldots e_y \ldots)$, that is, the current i_y through the element may depend on its own voltage e_y as well as various other variables. After addition of the capacitance ε the terminal relation becomes

$$i_y = \varepsilon \dot{e}_y + f_y(\ldots e_y \ldots)$$

Thus, if we now let i_y and e_y be vectors representing respectively the currents and voltages on all excess VCNR's then the effect of the augmentation is to add a term of the form $\varepsilon \dot{e}_y$ to all equations in which the currents i_y appear. It can be seen from Eq. (A-9e) that these currents appear in (and only in) Eq. (5e), so that the net result of this augmentation is to add the term $\varepsilon \dot{e}_y$ to the left-hand side of Eq. (5e) converting it to a differential equation. Similar reasoning follows for the other three types of augmentation illustrated in Fig. 2. The end result is that the constraint equations (5c-f) are now converted to the following differential form.*

$$\varepsilon \dot{q}_c = g_c(t, x, u) \tag{7a}$$

$$\varepsilon \dot{\lambda}_\ell = g_\ell(t, x, u) \tag{7b}$$

$$\varepsilon \dot{e}_y = g_y(t, x, u) \tag{7c}$$

$$\varepsilon i_z = g_z(t, x, u) \tag{7d}$$

Note that the order of the augmented system is equal to the order of the unaugmented system plus the number of excess elements. The augmentation with small stray elements described in Fig. 2 can be viewed as a physical justification for the conversion of Eq. (5c-f) to the form (7). However, once the procedure is understood, there

---

*It has been shown [1] that when no coupling is present, the augmentation procedure described is *minimal* in the sense that the number of augmenting elements is the least possible that will eliminate the constraints. When coupling is present other methods are sometimes required to produce a minimal augmentation.

is no need to interpret it in terms of network modification. One simply identifies the excess elements, formulates the constraints and adds the appropriate terms to the constraint equations. The canonical form (1) is thus changed by augmentation to

$$\dot{x} = f(t, x, u) \tag{8a}$$

$$\varepsilon \, \dot{u} = g(t, x, u) \tag{8b}$$

Returning to the example, we find that the left-hand side of the constraint equations (6b) must be augmented by the respective terms $\varepsilon\dot{q}_1$, $\varepsilon\dot{\lambda}_{13}$, $\varepsilon\dot{e}_{11}$, $\varepsilon\dot{i}_3$. Note that these terms correspond physically to the insertion of four elements of value ε shown dashed in Fig. 1a.

The principal justification for the augmentation procedure outlined above is the fact that every physical network element has associated with it some "stray" or "parasitic" energy storage or dissipation which is usually neglected in its mathematical model. Therefore our augmented network model should be a more realistic representation of the true behavior of the physical network than is the constrained model. The idea of augmentation is, of course, not new: it is commonly used to explain the discontinuous behavior of many types of nonlinear oscillators of the "relaxation" type.

1.3 *Limiting Solution of the Augmented Equations*

We must now relate some solution of the augmented system (8) to the behavior of the physical system originally described by (1). Since the parameter ε was not even present in the original network model, it is reasonable to attempt to minimize the differences between (8) and (1) by considering the *limiting* behavior of (8) as $\varepsilon \to 0+$ as a "true" representation of the physical network.

Thus, if $z(t, \varepsilon) = [x(t, \varepsilon), u(t, \varepsilon)]$ represents a solution of (8) for $\varepsilon>0$, we attempt to determine the limiting solution,

$$z^0(t) = [x^0(t), u^0(t)] \overset{\Delta}{=} \lim_{\varepsilon\to 0+} [x(t, \varepsilon), u(t, \varepsilon)] \tag{9}$$

Although systems of the form (8) have been the subject of considerable study (see, for example [2], [3], [4]), conditions under which $z^0(t)$ exists have not been determined in the most general case. Nevertheless, there is one result due to Tihonov [2], which is pertinent to our problem:

Let $z(t, 0)$ satisfy (8) for $\varepsilon = 0$ on some interval $[t_1, t_2]$, and let* $g_u(t, x(t, 0), u(t, 0))$ be stable** on that interval. If $z(t_1, \varepsilon) = z(t_1, 0)$ then

$$z^0(t) = z(t, 0) \text{ on } [t_1, t_2]$$

In other words, if g_u is *stable* along a solution of the constrained system then that solution is identical to a limiting solution of the augmented system starting at the same initial state. Generally, $z^0(t)$ is characterized by one or more continuous segments along which g_u is stable, separated by "critical points" at which g_u is not stable. At these points various pathological situations may arise in unusual cases. However, in the more common "well-behaved" cases the work of Pontriagin and Mishchenko [3],[4] offers a good guide to the behavior of $z^0(t)$. Let t_c represent a critical point at the end of an interval $[t_1, t_c]$ along which g_u is stable. In the "well-behaved" case $x(t)$ will be continuous at t_c and $u(t)$ will have at worst a simple discontinuity. If $u(t_c-)$ and $u(t_c+)$ denote respectively the left-hand and right-hand limits of $u(t)$ at t_c then $u(t_c+)$ will be given by

$$u(t_c+) = u_\infty = \lim_{\gamma\to\infty} u(\tau) \tag{10}$$

where $u(\tau)$ is the solution of

---

*Throughout these notes, subscript notation denotes Jacobian matrices. In this case, $[g_u]_{ij} = [\partial g_i/\partial u_j]$.

**We shall call a square matrix *stable* if all of its eigenvalues have negative real parts.

$$\frac{du}{d\tau} = g[t_c, x(t_c), u] \tag{11}$$

Intuitively one would expect that $u(0) = u(t_c-)$ would be the appropriate initial condition for (11). However, $u(t_c-)$ will generally be a singular point of (11) and would hence give the incorrect result, $u(t_c+) = u(t_c-)$. Instead, we choose $u(0)$ as follows: Let D^c denote the complement of the domain of attraction of $u(t_c-)$. In the "well-behaved" case, there will exist a neighborhood N of $u(t_c-)$ such that if $u(0) \varepsilon N \cap D^c$ then $u(t_c+)$ will be given by (10). Any $u(0)$ satisfying this condition can be used in (11).

Eqs. (10) and (11) can be justified intuitively by the fact that in the limit as $\varepsilon \to 0$, the auxiliary variables u move infinitely faster than the state variables x. Thus, at the critical points it is reasonable to suspend the movement of x and allow u to attain the steady state dictated by (11), henceforth to be termed the "fast equations". Fig. (3) illustrates the typical behavior of $z^0(t)$ at a critical point.

Thus, in the well-behaved cases,* $z^0(t)$ can be constructed by solving (1) along intervals of continuity (g_u stable) and applying equations (10), (11) at critical points (g_u not stable). Some numerical methods based on these ideas are discussed in Part II.

PART II. NUMERICAL SOLUTION OF THE NETWORK EQUATIONS

The digital computer would be a truly effective tool for engineering design of nonlinear networks if it were possible to compute the time response of a network of moderate complexity (say, 20-40 elements) to engineering

---

*In pathological cases, $z^0(t)$ may not exist. Then the only possible conclusions are either (1) the physical system on which (8) is based exhibits some type of structural instability, or (2) the mathematical model is faulty, in which case it must be abandoned for a more suitable one.

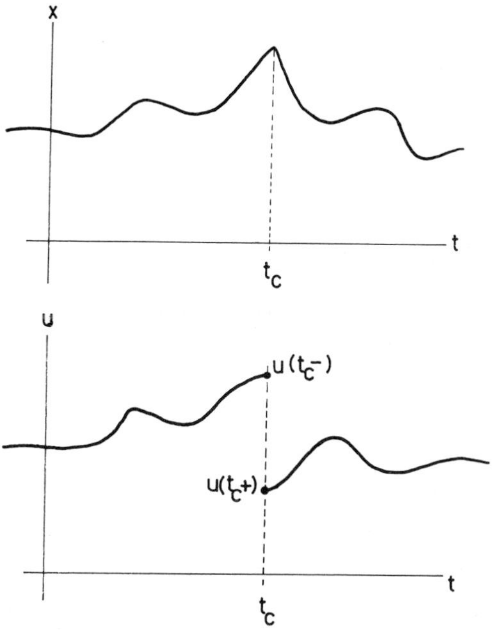

Fig. 3. Behavior of limiting solution at a critical point

accuracy (a few percent) in a matter of several seconds. Except in some unusual cases, this goal has not yet been achieved. Nevertheless, recent efforts to achieve it have produced considerable insight into the capabilities and limitation of various classes of numerical methods, so that much progress has been made toward designing efficient numerical methods for network problems. In this section we shall discuss some of these recent developments.

2.1 *General Characteristics of Numerical Solutions*

Initially we shall assume for simplicity that we are solving the autonomous unconstrained system

$$z = f(z) \qquad z(0) = z_0 \tag{12}$$

Generalization to the nonautonomous case is straightforward.

In order to solve (12) numerically, time must be discretized into a sequence $\{t_k\}$

$$t_0 = 0$$

$$t_{k+1} = t_k + h_k$$

where h_k is the *stepsize* for the k-th time step. (Generally, in solving highly nonlinear equations a variable stepsize is required.) We shall denote the computed approximation to the solution of (12) at time t_k by z_k, and the exact solution by $z(t_k)$. Some desirable features of any algorithm for solving (12) numerically are:

1) *Local errors*: Errors introduced at each step in the computation should be small.

2) *Stability*: Errors introduced at one step in the computation should not propagate through subsequent steps in an unstable manner; i.e., their effects should be attenuated rather than amplified at future steps.

3) *Stepsize*: h_k should be limited only by the desired time resolution of the solution.

4) *Computational Complexity*: The number of computations per step should be kept to a minimum, preferably no more than one function evaluation.

Items (1) and (2) above pertain to the validity and accuracy of an algorithm, while (3) and (4) deal with its efficiency.

Most of the "conventional" numerical integration procedures are not particularly suited to solving nonlinear network equations primarily because they are prone to instability (item (2) above) when solving "stiff" systems of differential equations. A system (12) is said to be stiff when the ratio of the modulus of the largest to the smallest eigenvalue of f_z (the "stiffness ratio") is large. Stiffness is a typical characteristic of the equations of many nonlinear networks: especially pulse and digital networks. It occurs because the time constants associated with the active elements, such as transistors, are usually orders of magnitude smaller than those associated with the rest of the network. In practical cases, stiffness ratios of the order of 10^6 are not uncommon.

Recently it has been pointed out by several investigators ([5],[6],[7]) that "implicit" methods are useful in overcoming the aforementioned stability problems. Because of this and because, in any case, we wish to solve systems subject to implicit constraints, we shall deal with a certain class of implicit algorithms together with some closely related explicit methods. In the remainder of this section we develop certain general relations which will be useful in analyzing the proposed algorithms.

Let a general method for integrating (12) be defined by

$$\Psi(Z, z; h) = 0 \tag{13}$$

where $z = z_k$ represents the computed state of (12) at time t_k

$h = h_k$ represents the stepsize at the k-th step of the computation

$Z = z_{k+1}$ represents the next computed state.

Clearly (13) must have the property

$$\Psi(z, z; 0) = 0 \quad \text{for all } z$$

If (13) can be solved explicitly for Z in closed form, it is said to be an explicit method; otherwise it is implicit. Methods of the above type are called *one-step* methods because computation of the next state depends only on the present computed state and no previous ones.

The *local truncation error* (LTE) incurred in using (13) to compute the next state of (12) is defined as

$$e(z; h) = Z - z(t_{k+1})$$

where $z(t_{k+1})$ is the exact solution of (12) when $z(t_k) = z$, and Z is the solution of (13).* Thus the LTE represents the error introduced in going from an *exact* present state to a computed next state.

A method (13) is said to be of order p (p a positive integer) if $e(z; h) = O(h^{p+1})$. Clearly, higher order methods should generate smaller truncation errors. However, it turns out that in going to higher order methods one often must sacrifice stability and computational simplicity.

To investigate the *propagation* of errors; i.e., stability, consider a sequence $\{Z_k\}$ generated by repeated application of (13) at *fixed* stepsize h. The Z_k satisfy

$$Z_0 = z_0$$

$$\Psi(Z_{k+1}, Z_{k;h}) = 0$$

Let

$$E_k \triangleq Z_k - z(t_k)$$

Using Taylor's formula we find that, to within second order in E_k, the propagation of the error E_k is governed by**

$$E_{k+1} = \Phi_k E_k + e_k \qquad (14)$$

---

*We shall assume throughout that (12) has unique solutions, all derivatives which are used exist and are continuous, and conditions for unique *local* solvability of (13) for Z are satisfied. In particular, when we refer to *the* solution of (13) we mean the unique solution for Z in a neighborhood of z.

**It will be helpful to fix ideas on the scalar case of (12). However, the development is written throughout for the vector case. Thus, for example, E_k, e_k are treated as column vectors, and Ψ_Z is a square matrix whose elements are $\partial\Psi_i/\partial Z_j$.

where

$$\Phi_k = \left[-\Psi_Z^{-1} \ \Psi_z \right]_{[z(t_k),\ z(t_k);\ h]} \tag{15}$$

and e_k represents the LTE introduced at the k-th step. (Rounding error is ignored here.)

We shall say that a method (13) is *stable* when solving (12) if bounded $\{e_k\}$ produce bounded $\{E_k\}$. A sufficient condition for stability of (13), subject to the assumption of small E_k, is that

$$||\Phi_k|| \leq c < 1 \quad \text{for all } k \tag{16}$$

Stability generally depends upon the particular equation being solved as well as the stepsize employed. Experience has shown that one of the fundamental difficulties in solving network equations is that most conventional algorithms become unstable unless h is kept extremely small relative to the total time interval over which the solution is computed. The reason for this is not hard to find. If λ_M represents the modulus of the largest eigenvalue of f_z in (12), then it can be shown by evaluation of (15) that most algorithms become unstable when $h \geq C/\lambda_M$, where the constant C varies from one method to another, but is generally ~ 1. Since the total solution interval will normally be of the order of the reciprocal of the modulus of the *smallest* eigenvalue of f_z, the number of computation steps required to insure stability will be of the order of the stiffness ratio for the system - a number that can easily be unacceptably large. Clearly, any method which has the above stability limitations is a very poor choice in problems involving stiff systems.

Recognizing this problem, Dahlquist has defined "A-stability". A method (13) is said to be A-stable if for *all* positive h, $z_k \to 0$ as $k \to \infty$, when solving the *linear* equation $\dot{z} = Az$, where A is any stable matrix. He also pointed out that among a certain commonly used class of methods (the linear multi-step methods) two necessary conditions for A-stability are (1) the method must be implicit, and (2) it must be of order no higher than two. A simple one-parameter family of A-stable methods in this category is

$$\Psi(Z,z;h) = Z - z - h[\rho f(z) + (1-\rho)f(Z)] \quad 0 \leq \rho \leq \tfrac{1}{2} \tag{17}$$

Eq. (17) is known as the trapezoidal rule for $\rho = 1/2$ and the Backward Euler method for $\rho = 0$.

Convergence: In choosing an implicit method of the form (13) one must consider how (13) is to be solved for Z. Some type of iteration is generally required. The simplest approach, direct substitution, when applied to the trapezoidal rule, for example, gives the iteration formula

$$Z^{(r+1)} = z + \tfrac{1}{2} h \, [f(z) + f(Z^{(r)})] \tag{18}$$

where $Z^{(r)}$ is the r-th approximation to the solution of (17). Since this method will fail to converge if h is too large, the A-stability property of (17) is nullified by lack of convergence of (18).* Noting this problem certain authors ([5],[6]) have proposed methods based on Newton-Raphson (N-R) iteration. The N-R formula applied to a general method of the form (13) gives

$$Z^{(r+1)} = Z^{(r)} - \left[\Psi_Z^{-1} \, \Psi \right]_{(Z^{(r)},\, z;h)} \tag{19}$$

---

*Strictly speaking, A-stability only applies to the linear case of (12), in which case (17) can be solved explicitly (with $\rho = 1/2$) to give

$$Z = \left(I - \tfrac{1}{2} hA\right)^{-1} \left(I + \tfrac{1}{2} hA\right) z$$

However, it is reasonable to assume that if a method is A-stable it should also perform well when solving the general case of (12) with f_Z a stable matrix. We shall return to this question in the next Section.

Assuming that the solution of (13) is sufficiently close to z and a sufficiently good initial trial is chosen, a sufficient condition for convergence of (19) is

$$||N|| < 1$$

where

$$N_{ij} = \left[\sum_k \left(\Psi_Z^{-1} \frac{\partial \Psi_Z}{\partial Z_j} \Psi_Z^{-1} \right)_{ik} (\Psi)_k \right]_{(z,z;h)} \tag{20}$$

In the scalar case this becomes

$$| \Psi_{ZZ} \Psi / \Psi_Z^2 | < 1$$

In particular, N-R applied to the trapezoidal rule in the scalar case gives

$$Z^{(r+1)} = Z^{(r)} - [1 - \tfrac{h}{2} f_z(Z^{(r)})]^{-1}$$
$$\times [Z^{(r)} - z - \tfrac{h}{2}(f(z) + f(Z^{(r)}))] \tag{21}$$

(Note that N-R iteration requires knowledge of the derivatives of f.)

Using (20) we find that (21) converges for

$$| 2f_{zz} f/f_z^2 | < 1$$

where $f_z < 0$.

Thus convergence of (21), and hence the validity of the implicit method (13) coupled with N-R iteration depends upon second derivatives of f, but does *not* depend (directly) on h.

Any implicit method (13) can be converted to an explicit one by applying a fixed number of iterations to it. In particular the general explicit formula

$$Z = z - \left[\Psi_Z^{-1} \Psi \right]_{(z,z;h)} \tag{22}$$

is derived by applying one N-R iteration to (13), taking $z^{(0)} = z$ as the initial trial.

The explicit formula (22) can serve as a basis for the development of a whole range of algorithms based on the evaluation of the right-hand side of (12) as well as its partial derivatives. If these methods are derived from implicit A-stable methods they should exhibit reasonably stable behavior when applied to stiff systems. Based on these observations we shall develop in the next two sections a class of algorithms for the solution of network equations with or without constraints.

2.2 *Algorithms based on Padé Approximants*

We develop in this section a class of A-stable explicit algorithms for solving (12).

Following Pope [8] we linearize (12) about $z(t_k)$ and integrate from t_k to t_{k+1} to give an approximation to the next state:

$$z(t_{k+1}) \approx z(t_k) + [e^S - I]\, S^{-1}\, hf[z(t_k)] \tag{24}$$

where $S = hf_z$.

Rather than using a series approximation for the matrix exponential in (24) as suggested by Pope, we use the Padé rational approximant:

$$e^S \approx E_{p,q}(S) = [d_{p,q}(S)]^{-1}\, [n_{p,q}(S)] \tag{25}$$

where

$$n_{p,q}(S) = \sum_{k=0}^{q} \frac{(p+q-k)!q!}{(p+q)!k!\ (q-k)!}\, S^k$$

$$d_{p,q}(S) = \sum_{k=0}^{p} \frac{(p+q-k)!p!}{(p+q)!k!\ (p-k)!}\, (-S)^k \tag{26}$$

and $E_{p,q}(S)$ agrees with the series expansion of e^S through terms of order p+q.

Substituting (25) into (24) we obtain a class of explicit methods to be designated as P(p,q), defined by

$$\Psi(Z,z;h) = Z - z - [D(S)^{-1}][N(S)]\ hf(z)$$

where

$$N(S) = n_{p,q}(S) - d_{p,q}(S) \quad \text{and} \quad D(S) = d_{p,q}(S)\ S$$

From the properties of $E_{p,q}(S)$ it is easily established that P(p,q) is A-stable for all $p \geq q$. When solving a nonlinear equation stability is determined by evaluating (15):

$$\Phi_k = \{[d_{p,q}(S)]^{-1}[n_{p,q}(S)] + hQ\}_{[z(t_k),z(t_k);h]} \qquad (27)$$

where Q is a matrix containing second partials of f. Restricting attention to the A-stable methods ($p \geq q$), we can establish from (27) that P(p,q) will be stable when applied to a solution along which f_z is always stable *provided that* the higher derivative term hQ is negligible. However, sufficiently large values of the second partials of f will eventually produce instability at large stepsizes. A counterpart of this stability problem occurs when one applies N-R iteration to an implicit method such as (17). Since convergence of N-R is dependent on the second partials of Ψ (see Eq. (20)) an implicit method with apparently good stability properties may be rendered useless by lack of convergence at large stepsizes.

It is interesting to note that for appropriate values of p and q, P(p, q) reduces to some of the commonly used implicit methods, coupled with one N-R iteration. For example, P(1, 0) corresponds to the Backward Euler method and P(1, 1) to the trapezoidal rule. Since P(p, q) is based on linearization of (23) one cannot expect to obtain an order greater than two for any equation except a linear one. The method P(1, 1) is of second order, and another second order method of some interest is P(2, 1) in which

$$N(S) = 6I - S$$

$$D(S) = 6I - 4S + S^2$$

This method has good stability properties when solving an equation in which f_z is stable, since the first term in (27) tends toward 0 as $h \to \infty$.

Thus, the family P(p, q) with $p \geq q$, appears to be a good candidate for solving network problems, because it requires only one evaluation of f and f_z per step, and its members are A-stable. In the next section we adapt these methods to the solution of constrained differential equations.

2.3 *Numerical methods for constrained equations*

As discussed in section 1.3, solution of the equations of nonlinear networks generally reduces to finding the limiting solution of (8). This in turn requires two different modes of solution:

Mode I: Continuous regime, valid when g_u is stable.
Mode II: Discontinuous regime, valid when g_u is unstable.

The basic structure of an algorithm for computing limiting solutions of (8), incorporating these two modes, is shown in Fig. 4. In the block labelled Mode II, any numerical method for solving unconstrained equations (for example, P(p,q)) can be used to integrate (11) to steady state; i.e., to a value u_∞.

In Mode I, a suitable time step h must be chosen and the state z must be updated in accordance with (1). To develop an algorithm for Mode I, we adapt the methods P(p,q) to the constrained system

$$\dot{x} = f(x, u) \tag{28a}$$

$$0 = g(x, u) \tag{28b}$$

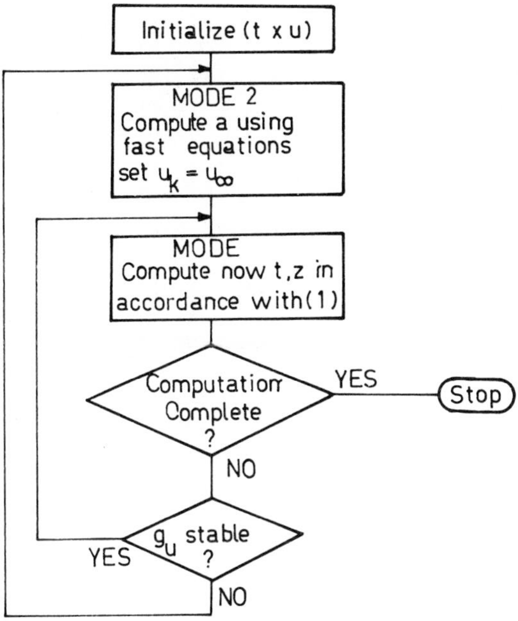

Fig. 4. Flow chart for computation of limiting solution of (8)

as follows:

Let z = (x, u) be the present computed state of (28), let Z = (X, U) be the next computed state, and let

$$u_1 = u - g_u^{-1} g$$

(In this section all functions are assumed to be evaluated at z.) Linearizing the equation

$$\dot{x} = f(x, u_1)$$

about z we obtain

$$\dot{X} \approx B(X - x) + f - f_u g_u^{-1} g \tag{29}$$

where $B = (f_x - f_u g_u^{-1} g_x)$. (Note that the constraint (28b) has been taken into account in the linearization.) Applying the method P(p,q) to (29) and including the constraint we obtain the implicit relation:

$$\Psi(Z,z;h) = \begin{pmatrix} X - x\ D^{-1}Nh[f - f_u g_u^{-1} g] \\ g(Z) \end{pmatrix} = 0 \qquad \begin{matrix} (30a) \\ (30b) \end{matrix}$$

where N and D are the appropriate polynomials in hB.

To derive an explicit formula from (30) we use an N-R iteration leaving (30a) unchanged and giving

$$\begin{aligned} X &= x + D^{-1}Nh[f - f_u g_u^{-1} g] \\ U &= u - g_u^{-1}[g_x D^{-1} Nhf + g] \end{aligned} \tag{31}$$

It will be observed that g_u will be nonsingular in Mode I.

Eq. (31) represents a class of explicit methods for computing solutions of (28) based on the Padé approximants and N-R iteration. It can be shown that these methods are A-stable for the variable x and are of second order if $p + q \geq 2$.* They require only one evaluation of f, g and their derivatives per step and thus should be quite efficient in solving network equations.

Several additional points must be considered in evaluating the algorithm described above. For example, some criterion for control of the step size h_k is required. One possible criterion could be based on $||g||$. If $||g||$ is too large after the completion of a step, this would indicate that the previous stepsize was too large and the computation should be repeated using a

---

*In determining A-stability we assume that (30a) and (30b) are linear and B is stable.

smaller stepsize. Another, more difficult problem is the test for stability of g_u, that determines which mode to enter. To do this correctly would require (1) determination of the characteristic polynomial of g_u and (2) some stability test such as Routh-Hurwitz on that polynomial. To perform these operations each step may consume considerable computation time. A possibly more efficient alternative might be to accept some sufficient condition for stability (for example row dominance of $-g_u$). This would mean that Mode II would be entered more often than necessary, but should not noticeably effect the results of the computation.

REFERENCES

1. OH, S.J., STERN, T.E. and MEADOWS, H.E. (April 12, 13, 14, 1966) On the Analysis of Nonlinear Irregular Networks, *Proceedings of the Symposium of Generalized Networks, Polytechnic Institute of Brooklyn.*

2. TIHONOV, A.N. (1952) Systems of Differential Equations Containing a Small Parameter as Coefficient, *Mat. Sb. 31 (73)* No. 3, pp. 575-586. (In Russian.)

3. PONTRIAGIN, L.S. (1961) Asymptotic Behavior of the Solutions of Systems of Differential Equations with a Small Parameter in the Higher Derivatives, *AMS Translations Ser. 2 vol. 18*, pp. 295f. (Translated from the Russian.)

4. MISHCHENKO, E.F. (1961) Asymptotic Calculation of Periodic Solutions of Systems of Differential Equations Containing Small Parameters in the Derivatives, *AMS Translations Ser. 2 vol. 18*, pp. 187f. (Translated from the Russian.)

5. LINIGER, W. and WILLOUGHBY, R.A. (Dec. 20, 1967) *Efficient Numerical Integration of Stiff Systems of Ordinary Differential Equations*, IBM Research Report RC 1970.

6. SANDBERG, I.W. and SHICHMAN, H. (1968) Numerical Integration of Stiff Nonlinear Differential Equations, *Bell System Tech. J. 47*, pp. 511-527.

7. DAHLQUIST, G.G. (1963) A Special Stability Problem for Linear Multistep Methods, *BIT. 3*, pp. 27-43.

8. POPE, D.A. (1963) An Exponential Method of Numerical Integration of Ordinary Differential Equations, *Communications of the ACM 6:* (8), pp. 491-493.

NONLINEAR NETWORK ANALYSIS

APPENDIX

DERIVATION OF THE AUGMENTED NETWORK EQUATIONS

A.1 *Partitioning of Variables*

Subject to the assumptions of Section 1.1, a fundamental loop matrix B_f based on any proper tree T will take the form

$$B_f = [I,F] = \tag{A-1}$$

		Y				Λ		Q				Z			
C	c	g	G	z	J	γ	Γ	S	s	V	y	R	r	ℓ	L
I_{CC}								F_{CS}							
	I_{cc}							F_{cQ}							
		I_{YY}						F_{YQ}		F_{YV}	F_{Yy}	F_{GR}			
				I_{zz}				F_{zQ}		F_{zV}	F_{zy}	F_{zZ}			
					I_{JJ}			F_{JQ}		F_{JV}	F_{JY}	F_{JZ}			
						$I_{\Lambda\Lambda}$		$F_{\Lambda Q}$		$F_{\Lambda V}$	$F_{\Lambda y}$	$F_{\Lambda Z}$		$F_{\Lambda \ell}$	$F_{\Lambda L}$

where the I's represent identity matrices of appropriate size and the empty partitions of B_f correspond to zero submatrices. (It can be shown that the priority ordering of elements of T always leads to a matrix B_f of this form. The rows and columns of B_f are partitioned into the sets C, c, ..., L, defined as follows:

Group	Symbol	Element	Class	Set
	C	LPC	Capacitors	Chords of T
	c (circled)	Nonlinear	Capacitors	Chords of T
Y	g	VCNR	Resistors	Chords of T
Y	G	LPR	Resistors	Chords of T
	z (circled)	CCNR	Resistors	Chords of T
	J	Independent current sources		Chords of T
Λ	γ	Nonlinear	Inductors	Chords of T
Λ	Γ	LPL	Inductors	Chords of T
Q	S	LPC	Capacitors	Branches of T
Q	s	Nonlinear	Capacitors	Branches of T
	V	Independent voltage sources		Branches of T
	y (circled)	VCNR	Resistors	Branches of T
Z	R	LPR	Resistors	Branches of T
Z	r	CCNR	Resistors	Branches of T
	ℓ (circled)	Nonlinear	Inductors	Branches of T
	L	LPL	Inductors	Branches of T

(The circled symbols indicate excess elements.)

The fundamental cutset matrix Q_f is derived from B_f:

$$Q_f = [-F^t, I] \tag{A-2}$$

and Kirchhoff's laws are expressed in terms of B_f and Q_f as

$$B_f e_N = 0 \qquad \text{(KVL)}$$

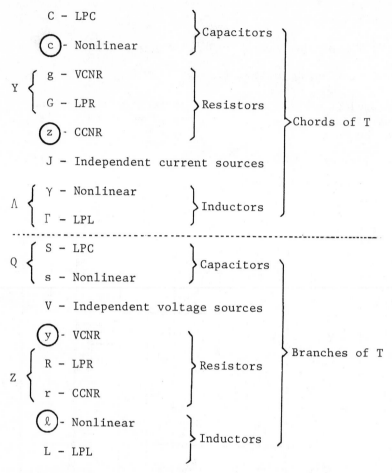

$$Q_f \; i_N = 0 \qquad \text{(KCL)}$$

where e_N and i_N represent the appropriately ordered branch voltages and currents respectively.

A.2 *The Constrained Network Equations*

In what follows we indicate variables associated with various subsets of elements by appropriate subscripts. For example, λ_ℓ is a vector (column matrix) whose elements are the flux linkages in each nonlinear tree branch inductor.

Subject to the above method of partitioning we rewrite the element terminal relations given in (2), (3), (4) as follows: (For economy of notation the argument t will be omitted from all functions.)

Capacitive elements:

$$e_S = S \; q_S \quad , \quad q_C = C \; e_C \qquad \text{(A-3a)}$$

$$e_s = f_s(q_s, q_c) \quad , \quad e_c = f_c(q_s, q_c) \qquad \text{(A-3b)}$$

Inductive elements:

$$i_\Gamma = \Gamma \; \lambda_\Gamma \quad , \quad \lambda_L = L \; i_L \qquad \text{(A-4a)}$$

$$i_\gamma = f_\gamma(\lambda_\gamma, \lambda_\ell) \quad , \quad i_\ell = f_\ell(\lambda_\gamma, \lambda_\ell) \qquad \text{(A-4b)}$$

Resistive elements:

$$i_G = G \; e_G, \qquad e_R = R \; i_R \qquad \text{(A-5a)}$$

$$i_g = f_g(e_g, e_y, i_r, i_z)$$

$$i_y = f_y(e_g, e_y, i_r, i_z) \qquad \text{(A-5b)}$$

$$e_r = f_r(e_g, e_y, i_r, i_z)$$

$$e_z = f_z(e_g, e_y, i_r, i_z)$$

Independent sources:

$$i_J = J \quad , \quad e_V = V \tag{A-6}$$

The state variables $x = (q, \lambda)$, are defined as the cutset charges,

$$q = q_Q - \begin{pmatrix} F^t_{CS} \\ \\ 0 \end{pmatrix} q_C - F^t_{cQ}\, q_c \tag{A-7a}$$

and loop flux linkages

$$\lambda = \lambda_\Lambda + \begin{pmatrix} 0 \\ \\ F_{\Lambda L} \end{pmatrix} \lambda_L + F_{\Lambda \ell}\, \lambda_\ell \tag{A-7b}$$

The auxiliary variables are

$$u = (q_c, \lambda_\ell, e_y, i_z) \tag{A-8}$$

From the Kirchhoff law relations implied by (A-1) and (A-2) we obtain the following canonical form of the network equations.

$$\dot{q} = f_q(x, u) = F^t_{YQ}\, I_Y + F^t_{zQ}\, i_z + F^t_{\Lambda Q}\, I_\Lambda + F^t_{JQ}\, J \tag{A-9a}$$

$$\dot{\lambda} = f_\lambda(x, u) = - F_{\Lambda Z}\, E_Z - F_{\Lambda y}\, e_y - F_{\Lambda Q}\, E_Q - F_{\Lambda V}\, V \tag{A-9b}$$

$$0 = g_c(x, u) = - E_c - F_{cQ}\, E_Q \tag{A-9c}$$

$$0 = g_\ell(x, u) = - I_\ell + F^t_{\Lambda \ell}\, I_\Lambda \tag{A-9d}$$

$$0 = g_y(x, u) = - I_y + F^t_{\Lambda y} I_\Lambda + F^t_{zy} i_z + F^t_{Yy} I_y + F^t_{Jy} J \tag{A-9e}$$

$$0 = g_z(x, u) = - E_z - F_{zQ} E_Q - F_{zy} e_y - F_{zZ} E_Z - F_{zV} V \tag{A-9f}$$

In (A-9) the lower case quantities q, λ, i_z, e_y all represent state or auxiliary variables. The upper case quantities I_Y, E_Z, I_Λ, E_Q, E_c, I_ℓ, I_y, E_z all represent explicit functions of the state and auxiliary variables defined as follows:

Let

$$F_{cQ} = [F_{cS}, F_{cs}], \; S_N = [S^{-1} + F^t_{CS} \, C \, F_{CS}]^{-1}, \quad q = \begin{pmatrix} q_2 \\ \\ q_1 \end{pmatrix}$$

(S_N is positive definite, q_1 is of the dimension of q_S, and q_2 of the dimension of q_s.)

Then, combining Kirchhoff's laws with (A-7a) and (A-3) and eliminating the linear terminal relations (A-3a) we obtain

$$E_Q(q, q_c) = \begin{pmatrix} E_S \\ \\ E_s \end{pmatrix} = \begin{pmatrix} S_N(q_1 + F^t_{cS} q_c) \\ \\ f_s(q_2 + F^t_{cs} q_c, q_c) \end{pmatrix}$$

$$E_c(q, q_c) = f_c(q_2 + F^t_{cs} q_c, q_c) \tag{A-10}$$

Let

$$F_{\Lambda\ell} = \begin{pmatrix} F_{\gamma\ell} \\ \\ F_{\Gamma\ell} \end{pmatrix}, \quad \Gamma_N = [\Gamma^{-1} + F_{\Lambda L} \, L \, F^t_{\Lambda L}]^{-1}, \quad \lambda = \begin{pmatrix} \lambda_1 \\ \\ \lambda_2 \end{pmatrix}$$

(Γ_N is positive definite, λ_1 is of the dimension of λ_γ and λ_2 of the dimension of λ_Γ.)

Then, combining Kirchhoff's laws with (A-7b) and (A-4) and eliminating the linear terminal relations (A-4a) we obtain

$$I_\Lambda(\lambda, \lambda_\ell) = \begin{pmatrix} I_\gamma \\ \\ I_\Gamma \end{pmatrix} = \begin{pmatrix} f_\gamma(\lambda_1 - F_{\gamma\ell}\,\lambda_\ell, \lambda_\ell) \\ \\ \Gamma_N(\lambda_2 - F_{\Gamma\ell}\,\lambda_\ell) \end{pmatrix}$$

$$I_\ell(\lambda, \lambda_\ell) = f_\ell(\lambda_1 - F_{\gamma\ell}\,\lambda_\ell, \lambda_\ell) \qquad \text{(A-11)}$$

Let

$$F_{YQ} = \begin{pmatrix} F_{gQ} \\ \\ F_{GQ} \end{pmatrix}, \quad F_{Yy} = \begin{pmatrix} F_{gy} \\ \\ F_{Gy} \end{pmatrix}, \quad F_{YV} = \begin{pmatrix} F_{gV} \\ \\ F_{GV} \end{pmatrix}$$

$$F_{\Lambda Z} = [F_{\Lambda R}, F_{\Lambda r}], \quad F_{zZ} = [F_{zR}, F_{zr}], \quad F_{JZ} = [F_{JR}, F_{Jr}]$$

Let

$$E_1 = -F_{gQ}\,E_Q - F_{gy}\,e_y - F_{gV}\,V$$

$$E_2 = -F_{GQ}\,E_Q - F_{Gy}\,e_y - F_{GV}\,V$$

$$I_1 = F^t_{\Lambda R}\,I_\Lambda + F^t_{zR}\,i_z + F^t_{JR}\,J$$

$$I_2 = F^t_{\Lambda r}\,I_\Lambda + F^t_{zr}\,i_z + F^t_{Jr}\,J$$

Let

$$G_N = [G^{-1} + F_{GR}\,R\,F^t_{GR}]^{-1} \quad \text{(positive definite)}$$

$$R_N = [R^{-1} + F^t_{GR}\,G\,F_{GR}]^{-1} \quad \text{(positive definite)}$$

Then, combining Kirchhoff's laws with (A-5) and eliminating the linear terminal relations (A-5a) we obtain

$$I_Y(x, u) = \begin{pmatrix} I_g \\ \\ I_G \end{pmatrix} = \begin{pmatrix} f_g(E_1, e_y, I_2, i_z) \\ \\ G_N(E_2 - F_{GR} R I_1) \end{pmatrix}$$

$$E_Z(x, u) = \begin{pmatrix} E_R \\ \\ E_r \end{pmatrix} = \begin{pmatrix} R_N(I_1 + F^t_{GR} G E_2) \\ \\ f_r(E_1, e_y, I_2, i_z) \end{pmatrix}$$

$$I_y(x, u) = f_y(E_1, e_y, I_2, i_z) \tag{A-12}$$

$$E_z(x, u) = f_z(E_1, e_y, I_2, i_z) \tag{A-13}$$

A.3 *Augmentation*

From Fig. 2 we note that when the excess elements are augmented, the expressions (A-10, 11, 12, 13) are modified as follows:

$$E_c \rightarrow \varepsilon \dot{q}_c + E_c$$

$$I_\ell \rightarrow \varepsilon \dot{\lambda}_\ell + I_\ell$$

$$I_y \rightarrow \varepsilon \dot{e}_y + I_y$$

$$E_z \rightarrow \varepsilon \dot{i}_z + E_z$$

Making these substitutions in (A-9) we obtain the augmented constraint equations:

$$\varepsilon \dot{q}_c = g_c(x, u) \qquad \varepsilon \dot{e}_y = g_y(x, u)$$

$$\varepsilon \dot{\lambda}_\ell = g_\ell(x, u) \qquad \varepsilon \dot{i}_z = g_z(x, u)$$

SOME BASIC GENERALIZED NETWORKS REPRESENTING NONLINEAR ELECTRICAL N-PORTS*

DIETER MÖNCH

Institut für Elektrische Nachrichtentechnik, Technische Hochschule Aachen, West Germany

1. INTRODUCTION

In this paper the electrical properties of a number of physical systems are investigated. These systems may be considered as "black boxes" which have a number of "ports" (or terminal pairs) for the exchange of electrical, mechanical and thermal quantities with the outside world. The types of physical systems under investigation are called "generalized networks" for two reasons:
(1) they consist of interconnections of lumped elements;
(2) they comprise electrical networks (more specifically, RLC networks) as special cases.

After introducing definitions of some generalized network elements, a number of simple networks constructed from these elements will be analyzed. In all cases, the aim of the analysis is to find out the relations between the electrical variables under the condition that the mechanical and thermal variables assume their free equilibrium values; in other words, electrical N-ports are investigated which do not contain purely electrical networks but, instead, electro-thermo-mechanical (N+M+L)-ports whose M mechanical ports and L thermal ports are

---

*The work reported in this paper was supported by the *Deutsche Forschungsgemeinschaft* under Grant Mo-162/1.

all connected to passive mechanical or thermal terminations.

Electrical N-ports constructed in this way from generalized networks are found to have quite surprising properties in spite of the simplicity of the physical phenomena that their elements are defined to represent. These N-ports are nonlinear in general, and may be small-signal active and small-signal nonreciprocal N-ports if suitably biassed by electrical d.c. sources.

Some remarks concerning the motivation of this work will be made at the end of the paper.

2. BASIC ASSUMPTIONS, TERMINOLOGY AND SYMBOLS

2.1. *Basic Assumptions*

The physical systems investigated in this paper satisfy the following assumptions.

A1. All variables occurring in the analysis of the systems are lumped variables (rather than densities or field variables).

A2. All system components are lumped elements, i.e. the relations between their port variables do not involve any nonzero delay time.

A3. The physical nature of the system variables is restricted to electrical, mechanical, and thermal variables, i.e. to (time derivatives or time integrals of) electrical voltages and currents, mechanical forces and coordinates, and heat currents and temperatures.

A4. The mechanical variables are restricted to translational movements in fixed directions.

2.2. *Terminology*

N-port: A "black box" defined exclusively by the externally observable relations between the pairs of variables at its N-ports.

Electrical N-port: N-port with exclusively electrical port variables (voltages, currents, charges, flux linkages).

Mechanical N-port: N-port with exclusively mechanical port variables (forces, velocities, momenta, coordinates).

Thermal N-port: N-port with exclusively thermal port variables (heat currents, temperatures).

Classical Electrical N-port: Electrical N-port whose port variable relations satisfy, under every admissible type of operation, the requirements of (1) linearity, (2) time invariance, (3) passivity and (4) reciprocity.

Corresponding definitions hold for Classical Mechanical N-ports and Classical Thermal N-ports. Note that for general (i.e. mixed variable type) N-ports the classification "reciprocal" (and, consequently, the term "classical") cannot be defined unambiguously [1].

Network: A structure consisting of a finite number of P-ports (network elements) and their mutual connections (and, possibly, their connections with N external ports).

Electrical Network: A network whose elements are all electrical P-ports.

Corresponding definitions hold for Mechanical Networks and Thermal Networks.

Converter: A 2-port with variables of different physical kinds at both ports.

(Electromechanical) Transducer: A converter with electrical and mechanical port variables.

Thermistor: A converter with electrical and thermal port variables.

Generalized Network: A network containing converters.

2.3. *Letter Symbols*

A_I, A_{II}	Numerical value of a capacitive transducer of type I or type II
B_I, B_{II}	Numerical value of an inductive transducer of type I or type II
C	Capacitance
E, E*	Energy, Coenergy (supplied to an N-port)
F	Mechanical force
G	Thermal conductance
h	Heat current
i, I, δi	Electrical current, current bias, current signal
K_P, K_N, K_0	Numerical value of a thermistor of type P, type N, or type 0
L	Inductance
q, Q, δq	Electrical charge, charge bias, charge signal
R	Resistance
S	Mechanical stiffness
T, T_a	Absolute temperature, ambient temperature
v, V, δv	Voltage, voltage bias, voltage signal
W, W*	Energy, coenergy (supplied to a network element)
x, $\bar{x}$	Mechanical coordinate, free coordinate
ϕ, Φ, $\delta\phi$	Flux linkages, flux bias, flux signal

Some additional auxiliary symbols will be explained where they are introduced.

2.4. *Graphical Symbols*

Symbol	Description	Port
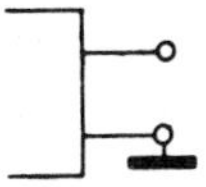	Electrical terminal Electrical reference terminal ("Ground")	Electrical port
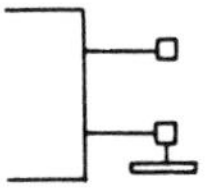	Mechanical terminal Mechanical reference frame	Mechanical port
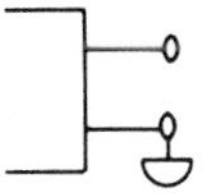	Thermal terminal Ambient temperature bath	Thermal port
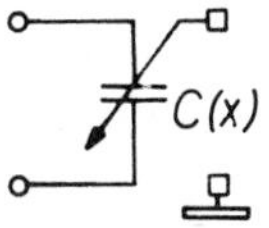	Capacitive transducer	
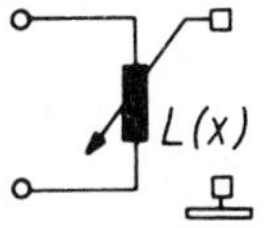	Inductive transducer	
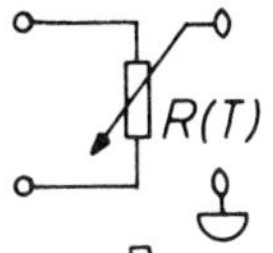	Thermistor	
	Mechanical spring (Stiffness S)	
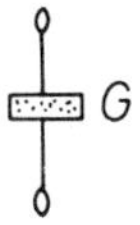	Heat conductor (Thermal conductance G)	

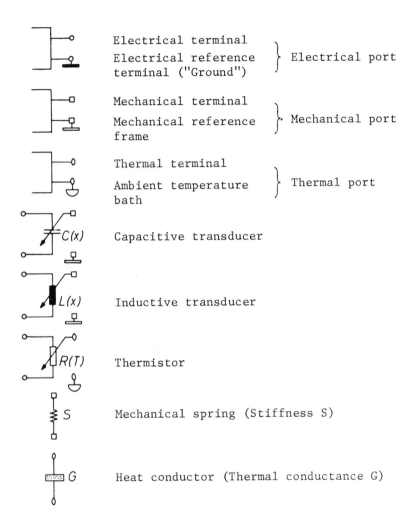

3. GENERALIZED NETWORK ELEMENTS

3.1. *Principles of Converter Elements*

The converters considered in this paper are electromechanical converters (transducers) and electrothermal converters (thermistors). Their physical principles may be considered as generalizations of those underlying the resistors, capacitors, and inductors of classical electrical networks, as explained by Figs. 1 and 2.

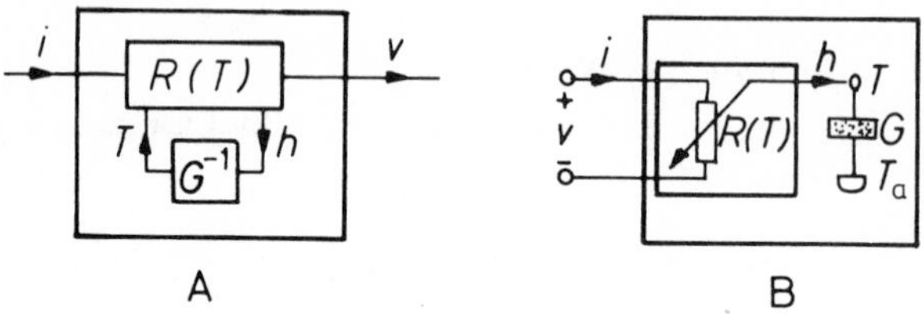

Fig. 1. Resistive electrical 1-port having an internal thermal degree of freedom.

A. Signal flow graph. B. Generalized network diagram.

If an electrical current flows through a resistor, two kinds of physical effects are produced by it: (1) the effect taken into account in electrical network theory is an electrical voltage; (2) the additional effect taken into account in generalized network theory is the heat current which equals the electrical power consumed by the resistor. If the resistor is temperature sensitive, its temperature may be considered as a second input signal in addition to the electrical current.

The essential step of generalization from an electrical resistor to an electrothermal converter of this type thus consists in redefining it as a double-input double-output system as in Fig. 1A; the actual physical component to which both the former and the new definitions apply may be the same. Even if the resistor is temperature insensitive so that the definition of the second input is not really meaningful, it may still be considered as an electrothermal converter since the heat current which it produces anyway may affect other temperature-sensitive components of a generalized network.

In the case of a temperature-sensitive resistor, an

additional condition must be met if the heat production associated with its power dissipation is to have an observable effect on its electrical properties: the heat current has to produce a nonzero temperature rise and, thereby, a resistance variation. Whether or not this is possible depends on the "thermal embedding" of the resistor, i.e. on the thermal properties of the resistive component itself and of its environment. In generalized network theory, this thermal embedding is represented by a corresponding thermal 1-port terminating the thermal port of the thermistor. In the simplest possible case (linear passive embedding system having instantaneous response), this terminating 1-port consists of only one element, a thermal conductance G (Fig. 1B). If G is infinitive (representing "isothermal embedding"), the electrical input properties correspond to those of a temperature-insensitive resistor. If G is finite (representing "generalized thermal embedding"), thermal feedback occurs, as represented by the loop in Fig. 1A.

The important feature of this feedback is its nonlinearity, which originates from the fact that the heat current is a quadratic function of the electrical current.

The problem of defining ideal thermistor elements based on the principle outlined above consists in selecting appropriate resistance-temperature functions and will be treated in the next section.

The capacitors and inductors of electrical network theory can be generalized in a similar way to obtain capacitive and inductive electromechanical transducers, as explained in Fig. 2 for the capacitive case.

If a voltage is applied to a capacitor, two kinds of physical effects are in general produced by it: (1) the effect taken into account in electrical network theory is an electrical charge; (2) the additional effect taken into account in generalized network theory is the mechanical force which usually occurs between the electrodes of the capacitor. If the capacitance is mechanically variable, the geometrical variation coordinate may be considered as a second input variable in addition to the voltage.

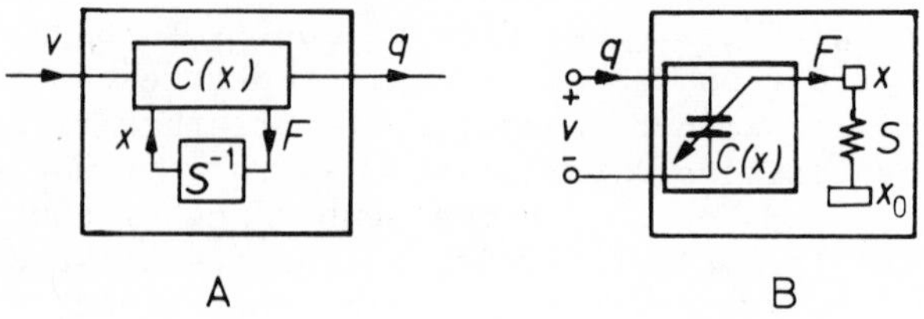

Fig. 2. Capacitive electrical 1-port having an internal mechanical degree of freedom.

A. Signal flow graph. B. Generalized network diagram.

The essential step of generalization from an electrical capacitor to a capacitive electromechanical transducer thus consists in redefining it as a double-input double-output system as in Fig. 2A; the actual physical component to which both the former and the new definitions apply may be the same.

An additional condition must be met if the mechanical force between the capacitor electrodes is to have an observable effect on its electrical properties: the mechanical force has to produce an electrode displacement and, thereby, a capacitance variation. Whether or not this is possible depends on the "mechanical embedding" of the capacitor electrodes. In generalized network theory, this mechanical embedding is represented by a corresponding mechanical 1-port terminating the mechanical port of the capacitive transducer. In the simplest possible case (linear passive embedding system having instantaneous response), this terminating 1-port consists of only one element, a spring with stiffness S (Fig. 2B). If S is infinite (representing "rigid mechanical embedding"), the electrical input properties of the transducer correspond to those of a constant capacitor. If S is finite (representing "generalized mechanical embedding"), mechanical feedback occurs as represented by the loop in Fig. 2A.

The important feature of this feedback is its nonlinearity, which originates from the fact that the mechanical force is a quadratic function of the voltage.

that the domain of admissible coordinate values is appropriately restricted. In terms of a system of two translationally movable electrodes, both types correspond to parallel-plate capacitors in the homogeneous-field-only approximation: the type I transducer has fixed plate distance and variable plate area; the type II transducer has fixed plate area and variable plate distance (Fig. 3A).

The domain of admissible coordinate values for which proportionality or inverse proportionality holds is restricted by the following assumption.

A9. For each admissible value of the voltage, the charge, and the mechanical coordinate of a capacitive transducer, its electrical energy, and thereby its capacitance, are nonnegative.

Thus, for negative values of the coordinate, the capacitance of the type I transducer is assumed zero (rather than negative); the capacitance of the type II transducer is undefined for negative coordinates (negative plate distances are considered as impossible). The restricted capacitance-coordinate functions are depicted in Fig. 3C. In some applications, the capacitance functions are additionally restricted by upper bounds of the capacitance values.

The relations between the electrical and mechanical port variables of the type I transducer are derived in Fig. 4. These relations may be derived either from an energy function $W(q,x)$ or from a coenergy function [3,5] $W^*(v,x)$. A_I or A_I^* is the transducer coefficient. Note that for negative coordinates $W(q,x)$ is not defined since charge and coordinate cannot be considered as independent variables if the capacitance is zero.

The dependent port variables are obtained from the energy (coenergy) functions using their conservation properties. The conservation equations also define the meaning of the sign of the mechanical force: it is positive (negative) if a displacement of the movable transducer electrode in the direction of increasing coordinates x by

A	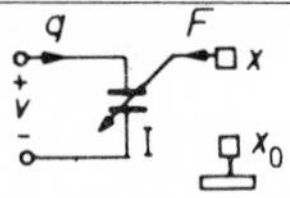	$C(x) \begin{cases} \sim x & (x>0) \\ =0 & (x \le 0) \end{cases}$
B	$W(q,x) = \begin{cases} \frac{q^2}{2C} = A_I \frac{q^2}{x} & (x>0) \\ \text{undefined} & (x \le 0) \end{cases}$ $A_I > 0$	$W^*(v,x) = \begin{cases} \frac{Cv^2}{2} = A_I^* v^2 x & (x>0) \\ 0 & (x \le 0) \end{cases}$ $A_I^* > 0$
	$A_I = 1/(4A_I^*)$	
C	$x>0$:	
	$dW = v\,dq + F\,dx$	$dW^* = q\,dv - F\,dx$
	$v(q,x) = \frac{\partial W(q,x)}{\partial q} = 2A_I \frac{q}{x}$	$q(v,x) = \frac{\partial W^*(v,x)}{\partial v} = 2A_I^* v x$
	$F(q,x) = \frac{\partial W(q,x)}{\partial x} = -A_I \frac{q^2}{x^2}$	$F(v,x) = \frac{\partial W^*(v,x)}{\partial x} = -A_I^* v^2$
D	q_r, x_r arbitrary > 0	v_r, x_r arbitrary > 0
	$v_r = 2A_I \frac{q_r}{x_r}, \quad F_r = A_I \frac{q_r^2}{x_r^2}$	$q_r = 2A_I^* v_r \cdot x_r, \quad F_r = A_I^* v_r^2$
	$v' = \frac{q'}{x'}, \quad F' = -\frac{q'^2}{x'^2}$	$q' = v' \cdot x', \quad F' = -v'^2$
E	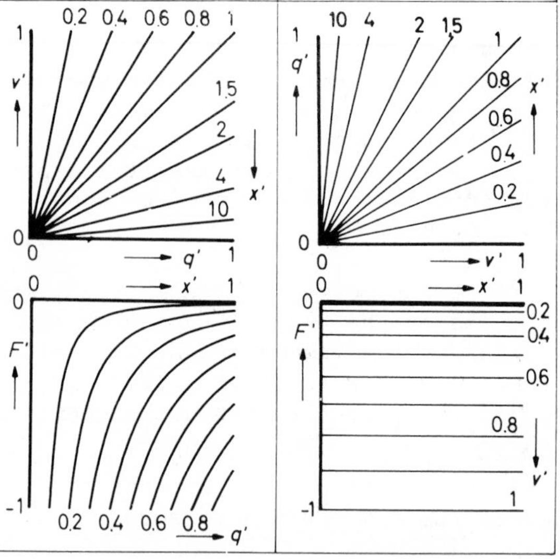	

Fig. 4. Port equations of type I ideal capacitive transducer element. A. Symbol and capacitance function. B. Energy and coenergy functions. C. Two forms of the port equations. D. Normalization for numerical analysis (reference values and normalized port equations). E. Graphical representation of transducer characteristics.

an external mechanical source supplies energy to (extracts energy from) the transducer. For the type I transducer, the case given in brackets applies; therefore the force is negative by the above definition.

In the applications, the transducer representation based on its coenergy function (with voltage and coordinate as independent variables) will be preferred since then the force is independent of the coordinate, and thus the equation to be solved for mechanical equilibrium will usually involve lower powers of the coordinate.

In Fig. 4E the characteristics of the electrical port are shown only in the first quadrant but really extend to the third quadrant symmetrically with respect to the origin.

The analysis of the type II transducer is presented in Fig. 5. A_{II} is the transducer coefficient if the analysis is based on the energy function, or A^*_{II} if it is based on the coenergy function $W^*(v,x)$.

In contrast to the type I transducer, the charge and coordinate will preferably be used as independent variables since then the force is independent of the coordinate x, and thus the equations to be solved for mechanical equilibrium will usually involve lower powers of x.

In Fig. 5E, the characteristics of the electrical port are shown only in the first quadrant, but they extend to the third quadrant symmetrically with respect to the origin.

Both types of capacitive transducers are nonlinear network elements because the equations for the dependent port variables involve products, ratios, and squares of the independent port variables.

3.3. *Ideal Inductive Transducer Elements*

Ideal inductive transducer elements of types I and II are defined as electrical duals of the ideal capacitive transducer elements of the corresponding types.

A

$C(x) \sim \dfrac{1}{x}$ $(x>0)$

$[x \leq 0 \text{ impossible}]$

(symbol: q, F, $\square x$, v, II, x_0)

B

$x > 0:$

$$W(q,x) = \dfrac{q^2}{2C} = A_{\mathrm{II}}\, q^2 x \qquad\Big|\qquad W^*(v,x) = \dfrac{C v^2}{2} = A_{\mathrm{II}}^* \dfrac{v^2}{x}$$

$$A_{\mathrm{II}} > 0 \qquad\qquad A_{\mathrm{II}} = 1/(4 A_{\mathrm{II}}^*) \qquad\qquad A_{\mathrm{II}}^* > 0$$

C

$$dW(q,x)=v\,dq+F\,dx \qquad\Big|\qquad dW^*(v,x)=q\,dv-F\,dx$$

$$v(q,x)=\dfrac{\partial W(q,x)}{\partial q}=2A_{\mathrm{II}}\,q\,x \qquad\Big|\qquad q(v,x)=\dfrac{\partial W^*(v,x)}{\partial v}=2A_{\mathrm{II}}^*\dfrac{v}{x}$$

$$F(q,x)=\dfrac{\partial W(q,x)}{\partial x}=A_{\mathrm{II}}\,q^2 \qquad\Big|\qquad F(v,x)=\dfrac{\partial W^*(v,x)}{\partial x}=A_{\mathrm{II}}^*\dfrac{v^2}{x^2}$$

D

$q_r,\ x_r$ arbitrary > 0 $\Big|$ $v_r,\ x_r$ arbitrary > 0

$$v_r = 2A_{\mathrm{II}}\,q_r\,x_r\ ,\quad F_r = A_{\mathrm{II}}\,q_r^2 \qquad\Big|\qquad q_r = 2A_{\mathrm{II}}^*\dfrac{v_r}{x_r}\ ,\quad F_r = A_{\mathrm{II}}^*\dfrac{v_r^2}{x_r^2}$$

$$v' = q'\cdot x'\ ,\quad F'=q'^2 \qquad\Big|\qquad q' = \dfrac{v'}{x'}\ ,\quad F'=\dfrac{v'^2}{x'^2}$$

E

(graphical representation: upper-left plot v' vs q' with curves labelled 10, 4, 2, 1.5, 1, 0.8, 0.6, 0.4, 0.2 (x'); upper-right plot q' vs v' with curves labelled 0.2, 0.4, 0.6, 0.8, 1, 1.5, 2, 4, 10 (x'); lower-left plot F' vs x' with curves labelled 1, 0.8, 0.6, 0.4, 0.2 (q'); lower-right plot F' vs x' with curves labelled 0.8, 0.6, 0.4, 0.2 (v'))

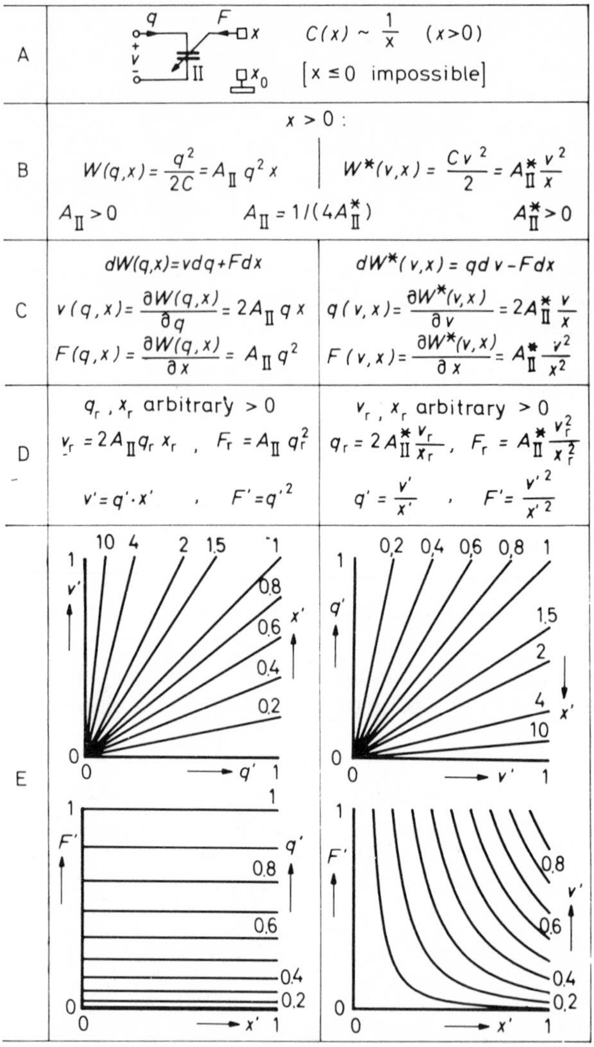

Fig. 5. Port equations of type II ideal capacitive transducer element. A. Symbol and capacitance function. B. Energy and coenergy functions. C. Two forms of the port equations. D. Normalization for numerical analysis (reference values and normalized port equations). E. Graphical representation of transducer characteristics.

Their properties are therefore obtained from those of the capacitive transducers by the dual correspondences listed in Table I.

TABLE I. Dual Correspondences Between Capacitive and Inductive Transducer Elements.

Capacitive transducers		Inductive transducers	
Capacitance-coordinate function	$C(x)$	$L(x)$	Inductance-coordinate function
Voltage	v	i	Current
Charge	q	ϕ	Flux linkages
Electric energy function	$W(q,x)$	$W(\phi,x)$	Magnetic energy function
Electric coenergy function	$W^*(v,x)$	$W^*(i,x)$	Magnetic coenergy function
Capacitive transducer coefficients	A_I, A_I^*, A_{II}, A_{II}^*	B_I, B_I^*, B_{II}, B_{II}^*	Inductive transducer coefficients

Ideal inductive transducer elements are thus required to satisfy A8 and the following assumption.

A9'. For each admissible value of the current, the flux linkages, and the mechanical coordinate of an inductive transducer, its magnetic energy, and thereby its inductance, are nonnegative.

The port equations of both types of ideal inductive transducer elements as obtained by Table I from Figs. 4 and 5 are listed in Table II.

Physical principles for the realization of the ideal inductive transducers are not as easily depicted as in Fig. 3A for capacitive transducers. If the "long coil" is taken as an analog of the parallel-plate capacitor, then the cross section of the coil corresponds to the overlapping area of the capacitor plates (and thus should be proportional to the mechanical coordinate for the type I transducer), and the length of the coil corresponds to the distance of the capacitor plates (and thus should be proportional to the mechanical coordinate for the type II transducer).

TABLE II. Port Equations of Ideal Inductive Transducer Elements.

	Type I	Type II
$i(\phi,x) =$	$2\ B_I \frac{\phi}{x} \quad (x > 0)$	$2\ B_{II}\ \phi\ x \quad (x > 0)$
$F(\phi,x) =$	$-B_I \frac{\phi^2}{x^2} \quad (x > 0)$	$B_{II}\ \phi^2 \quad (x > 0)$
$\phi(i,x) =$	$2\ B_I^*\ i\ x \quad (x > 0)$	$2B_{II}^* \frac{i}{x} \quad (x > 0)$
$F(i,x) =$	$-B_I^*\ i^2 \quad (x > 0)$	$B_{II}^* \frac{i^2}{x^2} \quad (x > 0)$

A difficulty of this analogy is related to the assumption that the field is homogeneous inside, and zero outside the capacitor (coil); this holds for $x \to 0$ in the case of the type II capacitive transducer but for $x \to \infty$ in the case of the type II inductive transducer.

This deficiency can be avoided by taking a different physical model using a pair of coupled coils which are rigid in themselves but movable with respect to each other.

3.4. *Ideal Thermistor Elements*

Ideal thermistor elements are required to satisfy the following restrictive assumptions.

A10. An ideal thermistor element is completely specified by a single coefficient.

A11. For each admissible value of the voltage, the current, and the temperature of a thermistor element, its power dissipation, and thereby its resistance, are non-negative.

Three types of resistance-temperature functions will be considered which satisfy both these requirements.

Type P thermistor (*positive* temperature coefficient of the resistance): Its resistance is proportional to its absolute temperature (in the domain of positive temperatures). Its physical model is a conducting material in which the availability of free charge carriers is temperature independent whereas their mobility decreases inversely proportional with the temperature. Its port equations are derived in Fig. 6.

Type N thermistor (*negative* temperature coefficient of the resistance): Its resistance is inversely proportional to its absolute temperature (in the domain of positive temperatures). Its physical model is a conducting material in which the availability of free charge carriers increases proportionally with the temperature, their mobility being temperature independent. Since the conductance of a type N thermistor is directly proportional to its absolute temperature, as is the resistance of a type P thermistor, it is obvious that both types are electrical duals of each other. The port equations of the type N thermistor are derived in Fig. 7.

Type 0 thermistor (*zero* temperature coefficient): Its resistance is temperature independent. In spite of this fact it may act as an electrothermal converter since the heat current which it produces may affect other temperature-sensitive elements of a generalized network. Its port equations are derived in Fig. 8.

The type 0 thermistor does not have a counterpart among the electromechanical transducers. If the capacitance (or inductance) of a transducer is independent of a mechanical coordinate, it does not produce a force in the direction of that coordinate, since the force is related to a capacitance variation (or inductance variation)

A	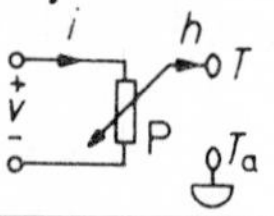 $R(T) \sim T \quad (T>0)$ $[T<0 \text{ impossible}]$	
B	$R(T) = K_P \cdot T \ , \ K_P > 0$	
	$v(i,T) = R(T)\cdot i = K_P \cdot i \cdot T$	$i(v,T) = \dfrac{v}{R(T)} = \dfrac{v}{K_P T}$
	$h(i,T) = R(T)\cdot i^2 = K_P \cdot i^2 \cdot T$	$h(v,T) = \dfrac{v^2}{R(T)} = \dfrac{v^2}{K_P T}$
C	$i_r \ , T_r$ arbitrary >0	$v_r \ , T_r$ arbitrary >0
	$v_r = K_P i_r T_r \ , \ h_r = K_P i_r^2 T_r$	$i_r = \dfrac{v_r}{K_P T_r} \ , \ h_r = \dfrac{v_r^2}{K_P T_r}$
	$v' = i' \cdot T' \ , \ h' = i'^2 \cdot T'$	$i' = \dfrac{v'}{T'} \ , \ h' = \dfrac{v'^2}{T'}$
D	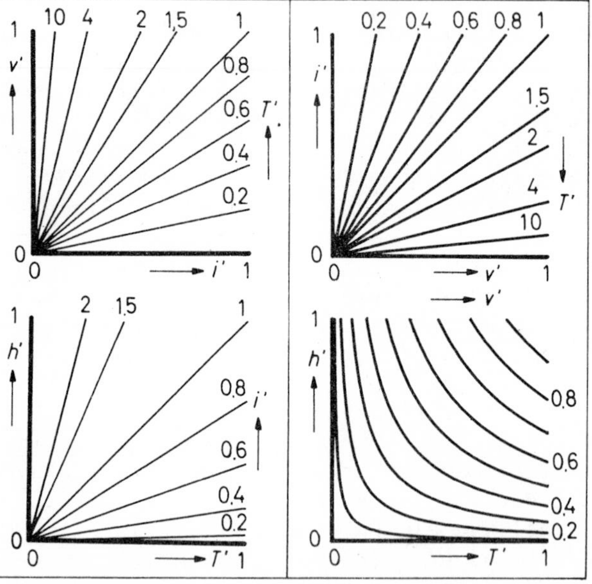	

Fig. 6. Port equations of type P ideal thermistor element. A. Symbol and resistance function. B. Two forms of the port equations. C. Normalization for numerical analysis (reference values and normalized port equations. D. Graphical representation of transducer characteristics.

A	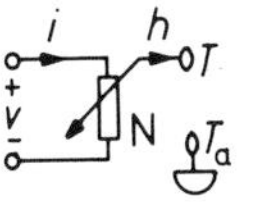	$R(T) \sim \frac{1}{T} \quad (T>0)$ $[T<0$ impossible$]$
B	$R(T) = \frac{K_N}{T}, \quad K_N > 0$ $v(i,T) = R(T)\cdot i = K_N \frac{i}{T}$ $h(i,T) = R(T)\cdot i^2 = K_N \frac{i^2}{T}$	$i(v,T) = \frac{v}{R(T)} = \frac{vT}{K_N}$ $h(v,T) = \frac{v^2}{R(T)} = \frac{v^2 T}{K_N}$
C	i_r, T_r arbitrary >0 $v_r = K_N \frac{i_r}{T_r}, \quad h_r = K_N \frac{i_r^2}{T_r}$ $v' = \frac{i'}{T'}, \quad h' = \frac{i'^2}{T'}$	v_r, T_r arbitrary >0 $i_r = \frac{v_r T_r}{K_N}, \quad h_r = \frac{v_r^2 T_r}{K_N}$ $i' = v' \cdot T', \quad h' = v'^2 T'$
D	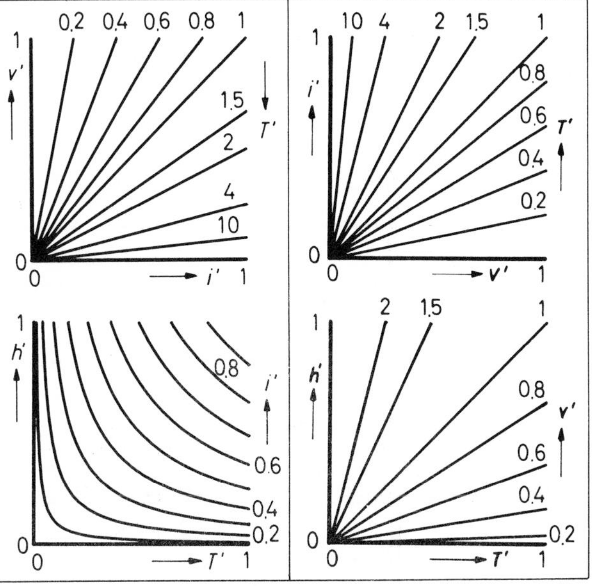	

Fig. 7. Port equations of type N ideal thermistor element. A. Symbol and resistance function. B. Two forms of the port equations. C. Normalization for numerical analysis (reference values and normalized port equations. D. Graphical representation of transducer characteristics.

A	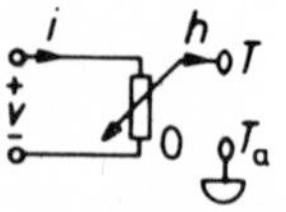	$R(T) = \text{const.}$
B	$R(T) = K_0, \quad K_0 > 0$	
	$v(i,T) = R(T)\cdot i = K_0 \cdot i$	$i(v,T) = \dfrac{v}{R(T)} = \dfrac{v}{K_0}$
	$h(i,T) = R(T)\cdot i^2 = K_0 \cdot i^2$	$h(i,T) = \dfrac{v^2}{R(T)} = \dfrac{v^2}{K_0}$
C	i_r, T_r arbitrary >0	v_r, T_r arbitrary >0
	$v_r = K_0 \cdot i_r, \quad h_r = K_0 \cdot i_r^2$	$i_r = v_r / K_0, \quad h_r = v_r^2/K_0$
	$v' = i', \quad h' = i'^2$	$i' = v', \quad h' = v'^2$
D	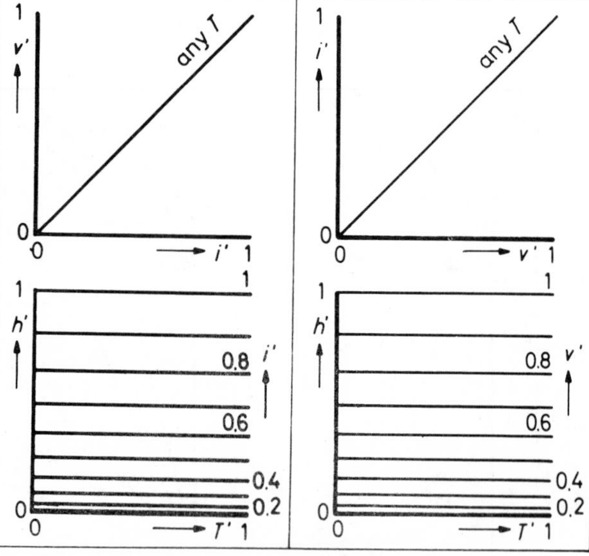	

Fig. 8. Port equations of type 0 ideal thermistor element. A. Symbol and resistance function. B. Two forms of the port equations. C. Normalization for numerical analysis (reference values and normalized port equations). D. Graphical representation of transducer characteristics.

rather than to the capacitance value (or inductance value) itself. The heat current produced by a thermistor, on the other hand, is related to resistance value itself rather than to the resistance variation with temperature.

This formal difference between thermistors and transducers can be expressed in other words as follows. The capacitive and inductive transducers have state function with conservation laws, e.g. their energy functions to which contributions can be made both at the electrical and at the mechanical ports, whereas the thermistors do not possess corresponding state functions having this property. This statement does not imply that each possible electromechanical device has a state function with a conservation law, and that each possible electrothermal device has none. As a wellknown example of an electromechanical device whose analysis cannot be based on a state function with a conservation property, consider the carbon microphone used in telephone sets, which is a mechanically variable resistor.

The three types of ideal thermistor elements are all nonlinear 2-ports, since the dependent variables are functions of products, ratios, and squares of the independent variables.

3.5. *Other Generalized Network Elements*

Only two additional elements are needed for the generalized networks considered in this paper.

Mechanical springs with finite positive stiffness are used as the simplest possible representations of non-rigid mechanical embeddings (or terminations) of electromechanical transducers.

Heat conductors with finite positive thermal conductance are used as the simplest possible representations of non-isothermal embeddings (or terminations) of thermistors.

In a somewhat more realistic approach to generalized networks, some additional types of elements would be required to express the fact that the responses of mechanical and thermal embedding systems are not usually instantaneous but involve transients: at least mechanical masses and friction elements, as well as heat capacitances.

Further, in a somewhat more general approach to generalized networks, mutual coupling between the mechanical and thermal degrees of freedom inside electrical N-ports would have to be taken into account by introducing thermomechanical converters.

The rather special nature of the types of generalized networks considered in this paper may be expressed by the following two additional assumptions.

A12. The generalized networks and electrical N-ports considered in this paper are described by algebraic (i.e. nondifferential) equations.

A13. No mutual coupling exists between internal mechanical and thermal degrees of freedom of the electrical N-ports.

When investigating the small signal properties of the electrical N-ports, the independent port variables will be supposed to consist of sums of a large constant (*d.c.*) value and a signal with comparatively small amplitude value. Since each part of the variables may be realized by an independent source, each electrical variable then splits into the series or parallel connection of a "signal port" and a "bias port". The "bias sources" (which may be constant voltages, constant currents, constant charges, or constant fluxes) may then be considered as belonging to the network, and thus as constituting another four types of elements of generalized networks.

4. EXAMPLES OF GENERALIZED NETWORKS REPRESENTING NON-LINEAR ELECTRICAL N-PORTS

4.1. *Capacitive 1-Ports*

The analysis of the network of Fig. 2 using type I and type II capacitive transducers is presented in Fig. 9.

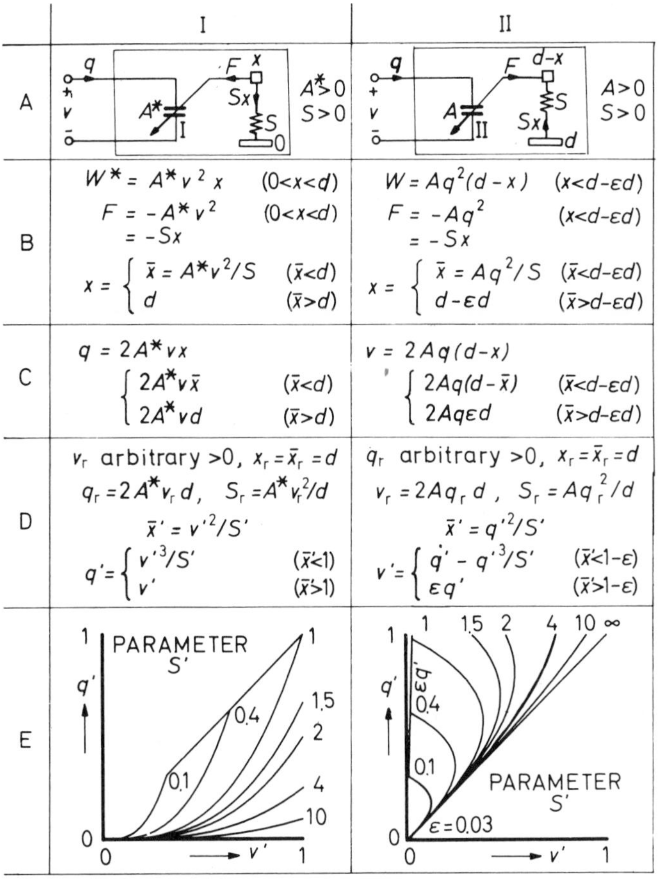

Fig. 9. Analysis of two generalized networks representing capacitive electrical 1-ports. A. Network diagrams. B. Mechanical equilibrium conditions. C. Resulting port equations. D. Normalized port equations. E. Graphical representation of port equations.

In both cases it is assumed that the capacitance values which the transducers may assume have a positive lower limit as well as an upper limit. In terms of the physical models of Fig. 3A the upper capacitance limit is motivated for type I by a limitation of the available plate area, and for type II by a minimum plate distance to avoid a short circuit.

For the type I transducer, voltage and coordinate are used as independent variables, and for the type II transducer charge and coordinate, in order to avoid the necessity of solving a third-order equation for the equilibrium coordinate.

The definition of force signs has been discussed in the comment on Fig. 4. Note that a discussion of force signs can be avoided here by calculating the total energy (the total coenergy) supplied to the 1-port as the sum (the difference) of the electric energy (coenergy) of the transducer and the potential energy (coenergy) of the spring [5] and searching for its minimum (maximum) with respect to the coordinate x [4], which gives its equilibrium value.

The mechanical stop associated with the maximum capacitance values of the transducers is taken account of by introducing the concept of a "free coordinate value $\bar{x}$" that the systems would assume if the stop were not present. If the free coordinate is outside the restricted domain of coordinates, the actual coordinate assumes the value of the domain boundary, as represented algebraically by the branching of the mechanical and resulting electrical equations, and as represented graphically by the breakpoints in the characteristic of Fig. 9E.

The characteristics are shown only in the first quadrant; they extend to the third quadrant symmetrically with respect to the origin.

In the limit of infinite stiffness of the spring (i.e. in the limit of rigid embedding where only purely electrical, not mechanical energy storage occurs), the electric input characteristics are those of linear capacitors (with zero capacitance in case I).

In case II, each finite stiffness leads to a characteristic with partly negative incremental capacitance (of the charge-controlled type), which is associated with limited energy storage capability of the 1-port.

The energy functions associated with the networks of Fig. 9 are represented in Fig. 10. By its definition, the total electrical energy (coenergy) supplied to a capacitive 1-port can be represented in its voltage-charge plane by the area between the 1-port characteristic and the charge axis (the voltage axis). Due to assumption A6, the electrical energy (coenergy) stored in the transducer can be represented by the area between a straight line (which passes through the corresponding point on the characteristic and through the origin) and the charge axis (voltage axis). Note that in linear 1-port storage elements energy and coenergy, though expressed by different functions, have the same value for each admissible pair of port variables; this reasoning applies both to the potential energy (coenergy) of the spring and, for each fixed mechanical coordinate value, to the electrical energy (coenergy) of the transducer. Therefore, the potential energy (coenergy) stored in the spring can be represented by the area between the electrical port characteristic and a straight line through the origin.

As soon as the breakpoint of the characteristic (corresponding to the mechanical stop) is reached, the potential energy storage capability of the spring is saturated, and all further energy storage is purely electrical. In case II, this further purely electrical energy storage may be negligible if ε (see Fig. 9) is sufficiently small; then the total energy storage capability of the 1-port is practically limited; as soon as the breakpoint is reached, the energy supplied to the 1-port is practically all stored in mechanical form.

This type of reasoning can also be used to derive necessary conditions for a given voltage-charge characteristic to be representable by an equivalent generalized network containing capacitive transducers and springs; investigations of this kind are, however, outside the scope of this paper.

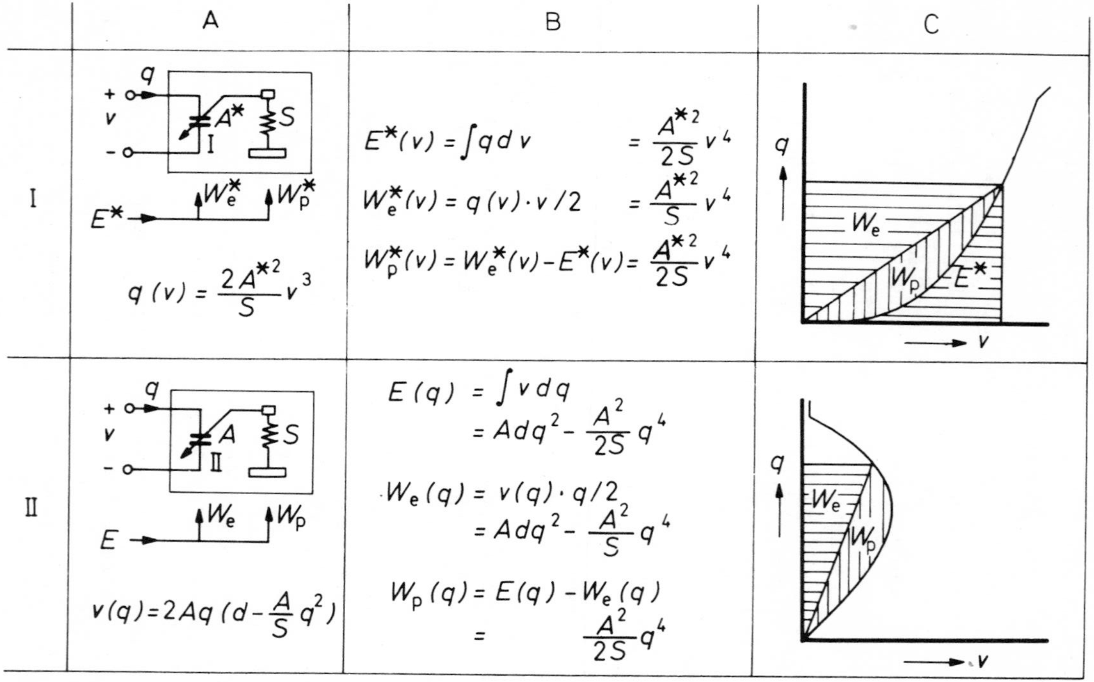

Fig. 10. Energy functions of the capacitive 1-ports of Fig. 9. A. Network diagrams, energy flow graphs, and electrical port equations. B. Energy (coenergy) of the 1-ports and of their elements. C. Representation by areas in the port variable planes.

4.2. *Inductive 1-Ports*

If, in the network of Fig. 9, the capacitive transducers are replaced by inductive transducers of the corresponding types, the resulting current-flux linkages relations correspond to the voltage-charge relations derived in Fig. 9 due to the fact that the capacitive and inductive transducers are electrical duals of each other, as expressed by the dual correspondences listed in Table I.

4.3. *Resistive 1-Ports*

The analysis of the network of Fig. 1 using type P and type N thermistors is presented in Fig. 11. Since both types are electrical duals of each other, a dual correspondence must also exist between the resulting electrical port relations. The analysis is, however, carried out independently for both cases in Fig. 11 in order to demonstrate the effects of different choices of independent variables used for the analysis. For the type P thermistor, the voltage is chosen as the independent variable; this results in a second-order equation for the equilibrium temperature. The same would hold true for the type N thermistor if the current were chosen as the independent variable; if, however, the voltage is again chosen as the independent variable as in Fig. 11B, a linear equation is seen to result for the equilibrium temperature. The former approach leads to a temperature equation which yields positive temperature values for each value of the independent variable whereas the latter does not. Thus, the domain of admissible voltages is restricted in case N (and, consequently, the domain of admissible currents is restricted in case P). In other words, the current in case P (the voltage in case N) tends to a limit or saturation value for increasing voltages (currents); this value depends on the thermal conductance.

The characteristics of Fig. 11E are shown only in the first quadrant, but extend to the third quadrant symmetrically with respect to the origin.

In the limit of infinite thermal conductance (corresponding to isothermal embedding), the electrical input characteristics are those of linear resistors.

	P	N
A	$K_P > 0$, $G > 0$	$K_N > 0$, $G > 0$
B	$h = v^2/(K_P T) = G(T - T_a)$ $T = \frac{1}{2} T_a (1 {\pm \atop (-)} \sqrt{1 + 4v^2/[K_P G]})$	$h = v^2 T/K_N = G(T - T_a)$ $T = T_a/(1 - v^2/K_N G)$
C	$i = v/(K_P T)$ $= \dfrac{2v}{K_P T_a (1 + \sqrt{1 + 4v^2/[K_P G]})}$	$i = vT/K_N$ $= \dfrac{v T_a}{K_N (1 - v^2/[K_N G])}$
D	v_r arbitrary > 0, $T_r = T_a$ $G_r = v_r^2/(K_P T_r^2)$, $i_r = v_r/K_P T_r$ $i' = 2v'/(1 + \sqrt{1 + 4v'^2/G'})$	v_r arbitrary > 0, $T_r = T_a$ $G_r = v_r^2/K_N$, $i_r = T_r\sqrt{G_r/K_N}$ $i' = v'/(1 - v'^2/G')$
E	PARAMETER G': ∞, 10, 4, 1, 0.4, 0.1, 0.04, 0.01	0.01, 0.04, 0.1, 0.4, 1, 4, 10, ∞ PARAMETER G'

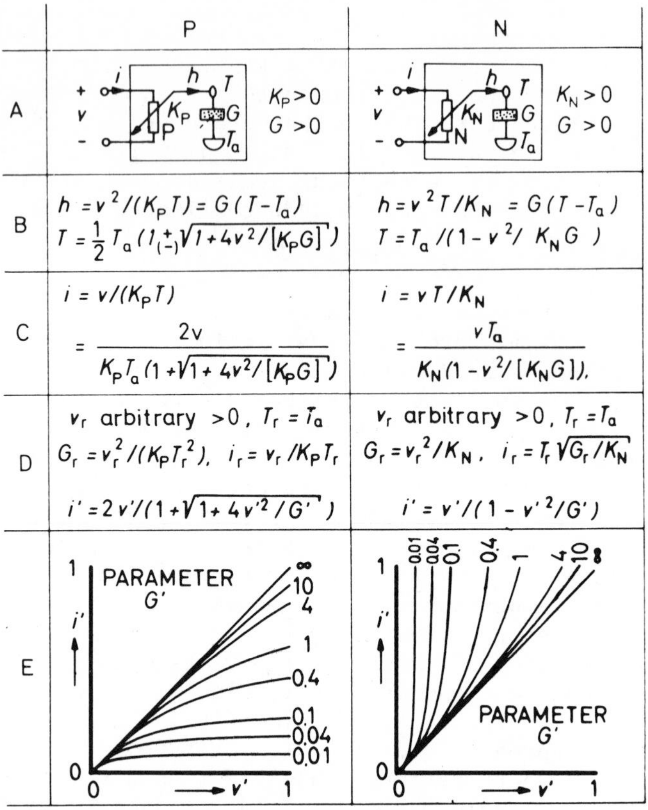

Fig. 11. Analysis of two generalized networks representing resistive electrical 1-ports. A. Network
diagrams. B. Thermal equilibrium conditions.
C. Resulting port equations. D. Normalized
port equations. E. Graphical representation
of port equations.

The type 0 thermistor is of no interest for resistive
1-ports since its electrical input characteristic corresponds to a linear resistor irrespective of its thermal
embedding (see Fig. 8).

4.4. *Reactive 2-Ports*

Nonlinear reactive (i.e. capacitive and/or inductive)

electrical 2-ports may be obtained by connecting capacitive and/or inductive transducers at their mechanical ports, and by terminating these connections in appropriate mechanical embeddings. Two examples of this kind will be analyzed.

Fig. 12 shows the mechanical connection of two capacitive transducers of types I and II terminated in a spring. The mechanical connection is supposed to be of the "push-pull" type, the forces exerted on the spring by both transducers having opposite directions.

The mechanical coordinate is supposed to have a positive quiescent value x_0 in order to avoid infinite quiescent capacitance of the type II transducer. For the sake of simplicity, an upper bound on the mechanical coordinate (such as in Fig. 9) is not taken into account here. The "free coordinate value $\bar{x}$" has the same meaning as in Fig. 9; if it is negative, the actual coordinate assumes its minimum possible value 0; all breakpoints occurring in the electrical characteristics (again only shown in the first quadrant but symmetrical with respect to the origin) correspond to this mechanical stop.

Fig. 13 shows the mechanical connection of a capacitive and an inductive transducer, both of type I, terminated in a spring. The mechanical connection is supposed here to be of the "pull-pull" type, the forces exerted on the spring by both transducers having the same direction. An upper bound d is assumed to exist for the mechanical coordinate, corresponding to maximum values of the capacitance at port 1 and of the inductance at port 2.

Due to the fact that the two transducers are electrical duals of each other, there is also a dual correspondence between the equations for both ports, as clearly exhibited by the identical shapes of both fields of characteristics (which are again only shown in the first quadrants in Fig. 13E but extend to the third quadrants symmetrically with respect to the origins).

4.5. *Resistive 2-Ports*

Nonlinear resistive electrical N-ports may be

A

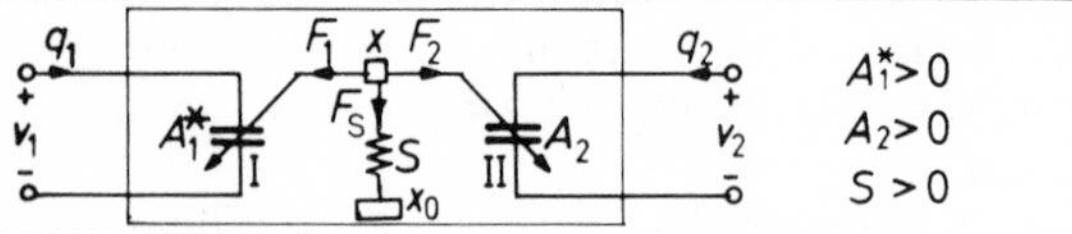

B

$$W_1^*(v_1,x)=A_1^* v_1^2 x \quad (x>0) \qquad W_2(q_2,x)=A_2 q_2^2 x$$

$$-F_S = F_1 + F_2 = -A_1^* v_1^2 + A_2 q_2^2 \quad (x>0)$$

$$= -S(x-x_0)$$

$$x = \begin{cases} \bar{x} = x_0 + \dfrac{A_1^*}{S} v_1^2 - \dfrac{A_2}{S} q_2^2 & (\bar{x}>0) \\ 0 & (\bar{x}<0) \end{cases}$$

C

$$q_1 = 2A_1^* v_1 x = \begin{cases} 2A_1 v_1 \bar{x} & (\bar{x}>0) \\ 0 & (\bar{x}<0) \end{cases} \qquad v_2 = 2A_2 q_2 x = \begin{cases} 2A_2 q_2 \bar{x} & (\bar{x}>0) \\ 0 & (\bar{x}<0) \end{cases}$$

D

$$x_r = \bar{x}_r = x_0$$

$$v_{1r} = \sqrt{Sx_0/A_1^*}\,, \; q_{1r} = 2x_0\sqrt{Sx_0A_1^*} \qquad q_{2r} = \sqrt{Sx_0/A_2}\,, \; v_{2r} = 2x_0\sqrt{Sx_0A_2}$$

$$\bar{x}' = 1 + v_1'^2 - q_2'^2$$

$$q_1' = \begin{cases} v_1' + v_1'^3 - v_1' q_2'^2 & (\bar{x}'>0) \\ 0 & (\bar{x}'<0) \end{cases} \qquad v_2' = \begin{cases} q_2' - q_2'^3 + v_1'^2 q_2' & (\bar{x}'>0) \\ 0 & (\bar{x}'<0) \end{cases}$$

E

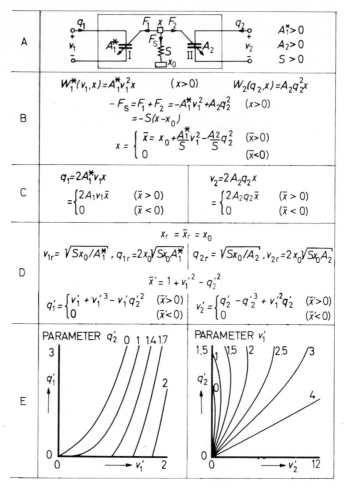

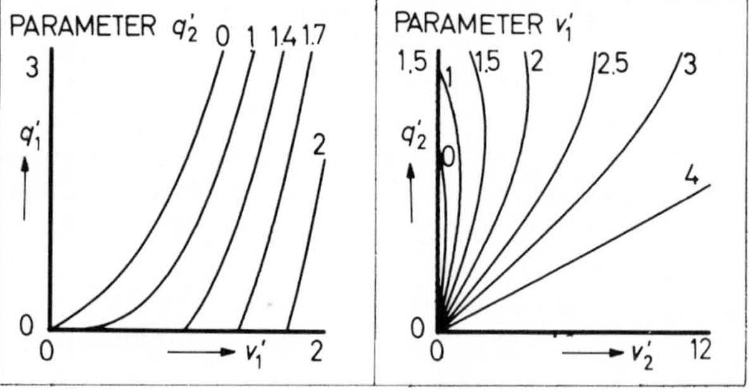

Fig. 12. Analysis of a generalized network representing a capacitive electrical 2-port. A. Network diagram. B. Mechanical equilibrium condition. C. Resulting electrical 2-port equations. D. Normalized 2-port equations. E. Graphical representation of 2-port equations.

obtained by connecting thermistors at their thermal ports, and by terminating these connections in appropriate

A

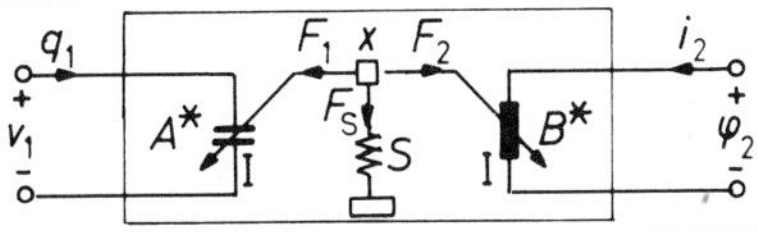

$A^* > 0$
$B^* > 0$
$S > 0$

B

$$W_1^*(v_1,x) = A^* v_1^2 x \qquad (0<x<d) \qquad W_2^*(i_2,x) = B^* i_2^2 x$$

$$-F_S = F_1 + F_2 = -A^* v_1^2 - B^* i_2^2 \qquad (0<x<d)$$
$$= -Sx$$

$$x = \begin{cases} \bar{x} = \dfrac{A^*}{S} v_1^2 + \dfrac{B^*}{S} i_2^2 & (\bar{x}<d) \\ d & (\bar{x}>d) \end{cases}$$

C

$$q_1 = 2A^* v_1 x = \begin{cases} 2A^* v_1 \bar{x} & (\bar{x}<d) \\ 2A^* v_1 d & (\bar{x}>d) \end{cases} \qquad \varphi_2 = 2B^* i_2 x = \begin{cases} 2B^* i_2 \bar{x} & (\bar{x}<d) \\ 2B^* i_2 d & (\bar{x}>d) \end{cases}$$

D

$$x_r = \bar{x}_r = d$$

$$v_{1r} = \sqrt{Sd/A^*}, \quad q_{1r} = 2d\sqrt{SdA^*} \qquad i_{2r} = \sqrt{Sd/B^*}, \quad \varphi_{2r} = 2d\sqrt{SdB^*}$$

$$\bar{x}' = v_1'^2 + i_2'^2$$

$$q_1' = \begin{cases} v_1'^3 + v_1' i_2'^2 & (\bar{x}'<1) \\ v_1' & (\bar{x}'>1) \end{cases} \qquad \varphi_2' = \begin{cases} i_2'^3 + v_1'^2 i_2' & (\bar{x}'<1) \\ i_2' & (\bar{x}'>1) \end{cases}$$

E

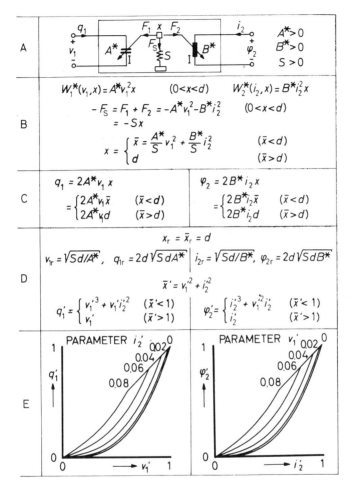

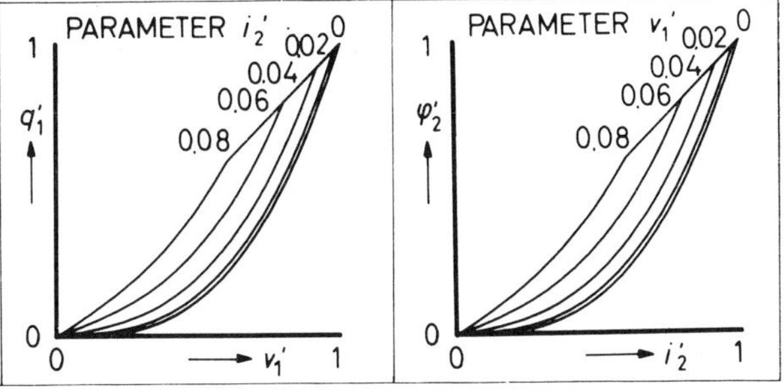

Fig. 13. Generalized network representing a reactive (capacitive/inductive) electrical 2-port. A. Network diagram. B. Mechanical equilibriu condition. C. Resulting electrical 2-port equ ticns. D. Normalized 2-port equations. E. Graphical representation of 2-port equation

thermal embeddings. Two examples of 2-ports of this kind will be analyzed. Fig. 14 shows the thermal connection c two thermistors of types P and N terminated in a heat

A	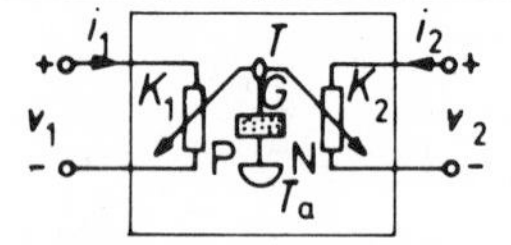 $K_1 > 0$, $K_2 > 0$, $G > 0$	
B	$h = h_1 + h_2 = K_1 i_1^2 T + \frac{v_2^2 T}{K_2} = G(T - T_a)$ $T = \frac{T_a}{1 - \frac{K_1}{G} i_1^2 - \frac{1}{K_2 G} v_2^2}$	
C	$v_1 = K_1 i_1 T = \frac{K_1 T_a i_1}{1 - \frac{K_1}{G} i_1^2 - \frac{1}{K_2 G} v_2^2}$	$i_2 = \frac{v_2 T}{K_2} = \frac{T_a v_2}{K_2 (1 - \frac{K_1}{G} i_1^2 - \frac{1}{K_2 G} v_2^2)}$
D	$v_{1r} = T_a \sqrt{G K_1}$ $\quad i_{1r} = \sqrt{G/K_1}$ $v_1' = \frac{i_1'}{1 - i_1'^2 - v_2'^2}$	$i_{2r} = T_a \sqrt{G/K_2}$ $\quad v_{2r} = \sqrt{G K_2}$ $i_2' = \frac{v_2'}{1 - i_1'^2 - v_2'^2}$
E	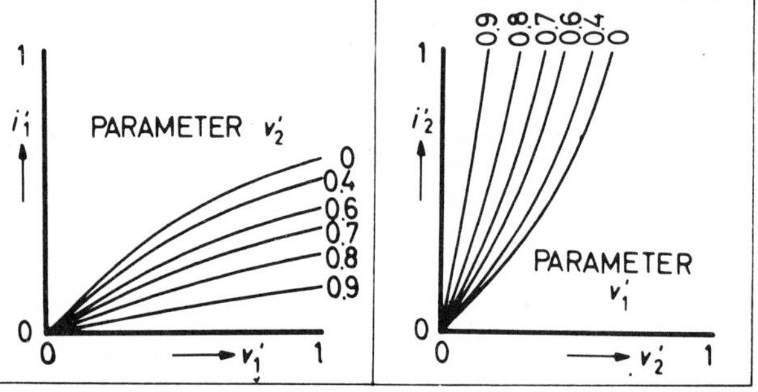	

Fig. 14. Generalized network representing a resistive electrical 2-port. A. Network diagram. B. Thermal equilibrium condition. C. Resulting electrical 2-port equations. D. Normalized 2-port equations. E. Graphical representation of 2-port equations.

conductor. In order to obtain a first-order equation for the equilibrium temperature, and simultaneously the limit values for the saturating variables, the current at port 1 and the voltage at port 2 are chosen as independent variables according to the comment given on Fig. 11. Due to the fact that the two thermistors are electrical duals of each other, there is also a dual correspondence between the equations for both ports, as clearly exhibited by the two fields of characteristics in Fig. 14E (which for simplicity are again only shown in the first quadrants).

Fig. 15 shows the thermal connection of two

A	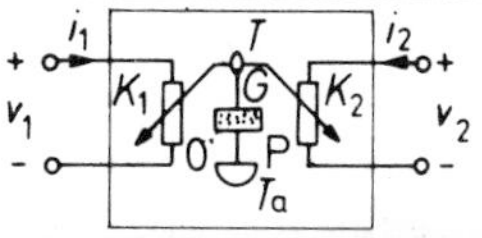	$K_1 > 0$, $K_2 > 0$, $G > 0$
B	$h = h_1 + h_2 = K_1 i_1^2 + K_2 i_2^2 T = G(T - T_a)$; $T = \dfrac{K_1 i_1^2 + G T_a}{G - K_2 i_2^2}$	
C	$v_1 = K_1 i_1$	$v_2 = K_2 i_2 T = K_2 i_2 \dfrac{K_1 i_1^2 + G T_a}{G - K_2 i_2^2}$
D	$v_{1r} = \sqrt{T_a G K_1}$, $i_{1r} = \sqrt{T G / K_1}$; $v_1' = i_1'$	$v_{2r} = T_a \sqrt{G K_2}$, $i_{2r} = \sqrt{G/K_2}$; $v_2' = \dfrac{1 + i_1'^2}{1 - i_2'^2} i_2'$
E	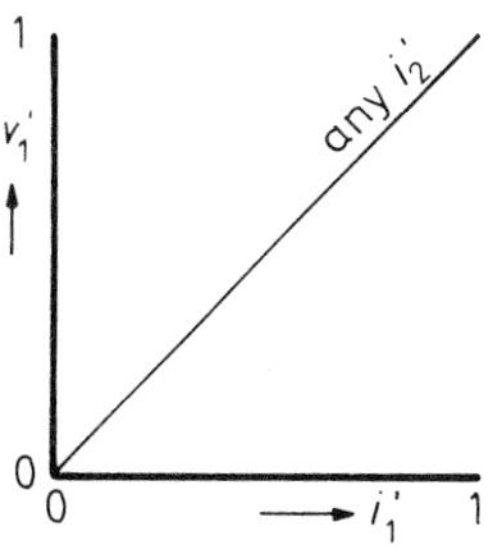	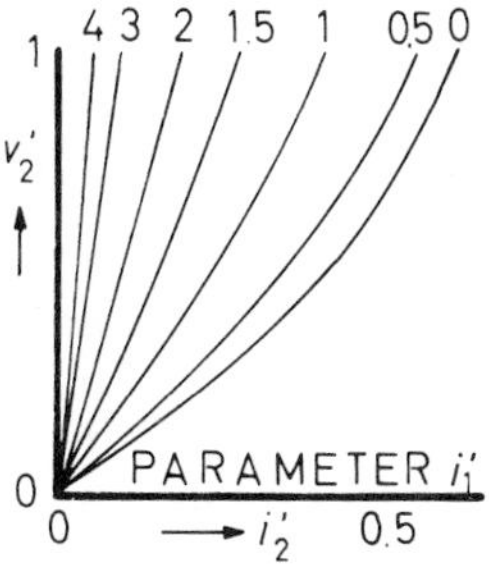

Fig. 15. Generalized network representing a resistive electrical 2-port. A. Network diagram. B. Thermal equilibrium condition. C. Resulting electrical 2-port equations. D. Normalized 2-port equations. E. Graphical representation of 2-port equations.

thermistors of types O and P terminated in a heat conductor. Its input characteristic at port 1 corresponds to a linear resistor irrespective of the current at port 2 due to the temperature insensitivity of the type O thermistor; the characteristics of port 2, however, depend on the current at port 1.

4.6. *Small-Signal 2-Port Properties*

The 2-ports of Figs. 12 an 15 are each represented by two algebraic equations for the dependent variables in terms of the two independent variables. If the independent variables are given some nonzero constant values ("bias values"), the dependent variables will also assume constant values ("quiescent values").

Suppose the 2-port equations are rewritten by expressing each variable as a deviation from its bias or quiescent value; the 2-port equations as expressed by these deviations (or "signals") are again in general nonlinear equations. Let the equations for the dependent signals be expressed as power series in the independent signals, and let only the first-order terms in both independent signals be retained; unless the bias values correspond to breakpoints of the characteristics or to regions of mechanical saturation or to inadmissible domains in the case of thermistor 2-ports, the linear equations obtained in this way define the "small-signal" properties of the 2-port. If the original 2-port equations are nonlinear, the coefficients of the small-signal equations are functions of the bias values.

The small-signal equations for the capacitive 2-port of Fig. 12 are given in Fig. 16B. They may be represented by an equivalent linear network as in Fig. 16C, consisting of two capacitors and an ideal transformer.

Within the restricted domain of admissable biasses, the capacitor at port 1 is always positive, but the capacitor at port 2 may assume negative capacitance values, corresponding to the regions of negative slope of the characteristics of port 2 in Fig. 12E.

The coupling between both ports, as represented by

A.

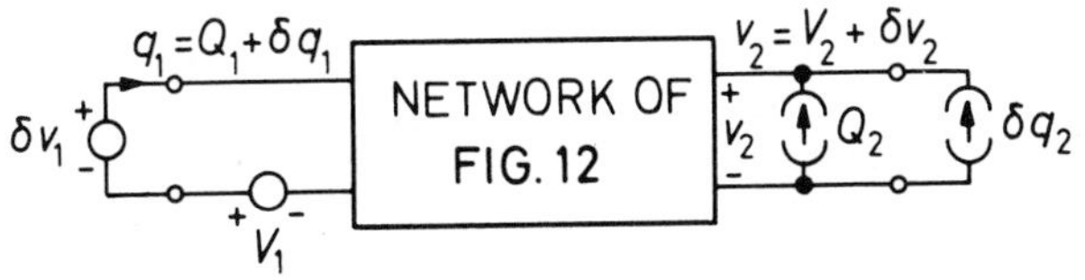

B.

$$\begin{bmatrix} \delta q_1' \\ \delta v_2' \end{bmatrix} = \begin{bmatrix} 1+3V_1'^2 - Q_2'^2 & -2V_1'Q_2' \\ 2V_1'Q_2' & 1-3Q_2'^2+V_1'^2 \end{bmatrix} \begin{bmatrix} \delta v_1' \\ \delta q_2' \end{bmatrix}$$

$$(1+V_1'^2-Q_2'^2>0)$$

C.

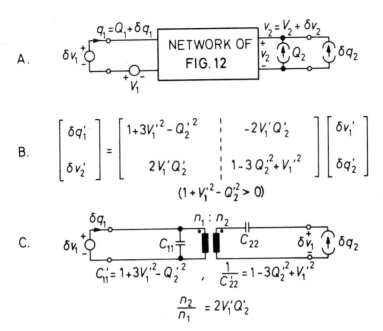

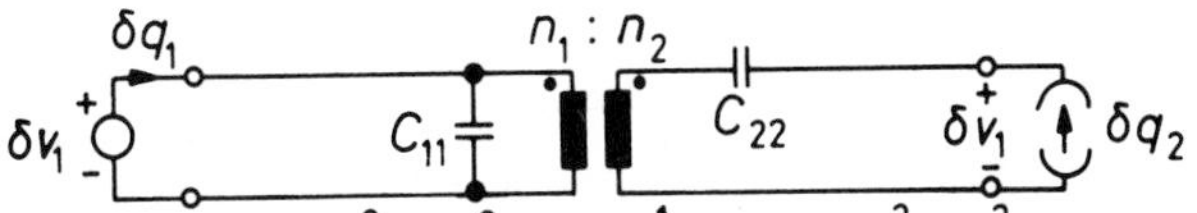

$$C_{11}' = 1+3V_1'^2 - Q_2'^2 \quad , \quad \frac{1}{C_{22}'} = 1-3Q_2'^2+V_1'^2$$

$$\frac{n_2}{n_1} = 2V_1'Q_2'$$

Fig. 16. Small-signal properties of the capacitive 2-port of Fig. 12. A. Circuit diagram. B. Small-signal equations. C. Small-signal equivalent circuit.

the ideal transformer, is proportional to the product of both biasses; the existence of nonzero signal transfer thus requires the presence of nonzero biasses both at the transmitting and at the receiving port. Their necessity is physically obvious: (1) Since the mechanical force and the equilibrium coordinate depend on the square of the electrical variable at the transmitting port, the first-order electromechanical transmission is zero for zero bias (2) since the mechanical displacement affects only a capacitance variation at the receiving port, its translation into a received signal requires again a bias at the receiving port.

The small-signal properties of the reactive (capacitive/inductive) 2-port of Fig. 13 are shown in Fig. 17. The coupling between both ports is represented by the

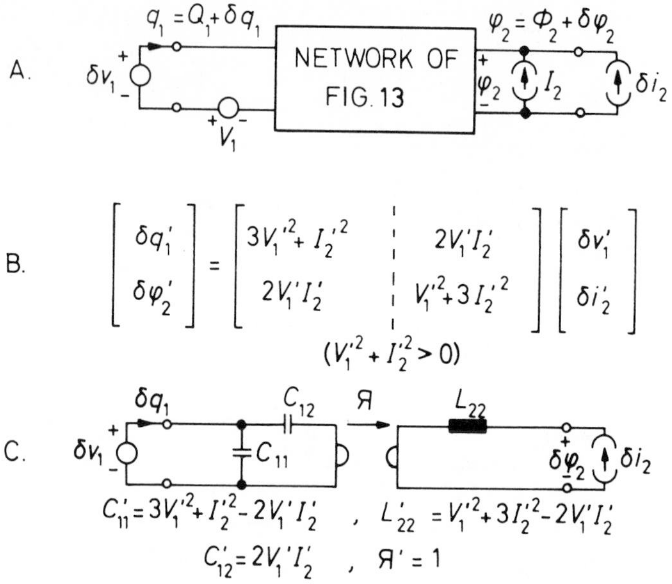

Fig. 17. Small-signal properties of the reactive (capacitive/inductive) 2-port of Fig. 13. A. Circuit diagram. B. Small-signal equations. C. Small-signal equivalent circuit.

cascade connection of a gyrator and a coupling capacitance. Thus, the coupling is antireciprocal; this property originates from the fact that the transducers at both ports are electrical duals of each other. Figs. 13 and 17 may be considered as a model representation of the fact (discovered about 25 years ago [6-8]) that lossless electrical 2-ports may violate the principle of reciprocity if they have, internally, mechanical degrees of freedom.

The coupling capacitance is zero unless nonzero biasses are present at both ports; the physical interpretation is nearly the same as in the case of the capacitive 2-port of Fig. 16.

The small-signal properties of the resistive 2-port

of Fig. 14 are shown in Fig. 18. The coupling between

A.

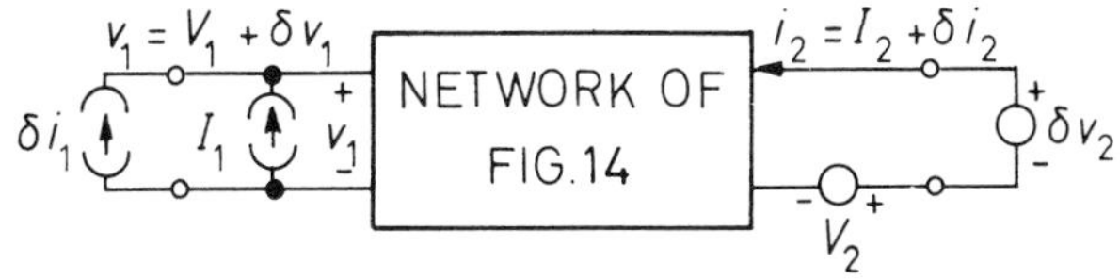

B.

$$\begin{bmatrix} \delta v_1' \\ \delta i_2' \end{bmatrix} = \begin{bmatrix} \dfrac{1 + I_1'^2 - V_2'^2}{(1 - I_1'^2 - V_2'^2)^2} & \dfrac{2 I_1' V_2'}{(1 - I_1'^2 - V_2'^2)^2} \\ \dfrac{2 I_1' V_2'}{(1 - I_1'^2 - V_2'^2)^2} & \dfrac{1 - I_1'^2 + V_2'^2}{(1 - I_1'^2 - V_2'^2)^2} \end{bmatrix} \begin{bmatrix} \delta i_1' \\ \delta v_2' \end{bmatrix}$$

$$(1 - I_1'^2 - V_2'^2 > 0)$$

C.

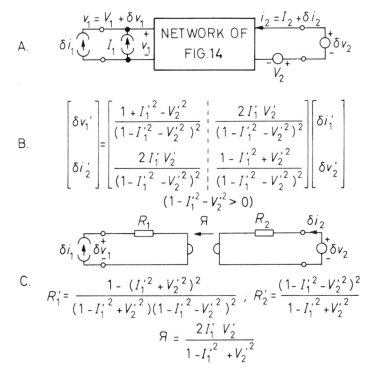

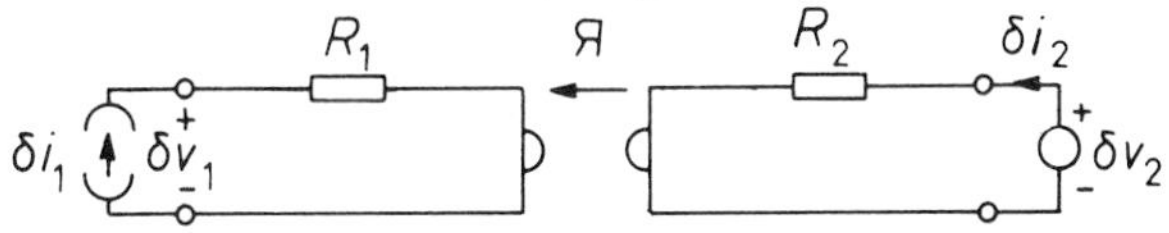

$$R_1' = \frac{1 - (I_1'^2 + V_2'^2)^2}{(1 - I_1'^2 + V_2'^2)(1 - I_1'^2 - V_2'^2)^2} \; ; \; R_2' = \frac{(1 - I_1'^2 - V_2'^2)^2}{1 - I_1'^2 + V_2'^2}$$

$$Я = \frac{2 I_1' V_2'}{1 - I_1'^2 + V_2'^2}$$

Fig. 18. Small-signal properties of the resistive 2-port of Fig. 14. A. Circuit diagram. B. Small-signal equations. C. Small-signal equivalent circuit.

both ports is again represented by a gyrator, whose phyical interpretation is here the electrically dual correspondence between the thermistors at both ports.

The gyration resistance is zero unless nonzero biasses are present at both ports; the physical interpretation of this fact is again very similar to that given above: (1) Since the heat current and the equilibrium temperature depend on the square of the electrical variable at the transmitting port, the first-order electrothermal transmission is zero for zero bias at the transmitting port; (2) since a temperature rise affects only a resistance variation at the receiving port, its translation into a received signal requires an additional bias to be present at the receiving port.

The small-signal properties of the resistive 2-port of Fig. 15 are shown in Fig. 19. The coupling between both ports is represented by the series connection of a gyrator and a resistor whose numerical value equals the gyration resistance for all admissable bias values; both elements together represent the unilateral coupling property of the 2-port, which physically originates from the fact that the thermistor at port 2 is temperature sensitive whereas the thermistor at port 1 is not.

This unilateral coupling again requires the presence of nonzero biasses at both ports for the same reasons as discussed in connection with Fig. 18.

The properties found in Figs. 18 and 19 may be used to generalize the statement contained in the comment on Fig. 17 as follows:

Electrical 2-ports may violate the principle of reciprocity if they have, internally, nonelectrical degrees of freedom.

(1) If the 2-ports are conservative (reactive, lossless), the nonelectrical degrees of freedom may be mechanical degrees of freedom. Lossless electrical 2-ports having internal mechanical degrees of freedom are either strictly reciprocal (as in the example of Fig. 16) or strictly antireciprocal (as in the example of Fig. 17) [9].

(2) If the 2-ports are dissipative (resistive), the nonelectrical degrees of freedom may be thermal degrees of freedom. Resistive electrical 2-ports having internal thermal degrees of freedom may be reciprocal (if thermistors of equal types are used; this is obvious for reasons of symmetry), or antireciprocal (if thermistors of dual types are used, as in the example of Fig. 18), or

A.

B

$$\begin{bmatrix}\delta i_1'\\ \delta i_2'\end{bmatrix} = \begin{bmatrix} 1 & 0 \\ \dfrac{2 I_1' I_2'}{1 - I_2'^2} & \dfrac{(1 + I_1'^2)(1 + I_2'^2)}{(1 - I_2'^2)^2} \end{bmatrix} \begin{bmatrix}\delta v_1'\\ \delta v_2'\end{bmatrix}$$

$$(1 - I_2'^2 > 0)$$

C.

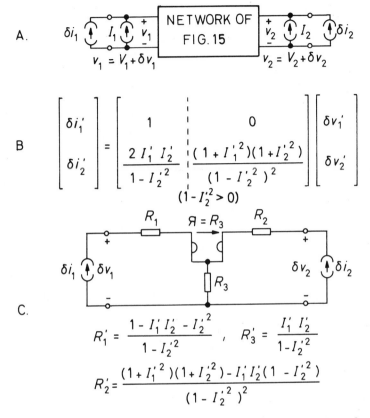

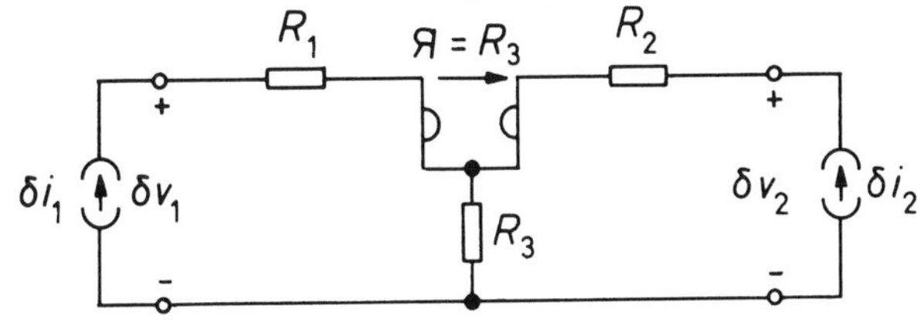

$$R_1' = \frac{1 - I_1' I_2' - I_2'^2}{1 - I_2'^2}, \quad R_3' = \frac{I_1' I_2'}{1 - I_2'^2}$$

$$R_2' = \frac{(1 + I_1'^2)(1 + I_2'^2) - I_1' I_2'(1 - I_2'^2)}{(1 - I_2'^2)^2}$$

Fig. 19. Small-signal properties of the resistive 2-port of Fig. 15. A. Circuit diagram. B. Small-signal equations. C. Small-signal equivalent circuit.

unilateral (if one thermistor of type 0 is used, as in the example of Fig. 19); the most general ratios of forward and backward transfer coefficients can in fact be obtained by combining the three former types of reciprocity behavior.

5. CONCLUSION

In this paper, some ideal physical system components connecting electrical variables with mechanical and thermal variables have been defined and used for some model considerations concerning the principle of reciprocity. Idealizing assumptions have been liberally used in order to obtain utmost mathematical simplicity of the models, without, however, losing sight of the possibility of physical interpretations. This simplicity is clearly reflected in the examples of the nonlinear systems analysed in this paper; the analysis was always finished within a few lines. Some intricacies associated with a number of geometrical limitations in the electromechanical systems may seem confusing at first but are quickly overcome.

It is felt that generalized networks of the kinds considered in this paper, since they are nonlinear, may prove as useful tools for the construction of examples and counter-examples whenever system properties are discussed such as the small-signal reciprocity properties treated above, where nonlinearities are explicitly or implicitly involved.

ACKNOWLEDGMENTS

Prof. V. Aschoff has stimulated the author's interest in the reciprocity theorems of network and field theory and in the principles of their violation; it is by studies on these theorems and principles that the author was led to turn his interest to generalized networks. Some useful critical comments on the first draft of this paper have been made by Prof. K.M. Adams (Delft), Prof. E. Schwartz, Dr. J. Mucha and Dr. H. Kindler. The author is also indebted to Mr. O. Fromhein for permission of, and assistance in, using his curve tracing program for producing the diagrams of Figs. 4 to 15.

REFERENCES

1. MILES, J.W. (1947) Coordinates and the reciprocity

theorem in electromechanical systems, *J. Acoust. Soc. Amer. 19*, pp. 910-913.

2. FISCHER, F.A. (1959) *Grundzüge der Elektroakustik* (2. Aufl.), Fachverlag Schiele und Schön, Berlin.

3. MEISEL, J. (1966) *Principles of Electromechanical Energy Conversion*, McGraw-Hill Book Co., New York.

4. FALK, G. (1968) *Theoretische Physik Band II (Allgemeine Dynamik, Thermodynamik)*, Springer Verlag, Berlin.

5. CHERRY, E.C. (1951) Some general theorems for non-linear systems possessing reactance, *Phil. Mag. (Ser. 7) 42*, pp. 1161-1177.

6. BRAUN, K. (1944) Elektroakustische Vierpole, *Telegr. Fernspr. Funk-Fernsehtechn. 33*, pp. 85-95.

7. JEFFERSON, H. (1945) Gyroscopic coupling terms, *Phil. Mag. (Ser. 7) 36*, pp. 223-224.

8. McMILLAN, E.M. (1946) Violation of the reciprocity theorem in electromechanical systems, *J. Acoust. Soc. Amer. 18*, pp. 344-347.

9. TELLEGEN, B.D.H. (1948) The gyrator, a new electric network element, *Philips Res. Rep. 3*, pp. 81-101.

CHARACTERIZATION OF NEGATIVE RESISTORS AND NEGATIVE CONDUCTORS

A.F. SCHWARZ

Department of Electrical Engineering,
Delft University of Technology, Delft, Netherlands

ABSTRACT

A unified characterization is given of the physical embodiments of the negative resistance: negative resistors (open-circuit stable) and negative conductors (short-circuit stable). These two classes of components will be characterized by a set of fundamental linear and nonlinear properties. Special attention will be paid to the conjoint occurrence of a specific linear model and a specific nonlinear characteristic associated with either a negative resistor or a negative conductor. This conjoint occurrence is explained using an equivalent one-port feedback network. To this end, two postulates pertinent to active components appear to be indispensable. The characterization as presented here enables one to explain stability properties of negative resistors and negative conductors and related components, such as negative-immittance converters and negative-immittance inverters.

1. INTRODUCTION

The continuing development of many new electronic devices with negative-resistance properties has led network theorists to introduce the *negative resistance* as a basic network element. The important point in this connection is that we should keep in mind the discrepancy between *ideal* negative resistances and *physical* negative resistances. The ideal negative resistance given by -R, where R is real and positive, is unrelated to physical reality and any analysis or synthesis based on this ideal element leads to impractical results.

It is well known that physical components with "negative-resistance" properties fall into two distinct classes according to their stability behaviour, usually referred to as the *open-circuit stable* class and the *short-circuit stable* class. In this paper, the terms *negative resistors* and *negative conductors*, respectively, will consistently be used to denote these classes. The collective term *negative resuctors* will be adopted to signify both classes.

Negative resuctors are frequently regarded as linear elements and treated as such. In other cases, they are described by a nonlinear voltage-current characteristic containing a negative-slope portion. In fact, negative resuctors have certain linear as well as nonlinear properties that by no means may be ignored. It appears that specific linear and specific nonlinear properties occur conjointly associated with either the negative resistor or the negative conductor.

When negative resuctors are to be used as basic components, it is indispensable to have a unified characterization that is simple enough to be tractable in analysis and synthesis, and at the same time giving a satisfactory approximation of physical reality.

The purpose of this paper is to characterize the negative resistor and the negative conductor in terms of their essential linear and nonlinear properties. In particular, the conjoint occurrence of specific linear and specific nonlinear properties will be explained in a simple manner.

2. CHARACTERIZATION OF NEGATIVE RESUCTORS

The characterizing linear and nonlinear properties of negative resistors and negative conductors are given in Table I. These properties will be discussed briefly.

No sources of infinite power exist. As a consequence, any voltage-current characteristic may exhibit a negative slope in a finite current or voltage region. The negative slope may change to a positive one by passing through zero or infinite differential resistance. Thus, we have two classes of nonlinear characteristics, namely, *current-controlled* and *voltage-controlled* characteristics. The characteristic of a negative resistor is current-controlled, that is, in the whole negative-slope region the voltage is a single-valued and continuous function of the current, but not vice versa. The negative conductor has a dual characteristic. The location of the negative-slope region in the V-I plane is not essential. We note that the characteristics as given in Table I are dc characteristics.

The linear characterization may be given in terms of poles and zeros. A negative resistor (a negative conductor) has impedance (admittance) zeros, but no impedance (admittance) poles in the right half-s-plane RHP. As a consequence, the negative resistor (negative conductor) is stable in open-circuited (short-circuited) condition, but unstable in short-circuited (open-circuited) condition. These stability properties have led to the well-known designations, *open-circuit stable* and *short-circuit unstable* in the case of a negative resistor, and *short-circuit stable* and *open-circuit unstable* in the case of a negative conductor. In the above, the expression "in open-circuited condition" or "in short-circuited condition" may be replaced by the expression "when driven by a current source" or "when driven by a voltage source", respectively.

For purposes of analysis, a negative resuctor is usually represented by a *linear model*. In Table I, linear models are introduced composed of three elements, one of which being a negative resistance. The other two

TABLE I

Characterization of Negative Resuctors

Nonlinear representation	
Negative resistor	Negative conductor
Current-controlled characteristic 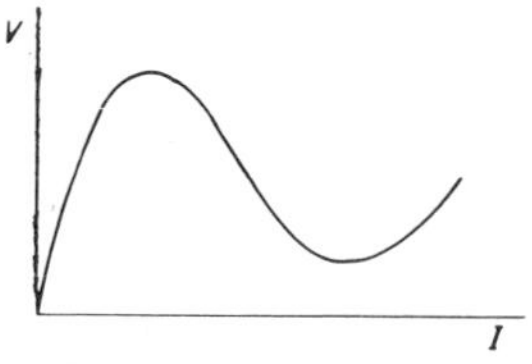	Voltage-controlled characteristic 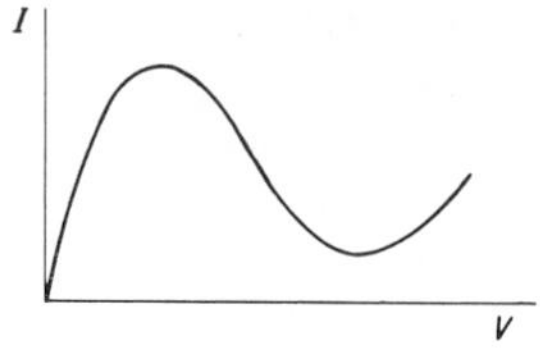
Linear representation	
Negative resistor	Negative conductor
Impedance zeros in RHP No impedance poles in RHP Open-circuit stable Short-circuit unstable Linear models 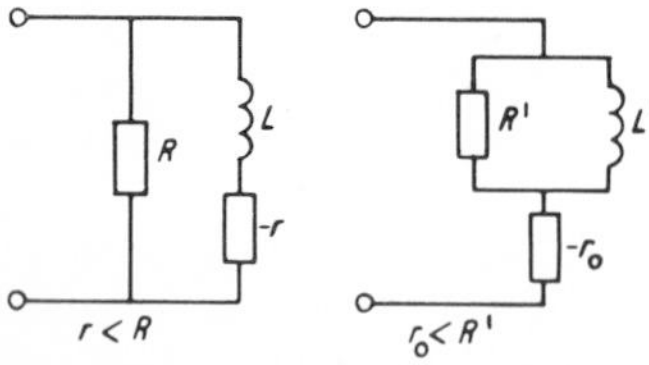 Impedance locus 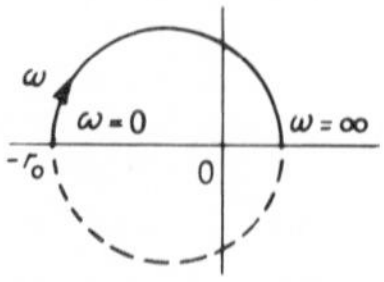	Admittance zeros in RHP No admittance poles in RHP Short-circuit stable Open-circuit unstable Linear models 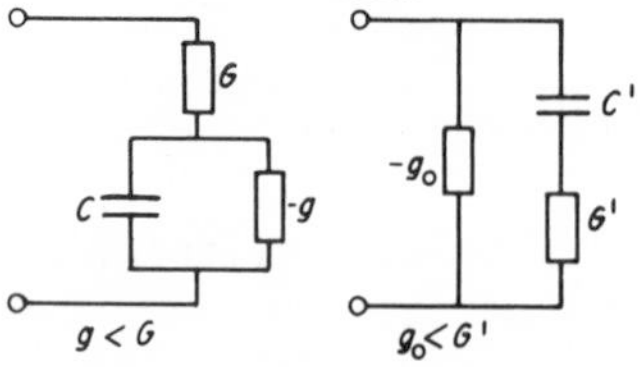 Admittance locus 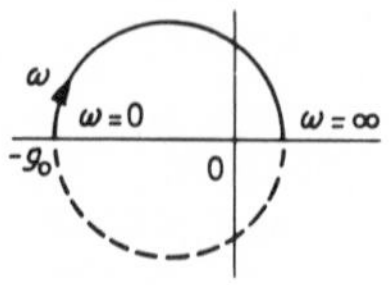

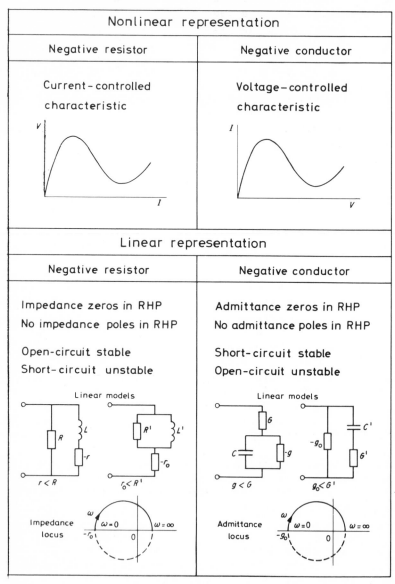

elements comprise a positive resistance and a reactive element. The corresponding impedance or admittance locus is also shown.

We note that the linear models shown in Table I differ from the models introduced first by Barkhausen [1]. No positive resistance is contained in the Barkhausen models. As a consequence, the real part of the impedance or admittance remains negative at all frequencies in contrast with experimental loci of physical devices. On that account, the Barkhausen models lead to unreliable conclusions, when the negative resuctor is terminated by a passive network. The inductance and the capacitance contained in the Barkhausen models have been introduced rather fortuitously. On the other hand, the linear models introduced in Table I can be derived in a straightforward manner as will be shown later. At any rate, the conclusions drawn in using these models are in good agreement with experimental results.

Any design of a negative resuctor should be aimed at attaining the closest possible approximation of the negative resistance -R in a largest possible frequency region. The frequency at which the real part of the immittance is zero, may be defined as the *cutoff frequency* of the negative resuctor.

It is readily seen from Table I that the negative resistor and the negative conductor have dual properties. What is most significant, specific linear properties and specific nonlinear properties occur conjointly, associated with either the negative resistor or the negative conductor, as follows:

Current-controlled characteristic Impedance zeros; no impedance poles in RHP	} Negative resistor
Voltage-controlled characteristic Admittance zeros; no admittance poles in RHP	} Negative conductor

There are no physical devices in existence that combine, e.g., a current-controlled characteristic and admittance zeros, no admittance poles in the right half-plane.

Since the conjoint occurrence of specific linear and non-linear properties have been observed with all negative resuctors (devices as well as circuits) known thus far, we shall give an explanation of this peculiar fact.

3. BASIC EQUIVALENT CIRCUITS

There is no reason to assume that linear and non-linear properties of any active one-port are directly related. For, when a linear model is given for a specified small region in the negative-slope portion of the V-I characteristic, no conclusions can be drawn as to the overall shape of the characteristic. In view of this, an explanation of the conjoint occurrence of specific linear and nonlinear properties is not possible, when the negative resuctor is represented by a black box. To overcome this problem we construct an equivalent one-port circuit model that exhibits the same driving-point properties as a negative resuctor. To this end, the basic circuits shown in Figs. 1a and 1b will be proposed as the equivalent circuit of a negative resistor and a negative conductor, respectively.

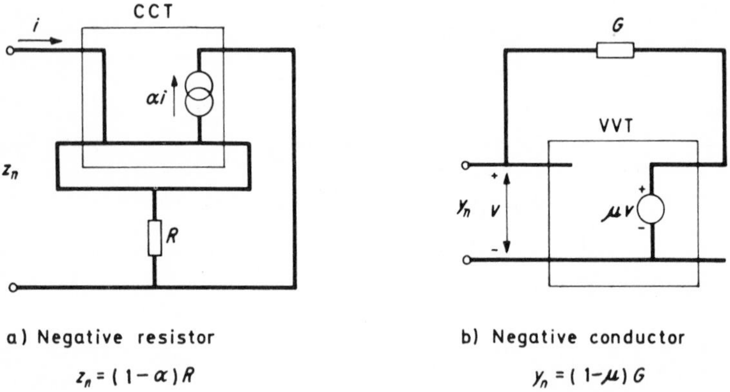

Fig. 1. Basic equivalent circuits

These circuits are composed of one transactor (controlled source considered as a basic two-port component) and one resistive component R or G.

The passive components R and G do not give any trouble. As is common usage in network theory, these components may be assumed to have a real and positive impedance. On the other hand, the introduction of an active component involves some fundamental physical limitations to be imposed on its active parameter, α or μ. The advantage of using a transactor is that its physical limitations are well known since transactors are idealizations of physical devices or circuits [2]. For example, the current-current transactor CCT as contained in Fig. 1 may be implemented by a pnpn transistor. The physical limitations to be imposed on α and μ will be formulated in terms of a set of postulates, namely, the Linearity Postulate and the Nonlinearity Postulate.

4. TWO FUNDAMENTAL POSTULATES

Due to finite nonzero transit times of charge carriers inside the transactor or the effect of stray reactances, both being inevitable in any physical device, the active parameters α and μ are complex rather than real. When given as a rational function A(s), where A stands for α or μ, at least one zero occurs at infinity. We shall ascribe a first-order function to A(s).

Linearity Postulate. The active parameter A of a transactor is given by the function

$$A = \frac{A_0}{1 + s/\omega_c} \tag{1}$$

A_0 denotes the magnitude of A at zero frequency, while ω_c is the cutoff frequency of the transactor.

The first-order function (1) appears to be adequate to account for the frequency behaviour of transactors.

In expressing the Linearity Postulate it is tacitly assumed that the active device is biased in the active region (operating region). As an illustration, Fig. 2a shows a typical family of output characteristics of a pnpn transistor which represents a current-current trans-

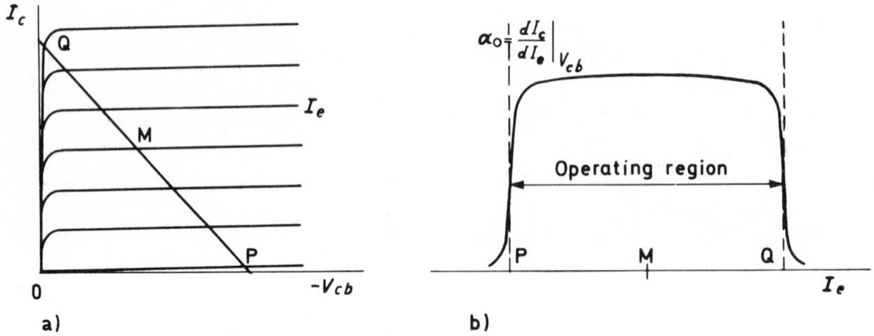

Fig. 2. Illustrating the limited operating region

actor. M is supposed to be the quiescent point. At this point, the current-amplification factor α_0 is greater than unity. When signal amplitudes are increased, the transistor will be driven to cutoff (point P) or to saturation (point Q). At points P and Q of the load line the factor α_0 drops to zero. The dependence of α_0 on bias current is sketched in Fig. 2b. This limitation on the operating region due to saturation effects as exemplified by the transistor is associated with all active devices. To account for this, we shall introduce the following postulate.

Nonlinearity Postulate. The magnitude of the parameter A of a transactor, which at zero frequency exceeds a given finite value, when the transactor is properly biased in a specified operating current or voltage region, drops to zero outside this region.

Let us now apply the Linearity Postulate and the Nonlinearity Postulate to the equivalent circuits which have been introduced earlier (Fig. 1). Let us restrict our discussion to the negative resistor. The impedance is given by

$$z_n = (1 - \alpha)R \tag{2}$$

Applying the Linearity Postulate in terms of α to (1) yields the following expression

$$z_n = \frac{s + (1 - \alpha_0)\omega_c}{s + \omega_c} R \tag{3}$$

It is readily seen that there is a zero but not any pole in the right half-plane. Hence, the one-port is open-circuit stable and short-circuit unstable. The linear models shown in Table I can be synthesized from (3). The elements contained in these models are related to parameters of the equivalent circuit of Fig. 1a as follows:

$$-r = \frac{1 - \alpha_0}{\alpha_0} R; \; L = \frac{R}{\alpha_0 \omega_c};$$

$$-r_0 = (1 - \alpha_0)R; \; L' = \frac{\alpha_0 R}{\omega_c}; \; R' = \alpha_0 R$$

The cutoff frequency of the negative resistor is given by

$$\omega = \sqrt{(\alpha_0 - 1)}\,\omega_c \tag{4}$$

The Nonlinearity Postulate should be applied at zero frequency, $s = 0$. Then, the differential resistance z_n is given by $(1 - \alpha_0)R$. In the operating region α_0 is greater than unity so that z_n is negative. However, as the instantaneous value of the bias current increases or decreases, α_0 ultimately drops to zero. As a consequence, at both sides of the operating region the differential resistance changes from negative to positive values by passing through zero resistance so that we have a current-controlled characteristic.

We have shown that the current-controlled characteristic and the properties of open-circuit stability and short-circuit instability are associated with the negative resistor in accordance with Table I.

Similar results are obtained in the dual case of a negative conductor. The conjoint occurrence of specific linear and specific nonlinear properties has been explained simply by applying the Linearity Postulate and the Nonlinearity Postulate to the equivalent circuit of a negative resistor and a negative conductor.

This result can also be derived in a more general manner using Blackman's formula for active one-ports [2]. When a negative resuctor is represented by an active one-port network with a *single feedback loop*, the conjoint occurrence of specific linear and specific nonlinear properties can be established. This leads to the following theorem [2].

Theorem. When a negative resuctor is realized using a single-loop feedback network consisting of one or more transactors and resistances, transformers and/or gyrators, either a negative resistor or a negative conductor is obtained, that is, specific linear and specific nonlinear properties occur conjointly.

This theorem is based on the result that application of the Linearity Postulate and the Nonlinearity Postulate leads to the same class of negative resuctor. As a consequence, when an analytical expression of a network realization is given as a function of the active parameter A, either the Linearity Postulate or the Nonlinearity Postulate applied to A suffices in determining the class of negative resuctor.

5. SINGLE-LOOP AND MULTIPLE-LOOP CIRCUITS

The restriction mentioned in the above theorem that the negative resuctor is realized using a single-loop feedback network is not of great significance, since this type of network is generally used in realizing negative resuctors. Moreover, the Linearity Postulate and the Nonlinearity Postulate can always be applied to the analytical expression of the immittance of any one-port, which may be a single-loop as well as a multiple-loop feedback network. In the case of *multiple-loop* feedback networks this expression can always be obtained using straightforward methods [3]. A simple example of a circuit using two transactors is shown in Fig. 3. We have

$$y_n = \frac{1 - \alpha_1\alpha_2}{R(1 + \alpha_1)(1 + \alpha_2)} \tag{5}$$

When the Linearity Postulate and the Nonlinearity Postulate are applied to both α_1 and α_2, it is found that the one-port has a voltage-controlled characteristic, while at the same time it is short-circuit stable and open-circuit unstable. Thus, the conjoint occurrence of specific nonlinear and specific linear properties is established.

The stability problem in multiple-loop feedback networks should not be underestimated. For example, when in Fig. 3 a phase reversal is introduced in the transactors,

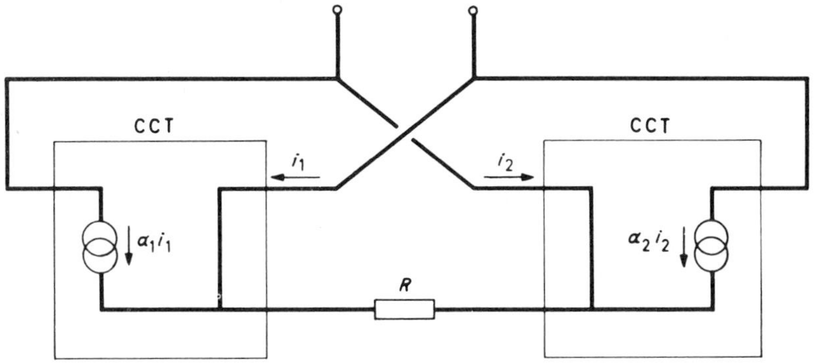

Fig. 3. Two-transactor circuit

that is, when α is replaced by $-\alpha$, we have a one-port that is unstable both in open-circuited and in short-circuited condition.

In the above, negative resuctors have been considered having a real and negative immittance at zero frequency. It is important to note that many one-ports can be synthesized having a real negative immittance at a nonzero finite frequency. In essence, the latter type of negative resuctors have similar linear and nonlinear properties as the former type on the understanding that zero frequency is replaced by the above-mentioned finite frequency [2].

The above results pertain to all negative resuctors, network realizations as well as physical devices, as long as they possess the fundamental linear and nonlinear properties as given in Table I.

6. APPLICATIONS

The characterization as given in Table I enables one to solve stability problems that arise, when negative resistors or negative conductors are used in active networks. A usual condition to be fulfilled by active networks is that system stability be ensured. It can easily be shown that neutralization of a positive resistance contained in a stable system requires either a negative resistor in series or a negative conductor in parallel with the positive resistance [2]. It can also be shown that imbedding a negative resistor in an RL network or a negative conductor in an RC network may lead to a stable system [2].

Synthesis methods using negative resistances have been proposed [4]. A negative resistor or a negative conductor must be used in the implementation of the negative resistance contained in the synthesized network. Using the proper linear model of the negative resuctor as shown in Table I, the stability behaviour of the system can be investigated.

Similar results as obtained with negative resuctors apply to related components such as *negative-immittance converters* and *negative-immittance inverters* [2]. A negative-immittance converter has invariably an open-circuit stable port and a short-circuit stable port, when the other port is terminated by a resistance. Which port is open-circuit stable and which short-circuit stable depends on the circuitry of the realization. A simple example is furnished by the basic circuits shown in Fig. 1. When the resistor R or conductor G is removed and the terminal-pair across which R or G has been connected is considered as a second port, a negative-immittance converter is obtained. The negative resuctor obtained at the second port, when the first port is terminated by a resistance, is dual with respect to the original negative resuctor given in Fig. 1.

7. CONCLUSIONS

In order that the network performance of negative resistors and negative conductors may be analyzed, a unique characterization of these components is indispensable. The characterization presented in this paper combines a close approximation of practical devices and circuits with tractability in network analysis and synthesis.

The introduction of the Linearity Postulate and the Nonlinearity Postulate pertaining to all physical devices enables one to explain the conjoint occurrence of specific linear and specific nonlinear properties of negative resuctors. Due to this conjoint occurrence it suffices to use either the linear or the nonlinear characterization of a negative resuctor, when its circuit behaviour is to be investigated.

In order to gain a better insight into the limitations and potential uses of negative resistors and negative conductors advantage can be taken of the characterization of negative resuctors as presented in this paper.

REFERENCES

[1] BARKHAUSEN, H. (15 Jan., 1926), Warum kehren sich die für den Lichtbogen gültigen Stabilitätsbedingungen bei Elektronenröhren um? *Physik Zeit. 27*, pp. 43 - 46.

[2] SCHWARZ, A.F. (26 Feb., 1969), *Negative Resistors and Negative Conductors* (Doct. Dissertation), Delft Univ. of Technology.

[3] SANDBERG, I.W. (March 1963), On the Theory of Linear Multi-Loop Feedback Circuits, *Bell Syst. Tech. Jour. 42*, pp. 355 - 382.
HAKIM, S.S. (Nov. 1963), Multiple-Loop Feedback Circuits, *Proc. IEE* (*London*) *110*, pp. 1955 - 1959.

[4] SU, K.L. (1965), *Active Network Synthesis*, McGraw-Hill, New York.

SOME ASPECTS OF N-PATH AND QUADRATURE MODULATION SINGLE SIDEBAND FILTERS

W. SARAGA

The General Electric Company Limited, Telecommunications Research Laboratories, Hirst Research Centre, Wembley, England.

SUMMARY

Some selected aspects of two types of periodically time-varying circuits, N-path filters and - to a smaller extent - quadrature modulation circuits are discussed. As background information an elementary description of these two types of circuits is given in engineering terms. This is followed by a discussion of two specific problems: the selection of an appropriate method of analysis for networks using ideal multipliers actuated by sinusoidal carrier oscillations; the concept and use of virtual transfer functions. Then a number of new results concerning N-path filters with sinusoidal carrier oscillations (novel passive, active and continuously variable N-path sections with biquadratic transfer functions) and *generalized* quadrature modulation circuits (i.e. quadrature modulation circuits not restricted to single sideband applications) are derived and discussed.

1. INTRODUCTION

Both N-path networks and quadrature modulation networks consist of time-invariant RLC (or RC, RL or C) networks combined with periodically time-varying modulators. The modulators are actuated by a local carrier oscillation or clock, the term "local" implying that this oscillation (characterized by its amplitude, frequency, phase and waveform) is considered to be a constituent of the modulator rather than an external signal applied to the modulator. We can therefore consider a modulator to have only one input signal, the information carrying signal.

The term "modulator" is sufficiently general to cover both ideal switches (Sw) and ideal multipliers (Mu). A Mu actuated by a switching function is equivalent to an Sw actuated by the same function; but if the actuating function is a sinusoid, then an Sw will still perform a switching action whereas a Mu will multiply the input signal by the instantaneous values of the actuating sinusoid. In this paper we shall only consider RLCMu circuits, each Mu being actuated by a sinusoidal oscillation of the form $\cos(\omega_c t + \theta)$*. The restriction of the discussion in this paper to RLCMu circuits does not imply that the results are necessarily invalid for RLCSw circuits (this would have to be investigated in each particular case).

From a practical engineering point of view, RLCSw circuits are of more immediate interest than RLCMu circuits, because microelectronic switches are cheap and reliable whereas multipliers have not yet reached this stage. However from a theoretical point of view RLCMu circuits, as an extension of time-invariant circuits, provide an interesting field of study, and some of the results obtained may stimulate the practical development of multipliers.

In the present paper we shall not cover the whole field of RLCMu networks. We shall take a special interest in RCMu circuits because of their suitability for microelectronic realization. Furthermore, in order to simplify our theoretical discussions, we shall exclude networks in

---

* In general, we shall use the normalized complex frequency variable $p = jx = jf/f_{ref}$ where f_{ref} is a suitably chosen reference frequency, and we shall therefore speak of the carrier frequency f_c (or ω_c) or x_c, where $f_c = \omega_c/2\pi = x_c f_{ref}$.

which feedback between output and input exists. However, there will be one exception: some aspects of active N-path sections will be discussed; this point will be clarified later.

2. ELEMENTARY DESCRIPTION OF N-PATH FILTERS AND QUADRATURE MODULATION CIRCUITS

2.1. *Basic N-path Filter Circuit*

Fig. 1a shows, as an example, the loss-frequency characteristic of a very narrow band pass filter. To realize such a characteristic by a conventional time-invariant filter high-grade components with extremely high Q-values and high constancy are required. Those requirements can be drastically reduced if the circuit shown in Fig. 1b is used. The input signal with frequencies f in the band f_a ... f_b is frequency translated by means of the first multiplier, with a carrier oscillation with a frequency f_c equal to the midband frequency f_m of the required pass band f_1 ... f_2. The resulting difference-frequency band can be represented (see Fig. 1c) either as $f-f_c$ or as the image band f_c-f, in each case covering both positive and negative frequencies (or, if only positive frequencies are used, as a *folded* band). It can be filtered by a comparatively simple low pass filter (with Q and constancy requirements reduced by about a factor $(f_2 - f_1)/2f_m$. Subsequently the signal is re-translated by means of a second multiplier into its original frequency position. However, the image band is translated into the same frequency position. In order to cancel the translated image band at the output terminals, a second path with carrier oscillations at quadrature to those in the first path (i.e. with a phase difference of $(\pm\pi/2)$ has to be provided (see Fig. 1d).

In switched N-path filters there is a bandwidth limitation. The input and output signal frequencies f are required to satisfy the inequality

$$f < Nf_c/2 . \quad (1)$$

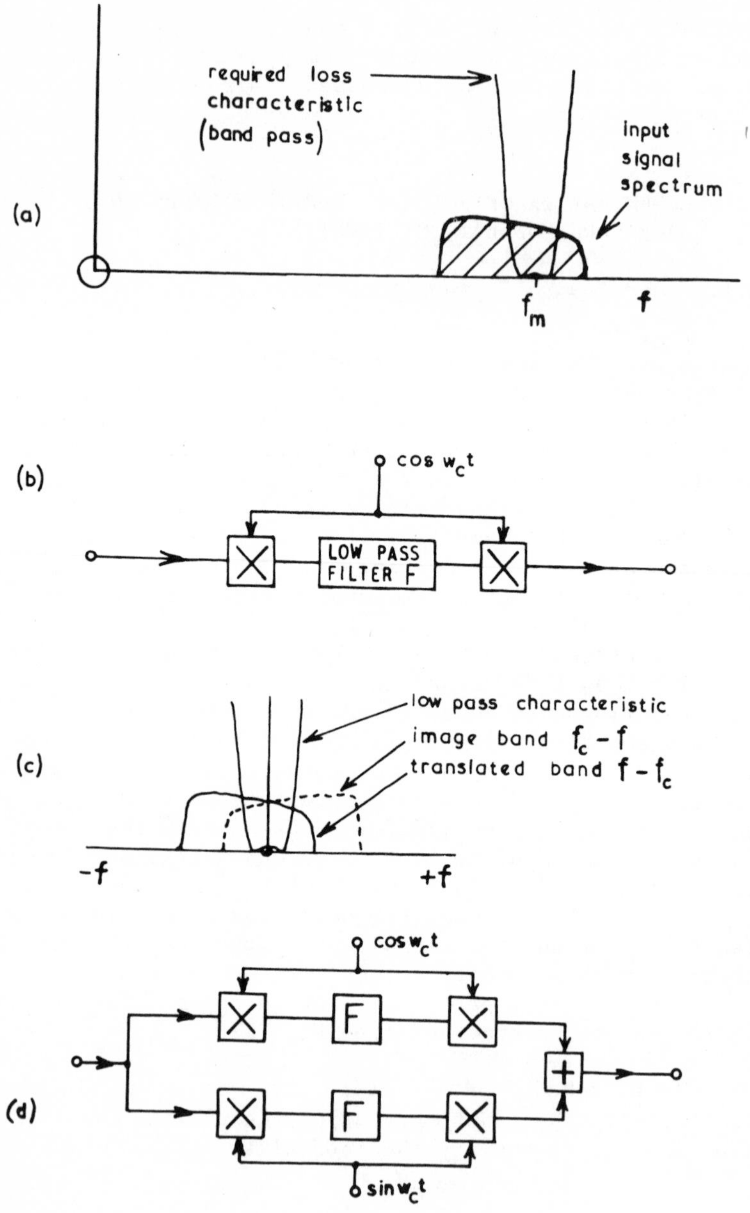

Fig. 1. Development of N-path filter concepts.

However, (1) does not apply if sinusoidal carrier oscillations are used. Therefore, for our present purposes, there is no advantage in considering high values of N, and we shall restrict our discussion to a physical two-path filter (degenerate case of $N = 4$). Formally (see Franks and Sandberg [1a]), the phase difference between corresponding carrier oscillations in adjacent paths is $2\pi/N$. In this case the formal theory as given in [1a] is valid only for $N \geq 3$. However, for even N-values degenerate forms with half the number of paths are possible, with phase shifts between adjacent paths equal to those in the non-degenerate case.

2.2. *Basic Quadrature Modulation Circuit*

Fig. 2a shows a conventional circuit for single side band (SSB) generation and reception. In the modulator M a double sideband (DSB) is produced, and by means of the SSB filter F a single sideband is selected. If the carrier frequency f_c is large compared with the lowest signal frequency f_1, then - as in the case of the narrow band pass filter considered in Fig. 1 - a practical realization of the filter F becomes difficult (or even impossible). This problem can be by-passed by using the quadrature modulation (QM) circuit shown in Fig. 2b [2, 3, 4]. Here the input signal s(t) is split into two versions $s_1(t)$ and $s_2(t)$, by two phase networks P_1 and P_2 which, over a finite signal frequency band f_1 to f_2 ($f_2/f_1 < \infty$, i.e. $f_2 < \infty$, $f_1 > 0$), have an approximate phase difference of $\pi/2$ (for the synthesis and design of such networks see [5] to [12]). Alternatively, tapped delay lines can be used which produce the required phase difference exactly but only an approximately flat amplitude characteristic (see [13, 14, 15]). $s_1(t)$ and $s_2(t)$ are multiplied by two carrier oscillations in quadrature, $c_1(t) = \cos \omega_c t$ and $c_2(t) = \sin \omega_c t$. Then the two products $s_1\ c_1$ and $s_2\ c_2$, both of which contain both sidebands, are obtained. The output signal z(t) is formed as the sum or the difference of the two products, and it can easily be shown that z(t) contains only one of the sidebands, the other one being suppressed by mutual cancellation. This method of suppression is, unfortunately, very sensitive to component variations and to relative amplitude and phase variations of the two carrier oscillations.

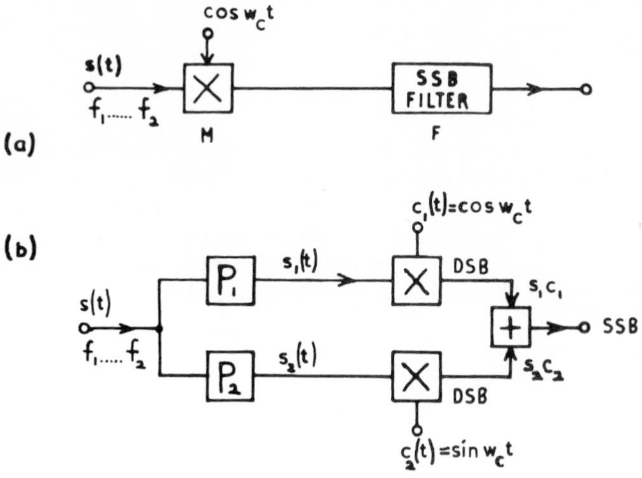

Fig. 2. Single sideband generators; (a) with sideband filter, (b) using quadrature modulation.

2.3. *SSB Generation Using Two Stages of Modulation*

One of the disadvantages of the circuit described in the preceding section is the need for high-precision wideband phase splitting networks. This can be avoided if $s_1(t)$ and $s_2(t)$ are produced in a preceding stage of modulation with carrier oscillations in quadrature. Since the output signals of these modulators are double sideband signals (see Fig. 3a), conventional SSB filters are in fact required after the first stage of modulation. However, if the first carrier frequency f_{c_1} is low compared with the carrier frequency required in a single-modulation system, then these filters are very much simpler than the SSB filters required in a conventional circuit of the type shown in Fig. 2a.

As far as its engineering purpose is concerned, the circuit in Fig. 3a is an SSB generating circuit which can be interpreted as a development of the basic QM circuit shown in Fig. 2b. From a circuit-theoretical point of view the circuit in Fig. 3a can be regarded as a generalization of the N-path circuit shown in Fig. 1d, the output signal being translated to a new frequency position (see, e.g., [16]).

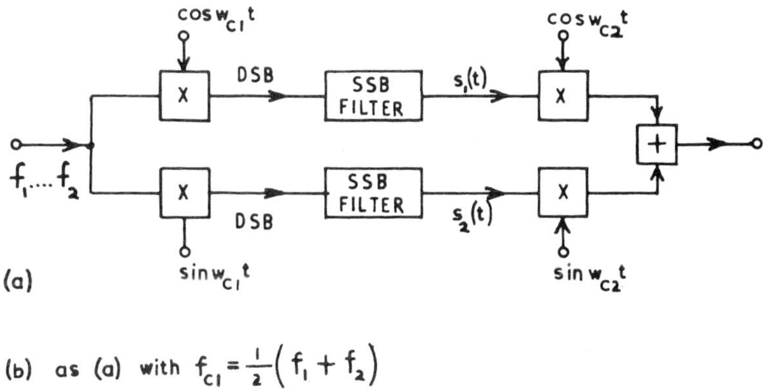

Fig. 3. Single sideband generator (quadrature modulation type), using two stages of modulation.

A special case of the circuit in Fig. 3a is Weaver's [17] circuit shown in Fig. 3b in which the first carrier oscillation is chosen to be equal to the mid-band frequency $f_m = (f_1 + f_2)/2$ of the input signal band (see also [18]). The advantage of this method is that those unwanted signal components, the suppression of which depends on mutual cancellation (i.e. not on the low pass filters) and is therefore particularly sensitive to component variations, constitute *in-band* distortion which is less serious than interference from and into other channels.

3. THE "NEW APPROACH"

Telecommunication engineers have used circuits incorporating modulators, such as those shown in Figs. 1b, 2b, 3a, b, for many decades, and the combination of filters and modulators is one of the basic features of frequency-division-multiplex systems. On the whole, telecommunication engineers feel quite capable of dealing with such systems and will normally call on the network theorist only for the solution of specific network problems, as for example for the design of 90°-phase-splitting networks. However, it can be very revealing and most fruitful to treat composite circuits as those shown in Figs. 1, 2, 3 not as combinations of separate networks but, as far as possible, as a single time-varying circuit having both time-invariant and time-varying components or sub-networks.

The advantages of such a "unified" approach can be manifold. Apart from its intellectual attraction, such an approach leads to the posing of questions which might not be asked within the framework of the conventional theory. The answers to such questions may reveal relationships previously not known and not suspected. More concise and more elegant ways of formulating statements about such systems may be found; it may become possible to carry out analysis and synthesis at a higher "hierarchical" level, with complex composite circuits as new "units."

4. SOME GENERAL PROBLEMS IN THE THEORY OF RLCMu NETWORKS

4.1. *Ideal Multipliers*

A multiplier has two input ports, with input signals s(t) and c(t), respectively, and one output port with output signal z(t); but, considered as a two-port with input signal s(t) and output signal z(t) (i.e. c(t) being regarded as part of the two-port), it is a *linear* network. For an ideal multiplier

$$z(t) = as(t)c(t) \tag{2a}$$

where a is a proportionality constant which is independent of time and of any of the port signals. With $c(t) = \cos \omega_c t$ and $s(t) = \cos \omega t$ we obtain (choosing $a = 1$)

$$z(t) = \cos\omega t \cos\omega_c t = \tfrac{1}{2}\cos(\omega_c + \omega)t + \tfrac{1}{2}\cos(\omega_c - \omega)t \; . \tag{2b}$$

The characteristic feature of (2b) is that there are two output frequencies for each signal input frequency ω, and both differ from ω. It follows that a required output frequency can be produced (using, of course, the same multiplier with the same carrier frequency ω_c) by two different input frequencies. Thus $\omega_2 = \omega_c + \omega_s$ can be produced either by an input frequency ω_s or by an input frequency $2\omega_c + \omega_s$ and $\omega_1 = \omega_c - \omega_s$ either by ω_s or by $2\omega_c - \omega_s$. The situation becomes clearer and logically more satisfactory if we consider both positive and negative frequencies, as shown in Fig. 4. Here input and output frequencies are represented along the abscissa axes, and the positive carrier frequency ω_c is represented along the ordinate axis. Then the projections of lines with positive and negative unit slope can be used to represent addition and subtraction of ω_c to and from the input and output signal frequencies.

4.2. *Representation of Signal and Carriers by Exponentials*

When we consider *linear* time-invariant circuits, it is convenient to represent all signals as the sum (or the integral) of elementary signals of the form e^{pt}. Since linearity still applies in the case of the time-varying circuits discussed in this paper, we certainly wish to continue the use of exponentials for signal representation, and we may be tempted to do the same for the carrier, i.e. to represent it by

$$e^{p_c t} = e^{jx_c t}.$$

Then we would obtain for a multiplier with carrier $e^{jx_c t}$ and input signal e^{jxt} the output signal $e^{j(x_c+x)t}$. Although we know that both x_c and x can take both positive and negative values so that $x_c + x$ represents, in a concise way, the four different expressions $x_c + x$, $x_c - x$, $-x_c + x$ and $-x_c - x$, this seems to be a situation where conciseness has been carried to the point of sacrificing clarity. The basic practical fact that in a wave analyzer we would get two separate output indications at frequencies $(x_c + x)$ and $x_c - x$ does not become evident from, but is rather obscured by, the single expression $e^{j(x_c+x)t}$.

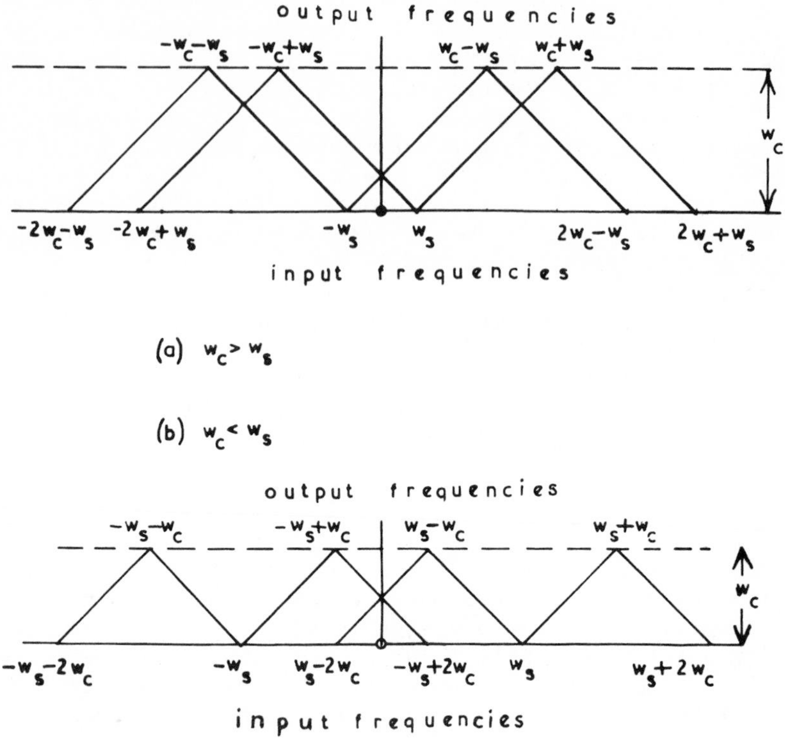

Fig. 4. Input and output frequencies of a multiplier.

The usual convention in this case is to represent the carrier by a sinusoid and the signal by an exponential; this is an admirable compromise. We preserve the signal notation used in the treatment of time-invariant networks (and avoid the clumsiness associated with the repeated use of the addition and multiplication theorems of trigonometric functions). On the other hand, for an input signal e^{jxt} to a multiplier actuated by the carrier oscillation $\cos(x_c t)$ we obtain now the output signal

$$z(t) = \frac{1}{2}e^{j(x_c+x)t} + \frac{1}{2}e^{j(-x_c+x)t} \tag{3a}$$

which clearly indicates the presence of two separate

output frequencies. Without loss of generality we can assume that both $x_c > 0$ and $x > 0$. Then $x_c + x$ is the sum frequency and $-x_c + x$ the difference frequency. In various circumstances it may be preferable to write the difference frequency in the form $x_c - x$. In this case we recall that (as in the case of time-invariant networks) z(t) as given by (3a) must be associated with

$$\bar{z}(t) + \tfrac{1}{2}e^{-j(x_c+x)t} + \tfrac{1}{2}e^{-j(-x_c+x)t} \tag{3b}$$

However, this is not different from what we are accustomed to in the case of time-invariant networks, viz., that each phasor is associated with another phasor rotating in the opposite sense.

In this context one other point may be worth mentioning. Although we have just stated that a multiplier (actuated by a sinusoidal carrier oscillation provides two different single-frequency output signal components for each single-frequency input component, it would not be correct to conclude that if the input signal contains m different single-frequency input components, the output signal will contain 2m components. Two output frequencies produced by different input frequencies may coincide. In such a case the number of output frequencies may be reduced by one, due to component merger, or by two, due to mutual cancellation of components.

4.3. *Multipliers and Time-Invariant Two-Ports in Cascade*

In this section we wish to consider one of the simplest types of composite circuits involving time-invariant networks and multipliers: cascade connections between a multiplier Mu and a linear two-port N. There are two such circuits; Fig. 5a shows a two-port N_1 *preceding* Mu, and Fig. 5b shows a two-port N_2 *following* Mu. We apply to both circuits the same input signal e^{pt} and consider a hypothetical situation in which we obtain the same output signal in both circuits, so that the two circuits are equivalent, i.e.,

$$v_1(t) = v_2(t). \tag{4a}$$

In each circuit the output signal will contain two

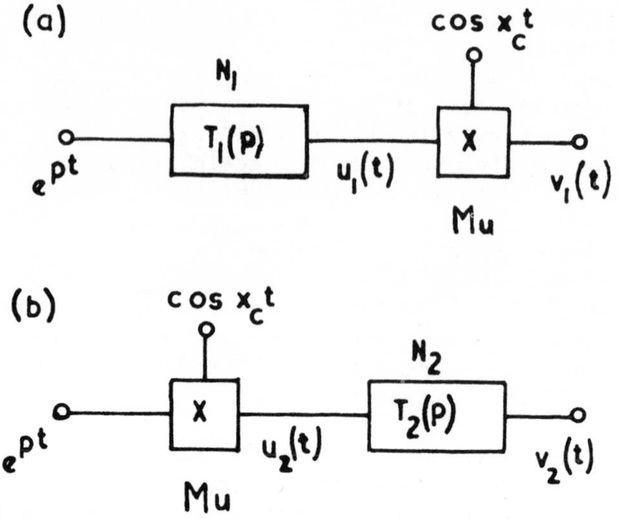

Fig. 5. Time-invariant circuit (a) preceding and (b) following multiplier stage.

output components, $e^{(p+jx_c)t}$ and $e^{(p-jx_c)t}$. If we consider real x-values then a band-limited input signal, for which $-x_0 < x < +x_0$ will produce an output signal band limited to $x_c - x_0 < x < x_c + x_0$. Let us now assume that an input signal with a wider frequency spectrum, extending beyond $-x_0$ and $+x_0$, is given and that $x_0 \ll x_c$. Then it is obvious, without any formal analysis, that an output signal band limited as before could be obtained in two different ways: by means of the circuit in Fig. 5a, making N_1 a low pass filter with pass band $-x_0$ to $+x_0$ or by means of the circuit in Fig. 5b making N_2 a narrow band pass filter with pass band $x_c - x_0$ to $x_c + x_0$ (and $-x_c + x_0$ to $-x_c - x_0$). If we derive N_2 from N_1 by means of a conventional low-pass to band-pass transformation, Eq. (4a) would be approximately satisfied.

We shall now investigate the implications of this equation in a more formal way. In Fig. 5a we find

$$u_1(t) = T_1(p)e^{pt}$$

$$v_1(t) = \tfrac{1}{2}T_1(p)e^{(p+jx_c)t} + \tfrac{1}{2}T_1(p)e^{(p-jx_c)t}, \tag{4b}$$

and in Fig. 5b

$$u_2(t) = \tfrac{1}{2}e^{(p+jx_c)t} + \tfrac{1}{2}e^{(p-jx_c)t}$$

$$v_2(t) = \tfrac{1}{2}T_2(p + jx_c)e^{(p+jx_c)t} + \tfrac{1}{2}T_2(p - jx_c)e^{(p-jx_c)}. \tag{4c}$$

Thus, if (4a) is to be satisfied, we must have (using (4b) and (4c))

1. for the component $e^{(p+jx_c)t}$: $T_1(p) = T_2(p + jx_c)$ (4d)

2. for the component $e^{(p-jx_c)t}$: $T_1(p) = T_2(p - jx_c)$ (4e)

For some purposes it is convenient to replace the second condition by

2a. for the component $e^{(jx_c-p)t}$: $T_1(-p) = T_2(-p + jx_c)$ (4f)

For real signal frequencies x we obtain from (4d) and (4f)

for the component $e^{j(x_c+x)t}$: $T_1(jx) = T_2[j(x_c + x)]$ (4g)

for the component $e^{j(x_c-x)t}$: $T_1(-jx) = T_2[j(x_c - x)]$ (4h)

Since $T_1(jx)$ is symmetrical about $x = 0$, i.e.

$$|T_1(jx)| = |T_1(-jx)| \text{ and } \arg[T_1(jx)] = -\arg[T_1(-jx)] \tag{4i}$$

(4g) and (4h) require that $T_2[j(x_c + x)]$ shows a similar symmetry about $x = x_c$. This can be achieved, at least approximately, by a suitable design of N_2 for "arithmetical symmetry" (see Szentirmai [19]).

Thus, if N_1 is given, in a circuit of the type shown in Fig. 5a, then it is possible to find a network N_2 which, in the circuit of Fig. 5b, would produce *approximately* the same output signal as that produced in Fig. 5a. On the other hand, if a network N_2 in the circuit in Fig. 5b is given, with a transfer function $T_2(p)$ which does not show any symmetry about $x = x_c$ (e.g. a band pass filter with a dissymmetrical loss-frequency curve or a narrow symmetrical band pass filter with a mid-band frequency significantly different from x_c), then it is obviously not only physically but also conceptually impossible to find a network N_1 which would enable the circuit in Fig. 5a to produce the same output signal as the circuit in Fig. 5b. We shall return to this point later. However, even in the first situation, where N_1 is given, a network N_2 satisfying (4a) *accurately* (as opposed to approximately, over a limited frequency range) is, as we shall now show, non-physical.

To demonstrate this, we shall use a different frequency variable. We note that the equivalence conditions (4d) and (4f) are written in terms of the input frequency p to the composite circuits in Figs. 5a and 5b. As far as N_2 is concerned, it is more appropriate to write them in terms of the output frequency of the composite circuits which is at the same time the input and output frequency of N_2. As there are two output frequencies, we introduce two new variables:

$$p_+ = jx_c + p, \; p = p_+ - jx_c \tag{4j}$$

$$p_- = jx_c - p, \; p = -p_- + jx \tag{4k}$$

Then,

$$\text{for} \quad e^{p_+t}: \quad T_2(p_+) = T_1(p_+ - jx_c) \tag{5a}$$

$$\text{for} \quad e^{p_-t}: \quad T_2(p_-) = T_1(p_- - jx_c). \tag{5b}$$

If we use p' as a common symbol for all output frequencies, p_+ or p_-, we can combine (5a) and (5b) into one equation

$$T_2(p') = T_1(p' - jx_c)^*. \tag{5c}$$

---

* There is also an "associated" equation

$$T_2(-p') = T_1(-p' - jx_c). \tag{5d}$$

Since N_1 is a given physical network, $T_1(p)$ is a *real* function of p (it is also a rational function if N_1 consists of a finite number of lumped components). This means that $T_1(p' - jx_c)$ cannot be a real function of p', and therefore (see (5c)) $T_2(p')$ is not a real function of p'. This may be regarded as an additional degree of freedom or as an additional constraint, depending on circumstances.

The question arises whether this is just an interpretation and of purely formal interest, or whether there are any practical consequences of $T_2(p')$ being non-real. The answer is that the non-real character of $T_2(p')$ can be demonstrated by laboratory experiments. Let us consider a physical arrangement as shown in Fig. 5a, where N_1 is a low pass filter with pass band limits $-x_b$ to x_b with x_b roughly equal to x_c. We apply a signal with frequency $x = x_1$ to the input of the composite circuit where x_1 is small compared with x_c but large enough to enable us to separate the two output signal components with frequencies $x_c - x_1$ and $x_c + x_1$ from each other. We tune the detector to the lower frequency, i.e. to $x_c - x_1$ and measure the amplitude of the output signal. Then we increase the input frequency x continuously and record the output amplitude as function of $x' = x_c - x$, the detector tuning being altered correspondingly so that we continue to measure the amplitude of the difference-frequency component. The amplitude/frequency characteristic obtained in this way represents the function $|T_2(jx')|$ (see Fig. 6a)*. It is obviously not constrained to be an even function of the output frequency x' since the behavior of the curve in the neighborhood of $x' = 0$ depends only on the behavior of the characteristic of the low pass filter at an input frequency x of a value near to that of the carrier frequency x_c. Since the low pass filter "cannot know" the value of the carrier frequency applied to its *output* signal it is obvious that we cannot expect any other result. If, however, the same experiment is repeated with a circuit of the type shown in Fig. 5b, then the recorded characteristic is in fact

---

* The presentation in Fig. 6a is open to criticism. This has been done deliberately and will be discussed presently.

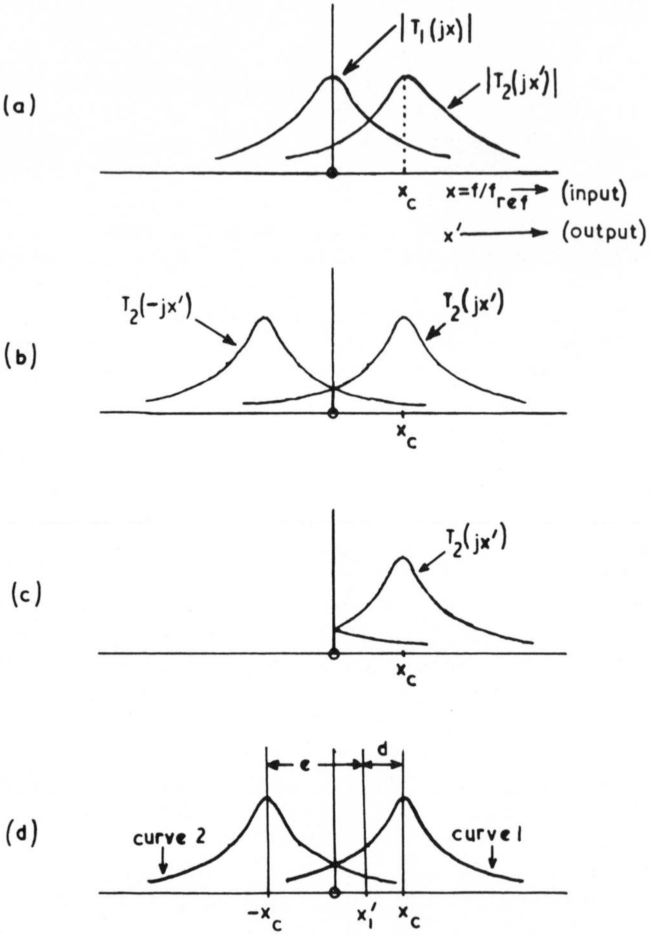

If output frequency $x'=x_1'$ is produced by input frequency |d|, then use curve 1 for output amplitude; if the same output frequency is produced by input frequency |e|, then use curve 2.

Fig. 6. Response characteristics associated with circuit in Fig. 5a.

the characteristic of the network N_2 and will therefore be an even function of the output frequency x' (see Fig. 7). If N_2 is a band pass filter, it can be designed to give an output-amplitude/frequency characteristic which over a wide frequency band coincides with that obtained in the first experiment with the circuit of Fig. 5a; but near x' = 0 the two output characteristics will necessarily differ. The difference between the results obtained in the two experiments shows that if the two circuits shown in Figs. 5a and 5b were presented to us as physical "black boxes", we could identify each circuit by means of the measurements described.

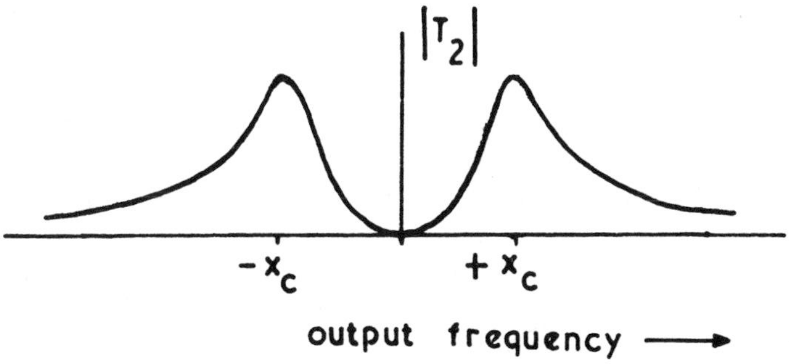

Fig. 7. Response characteristics associated with circuit in Fig. 5b.

In the discussion of the mathematical form of $T_2(p')$ as given by (5c) we have pointed out that it is a *non-real* function of p. In the discussion of the corresponding measurement of $|T_2(p')|$ we have commented on the fact that it is not an *even* function of p'. A few comments on the relationship between the two statements are relevant. In elementary electrical engineering we may, for example, consider a real impedance R and an imaginary impedance $j\omega L$ and define the total impedance as $Z = R + j\omega L$, its magnitude by $|Z| = (R^2 + \omega^2 L^2)^{\frac{1}{2}}$ and its phase angle ψ by $\tan\psi = \omega L/R$. However, this decomposition of Z into the particular real and imaginary components given above is only valid on the imaginary p-axis. At a higher level of sophistication we recognize that any function F(p) can be decomposed into an even part F_e and an odd part F_0, and that this decomposition is valid anywhere in the complex p-plane. On the imaginary axis the even part coincides with the real part and the odd part with the imaginary part. We can therefore express $|F|$ and $\tan\psi$ by means of $|F|^2 = F_e^2 - F_0^2$ and $\tan\psi = -jF_0/F_e$, to be evaluated on the imaginary p-axis. From this point of view we would consider "being even" and "being odd" as more fundamental and more general characteristics than "being real" and "being imaginary." Now, in discussing functions such as $T_2(p')$ the wheel has come full circle. In order to find the magnitude and the phase angle of a non-real function we must know its real and its imaginary part.

The presentation in Fig. 6a can be criticized because the low pass characteristic is shown both for positive and for negative frequencies whereas only one of the shifted characteristics is shown (although it is continued into the negative frequency range). The best presentation is probably that given in Fig. 6b, but that given in Fig. 6c which is restricted to positive frequencies gives all the relevant information (in Fig. 6b this information is duplicated). The relevant feature in both figures is that for any abscissa value there are *two* curves, both representing the output characteristic. How do we know which of these curves to choose?

For a discussion of this question the presentation in Fig. 6b is probably the most suitable one. The decisive point is that we are discussing a network characteristic as a function of the *output* frequency. In the case of the circuit shown in Fig. 5b where the network, N_2, follows the modulator, the frequency of a signal component at the output of the composite circuit is identical with the frequency of the corresponding signal component processed by N_2. Therefore no ambiguity arises. The situation is quite different when a circuit as shown in Fig. 5a is given. Any particular output frequency under consideration may have been produced by either of two different input frequencies. The output amplitude depends on the frequency of the input component which is processed by N_1. This is explained in Fig. 6d.

4.4. *Virtual Transfer Functions*

A transfer function $T(p_0)$ (where p_0 is an arbitrary but specific value of p) can only be defined for a given circuit, if any input signal component $e^{p_0 t}$ applied to this circuit produces a corresponding output component of the form $T(p_0)e^{p_0 t}$ and no output components with different p-values, and if an output component with the value $p = p_0$ cannot be produced by any input component with a different p-value. Neither of these conditions is satisfied by the circuits in Figs. 5a and 5b. However, there is a significant difference between these two circuits. We have shown that the output/input characteristic of the circuit in Fig. 5b can be described directly in terms of the transfer characteristic of the time-invariant network. This is not possible with the same degree of simplicity in the case of the circuit in Fig. 5a; we have shown that it is in fact convenient to explain the operation of the circuit in Fig. 5a in terms of an equivalent circuit of the type shown in Fig. 5b, i.e. in terms of a *fictitious* network N_2 contained in this equivalent circuit. We shall refer to the transfer characteristic of this network as a "virtual transfer function" (see [20], and also [16]).

The frequency selectivity at the output port of a circuit of the type shown in Fig. 5a, containing a low pass filter N_1, can be very high if the ratio of carrier frequency to low-pass band-width is sufficiently high. In such circumstances an RC network N_1 having a very low frequency-selectivity may produce a virtual transfer function $T_2(p')$ with such high selectivity that the corresponding fictitious network N_2 would for its approximate physical realization require RLC or crystal components. The poles p_i of the transfer functions of RC networks are restricted to the negative real p-axis, i.e.

$$T_1(p) \to \infty, \text{ if } p = p_i = -\sigma_i, \text{ all } \sigma_i > 0; \qquad (6a)$$

whereas the poles p_j of RLC networks can occur anywhere in the left half of the p-plane in conjugate complex pairs: $p_j = -\sigma_j \pm jx_j$, all $\sigma_j > 0$. We find that this "derestriction" of poles also occurs when we consider, instead of $T_1(p)$, the corresponding virtual transfer function $T_2(p')$: from (5c) and (6a) it follows that

$$T_2(p') \to \infty \text{ if } p' = -\sigma_i + jx_c\,. \qquad (6b)$$

We note that in this case the poles do not occur in negative conjugate pairs; since $T_2(p')$ is a non-real function, conjugateness is not required.

5. N-PATH FILTERS

5.1. *General*

There is a vast literature on N-path filters and, as this is not a survey paper, no attempt will be made to do justice to the many papers published on this subject. Franks and Sandberg, in their classical paper [1] mentioned the possibility of RC realization of stable transfer functions by means of elementary N-path filters which use pure multipliers and are driven by sinusoidal carrier oscillations; but nobody seems to have taken up their suggestion. In the same paper it was stated clearly that the input and output carrier oscillations applied to the same path need not have zero phase difference; again, little notice seems to have been taken of the availability of this phase difference (which has to be the same for all paths) as a design parameter. In the present paper these two omissions will be made good.

5.2. *The Basic Transfer Functions*

Although the basic transfer function is given in [1a], it is derived there as a special case in the course of a very general investigation. As this transfer function is the starting point of the considerations in this section, it is thought that it will be useful to give a brief derivation of this function here, particularly as this derivation employs the elementary mathematical methods used throughout this paper.

We consider the circuit shown in Fig. 8. For an input signal e^{pt} we obtain

$$q_1(t) = \tfrac{1}{2}e^{(p+j\omega_c)t} + \tfrac{1}{2}e^{(p-j\omega_c)t}$$

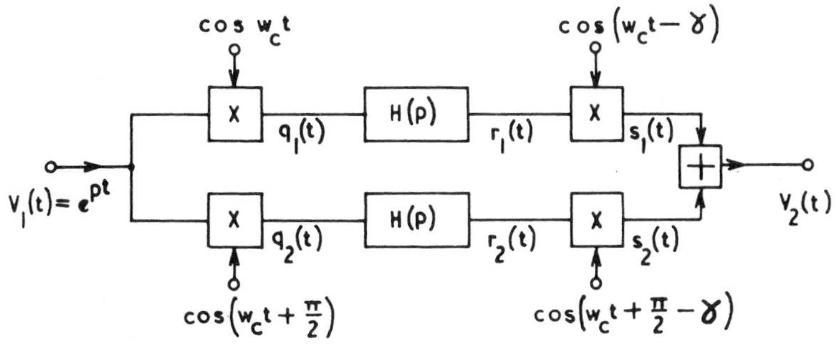

Fig. 8. Two-path filter.

$$q_2(t) = \tfrac{1}{2}e^{(p+j\omega_c)t+j\frac{1}{2}\pi} + \tfrac{1}{2}e^{(p-j\omega_c)t-j\frac{1}{2}\pi}$$

$$r_1(t) = \tfrac{1}{2}H(p+j\omega_c)e^{(p+j\omega_c)t} + \tfrac{1}{2}H(p-j\omega_c)e^{(p-j\omega_c)t}$$

$$r_2(t) = \tfrac{1}{2}H(p+j\omega_c)e^{(p+j\omega_c)t+j\frac{1}{2}\pi} + \tfrac{1}{2}H(p-j\omega_c)e^{(p-j\omega_c)t-j\frac{1}{2}\pi}$$

$$s_1(t) = \tfrac{1}{4}H(p+j\omega_c)[e^{(p+j2\omega_c)t-jy}+e^{pt+jy}] +$$
$$\tfrac{1}{4}H(p-j\omega_c)[e^{pt-jy}+e^{(p-j2\omega_c)t+jy}]$$

$$s_2(t) = \tfrac{1}{4}H(p+j\omega_c)[-e^{(p+j2\omega_c)t-jy}+e^{pt+jy}] +$$
$$\tfrac{1}{4}H(p-j\omega_c)[e^{pt-jy}-e^{(p-j2\omega_c)t+jy}]$$

$$v_2(t) = s_1(t)+s_2(t) = \tfrac{1}{2}[e^{jy}H(p+j\omega_c)+e^{-jy}H(p-j\omega_c)]e^{pt}$$

We see that all components with frequencies different from the input frequency p are cancelled, for any function H(p). Therefore a transfer function T(p) exists and is given by (replacing ω_c by x_c)

$$T(p) = \frac{v_2(t)}{v_1(t)} = \tfrac{1}{2}[e^{jy}H(p + jx_c) + e^{-jy}H(p - jx_c)]^{*} \quad (7a)$$

5.3. *N-Path Filters with RC Path Networks*

Consider the important practical problem of the

---

* Note that if $H(p) \to \infty$ for $p = -\sigma_i$ (RC network),

$$T(p) \to \infty \text{ for } p = -\sigma_i \pm jx_c . \quad (7b)$$

microelectronic realization (i.e. the inductor-less realization) of arbitrary stable transfer functions. Since the poles of transfer functions of passive time-invariant RC networks are restricted to the negative real p-axis, it is necessary to use active, non-reciprocal or time-varying components to remove this constraint. If we use a single N-path filter with passive time-invariant RC path networks, then the poles are restricted to two parallels to this axis at distances $+x_c$ and $-x_c$ (see equation (7a)). Although this is sometimes adequate for the synthesis of narrow band pass filters, in the general case this restriction is not acceptable. Franks and Sandberg [1a] have proposed the realization of a general transfer function by connecting m different N-path filters, with different carrier frequencies, in parallel (in some cases in parallel with a constant-gain path). In the general case each N-path filter is a second-order low pass filter with two transmission zeros at infinite frequency. The individual low pass filter transfer functions are identified by partial fraction expansion of the required function T(p) as follows:

$$T(p) = a_0 + \sum_{i=1}^{m} T_i(p) = a_0 + \sum_{i=1}^{m} \left(\frac{a_i}{p-p_i} + \frac{a_i^*}{p-p_i^*} \right)$$
$$= a_0 + \sum_{i=1}^{m} \left(\frac{a_i}{p+\sigma_i + jx_i} + \frac{a_i}{p+\sigma_i - jx_i} \right) . \tag{8a}$$

If all m values of x_i are different, then each second-order N-path low pass filter has a transfer function T_i as defined in (8a) and is produced with 1st order RC path networks with transfer functions

$$H_i(p) = \frac{2|a_i|}{p+\sigma_i}, \quad x_c = x_i \text{ and } \gamma_i = \arg(a_i). \tag{8b}$$

Transmission zeros of the overall transfer function T(p) are produced by the interaction of all separate elementary N-path filters. This is usually undesirable, as it makes adjustment difficult and leads to a high sensitivity of the performance characteristic (at least in the neighborhood of the transmission zeros) to component variations. Let us therefore consider - following the usual (although not exclusive) practice in active RC network design - a cascade arrangement of elementary N-path filter sections with biquadratic transfer functions T_j; note that for a cascade connection low pass sections alone would not be sufficient.

For the purpose of cascade connection we decompose a given transfer function T(p) in accordance with

$$T(p) = (BLF) \prod_j \frac{A_{0j} + A_{1j}p + A_{2j}p^2}{1 + B_{1j}p + B_{2j}p^2} \tag{9a}$$

where (BLF) indicates bilinear factors which can be realized by passive time invariant RC networks. All coefficients in (9a) are real and B_{1j} and B_{2j} are always > 0. As we shall deal with single biquadratic factors, we shall drop the suffix j. Furthermore it is convenient to renormalize the variable p for each factor so that T(p) is of the form

$$T(p) = \frac{A_0 + A_1p + A_2p^2}{1 + Dp + p^2} \tag{9b}$$

where Q = 1/D is the Q-factor of the denominator.

5.4. *N-Path Filters with Biquadratic Transfer Functions*

Our aim is to find a suitable 1st order RC path function

$$H(p) = \frac{C_0 + C_1p}{1 + Ep} \tag{10a}$$

and suitable values of x_c and γ, so that when H(p), x_c and γ are substituted in (7a), we obtain a function T(p) proportional to the required function T(p) defined in (9b). We can only expect proportionality (not equality) because the coefficients of H(p) are constrained by the requirement of passivity. For given H(p), x_c and γ we can find the *actual* function T(p), but if a required T(p) is specified, it is more convenient to consider as *given*, in addition to the denominator coefficients D and unity (for p^2), the ratios of the numerator coefficients A_0/A_2 and A_1/A_2. Similarly we shall regard as our unknown parameters the values of x_c, γ and the ratio C_0/C_1.

From (10a) and (7) we obtain

$$T(p) = \frac{(1+E^2x_c^2)^{-1}[(C_0+C_1Ex_c^2)\cos\gamma + x_c(C_0E-C_1)\sin\gamma]}{1 + 2E(1+E^2x_c^2)^{-1}p + E^2(1+E^2x_c^2)^{-1}p^2}$$

$$+ \frac{(1+E^2x_c^2)^{-1}\{[(C_1+C_0E)\cos\gamma]p + [C_1E\cos\gamma]p^2\}}{1 + 2E(1+E^2x_c^2)^{-1}p + E^2(1+E^2x_c^2)^{-1}p^2} \qquad (10b)$$

By comparing the denominator coefficients in (9b) and (10b), we find

$$E = 2D^{-1} = 2Q \qquad (10c)$$

$$x_c = +(1 - \tfrac{1}{4}D^2)^{\frac{1}{2}} \qquad (10d)$$

$$(\text{and } 1 + E^2x_c^2 = 4Q^2) \qquad (10e)$$

Substituting these expressions for E and x_c in (10b) and comparing the numerator coefficients in (9b) and 10b) we obtain

$$A_0 = \tfrac{1}{2}D[(C_1x_c^2 + \tfrac{1}{2}C_0D)\cos\gamma + x_c(C_0 - \tfrac{1}{2}C_1D)\sin\gamma] \quad (11a)$$

$$A_1 = \tfrac{1}{2}D(C_0 + \tfrac{1}{2}C_1D)\cos\gamma \qquad (11b)$$

$$A_2 = \tfrac{1}{2}DC_1\cos\gamma \qquad (11c)$$

i.e.

$$A_0/A_2 = [x_c^2 + \tfrac{1}{2}D(C_0/C_1)] + x_c[(C_0/C_1) - \tfrac{1}{2}D]\tan\gamma \quad (12a)$$

$$A_1/A_2 = (C_0/C_1) + \tfrac{1}{2}D \qquad (12b)$$

From (12b) we find

$$C_0/C_1 = (A_1/A_2) - \tfrac{1}{2}D \tag{12c}$$

and from (11a) and (12c)

$$\tan\gamma = \frac{A_0 - \frac{1}{2}DA_1 - (1 - \frac{1}{2}D^2)A_2}{x_c(A_1 - DA_2)} \quad . \tag{12d}$$

If $C_0/C_1 \neq 0$, we shall normalize $C_0 = 1$. (12e)

If $C_0/C_1 = 0$, then we shall normalize $C_1 = E = 2Q$. (12f)

5.5. *Low Pass Filter*

We require $A_1 = 0$ and $A_2 = 0$, A_0 as large as possible. From (12d) we obtain $\tan\gamma \to \infty$, but from (12c) we obtain an indeterminate result of the form 0/0. We investigate therefore equations (11a, b, c) for $\tan\gamma \to \infty$, i.e. $\cos\gamma = 0$, $\sin\gamma = \pm 1$. We find that $A_1 = 0$ and $A_2 = 0$, independent of C_1 and C_2. For A_0 we obtain $A_0 = \pm \frac{1}{2}Dx_c(C_0 - \frac{1}{2}C_1D)$. Thus $|A_0|$ becomes a maximum if $C_0 = 1$ and $C_1 = -2Q$: $A_0 = \pm Dx_c$. In this case $H(p) = (1 - 2Qp)/(1 + 2Qp)$, i.e. an all-pass function. The sets of values $C_0 = 1$, $C_1 = 0$ (low pass) and $C_0 = 0$, $C_1 = +2Q$ (high pass) provide "local" maxima $A_0 = \pm \frac{1}{2}Dx_c$ and $A_0 = \mp \frac{1}{2}Dx_c$, respectively (if we make $C_0 = 1$, $C_1 = +2Q$, i.e. $H(p) = 1$, we obtain $A_0 = 0$)*.

5.6. *Band Pass Filter*

Ideally we require $A_0 = 0$, $A_2 = 0$; A_1 to be maximized. From (12d) we obtain $\tan\gamma = -\frac{1}{2}D/x_c$, i.e. $\cos\gamma = x_c$, $\sin\gamma = -\frac{1}{2}D$. Normalizing $C_0 = 1$, we find $C_1 = 0$ and $A_1 = \frac{1}{2}Dx_c$. For small values of D we find $\gamma \simeq 0$. Therefore it is of interest to consider the case $\gamma = 0$. Then from (11a, b, c) we obtain $A_0 = \frac{1}{2}D(C_1x_c^2 + \frac{1}{2}C_0D)$, $A_1 = \frac{1}{2}D(C_0 + \frac{1}{2}C_1D)$ and $A_2 = \frac{1}{2}DC_1$.

Two sets of values for C_0 and C_1 seem to be of interest: $C_0 = 1$, $C_1 = 0$ and $C_0 = 1$, $C_1 = -\frac{1}{2}D/x_c^2$. In the first

---

* This invariance of T(p) to H(p) is discussed, within a much wider context and in greater detail, in [21].

case we obtain $A_0 = \frac{1}{4}D^2$, $A_1 = \frac{1}{2}D$, $A_2 = 0$ (i.e. $A_0/A_1 = \frac{1}{2}D$ instead of 0) and in the second case $A_0 = 0$, $A_1 = \frac{1}{2}D(1 - \frac{1}{4}D^2/x_c^2)$, $A_2 = -\frac{1}{4}D^2/x_c^2$ (i.e. $A_2/A_1 \simeq -\frac{1}{2}D$ instead of 0). For small values of D these are small deviations from the ideal characteristic which may be acceptable in order to make unnecessary the provision of a small phase shift between input and output carrier.

5.7. *High Pass Filter*

We require $A_0 = 0$, $A_1 = 0$; A_2 to be maximized. From (12d) we obtain $\tan\gamma = Q(1 - \frac{1}{2}D^2)/x_c$, $\cos\gamma = Dx_c$, $\sin\gamma = 1 - \frac{1}{2}D^2$, and from (12c) $C_0/C_1 = -\frac{1}{2}D$. Thus for $C_0 = 1$ we have $C_1 = -2Q$ and $H(p) = (1 - 2Qp)/(1 + 2Qp)$, an all-pass function; $A_2 = -Dx_c$. Negative values of C_0/C_1 cannot be realized by passive three-terminal RC networks. Therefore, we should like to avoid such functions as H(p). For this purpose we investigate the case where $A_1 = bA_2$ (instead of $A_1 = 0$). Then we obtain $\tan\gamma = (-1 + \frac{1}{2}D^2 - \frac{1}{2}bD)/x_c(b - D)$, $C_0/C_1 = b - \frac{1}{2}D$. Thus, if we choose $b = +\frac{1}{2}D$, we obtain $C_0 = 0$, $C_1 = 2Q$ (high pass filter) and $\tan\gamma = 2Qx_c$, $\cos\gamma = \frac{1}{2}D$, $\sin\gamma = x_c$; $A_2 = \cos\gamma = \frac{1}{2}D$.

5.8. *All-Pass Network*

We require $A_0 = A_2$ and $A_1 = -DA_2$. Then $\tan\gamma = -\frac{1}{2}D/x_c$, $\cos\gamma = x_c$, $\sin\gamma = -\frac{1}{2}D$, as in the ideal band pass case. From (12c) we obtain $C_0/C_1 = -\frac{3}{2}D$; we choose therefore $C_0 = 1$, $C_1 = -\frac{2}{3}Q$. If we leave C_0 and C_1 unchanged, but choose $\gamma = 0$, we obtain $A_2 = -\frac{1}{3}$ and $A_1/A_2 = -D$, as required, but $A_0/A_2 = 1 - D^2$ (instead of 1).

5.9. *Low and High Pass Filters with Transmission Zero at* $x = x_\infty$

We require $A_1 = 0$ and $A_0 = A_2x_\infty^2$. Then

$\tan\gamma = Q(1 - x_\infty^2 - \frac{1}{2}D^2)/x_c$, $C_0/C_1 = -\frac{1}{2}D$, i.e. $C_0 = 1$, $C_1 = -2Q$. In order to avoid negative values of C_0/C_1, let us consider $A_1 = bA_2$ (instead of $A_1 = 0$). Then $\tan\gamma = (x_\infty^2 - 1 + \frac{1}{2}D^2 - \frac{1}{2}bD)/x_c(b - D)$ and $C_0/C_1 = b - \frac{1}{2}D$. For $b = \frac{1}{2}D$ we obtain $C_0/C_1 = 0$, i.e. $C_0 = 0$, $C_1 = 2Q$. Then $\tan\gamma = 2Q(x_c^2 - x_\infty^2)/x_c$ and $A_2 = \cos\gamma$.

5.10. *Path Networks*

An important part of the theory and practice of N-path networks is concerned with the interaction of switching and path networks. In this paper we shall ignore this interaction (we can assume that where necessary buffer amplifiers are inserted to prevent it). On this basis we shall regard H(p) as the output/input voltage ratio of the unterminated path network.

For $C_0/C_1 \to \infty$ and for $C_0/C_1 = 0$ the simple path networks shown in Fig. 9a and 9b, respectively, can be used. For negative values of C_0/C_1 we can use the passive 4-terminal network shown in Fig. 9c or the half-lattice network driven by a paraphase amplifier shown in Fig. 9d (see [22]). The necessary design information is given in these figures.

5.11. *Additional Constant-Gain Path*

There is a different method by which the use of 4-terminal passive or equivalent active 3-terminal networks can be avoided: by the provision of an additional constant gain path, with transfer constant K, in parallel with the elementary N-path filter (see also [1a]). A detailed analysis shows that the expression for $\tan\gamma$ (Eq. 12d) remains unchanged; this means that the same phase angle γ is required for a specified function T(p), whether or not an additional parameter K is provided (this is most easily seen by replacing in (12d) A_0 by $A_0 - K$, A_1 by $A_1 - DK$ and A_2 by $A_2 - K$). However the new expression for C_0/C_1 depends on K:

$C_0/C_1 = [(A_1 - K)/(A_2 - K)] - \frac{1}{2}D$. Therefore K can be so chosen that C_0/C_1 can be made 0 or ∞, as desired.

(a) $H(p) = \dfrac{V_2}{V_1} = \dfrac{1}{1+2Qp}$

if $bc = 2Q$

(b) $H(p) = \dfrac{V_2}{V_1} = \dfrac{2Qp}{1+2Qp}$

if $bc = 2Q$

(c) $H(p) = \dfrac{V_2}{V_1} = \dfrac{1-bcp}{2(1+b)(1+cp)}$

$= \dfrac{1-2Qp}{1+2Qp}$ if $b=1, c=2Q$

$= \dfrac{1}{2(1+b)} \cdot \dfrac{1-\frac{2}{3}Qp}{1+2Qp}$

if $b = \dfrac{1}{3}, c = 2Q$

(d) $H(p) = \dfrac{V_2}{V_1} = \dfrac{1+(d-b)cp}{1+(d+b)cp}$

$= \dfrac{1-2Qp}{1+2Qp}$ if $d=0$
                          $bc = 2Q$

$= \dfrac{1-\frac{2}{3}Qp}{1+2Qp}$

if $dc = \dfrac{2}{3}Q, bc = \dfrac{4}{3}Q$

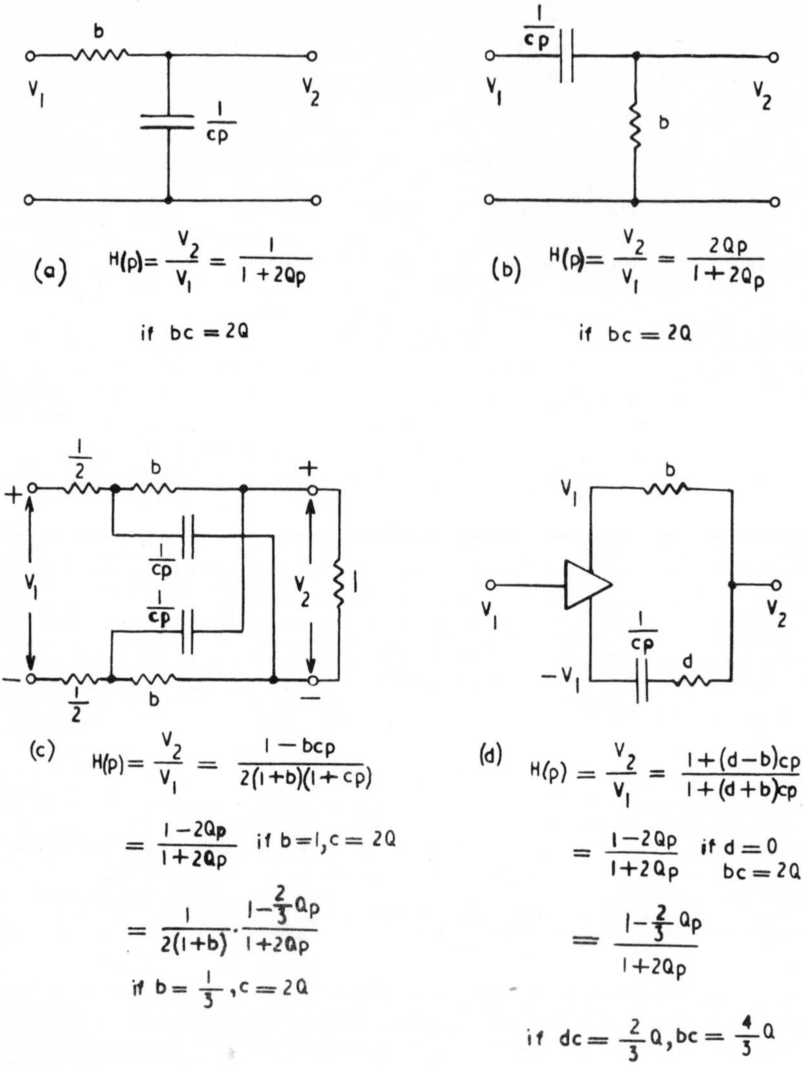

Fig. 9.   Path-networks for two-path filter.

Instead of continuing this detailed discussion we shall give a more elegant analysis and synthesis discussion. It is based on a simple relationship concerning bilinear functions and on a particular feature of Eq. (7a) which connects T(p) and H(p). Starting with the bilinear function

$$H(p) = \frac{C_0 + C_1 p}{1 + Ep} \qquad (10a)$$

we note that by addition of a real constant k we can form a new function, H'(p), which is proportional to any bilinear function with the same denominator 1 + Ep but with an arbitrary numerator $C_0' + C_1'p$.

$$H'(p) = k + H(p) = k + (C_0 + C_1 p)/(1 + Ep)$$

$$= m(C_0' + C_1'p)/(1 + Ep) \qquad (13a)$$

$$\text{if} \quad m = (C_1 - EC_0)/(C_1' - C_0'E)$$

$$\text{and} \quad k = (C_0'C_1 - C_1'C_0)/(C_1' - C_0'E) . \qquad (13b)$$

We note that $m \to \infty$ and $k \to \infty$, if H'(p) is a constant, and $m \to 0$ if H(p) is a constant (k = 0 occurs if the two bilinear functions are proportional to each other). It follows that we can obtain a low pass function $H'(p) = m/(1 + 2Qp)$ (or a high pass function $H'(p) = m2Qp/(1 + 2Qp)$) from any function H(p) required in accordance with the formulae derived in Sections 5.4 to 5.9 by adding an appropriately chosen constant k to H(p).

Let us now consider (using Eq. (7)) the effect on T(p) of adding a constant k to H(p). If we define T'(p) by

$$T'(p) = \tfrac{1}{2}[e^{j\gamma}H'(p + jx_c) + e^{-j\gamma}H'(p - jx_c)] \qquad (13c)$$

then

$$T'(p) = T(p) + \tfrac{1}{2}[e^{j\gamma} + e^{-j\gamma}]k = T(p) + k\cos\gamma$$

and

$$T(p) = T'(p) - k\cos\gamma = T'(p) + K; \; K = -k\cos\gamma \qquad (13d)$$

(13d) can be interpreted as follows: We generate $T'(p)$ by means of *simple* path networks with transfer function $H'(p)$. We obtain the required $T(p)$ by means of an additional path with transfer constant $K = -k\cos\gamma$ (connected in parallel), where k is defined by $k = H'(p) - H(p)$. We note that if $T(p)$ is a low pass function, for which $\gamma = \pm\pi/2$, then any constant k can be added to $H(p)$ without affecting $T(p)$, because $\cos\gamma = 0$.

We shall consider one example. For a high pass filter ($A_0 = 0$, $A_1 = 0$) we found in Section 5.7 that $H(p) = (1 - 2Qp)/(1 + 2Qp)$, $\cos\gamma = Dx_c$, is required. By adding $k = \frac{1}{2}$ to $\frac{1}{2}H(p)$ we obtain $H'(p) = 1/(1 + 2Qp)$. Therefore we can obtain the required function $T(p)$ with the simpler path network for $H'(p)$, if we add a constant gain path with $K = -\frac{1}{2}\cos\gamma = -\frac{1}{2}Dx_c$. Check: for $C_0 = 1$, $C_1 = 0$, $\cos\gamma = Dx_c$, $\sin\gamma = 1 - \frac{1}{2}D^2$, we obtain $A_0 = \frac{1}{2}Dx_c$, $A_1 = DA_0$, $A_2 = 0$. Thus

$T'(p) = \frac{1}{2}Dx_c(1 + Dp)/(1 + Dp + p^2)$ and

$T'(p) - \frac{1}{2}Dx_c = -\frac{1}{2}Dx_c p^2/(1 + Dp + p^2)$, as required.

6. CONTINUOUSLY ADJUSTABLE BIQUADRATIC N-PATH FILTER SECTIONS

The results obtained from the analysis and synthesis of passive biquadratic N-path filter sections show clearly that the character of the transfer function $T(p)$ depends to a very large extent on the value of the phase angle γ. Thus, with $H(p) = 1/(1 + 2Qp)$ we obtain

$T(p) = \frac{1}{2}Dx_c/(1 + Dp + p^2)$ for $\gamma = \frac{1}{2}\pi$, but

$T(p) = \frac{1}{2}Dx_c p/(1 + Dp + p^2)$ for $\sin\gamma = -\frac{1}{2}D$. If we choose $H(p) = (1 - 2Qp)/(1 + 2Qp)$ we obtain again a low pass function $T(p) = Dx_c/(1 + Dp + p^2)$ for $\gamma = \frac{1}{2}\pi$ but $T(p) = -Dx_c p^2/(1 + Dp + p^2)$ for $\sin\gamma = 1 - \frac{1}{2}D^2$. This suggests that a continuous variation of $T(p)$ by means of continuously varying γ may be of practical interest.

The relationship between T(p) and γ can be brought into clearer evidence if we derive special expressions for for A_0, A_1, A_2 (see (11a, b, c)) for the 3 cases:

Case 1. $H(p) = 1/(1 + 2Qp)$; i.e. $C_0 = 1$, $C_1 = 0$: *low-pass*

Case 2. $H(p) = 2Qp/(1 + 2Qp)$; i.e. $C_0 = 0$, $C_1 = 2Q$: *high-pass*

Case 3. $H(p) = (1 - 2Qp)/(1 + 2Qp)$, i.e. $C_0 = 1$, $C_1 = -2Q$: *all-pass*

The final expressions become particularly simple if we introduce an angle δ defined by $\sin\delta = \frac{1}{2}D$. Then $\cos\delta = (1 - \frac{1}{4}D^2)^{\frac{1}{2}} = x_C$, $\sin 2\delta = Dx_C$ and $\cos 2\delta = 1 - \frac{1}{2}D^2$ (see also Fig. 10). With the help of these relationships we obtain the following results:

Case 1

$$A_0 = \sin\delta \sin(\gamma + \delta)$$

$$A_1 = \sin\delta \cos\gamma$$

$$A_2 = 0$$

$$T(p) = \sin\delta \frac{\sin(\gamma + \delta) + (\cos\gamma)p}{1 + (2\sin\delta)p + p^2}$$

$T(p) = 0$ for two values of p, $(p_\infty)_1$ and $(p_\infty)_2$. $(p_\infty)_1 = \infty$ for any value of γ (because of the higher degree of the denominator).

$$(p_\infty)_2 = -\sin(\gamma + \delta)/\cos\gamma = -[\tan\gamma\cos\delta + \sin\delta]$$

$$\tan\gamma = -[(p_\infty)_2 + \sin\delta]/\cos\delta$$

We obtain a low pass characteristic, i.e. $(p_\infty)_2 \to \infty$, if $\tan\gamma \to \infty$ and a band pass characteristic, i.e. $(p_\infty)_2 \to 0$ if $\gamma = -\delta$. If, for the sake of simplicity, we choose $\gamma = 0$, we obtain $A_0/A_1 = \tan\delta$ (instead of $A_0/A_1 = 0$ for a perfect band pass filter).

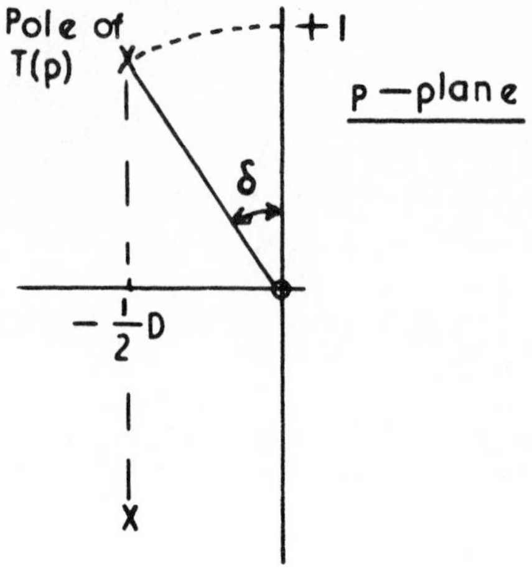

Fig. 10. Definition of design parameter δ.

Case 2

$A_0 = \cos\delta\cos(\gamma + \delta)$

$A_1 = \sin\delta\cos\gamma$

$A_2 = \cos\gamma$

$$T(p) = \cos\gamma \frac{\frac{\cos\delta}{\cos\gamma}\cos(\gamma+\delta) + (\sin\delta)p + p^2}{1 + (2\sin\delta)p + p^2}$$

$$T(p) = 0 \text{ if } (p_\infty)_{\frac{1}{2}} = -\tfrac{1}{2}\sin\delta \pm j\left(\frac{\cos\delta\cos(\gamma+\delta)}{\cos\gamma} - \tfrac{1}{4}\sin^2\delta\right)^{\frac{1}{2}}$$

As before, we obtain a low pass filter, i.e. $(p_\infty)_{\frac{1}{2}} \to \infty$ if $\cos\gamma = 0$. We note that for *complex* values of $(p_\infty)_{\frac{1}{2}}$ the real component is independent of γ, i.e. if γ is varied the zeros of T(p) move along a parallel to the imaginary axis with real value $-\frac{1}{2}\sin\delta = -\frac{1}{4}D$ (note that the *poles* have the real component $-\frac{1}{2}D$).

Case 3

$A_0 = -\cos(\gamma + 2\delta)$

$$A_1 = 0$$

$$A_2 = -\cos\gamma$$

$$T(p) = -\frac{\cos(\gamma + 2\delta) + (\cos\gamma)p^2}{1 + Dp + p^2}$$

$$T(p) = 0 \text{ for } p^2 = p_\infty^2 = -\frac{\cos(\gamma + 2\delta)}{\cos\gamma}$$

$$x_\infty^2 = +\frac{\cos(\gamma + 2\delta)}{\cos\gamma}$$

For $\gamma = -\frac{1}{2}\pi$, $T = 0$ for $x_\infty^2 \to \infty$ (low pass)

For $\gamma = \frac{1}{2}\pi - 2\delta$, $T = 0$ for $x_\infty^2 \to 0$ (high pass)

For intermediate values of γ, i.e. in the range $-\frac{\pi}{2} \ldots 0 \ldots +\frac{\pi}{2} - 2\delta$, we obtain a transmission zero at a finite real frequency. If x_∞^2 is specified then we require

$$\tan\gamma = \cot 2\delta - x_\infty^2/\sin 2\delta .$$

7. ACTIVE N-PATH FILTERS

In view of the fact that the poles of the transfer function of a single N-path filter with passive RC path networks are restricted to two parallels with the real axis, the use of *active* RC path networks has been proposed [23, 24]; but even then the pole pattern is restricted to being symmetrical about $+jx_c$ and $-jx_c$. On the other hand the multi-N-path methods proposed in [1] and in Section 5.3 of this paper have the disadvantage of needing a large number of different carrier frequencies. In this section a new method of using activity in conjunction with N-path filters will be described which does not impose any restrictions on the pole pattern of T(p). It can be used to reduce the number of different carrier frequencies required to generate a function T(p) (either by the cascade method described in Section 5.3 of this paper or by the parallel method proposed in [1a]). This is achieved by modifying the carrier frequency, required to generate a specified biquadratic transfer function, by the application of feedback. Thus it should become possible to generate a number of biquadratic functions which have poles with *different* imaginary components with the *same* carrier frequency (note that only the number of different carrier frequencies is reduced, but not the number of multipliers).

The introduction of activity into N-path filters seems to contradict the aims of the policy of using passive N-path filters: the reduction of sensitivity as compared with that of active filters. However, whereas in active RC filters activity can be said to be used to move the transfer function poles away from the real axis, in the new method proposed here the poles have already non-zero imaginary components due to the action of the passive N-path filter, and activity is only used to move these poles in the complex plane by a certain amount. It is therefore to be expected that the sensitivity of the performance characteristic to component variations will be much smaller and will decrease if the amount of movement required decreases. Thus the method should be particularly attractive in the case of narrow band pass filters where the imaginary components of all poles are roughly equal. In such a case it might become possible to use only one single carrier frequency. On the other hand, in the proposed new circuit the active elements have to work in the frequency band of the input and output signal, whereas in the arrangements proposed in [23, 24] the active elements operate in a different frequency band which may (but need not) be far removed from the original signal frequency band and may therefore be more practicable.

Consider again a network with a biquadratic transfer function

$$T(p) = (A_0 + A_1 p + A_2 p^2)/(1 + Dp + p^2)\,. \qquad (9b)$$

If this network forms part of a feedback circuit as shown in Fig. 11a, then we obtain the new transfer function

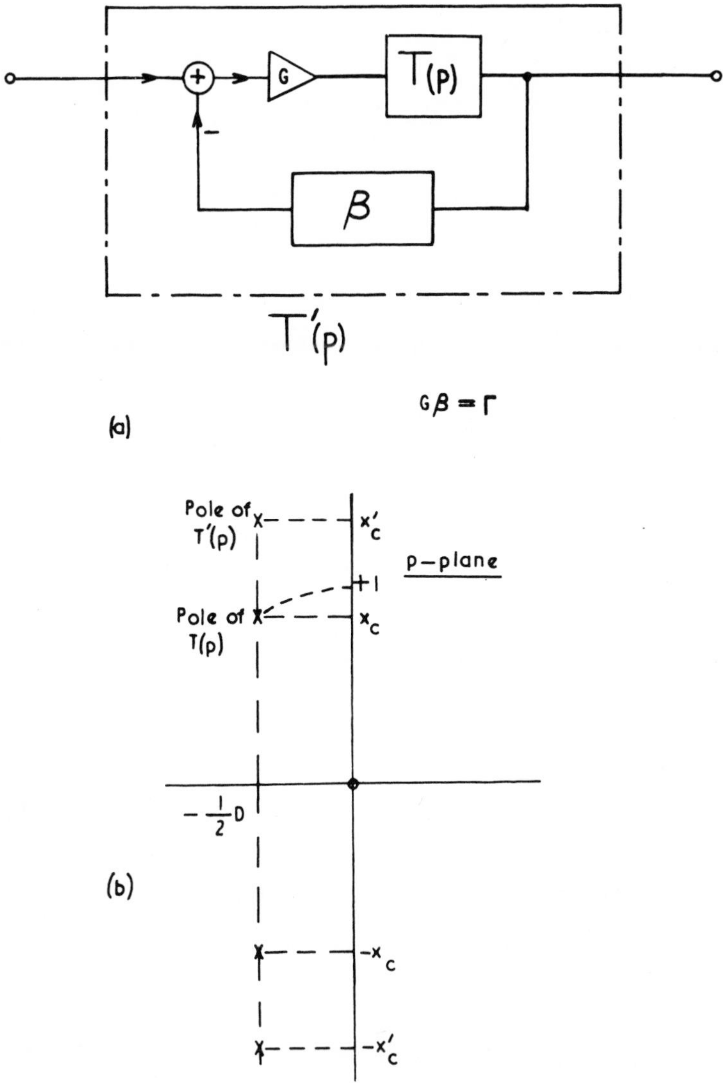

Fig. 11. Active N-path filter; (a) block diagram, (b) pole patterns for active and passive circuit.

$$T'(p) = \frac{GT(p)}{1 + G\beta T(p)}$$

$$= \frac{G(A_0 + A_1p + A_2p^2)}{(1 + \Gamma A_0) + (D + \Gamma A_1)p + (1 + \Gamma A_2)p^2} \qquad (14a)$$

where

$$\Gamma = G\beta \qquad (14b)$$

If it is required to realize T'(p) by means of a *passive* N-path filter with RC path networks, then the required carrier frequency is

$$x_c' = \left\{\frac{1 + \Gamma A_0}{1 + \Gamma A_2} - \frac{1}{4}\left(\frac{D + \Gamma A_1}{1 + \Gamma A_2}\right)^2\right\}^{\frac{1}{2}} \qquad (14c)$$

If we use the *active circuit* in Fig. 11a but realize the network with transfer function T(p) by a passive N-path filter, then the required carrier frequency is

$$x_c = (1 - \tfrac{1}{4}D^2)^{\frac{1}{2}} \qquad (10d)$$

(note that $x_c' \to x_c$ for $\Gamma \to 0$). Thus, by applying feedback we have changed the required carrier frequency by the ratio

$$\eta = x_c/x_c' \text{ (see Fig. 11b)} \qquad (14d)$$

The method described here can be used for any biquadratic function. However, in this paper we shall discuss only the case of a low pass filter with both transmission zeros at infinite frequency, i.e. $A_1 = 0$, $A_2 = 0$. Then

$$x_c' = (1 + \Gamma A_0 - \tfrac{1}{4}D^2)^{\frac{1}{2}} \qquad (14e)$$

$$\eta^{-2} = (1 - \tfrac{1}{4}D^2 + \Gamma A_0)/(1 - \tfrac{1}{4}D^2) \qquad (14f)$$

$$T'(p) = GA_0/[(1 + \Gamma A_0) + Dp + p^2] \qquad (14g)$$

The poles of T' lie at $-\frac{1}{2}D \pm jx'_c$. We note that in this case (low pass filter with $A_1 = 0$, $A_2 = 0$) the real component of the poles is unaffected by the application of feedback. We shall also need the Q-factor of the denominator of T' which we designate by Q'; we obtain

$$Q' = 1/D' = Q(1 + \Gamma A_0)^{\frac{1}{2}} \tag{14h}$$

From (14f) we obtain

$$\Gamma A_0 = (\eta^{-2} - 1)(1 - \tfrac{1}{4}D^2) \tag{14i}$$

From our discussion of passive networks in Section 5.5 we know that for $C_0 = 1$ and $C_1 = -2Q$ we have $A_0 = Dx_c$. Therefore

$$\Gamma = Q(\eta^{-2} - 1)(1 - \tfrac{1}{4}D^2)^{\frac{1}{2}} \tag{14j}$$

Using (14h), we can express Γ as a function of Q' and D' rather than of Q and D. Then, after some manipulation, we obtain

$$\Gamma = Q'(\eta^{-1} - \eta)(1 - \tfrac{1}{4}D'^2)^{\frac{1}{2}} \tag{14k}$$

(14k) gives the closed loop gain G(multiplied by the feedback factor β) which is required to change the carrier frequency be a factor η.

We are interested in the sensitivity of the transfer function T'(p) to component variations. In the first instance we are interested in the sensitivity to variations of the closed loop gain G and the open loop gain μ of the amplifier, where

$$G = \frac{\mu}{1 + \beta_F\mu} \tag{15a}$$

and β_F is the feedback ratio of the amplifier (to be distinguished from the feedback ratio β of the active N-N-path filter). We shall only need the partial derivative

$$\frac{\partial G}{\partial \mu} = \frac{G^2}{\mu^2} \tag{15b}$$

in which β_F does not appear explicitly.

In this paper we shall only discuss the Q-sensitivity, i.e. the sensitivity $S_G^{Q'}$ of Q' to G and the sensitivity $S_\mu^{Q'}$ to μ. We shall derive $S_G^{Q'}$ first and then obtain $S_\mu^{Q'}$ by means of the relationship

$$S_\mu^{Q'} = S_G^{Q'} \cdot S_\mu^G \tag{15c}$$

In order to find $S_G^{Q'}$ we assume that the passive N-path network with parameter Q (and $D = Q^{-1}$) is given and Q' is related to Q and G by means of (14h). Now

$$S_G^{Q'} = \frac{\partial Q'}{\partial G}\frac{G}{Q'} \tag{15d}$$

$$\frac{\partial Q'}{\partial G} = \frac{\partial Q'}{\partial \Gamma}\frac{\partial \Gamma}{\partial G}$$

Therefore

$$S_G^{Q'} = GD'\beta\tfrac{1}{2}QA_0(1 + \Gamma A_0)^{-\frac{1}{2}} = \tfrac{1}{2}D'\Gamma(1 - \tfrac{1}{4}D^2)^{\frac{1}{2}}\ (1 + \Gamma A_0)^{-\frac{1}{2}}$$

$$S_G^{Q'} = \tfrac{1}{2}\eta\,(\eta^{-1} - \eta)\ (1 - \tfrac{1}{4}D'^2) \tag{15e}$$

and

$$S_\mu^{Q'} = \tfrac{1}{2}Q'\eta(\eta^{-1} - \eta)^2(1 - \tfrac{1}{4}(D')^2)^{\frac{3}{2}}\mu^{-1}\beta^{-1} \tag{15f}$$

The stability of the circuit, shown in Fig. 11a, under practical conditions (bandwidth limitation of the amplifier, non-zero input admittance and output impedance, etc.) requires further investigation. This problem would, of course, arise with any active circuit; but in the circuit under consideration an additional specific problem arises, due to the use of time-varying methods which has been ignored in the preceding discussion. The analysis of this circuit given so far can be summarized as follows:

1. By application of feedback the position of the poles of a second-order network is modified; this part of the discussion is independent of the type of second-order network (e.g. passive and time-invariant; active;

N-path).

2. Assuming that we are considering N-path realizations for the second-order network, either with or without feedback, we need different carrier frequencies in the two cases since the carrier frequency is in each case equal to the imaginary parts of the poles which are modified by the application of feedback.

From a formal point of view, this analysis is correct (with the reservations mentioned above concerning non-ideal amplifiers). However, we have assumed perfect N-path sections. In physical reality there is one type of imperfection which may be of crucial importance in practice: this is the occurrence of unwanted frequency components at the output of the passive N-path filter. If these unwanted components are fed back to the input, they will produce further unwanted components, and so on. Obviously this point needs further investigation.

8. QUADRATURE MODULATION

8.1. *The Conventional Circuit*

Consider the circuit in Fig. 12a. With an input signal e^{pt} we obtain

$$u_1(t) = T_1(p)e^{pt};\; u_2(t) = T_2(p)e^{pt}$$

$$v_1(t) = \tfrac{1}{2}T_1(p)e^{(p+jx_c)t} + \tfrac{1}{2}T_1(p)e^{(p-jx_c)t}$$

$$v_2(t) = \tfrac{1}{2}T_2(p)e^{(p+jx_c)t+j\frac{1}{2}\pi} + \tfrac{1}{2}T_2(p)e^{(p-jx_c)t-j\frac{1}{2}\pi}$$

$$z(t) = v_1(t) + v_2(t) = \tfrac{1}{2}[T_1(p) + jT_2(p)]e^{(p+jx_c)t} + \tfrac{1}{2}[T_1(p) - jT_2(p)]e^{(p-jx_c)t} \qquad (16a)$$

Thus considering the two components of the output signal z(t) separately, we obtain:

$$\text{for}\quad e^{(p+jx_c)t}:\quad \tfrac{1}{2}[T_1(p) + jT_2(p)] \qquad (16b)$$

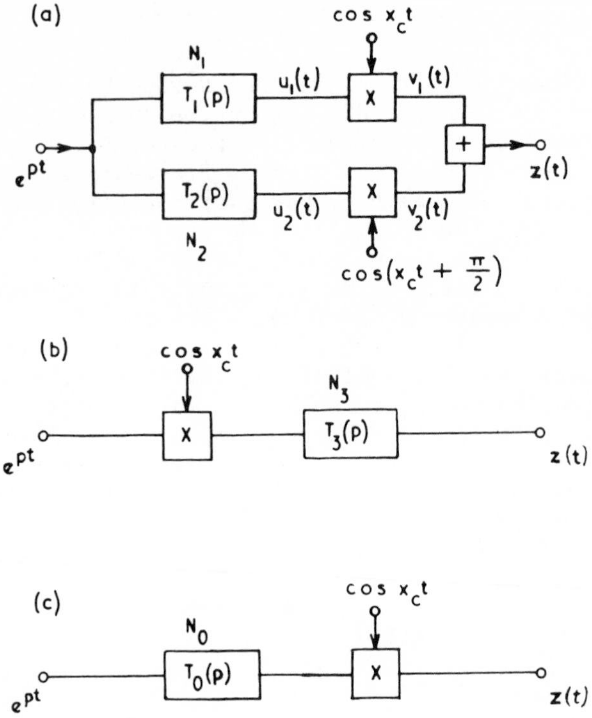

Fig. 12. Single sideband modulator (quadrature modulation type); (a) block diagram, (b) fictitious equivalent circuit, (c) alternative equivalent circuit.

$$\text{for} \quad e^{(p-jx_c)t}: \quad \tfrac{1}{2}[T_1(p) - jT_2(p)] \tag{16c}$$

As in the discussion in Section 4.3, it is often more useful to consider, instead of $e^{(p-jx_c)t}$, the component $e^{(-p+jx_c)t}$. Then we can replace (16c) by:

$$\text{for} \quad e^{(-p+jx_c)t}: \quad \tfrac{1}{2}[T_1(-p) + jT_2(-p)] \tag{16d}$$

and (16b) can be considered as including also (16d). If we want to rewrite (16b) in terms of the output frequency p', we can - as in section 4.3 - introduce p' by

$$p' = p + jx_c \tag{16e}$$

Then we define

$$T_3(p') \equiv \tfrac{1}{2}[T_1(p' - jx_c) + jT_2(p' - jx_c)] \tag{16f}$$

$T_3(p')$ can be regarded as a virtual transfer function, considering the circuit in Fig. 12b with a fictitious post-multiplier network N_3 as equivalent to the circuit in Fig. 12a; $T_3(p')$ is obviously a non-real function of p'. However, whereas $T_2(p')$ in (5c) is non-real because it is equal to a real function T_1 of a complex variable $(p' - jx_c)$ obtained by a frequency translation, the function $T_3(p')$ considered here corresponds to a function $\frac{1}{2}[T_1(p) + jT_2(p)]$ before frequency transformation which is itself non-real. This is due to the occurrence of the term j in (16b, c). We can therefore consider the equivalence shown in Fig. 12a, b, c, where N_1 and N_2 are physical networks, but both N_3 and N_0 are fictitious; N_0 has the transfer function

$$T_0(p) \equiv \tfrac{1}{2}[T_1(p) + jT_2(p)] \tag{16g}$$

If N_1 and N_2 are all-pass networks with

$$T_1(p) = e^{-j\beta_1}, \quad T_2(p) = e^{-j\beta_2}, \tag{17a}$$

then, writing T_0 in the form

$$T_0(p) = |T_0|e^{-j\beta_0}, \tag{17b}$$

we obtain

$$|T_0|^2 = \tfrac{1}{2}[1 + \sin(\beta_2 - \beta_1)]$$

$$\tan\beta_0 = \frac{\sin\beta_1 - \cos\beta_2}{\cos\beta_1 + \sin\beta_2}$$

With

$$\beta_1 = \beta_2 + \tfrac{1}{2}\pi + \delta \tag{17c}$$

$$|T_0|^2 = \cos^2(\tfrac{1}{2}\delta) = 1/[1 + \tan^2(\tfrac{1}{2}\delta)] \tag{17d}$$

$$\beta_0 = \beta_1 + \tfrac{1}{2}\delta \tag{17e}$$

For small δ,
$$|T_0|^2 \simeq 1 \tag{17f}$$

These results (17c, d) are only valid for *positive* values of x (which become the upper side band after frequency translation). For *negative* values of x (lower sideband) we have to replace $+\beta_1$ by $-\beta_1$ and $+\beta_2$ by $-\beta_2$, so that

$$|T_0|^2 = \sin^2(\tfrac{1}{2}\delta) = 1/[1 + \cot^2(\tfrac{1}{2}\delta)] \tag{17g}$$

$$\beta_0 = \beta_1 + \tfrac{1}{2}\delta - \tfrac{1}{2}\pi \tag{17h}$$

Thus, for small δ,
$$T_0 \to 0. \tag{17i}$$

This derivation, where the signs of β_1 and β_2 have to be changed, is - although formally correct - not very satisfactory. We obtain a clearer picture if we express $T_1(p)$ and $T_2(p)$ in terms of driving point impedance functions. In the case of all-pass networks we can write (instead of (17a))

$$T_1(p) = \frac{1 - jX_1}{1 + jX_1}, \quad T_2(p) = \frac{1 - jX_2}{1 + jX_2} \tag{17j}$$

where jX_1 and jX_2 are normalized reactance functions of p. Then we obtain for $|T_0|^2$ (we shall *not* discuss β_0)

$$|T_0|^2 = \left\{1 + \left(\frac{1 - (X_2 - X_1)/(1 + X_1X_2)}{1 + (X_2 - X_1)/(1 + X_1X_2)}\right)^2\right\}^{-1} \tag{17k}$$

Although the result given in (17k) is the same as that stated in (17d) (since $X_2 = \tan(\beta_2/2)$, $X_1 = \tan(\beta_1/2)$), Eq. (17k) demonstrates explicitly that for negative x we must replace X_1 by $-X_1$ and X_2 by $-X_2$. The difficulties, if any, in appreciating the need for this step stems from the non-realness of the function $T_0(p)$. If $T_0(p)$ were real, then $|T_0|$ would remain the same if p were replaced by -p.

If we define K by

$$|T_0|^2 = \frac{1}{1 + K^2}, \quad K^2 = \frac{1 - |T_0|^2}{|T_0|^2} \tag{17m}$$

we observe that in (17k)

$$K \to K^{-1} \text{ for } x \to -x \tag{17n}$$

This means that over the frequency range $-x_2 < -x_1 < 0 < x_1 < x_2$, if δ is small for $x_1 \leq x \leq x_2$, the function $T_0(p)$ is a fictitious "high-pass function" with "stop-band" $-x_2 \ldots -x_1$, "transition-band" $-x_1 \ldots +x_1$ and "pass-band" $+x_1 \ldots +x_2$.

This fictitious high-pass characteristic illustrates the point made above; whereas, in the equivalence between the circuits in Figs. 5a and b, N_1 was a physical network, in the equivalence between the circuits in Figs. 12a, b and c, N_0 is an extremely fictitious network. However, this network is conceptually very useful because a low-pass to band-pass transformation of its fictitious transfer characteristic gives the physical output characteristic of a quadrature modulation SSB generator for an upper sideband.

If in the circuit in Fig. 12a we change the (+) to (-) at the junction of the two paths or, alternatively, if we change the phase difference between the carrier oscillations applied to the paths from $+\pi/2$ to $-\pi/2$, then we transform

$$T_1(p) + jT_2(p) \to T_1(p) - jT_2(p) \tag{17p}$$

We find that as far as $|T_0|^2$ is concerned this transformation leads to the same result as the transformation $p \to -p$, i.e. to the interchange of pass-band and stop-band (as far as the phase is concerned, $\beta_0 \to -\beta_0$). Thus the transformation creates a fictitious low-pass filter from the fictitious high-pass filter; the frequency translation (by x_c) of such a low pass filter characteristic would lead to a quadrature modulation SSB output characteristic for a lower sideband.

8.2. *Generalization*

There is no reason to restrict the circuit shown in Fig. 12c to fictitious low-pass or high-pass transfer characteristics. Any characteristic $T_0(p)$, dissymmetrical with respect to zero-frequency, which can be expressed in the form $T_0(p) = T_1(p) + jT_2(p)$ where $T_1(p)$ and $T_2(p)$ are physical transfer characteristics can be obtained by means of the circuit in Fig. 12a. Let us, for example, assume that N_1 and N_2 are no longer required to be all-pass networks, but that we wish to retain the relation (17n). For imaginary values of p, we can write

$$T_1 = A_1(x) + jB_1(x), \quad T_2 = A_2(x) + jB_2(x) \qquad (18a)$$

where A_1, A_2 are real and even and B_1 and B_2 real and odd functions of x. Then substituting (18a) in (16g) and (17m), we find

$$K = \frac{4 - |T_1|^2 - |T_2|^2 + 2(A_1B_2 - A_2B_1)}{|T_1|^2 + |T_2|^2 - 2(A_1B_2 - A_2B_1)} \qquad (18b)$$

Thus, to satisfy (17n) we require

$$|T_1|^2 + |T_2|^2 = 2 \qquad (18c)$$

If we choose $T_1 = (a_1 + b_1p)/(1 + p)$, $T_2 = (a_2 + b_2p)/(1 + p)$, then (18c) leads to $a_1^2 + a_2^2 = 2$, $b_1^2 + b_2^2 = 2$. These conditions are, for example, satisfied by the sets ($a_1 = 1$, $a_2 = 1$, $b_1 = \sqrt{2}$, $b_2 = 0$); ($a_1 = \sqrt{2}$, $a_2 = 0$, $b_1 = 1$, $b_2 = 1$); ($a_1 = 0$, $a_2 = \sqrt{2}$, $b_1 = \sqrt{2}$, $b_2 = 0$). If we choose $T_1 = (a_1 + b_1p + c_1p^2)/(1 + Dp + p^2)$, $T_2 = (a_2 + b_2p + c_2p^2)/(1 + Dp + p^2)$, we obtain for the coefficients the conditions $a_1^2 + a_2^2 = 2$; $2(a_1c_1 + a_2c_2) - (b_1^2 + b_2^2) = 2(2 - D^2)$; $c_1^2 + c_2^2 = 2$. A set of coefficient values satisfying these three equations, for $D^2 = 1/32$, is ($a_1 = a_2 = c_1 = c_2 = 1$,

$b_1^2 = 1/64$, $b_2^2 = 3/64$).

Dropping condition (17n), we arrive at completely general characteristics of the form $T_0 = \frac{1}{2}(T_1 + jT_2)$. Even a simple case like $T_1 = 1/(1 + p)$, $T_2 = p/(1 + p)$ may be of some interest. We obtain

$T_0(jx) = (1 - x)/2(1 + jx)$. When frequency-translated by x_c, this represents a characteristic which in the upper sideband has a zero for $x' = x_c + 1$; but in the lower sideband for the corresponding frequency $x' = x_c - 1$ we obtain $T_0 = -1/(1 - jx)$.

When $T_0(jx)$ discriminates strongly between positive and negative x-values, we can use this function for SSB generation; but in practice we suffer from the sensitivity, mentioned in Section 2.2, of suppression by cancellation. In applications where less severe discrimination is required this sensitivity problem would not arise. There may, for example, be cases where we would be willing and able to provide a physical SSB filter for the suppression of the unwanted sideband. Then, an additional quadrature modulation circuit producing the function $T_0(jx)$ could be used (when shifted by x_c) for equalizing the wanted sideband by means of such rather unusual non-real functions.

Another problem where the generalized quadrature modulation circuit proposed here may possibly be of use arises when it is required to minimize the time-bandwidth product of band pass filters for double sideband signals*. These filters have a dissymmetrical frequency characteristic but provide only mild discrimination between the two sidebands [25].

8.3. *Some Comments on Multi-Quadrature Modulation Circuits*

We can classify circuits with reference to the number of *different* frequency changes taking place

---

* Private communication by Dr. Stehle, University of Karlsruhe, Germany.

(irrespective of the number of multipliers used). From this point of view we would designate the circuit in Fig. 12a as a single-change circuit, but the circuits in Figs. 8 and 13a as double-change circuits. Then we can argue in the following way (assuming that there are no feedback connections): starting with an input frequency x, we obtain in the first stage of modulation, with carrier frequency x_{c1}, two new frequencies: $x_{c1} + x$ and $x_{c1} - x$; if only one output frequency is required, we have to suppress the second one. In a second stage of modulation, with carrier frequency x_{c2}, we obtain in the general case four new frequencies $x_{c2} + x_{c1} + x$, $x_{c2} + x_{c1} - x$, $x_{c2} - x_{c1} - x$, $x_{c2} - x_{c1} + x$. If $x_{c1} = x_{c2}$, two of these frequency components (with output frequency x equal to input frequency x) merge. Thus if only one of these 4 (or 3) output frequencies is wanted, we have to suppress the other 3 or 2 unwanted frequencies.

In an SSB generator with SSB filter two output frequencies are produced, one of which is suppressed by this filter. In a quadrature modulation circuit as shown in Fig. 12a again two frequencies are produced, one of which is eliminated by mutual cancellation (of two equal-frequency components)*. In the 2-path filter as shown in Fig. 8, $x_{c1} = x_{c2}$; therefore there are two wanted components of equal frequency, but with different phase angles. The two unwanted components are suppressed by mutual cancellation. One of the unwanted output frequency components may already have been suppressed before the second modulation, by the network with transfer function H(p). However, it is obviously quite wrong to explain the operation of an N-path filter in general in terms of the specific function H(p) in a particular case.

In circuits of the type shown in Fig. 13a one of the

---

* It is possible to argue that in some quadrature modulation circuits the unwanted sideband is never produced but rather produced and suppressed at the same point in time and space (see [3, 4]); but this does not affect the present argument.

frequency components obtained after the first modulation is suppressed by a filter. Thus only two output frequency components appear at the final output, one of which is suppressed by mutual cancellation (see Section 8.4).

In this context of multi-quadrature modulators it is relevant to point out that a different type of multi-quadrature modulators has been described in the literature (see [26, 27]). These multi-quadrature modulators consist of two completely separate quadrature modulators*. There are different types of quadrature modulation circuits. Considering two different input frequencies, $x_a < x_c$ and $x_b > x_c$, we can use quadrature modulation filters in four different ways: to obtain only sum frequencies $x_c + x_a$, $x_c + x_b$; to obtain only difference frequencies $x_c - x_a$, $x_c - x_b$; to obtain only positive difference frequencies $x_c - x_a$; to obtain only negative difference frequencies $x_c - x_b$. By suitable interconnection of two such circuits various filter characteristics can be obtained**.

8.4. *SSB Generators with Two Stages of Quadrature Modulation*

We consider now a circuit of the type shown in Fig. 13a. The signals $q_1(t)$, $q_2(t)$, $r_1(t)$ and $r_2(t)$ will be identical with those obtained in the 2-path filter shown in Fig. 8 and given in Section 5.2. The

---

* Unfortunately the statement that there are two completely separate quadrature modulation circuits (connected in cascade) is not as unambiguous as it may sound; in [26] a circuit of this type is shown to have an equivalent circuit in which a distinction between separate quadrature modulation circuits is no longer possible.

** In this context an N-path type circuit consisting of an SSB quadrature demodulator, a modulator of the same type and one time invariant filter, which is described in [28], is also of interest.

only difference between the circuits in Fig. 13 and Fig. 8 is that in the circuit in Fig. 13 the carrier frequency x_{c_2} differs from the carrier frequency x_{c_1}. Therefore we obtain the following output signals. For ease of reference, we repeat $r_1(t)$ and $r_2(t)$ (with x_{c_1} instead of ω_c or x_c).

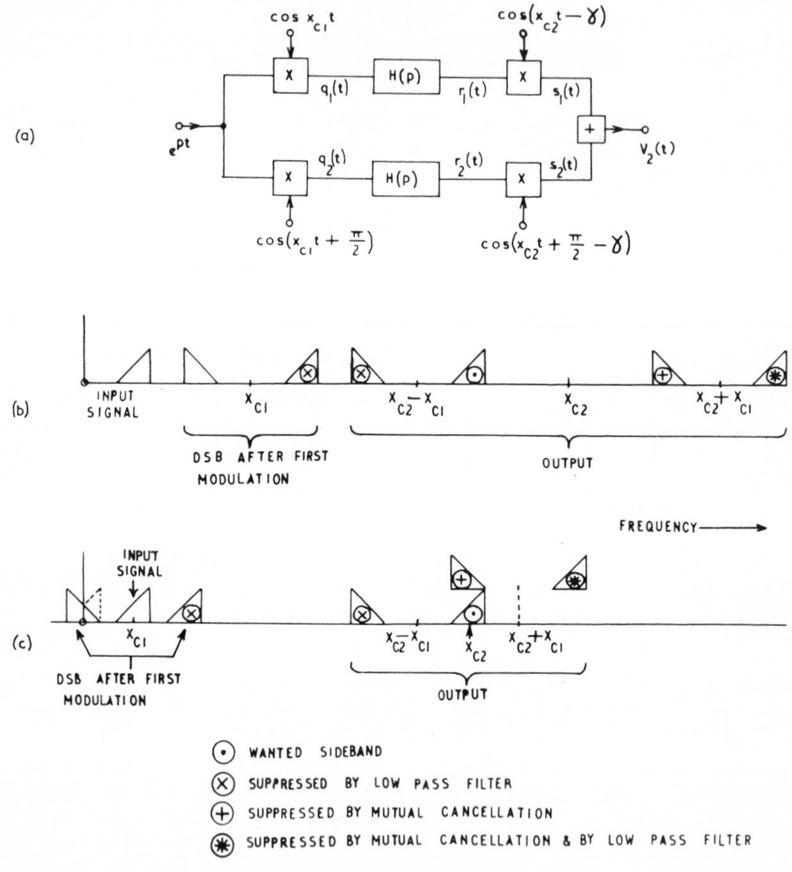

Fig. 13. Single sideband generator (quadrature modulation type) using two stages of modulation; (a) block diagram, (b) analysis of sidebands, (c) as (b), for x_{c_1} at centre of input frequency band.

$$r_1(t) = \tfrac{1}{2}H(p + jx_{c1})e^{(p+jx_{c1})t} + \tfrac{1}{2}H(p - jx_{c1})e^{(p-jx_{c1})t}$$

$$r_2(t) = \tfrac{1}{2}H(p+jx_{c1})e^{(p+jx_{c1})t+j\frac{1}{2}\pi}+\tfrac{1}{2}H(p-jx_{c1})e^{(p-jx_{c1})t-j\frac{1}{2}\pi}$$

$$s_1(t) = \frac{H(p+jx_{c1})}{4}[e^{[p+j(x_{c1}+x_{c2})]t-j\gamma}+e^{[p+j(x_{c1}-x_{c2})]t+j\gamma}]$$

$$+ \frac{H(p-jx_{c1})}{4}[e^{[p-j(x_{c1}-x_{c2})]t-j\gamma}+e^{[p-j(x_{c1}+x_{c2})]t+j\gamma}] \tag{19a}$$

$$s_2(t) = \frac{H(p+jx_{c1})}{4}[-e^{[p+j(x_{c1}+x_{c2})]t-j\gamma}+e^{[p+j(x_{c1}-x_{c2})]t+j\gamma}]$$

$$+ \frac{H(p-jx_{c1})}{4}[e^{[p-j(x_{c1}-x_{c2})]t-j\gamma}-e^{[p-j(x_{c1}+x_{c2})]t+j\gamma}] \tag{19b}$$

$$v_2(t) = s_1(t) + s_2(t) = \tfrac{1}{2}e^{j\gamma}H(p + jx_{c1})e^{[p+j(x_{c1}-x_{c2})]t}$$

$$+ \tfrac{1}{2}e^{-j\gamma}H(p - jx_{c1})e^{[p-j(x_{c1}-x_{c2})]t} \tag{19c}$$

We note that there are four different output frequency components (see Fig. 13b) in (19a) and (19b). The final output signal, which is the *sum* of the output signals from each path, contains only two different frequency components because the two sideband components of $x_{c1} + x_{c2}$ have been eliminated by cancellation. The two sideband components of $x_{c2} - x_{c1}$ have different *virtual* transfer constants $H[j(x + x_{c1})]$ and $H[j(x - x_{c1})]$, frequency translated by x_{c2}. If $H[j(x - x_{c1})]$ is a low pass function, efficiently discriminating between the lower and the upper sideband of x_{c1}, then the upper sideband of x_{c1}, and consequently the lower sideband of $x_{c2} - x_{c1}$ will be suppressed. The same considerations apply when x_{c1} is the mid-band frequency of the signal input band (see [17, 18]); this is illustrated in Fig. 13c.

9. CONCLUDING REMARKS

Although the discussions in this paper have been restricted to circuits containing ideal multipliers activated by sinusoidal carrier oscillations, the occurrence, in practical circuits, of non-sinusoidal carriers and non-ideal multipliers does not necessarily invalidate the results obtained. Thus an SSB generator with x_c large compared with the input signal frequencies will work in broad agreement with the theory given here when driven by a switching carrier. On the other hand, an N-path section with *high*-pass path networks will not operate as described, if switching harmonics are present.

Thus the restriction to ideal multipliers and sinusoidal carriers may be regarded either as a weakness of the paper (results are obtained which may not yet be applicable in present-day practice) or as a source of strength: the new results obtained here for passive, active and adjustable N-path sections with biquadratic transfer functions, which are regarded as being both of theoretical and of potential practical interest, would scarcely have been obtained within the context of an investigation emphasizing the switching and sampling aspects of N-path circuit operation.

As to virtual transfer functions, the writer is very conscious of the non-rigorous way in which this concept has been introduced and used in this paper. However, he feels that in focussing attention on the non-realness of certain relevant performance functions, the new concept performs a useful service and suggests some generalizations which seem to be of interest. The explicit expressions for the output signal obtained in quadrature modulation SSB circuits (including Weaver's circuit) may occasionally find application.

ACKNOWLEDGEMENTS

I wish to thank my colleagues J.T. Lim and D.G. Haigh and my former colleague R.F. Hoskins for many helpful discussions.

REFERENCES

1a. FRANKS, L.E. and SANDBERG, I.W. (Sept. 1960) An alternative approach to the realization of network transfer functions: the N-path filter, *Bell System Tech. J. 39*, p. 1321.
See also:

b. MACDIARMID, I.F. and TUCKER, D.G. (Sept. 1950) Polyphase modulation as a solution of certain filtration problems in telecommunication, *Proc. IEE, Part III*, p. 349.

2. NORGAARD, D.E. (Dec. 1956) The phase-shift method of single sideband signal generation, *Proc. IRE*, p. 1718; The phase-shift method of single sideband signal reception, ibid., p. 1735.

3. SARAGA, W. (May 1962) Single sideband generation - a new principle, *Electronic Technology (London)*, p. 168.

4a. SARAGA, W. (29th December 1960) Brit. Patent Spec. 941,619.

b. SARAGA, W. and GALPIN, R.K.P. (1964) Hall effect vector-product generators (Paper presented at IEEE Electron Devices Meeting, November 1st 1963, at Washington D.C.), *Solid State Electronics*, Vol. 7, Pergamon Press, p. 335.

5. ROWLANDS, R.O. (1949) Constant phase shift networks, *Wireless Engineer*, p. 283.

6. BAUMANN, E., (1950) Weber Scheinwiderstände mit vorgeschriebenem Verhalten des Phasenwinkels, *Z. Angew. Math. Phys. 1*, p. 43.

7. DARLINGTON, S. (1950) Realization of a constant phase difference, *Bell System Tech. J.*, p. 94.

8. ORCHARD, H.J. (1950) Synthesis of wide-band two-phase networks, *Wireless Engineer*, p. 72.

9a. SARAGA, W. (1950) The design of wide-band phase-

splitting networks, *Proc. IRE*, p. 754.
b. SARAGA, W. (1951) Wideband two-phase networks, *Wireless Engineer*, p. 30.

10. BEDROSIAN, S.D. (June 1960) Normalized design of 90° phase difference networks, *IRE Trans. on Circuit Theory*, p. 128.

11. HERRMANN, O. (1969) Quadraturfilter mit rationalem Übertragungsfactor, *Archiv der Elektrischen Übertragung 23*, p. 77.

12. ALBERSHEIM, W.J. and SHIRLEY, F.R. (May 1969) Computation methods for broadband 90° phase difference networks, *IEEE Trans. on Circuit Theory, CT-16: (2)*, p. 189.

13. POWERS, K.H. (August 1960) The compatibility problem in single-sideband transmission, *Proc. IRE*, p. 1431.

14. GOURIET, G.G. and NEWELL, F.G. (Oct. 1959) A quadrature network for generating vestigial-sideband signals, *Proc. IEE, Part B*, p. 253.

15. SCHNEIDER, W.H., (1967) *Telefunken Zeitung*, p. 107.

16. SARAGA, W. (1968) Multipath selective systems, (Paper presented at Summer School for Circuit Theory, Prague).

17. WEAVER, D.K., Jr. (Dec. 1956) A third method of generation and detection of single-sideband signals, *Proc. IRE*, p. 1703.

18. LEUTHOLD, P. (Feb. 1969) Das Abtasttheorem als Hilfsmittel zur anschaulichen Darstellung der Theorie der Einseitenband-modulation, *NTZ*, p. 66.

19. SZENTIRMAI, G. (1963) The design of arithmetically symmetrical band-pass filters, *IEEE Trans. on Circuit Theory*, p. 367.

20. SARAGA, W. (1965) A note on quadrature modulation circuits for single sideband generation and some associated curve approximation problems (Paper read at "Symposium on Network Theory," Cranfield, England).

21. SARAGA, W. (1969) Realizability conditions and associated invariance relations for N-path networks with sinusoidal carrier oscillations (Paper accepted for presentation at 1969 International Circuit Theory Symposium, Dec. 8-10, San Francisco).

22. GALPIN, R.K.P., HAWKES, P., SARAGA, W. and TARBIN, F.G. (March 1967) A practical tantalum thin-film single-sideband demodulator (1 Mc/s to voice frequency) using RC time varying and active networks (*Proceedings of 2nd International Symposium on Microelectronics*, Munich (1966) and *IEEE Journal of Solid-State Circuits SC-2*,

23. MÖHRMANN, K. and HEINLEIN, W. (Dec. 1967) N-Pfad Filter hoher Selektivität mit spulenlosen Schwingkreisen, *Frequenz*, p. 369.

24a. LANGER, E. (March 1968) Ein neuartiges N-Pfad Filter mit zwei konjugiert komplexen Polparen, *Frequenz*, p. 90.

b. OWENS, A.R. (17th Jan. 1966) IEE Colloquium on active filters, London.

25. STEHLE, Q. (1968) Über den Entwurf von Bandpässen mit minimalem Zeit-Bandreite Product, Dr. Ing. Thesis, University of Karlsruhe.

26. SARAGA, W. (April 1967) New class of time-varying filters, *Electronics Letters 3: 4*.

27. MILLER, A.K.H. (10th July 1969) New type of time-varying octave filter, *Electronics Letters*, p. 310.

28. GALPIN, R.K.P. (3rd May 1968) Narrow-bandpass filtering with modulation, *Electronics Letters*.

ANALYSIS TECHNIQUES FOR THE EVALUATION OF PARAMETRIC CIRCUITS

D. P. HOWSON

University of Bradford, Postgraduate School of Studies in Electrical and Electronic Engineering

1. INTRODUCTION

The Circuits group of the University of Bradford has as its major preoccupation the development of improved frequency changers, mixers, and parametric amplifiers in cooperation with industry. Analytic techniques have been used and developed to evaluate the performance of the circuits as exactly as possible, to optimise where possible, and to try to gain an understanding of the fundamental mechanisms involved in producing the observed performance. The background of the principal members of the group has been experience of Tucker's [1] work on circuits with periodically-varying parameters, an approach which is still found to be of use in the prediction of a number of aspects of performance of circuits of these types. Several more sophisticated techniques in the literature have appeared disappointing in being inapplicable to circuits of practical importance. Recently, however, the new approaches to non-linear analysis that have become available have led to several methods of value, and indications are that computer-aided design procedures will eventually solve - possibly at an economic cost - most of our outstanding problems.

The circuits under discussion use from one to four severely non-linear elements - see Figs. 1 and 2 - the non-linearity of which is largely controlled by a dominant vol-

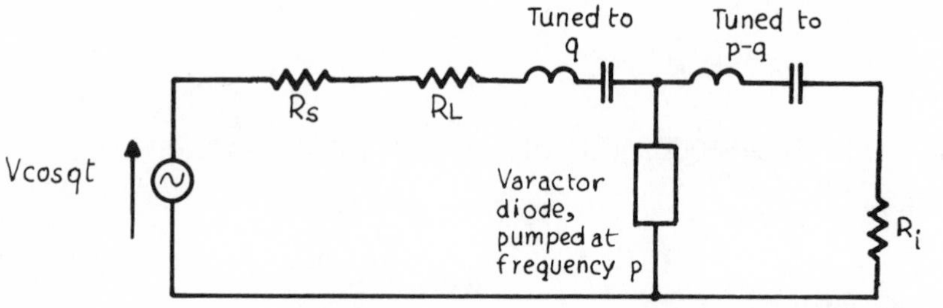

FIG. 1. Parametric amplifier

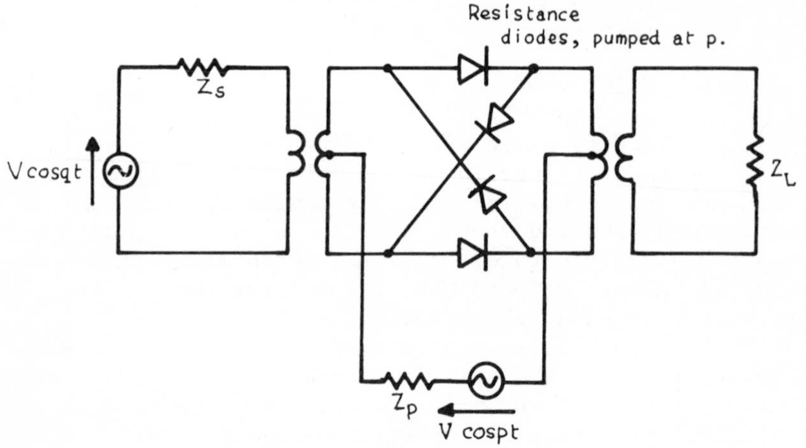

FIG. 2. Ring modulator

tage or current source termed the local oscillator, or the carrier, or the pump, depending on the application. There are usually significant currents or voltages in the circuit at at least three frequencies - signal, q, pump, p, and a *modulation product frequency*, $q \pm np$, usually $q + p$ or $q - p$. The circuit operates as a frequency changer or, if a non-linear reactance is present, possibly as an amplifier. The results that must be evaluated for any circuit will be included in the two lists to follow.

List 1:

a) Conversion (power) efficiency.
b) Power amplification.
c) System and circuit noise figure.
d) Relative magnitude of frequency components of unwanted signal and/or carrier leaks.
e) Relative magnitude of unwanted modulation products.
f) Sensitivity of any or all of the above to change in circuit parameters or applied voltages.

These properties of the circuits are designated *first order*, since they can usually be adequately investigated by analysis based on the representation of the non-linear elements as time-varying elements whose variation is completely controlled by the pump.

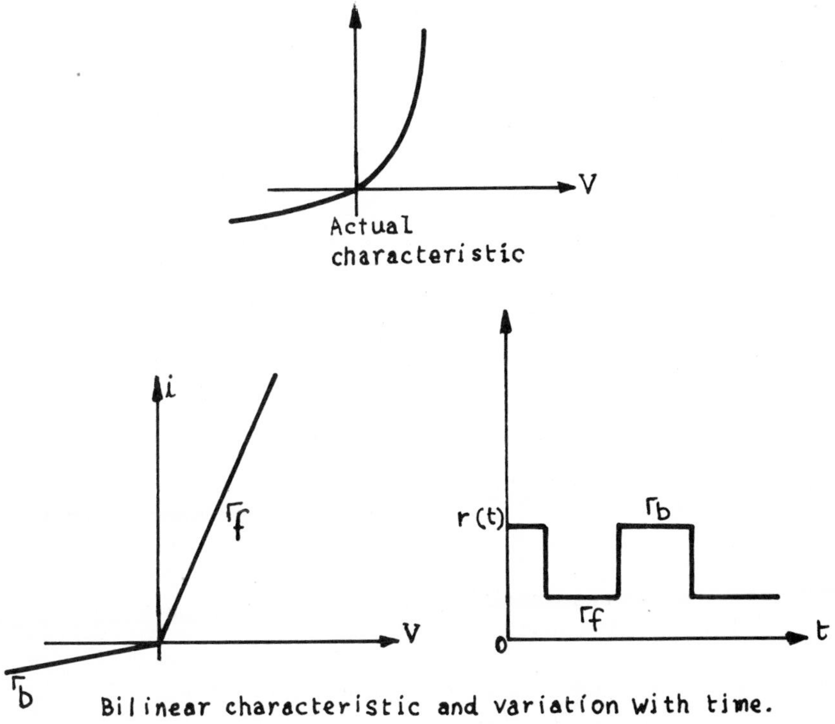

FIG. 3. Bilinear model of resistance diode

Complete algebraic analysis is practicable if (i) the time-varying elements are assumed to have simple models - see Fig. 3, (ii) any tuned circuits are idealised to have either zero or infinite impedance at all but a few frequencies, (iii) the number of frequencies of interest is restricted to three or four, at most. (There are exceptions to this rule.) When computer analysis is used these conditions can and have been relaxed, but at the cost of solving a particular circuit rather than investigating the interplay of parameters and - hopefully - proving a general principle.

In addition to the first-order properties it is necessary in most cases to know the magnitude of some or all of the parameters listed below, before a final decision between competing circuits can be made.

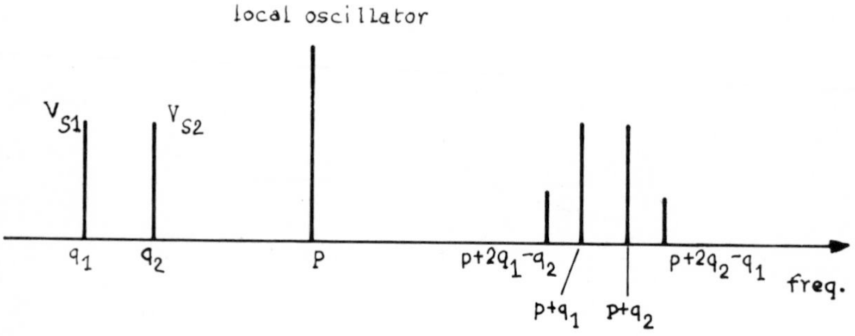

FIG. 4. Two-tone intermodulation test

List 2:

g) Intermodulation properties. A standard two-tone intermodulation test used for frequency changers is given in Fig. 4, and it is often necessary for the intermodulation products to be 60dB or more below the level of the wanted output product.

h) Cross-modulation properties. Measurement or calculation of the level of interfering signal (30% modulated) relative to the level of wanted signal for the wanted output sideband to consist of exactly 3% spurious modulation transferred from the interfering signal is an important property of a device used at the input of a radio receiver.

i) Overload properties, when the input signal becomes comparable to the pump. Often the worst effect is shown up under g).

j) Dynamic range - which may be a combination of i) and c).

These properties are termed *second-order*, and normally require a more sophisticated non-linear analysis for their evaluation.

In some circuits where it is important for the intermodulation components to be 100dB or more down on the wanted output, which may in turn be 30dB down on the pump signal, most general non-linear computer techniques that have been encountered to date cannot distinguish the smaller components from computer noise. There are a number of analysis techniques available for particular circuits which will be discussed later.

2. FIRST-ORDER CIRCUIT PROPERTIES

Turning once again to first-order circuit properties, for which treatment of the circuit as having an element or elements with periodically-varying parameters is appropriate, there are two principal facets to the work.

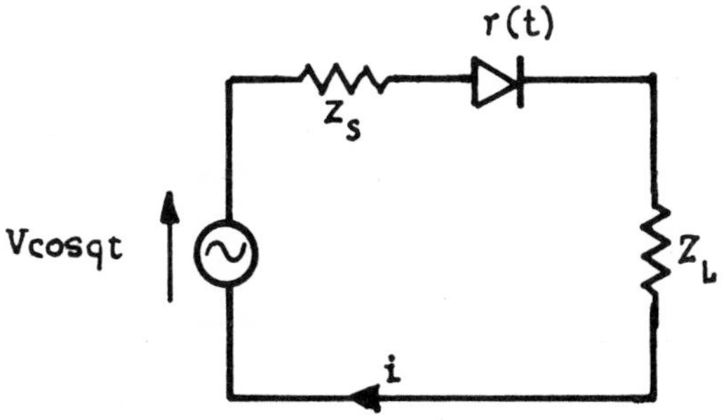

FIG. 5. Series modulator

The first consists of an algebraic approach, aimed at establishing fairly general results for use as guide lines for designers - often this has led to earlier results in the literature being expanded and generalised. For example, attention has been focussed on the simple single-diode modulator of Fig. 5. The circuit equation has been developed.

$$V \cos qt = [Z_s + Z_L + r(t)]i \qquad (1)$$

$$r(t) = r_0 + 2 \sum_{n=0}^{\infty} r_n' \cos (npt + \theta_n) \qquad (2)$$

$$Z = Z_s + Z_L \qquad (3)$$

Eq. (2) has been split into a set of equations [2] at all modulation-product frequencies (including the signal frequency), which is valid provided the circuit is non-oscillatory, by the frequency-balance technique.

$$\begin{pmatrix} \vdots \\ 0 \\ 0 \\ V \\ 0 \\ 0 \\ \vdots \end{pmatrix} = \begin{pmatrix} & \vdots & \vdots & \vdots & \\ \cdots & (Z_{-1} + r_0) & r_1 & r_2 & \cdots \\ \cdots & r_1^* & (Z_0 + r_0) & r_1 & \cdots \\ \cdots & r_2^* & r_1^* & (Z_{+1} + r_0) & \cdots \\ & \vdots & \vdots & \vdots & \end{pmatrix} \begin{pmatrix} \vdots \\ i_{-1} \\ i_0 \\ i_{+1} \\ \vdots \end{pmatrix} \qquad (4)$$

where $Z_{\pm n}$, $i_{\pm n}$ are at frequencies $q \pm np$ and r_n^* is the conjugate of r_n, where $r_n = r_n' \exp(\pm j\theta_n)$.

An examination of the square matrix allows several circuit properties of importance to be deduced - reciprocity relationships [2], the minimum loss of a large class of mixers [3], the relationship between optimum source and load resistance in frequency changers [4], for example. For specific circuits, where only a few currents (or voltages, in the dual case) are significant, it is straightforward to continue the analysis and solve for the output products, circuit efficiency also following.

The second facet of the analysis of circuits with periodically-varying parameters has been the development of optimization programmes for use on (4). This has allowed the efficiency of circuits with up to sixteen significant frequencies present to be investigated [5,6], subject to the restriction that the model of the non-linear element was simple, and that the circuit was appropriately terminated with ideal filters. Mixer circuits have also been examined, in which the termination at the image frequency, 2p - q, is the same as that at the signal frequency, q - which complicates the optimization procedure.

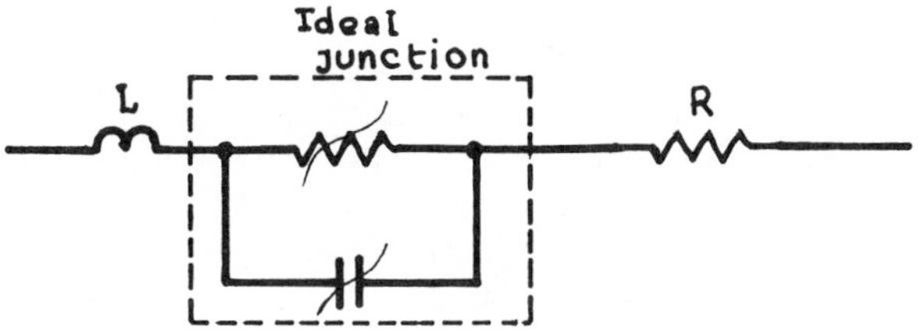

FIG. 6. Diode model

More complex diode models have now been incorporated, see Fig. 6, and the requirement for ideal filters has been removed at the cost of increased computer time. To sum up, this type of circuit analysis seems suitable for the rapid - and cheap - evaluation and optimization of properties such as conversion loss, which is normally not very sensitive to minor imperfections in diode or filter models.

3. SECOND-ORDER CIRCUIT PROPERTIES

Turning now to second-order circuit properties, the intermodulation and cross-modulation performance of simple

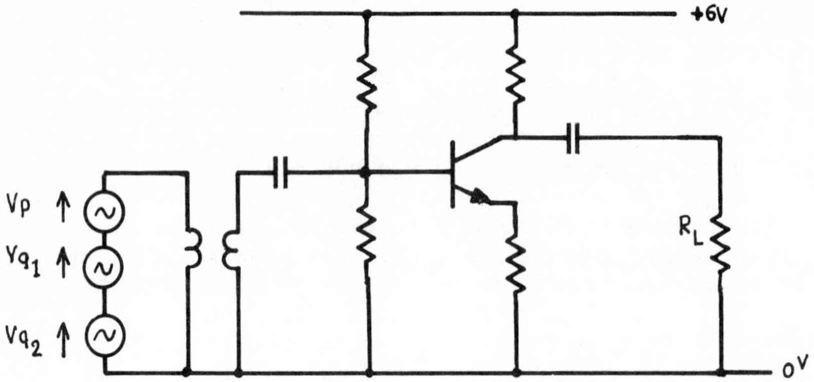

FIG. 7. Transistor mixer

switching transistor mixers has been evaluated using the bilinear diode model of Fig. 3, as a representation of the base-emitter junction of the transistor of Fig. 7. The effect of the strong signals is to phase-modulate the square-wave resistance function shown in Fig. 3, as pointed out by previous workers. Satisfactory agreement between theory and practice was observed for intermodulation products in the range 20-70 dB below the wanted output sideband. Cross-modulation and conversion gain also showed good agreement between theory and practice [7]. This work is now being extended to other forms of transistor mixers.

In attempting to predict the intermodulation distortion in the ring modulator using hot carrier diodes - see Fig. 2 - however, the simple model used previously [8] was found to be overpessimistic for small signals. This was accordingly modified to the form shown in Fig. 8, incorporating an offset voltage. Application of the new model gave improved agreement between theory and practice for intermodulation products between 35 and 80 dB below the wanted sideband [9], correctly predicting an improved performance of some 20 dB for small signals.

Neither of the models discussed above has taken into account the actual shape of the diode i-v characteristic, and it is very difficult to extend the approach outlined to do so. However, use of Savin's methods [10-12] seems

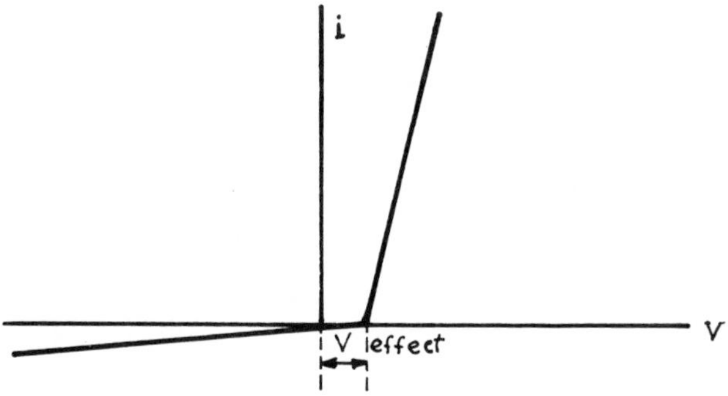

FIG. 8. Modified diode model

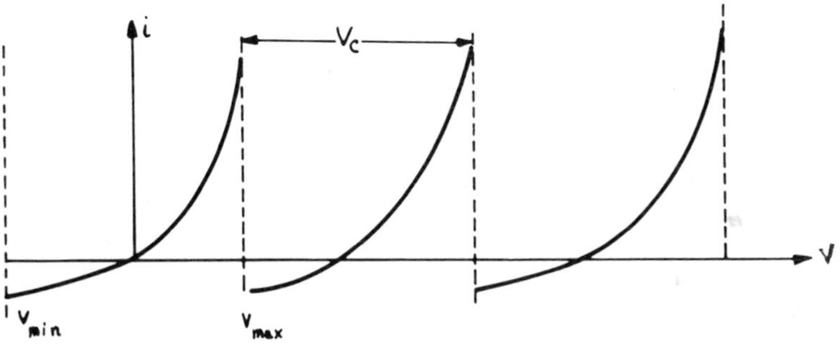

FIG. 9. Savin's model

to offer this possibility, and this approach will now be described.

Savin considers the i-v characteristic of the diode to be a repetitive function of applied voltage, V_D - see Fig. 9. He expands the diode current as a Fourier series with period $2\pi/V_R$ as follows:

$$i = a_0 + \sum_{n=1}^{\infty} (a_n \cos n \frac{2\pi}{V_R} V_D + \text{sine term}). \qquad (5)$$

V_D can consist of a number of applied voltages of different frequencies.

$$V_D = V_B + \sum_{K=1}^{r} V_{SK} \cos \omega_K t \tag{6}$$

$$= V_{S1}(\cos \omega_1 t - \cos \theta) + \sum_{K=2}^{r} V_{SK} \cos \omega_K t \tag{7}$$

where cos θ is introduced for the d.c. term.
Then

$$i = a_0 + \sum_{n=1}^{\infty} \{a_n \cos[n\lambda_{M1}(\cos\omega_1 t - \cos\theta) + n \sum_{K=2}^{r} \lambda_{MK}\cos\omega_K t] + \text{sine term}\} \tag{8}$$

where

$$\lambda_{MK} = \frac{2\pi}{V_R} V_{SK}$$

Using the Bessel expansions such as

$$\cos(x \cos y) = J_0(x) + 2 \sum_{n=1}^{\infty} (-1)^n J_{2n}(x) \cos 2ny \tag{9}$$

i may be separated into a spectrum of currents at modulation and intermodulation product frequencies. For example, if

$$i_{g_1,g_2,\ldots,g_r} = I_{g_1,g_2,\ldots g_r} \cos[g_1\omega_1 t + g_2\omega_2 t + \ldots + g_r\omega_r t] \tag{10}$$

and $g_1 + g_2 + \ldots + g_r$ is even

$$I_{g_1,g_2,\ldots g_r} = 2 \sum_{n=1}^{\infty} [a_n \cos(n\lambda_{M1}\cos\theta) - \text{sine term}] \times J_{g_1}(n\lambda_{M1}) \cdot J_{g_2}(n\lambda_{M2}) \ldots J_{g_r}(n\lambda_{Mr})$$

From this and similar results the intermodulation distortion may be calculated. Application appears limited to circuits in which a single resistive diode is present plus linear passive elements, because of the difficulty of calculating V_D. This, however, represents a significant advance on present theory for the series and shunt modulators. For the ring modulator, however, as pointed out by Belevitch in discussion, it appears virtually impossible to evaluate V_D.

4. CONCLUSION

Analysis methods based upon the representation of the non-linear element(s) in a parametric circuit as a time-varying element(s) controlled entirely by the pump allow a simple and rapid evaluation of a number of important circuit properties. These are designated 'first-order' in this paper, and include conversion efficiency, noise figure, and sensitivity. However, for a realistic assessment of the work of a parametric circuit, in most applications some of the properties here termed 'second-order' need to be evaluated. These include inter- and cross-modulation levels, overload properties and dynamic range. More sophisticated, non-linear analysis is needed to calculate the 'second-order' properties, and usually special methods have to be devised for each circuit in question.

Although general non-linear analysis by digital computer is becoming a possibility, the particular problems posed by the 'second-order' properties of non-parametric circuits appear soluble at the present time only by special-purpose, approximate techniques such as are described in Section 3 of this paper.

5. ACKNOWLEDGMENT

I should like to acknowledge the help of all my colleagues in the Circuit research group, University of Bradford, U.K., who have helped me to prepare this summary of our work. In particular Dr. J.G. Gardiner, who is primarily responsible for the success of the group, and whose distortion studies formed the basis for Section 3.

REFERENCES

1. TUCKER, D. G., (1964) *Circuits with Periodically-Varying Parameters*, Macdonald, London.

2. HOWSON, D. P. and TUCKER, D. G., (September 1965) Reciprocity in Parametric Circuits and Polyphase Modulation, *Symposium on Network Theory*, Cranfield.

3. HOWSON, D. P., (March 1969) Minimum Loss of a General Broadband Mixer, *Electronics Letters 5*, p. 123.

4. HOWSON, D. P., (1969) Terminating Impedances in Diode Modulators and Mixers, *Radio and Electronic Engineer 38*, p. 147.

5. HOWSON, D. P., (1966) Low-loss Shunt or Series Rectifier Modulators, *Electronics Letters 2*, 225.

6. HOWSON, D. P., (1968) The Analysis of Mixers and Frequency Changers with Tuned Terminations, *Summer School on Circuit Theory*, Prague.

7. GARDINER, J. G. and SURANA, D. C., Distortion Phenomena in a Switching Transistor Mixer, to be published in *Proc. IEE.*

8. BELEVITCH, V., (April 1950) Non-linear Effects in Rectifier Modulators, *Wireless Engineer 27*, p. 130.

9. GARDINER, J. G., Intermodulation Phenomena in the Ring Modulator, to be published in *Radio and Electronic Engineer.*

10. SAVIN, S. K., (1967) Harmonic Analysis of Non-Inertial Non-Linear Systems, *Radiotekhnika 22.*

11. SAVIN, S. K., (1968) Response of a Non-Linear Unilaterally Conducting Resistance to the Sum of Sinusoidal Oscillations, *Radio Engineering 23*, p. 65.

12. GARDINER, J. G., SURANA, D. C. and YOUSIF, A. M., (December 1969) Distortion Analysis for Mixer Diodes Possessing Discontinuous Resistance Voltage Characteristics (To be read at IEEE Symposium on Circuit Theory, San Francisco).

Other Related Publications

13. GARDINER, J. G., (November 1968) The Relationship between Cross-Modulation and Intermodulation Distortions in the Double-Balanced Modulator, *Proc.*

IEEE (Letters) 56, pp. 2069-2071.

14. HOWSON, D. P., (November 1968) High Frequency Mixers Using Square-law Diodes, *Radio and Electronic Engineer 36*, No. 5, pp. 311-316.

15. GARDINER, J. G., (May 1969) Single-Balanced Modulator Using Square-law Resistors (S.C.L. Diodes), *Radio and Electronic Engineer 37*, No. 5, pp. 305-314.

16. GARDINER, J. G., (June 1969) Cross-Modulation and Intermodulation Distortions in the Tuned Square-law Diode Modulator, *Radio and Electronic Engineer 37*, No. 6, pp. 353-363.

17. GARDINER, J. G., (April 1969) Distortion Rejection in Square-law Diode Mixer, Technical Report No. 19.

18. HOWSON, D. P., (July 1969) Unwanted Modulation Products in the Ring Modulator, Technical Report No. 23.

19. SURANA, D. C. and GARDINER, J. G., (November 1969) Mixing Coefficients for Two and Three-Terminal Non-Linear Resistance, Technical Report No. 24.

20. YOUSIF, A. M. and GARDINER, J. G., (November 1969) Distortion phenomena in Single-Balanced Diode Mixers, Technical Report No. 25.

21. GARDINER, J. G. and YOUSIF, A. M., Local-Oscillator Noise Mixing Effects in the Ring Modulator, in preparation.

POLYNOMIAL RECURSIVE DIGITAL FILTERS APPROXIMATING THE IDEAL LOW-PASS CHARACTERISTIC

J.P. THIRAN

M.B.L.E. Reseach Laboratory, Brussels, Belgium

1. INTRODUCTION

From the standpoint of approximation, digital filters are divided into two classes, viz., recursive and nonrecursive filters. As regards their respective advantages, one can mention [1], on the one hand, the strictly linear phase of nonrecursive filters which results from most of their design methods, on the other hand, the capability of recursive filters to meet high selectivity requirements to a reasonable degree. For the latter, which have the non-ideal phase of the continuous filters, there thus arises the problem of performing some approximation of a linear phase together with a selective magnitude function. Recently [2] a method has been suggested for direct approximation of the ideal low-pass characteristic by continuous filters: by using a complex norm, one can meet both phase and amplitude requirements. The aim of this paper is to extend this approach to "polynomial" recursive digital filters, i.e. whose transfer functions are the reciprocals of polynomials.

From the definition of a Chebyshev error norm, which includes the maximally flat approximation as a particular case, one obtains the general form of the optimal error and, consequently, the algorithm for finding the solution which achieves such an optimum. Finally, the implementation of this procedure enables one to analyze the frequency characteristics and the stability property of the optimal filters.

APPROXIMATION PROBLEM

2.1. Statement of the problem

The ideal low-pass digital filter is defined by the transmittance

$$H(e^{-j\omega}) = \begin{cases} e^{-j\omega\tau} & |\omega| < \omega_c \\ 0 & \omega_c < |\omega| < \pi \end{cases}$$

where a unit sampling period has been obtained by time normalization.

As a consequence of this definition, the ideal filter transfer function depends on two parameters, the cutoff angular frequency ω_c and the delay τ. In the following, one degree of freedom will be frozen by choosing $\tau = 1$, thus the shortest delay such that the problem can be stated in terms of rational functions in the variable $z = \exp(j\omega)$.

As $H[\exp(-j\omega)]$ is assumed to be the reciprocal of a polynomial of degree n, $P[\exp(-j\omega)]$, the Chebyshev error norm, is

$$\sup \left|e^{-j\omega} - P^{-1}(e^{-j\omega})\right| \qquad |\omega| < \omega_c$$

Since $\left|P[\exp(-j\omega)]\right| \cong 1$ in the approximation interval, this norm is practically equal to

$$\sup \left|e^{j\omega} - P(e^{-j\omega})\right| \qquad |\omega| < \omega_c$$

or

$$\sup \left|e^{j\omega(n+1)} - Q(e^{j\omega})\right| \qquad |\omega| < \omega_c$$

where $Q[\exp(j\omega)] = \exp(jn\omega)\, P[\exp(-j\omega)$ is a polynomial of degree n in $\exp(j\omega)$.

The approximation problem thus consists in minimizing the modulus of a monic polynomial of degree n+1 over an interval C of the unit circle. Such an optimal polynomial is frequently called the Chebyshev polynomial for C [3, p. 264] [4, p. 146].

2.2. *Characterization of the optimum*

The following two properties will help to characterize the optimal solution.

Property 1. [3] [4] Let S be a closed and bounded set in the complex plane that contains at least n+1 points. The Chebyshev polynomial of degree n for S

a) exists,
b) is unique,
c) is such that the modulus reaches its maximum value at least at n+1 distinct points of S.

Property 2. [5] The squared modulus of a polynomial of degree n whose argument is equal to $\exp(j\omega)$ is a polynomial of degree n in $\cos\omega$.

Therefore, denoting by $\varepsilon[\exp(j\omega)]$ the complex error

$$e^{j\omega(n+1)} - Q(e^{j\omega}), \tag{1}$$

one deduces from Property 1 that $|\varepsilon[\exp(j\omega)]|^2 = M(\cos\omega)$ reaches its maximum value at least at n+2 distinct points on $[-\omega_c,+\omega_c]$ if it is optimal.

1) For $n = 2\ell$, by Property 2, $M(\cos\omega)$ has at most ℓ distinct maxima at $\cos\omega_i$ $(i = 1,2,\ldots,\ell)$ interlaced with ℓ minima. On the interval $[\cos\omega_c,1]$, the error can be equalized at $\ell+1$ points: the ℓ abscissae $\cos\omega_i$ and one boundary. According to whether the latter is $\cos\omega_c$ or 1, the error will have $2\ell+2 = n+2$ or $2\ell+1 = n+1$ equal magnitude maxima on $[-\omega_c,+\omega_c]$. The optimal error characteristic is thus necessarily the one plotted in Fig.1(a).

2) For $n = 2\ell+1$, on the $\cos\omega$-axis $M(\cos\omega)$ has at most $\ell+1$ distinct maxima interlaced with ℓ minima or conversely. In the first case, $M(\cos\omega)$ can be equalized at

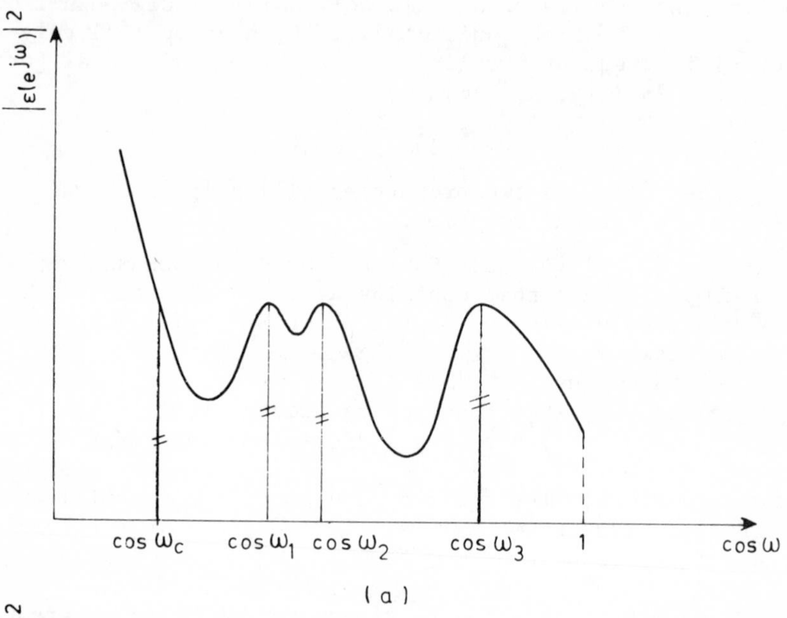

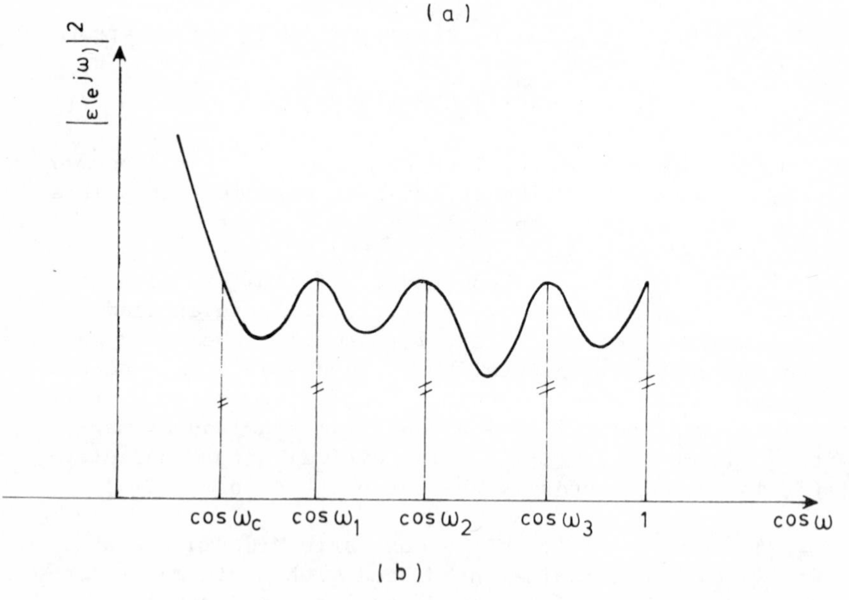

Fig. 1. Optimal error characteristics (a) for even n
(b) for odd n

the $\ell+1$ abscissae of the maxima, thus at $2\ell+2$ or $n+1$ distinct points on $[-\omega_c,+\omega_c]$. On the other hand, if $M(\cos\omega)$ has ℓ maxima at $\cos\omega_i$ $(i = 1,2,\ldots,\ell)$, it can be equalized on $[\cos\omega_c,1]$ at $\ell+2$ points: the ℓ abscissae $\cos\omega_i$ and the two boundaries $\cos\omega_c$ and 1. Therefore, on $[-\omega_c,+\omega_c]$, the total number of equal magnitude maxima is $2\ell+3$ or $n+2$, i.e. the required number [Fig. 1(b)].

In order to find the parameters p_i $(i = 1,2,\ldots,n+1)$, coefficients or roots, which define $\varepsilon(z)$, one thus has to minimize $M(\cos\omega_c)$ with the constraints

$$\begin{cases} M(\cos\omega_c) = M(\cos\omega_j) & j = 1,2,\ldots,\ell \\ M(\cos\omega_c) = M(1) & \text{for odd } n \end{cases} \tag{2}$$

Using Lagrange multipliers, one obtains the set of equations

$$\frac{\partial M(\cos\omega_c)}{\partial p_i} + \sum_{j=1}^{\ell} \lambda_j \left(\frac{\partial M(\cos\omega_c)}{\partial p_i} - \frac{\partial M(\cos\omega_j)}{\partial p_i}\right) + \partial\lambda_0 \left(\frac{\partial M(\cos\omega_c)}{\partial p_i} - \frac{\partial M(1)}{\partial p_i}\right) = 0, \quad i = 1,2,\ldots,n+1 \tag{3}$$

where δ is 0 or 1 for even or odd n respectively.

These relations, together with the conditions (2), enable one to determine the parameters p_i and the multipliers λ_j.

2.3. *Maximally flat approximation*

If $\omega_c = 0$, in the sense of approximation theory, the Chebyshev polynomial is not unique for nonzero n. Its general expression is indeed $(z-1)q(z)$, where $q(z)$ is any monic polynomial of degree n. However, the optimal family described in section 2.2. tends to the particular solution

$$\varepsilon(z) = (z - 1)^{n+1} \tag{4}$$

Indeed, all extrema of $|\varepsilon[\exp(j\omega)]|^2 = M(\cos\omega)$ are located in the interval $[\cos\omega_c, 1]$ so that for $\omega_c = 0$ the error has the form

$$M(\cos\omega) = K(\cos\omega - 1)^{n+1}$$

and cancels at $\omega = 0$ with its derivatives up to order n according to a maximally flat criterion.

2.4. *Critical solution for ω_c approaching π*

For the whole unit circle, the Chebyshev polynomial of degree n+1 is [4]

$$\varepsilon(z) = z^{n+1} \tag{5}$$

and, from the general expression (1) for the error $\varepsilon(z)$, one deduces that the approximating polynomial $Q(z)$ vanishes. In fact, this critical solution will occur for $\omega_0 \le \omega_c \le \pi$, where the lower bound ω_0 is the cutoff angular frequency, for which (5) is a solution of equations (2) and (3). From the expression

$$\varepsilon(z) = \sum_{i=0}^{n+1} a_i z^{n+1-i} \quad \text{with } a_0 = 1$$

it follows that

$$\lim_{a_i (i=1,2,\ldots,n+1)\to 0} \frac{\partial M(\cos\omega)}{\partial a_k} = 2\cos k\omega, \qquad k=1,2,\ldots,n+1$$

Therefore, the system (3) becomes

$$\left(1 + \sum_{j=1}^{\ell} \lambda_j + \delta\lambda_0\right) \cos k\omega_0 - \sum_{j=1}^{\ell} \lambda_j \cos k\omega_j - \delta\lambda_0 = 0, \qquad k=1,2,\ldots,n+1. \tag{6}$$

The solution is readily found for the lowest degrees, and can be generalized in the form

$$\begin{aligned} \omega_j &= \frac{n+1-2j}{n+2}\pi & j&=0,1,2,\ldots,\ell \\ \lambda_j &= -\frac{2}{n+2} & j&=1,2,\ldots,\ell \\ \lambda_0 &= -\frac{1}{n+2} \end{aligned} \tag{7}$$

Indeed, the insertion of (7) into (6) yields

for even n $$\sum_{j=0}^{n/2} \cos\left(\frac{n+1-2j}{n+2} k\pi\right) = 0, \quad k=1,2,\ldots,n+1$$

for odd n $$\sum_{j=0}^{(n-1)/2} \cos\left(\frac{n+1-2j}{n+2} k\pi\right) + \frac{1}{2} = 0, \ k=1,2,\ldots,n+1$$

Now, from the series [6, p. 86]

$$\sum_{j=0}^{n-1} \cos(\alpha+j\delta) = \frac{\cos[\alpha+(n-1)\delta/2]\ \sin(n\delta/2)}{\sin(\delta/2)}$$

one easily verifies

for even n $$\sum_{j=0}^{n/2} \cos\left(\frac{n+1-2j}{n+2} k\pi\right) = \frac{\sin k\pi}{2\sin[k\pi/(n+2)]} = 0, \quad k=1,2,\ldots,n+1$$

for odd n $$\sum_{j=0}^{(n-1)/2} \cos\left(\frac{n+1-2j}{n+2} k\pi\right) = -\frac{1}{2} + \frac{\sin k\pi}{2\sin[k\pi/(n+2)]} = -\frac{1}{2}, \ k=1,2,\ldots,n+1$$

To summarize: for fixed n, the approximation problem stated in Section 2.1 will have a meaningless solution when ω_c belongs to the interval $[\pi(n+1)/(n+2),\pi]$.

2.5 Analytical solution for n = 0 *and* 1.

For n=0, from $\varepsilon(z) = z + a_1$, one computes

$$M(\cos\omega) = 2a_1\cos\omega + 1 + a_1^2$$

Since $M(\cos\omega)$ must exhibit the optimal behaviour illustrated in Fig. 1(a), a_1 is necessarily negative. Its value is determined by the condition $\delta M(\cos\omega_c)/\partial a_1 = 0$. Thus, $a_1 = -\cos\omega_c$ and ω_c must be smaller than $\pi/2$ to select the negative values of a_1. To summarize, the Chebyshev polynomial of degree 1 is

$$\varepsilon(z) = \begin{cases} z - \cos\omega_c & \omega_c \leq \pi/2 \\ z & \pi/2 \leq \omega_c \leq \pi \end{cases}$$

For n=1, one has, successively,

$$\varepsilon(z) = z^2 + a_1 z + a_2$$
$$M(\cos\omega) = 4a_2\cos^2\omega + 2a_1(1+a_2)\cos\omega + (a_2-1)^2 + a_1^2$$

and a_2 must be positive in order to satisfy the optimal criterion illustrated in Fig.1(b). The set of equations (2) and (3) becomes

$$\begin{cases} 2a_2\cos^2\omega_c + a_1(1+a_2)\cos\omega_c - 2a_2 - a_1(1+a_2) = 0 \\ (1+a_2)\cos\omega_c + a_1 + \lambda_0(\cos\omega_c-1)(1+a_2) = 0 \\ 2\cos^2\omega_c + a_1\cos\omega_c + a_2 - 1 + \lambda_0(\cos\omega_c-1)(2\cos\omega_c+2+a_1) = 0 \end{cases}$$

From the first equation, a_1 is expressed in terms of a_2 as

$$a_1 = \frac{4a\cos^2(\omega_c/2)}{1 + a_2} \tag{8}$$

where elimination of λ_0 from the last two equations yields

$$(a_1 + a_2 + 1)(a_1 - a_2 + 4\cos^2(\omega_c/2) - 1) = 0 \tag{9}$$

By inserting (8) into (9), one verifies that the only real value of a_2 which is not negative for all ω_c ranging from 0 to π is $a_2 = -1 + 2\cos(\omega_c/2)$. Whence $a_1 = 2\cos(\omega_c/2)[1 - 2\cos(\omega_c/2)]$ and the condition to have a positive value for a_2 reduces to $\omega_c \leq 2\pi/3$.

Finally, the optimal solution is

$$\varepsilon(z) = \begin{cases} z^2 + 2\cos(\omega_c/2)[1-2\cos(\omega_c/2)]z + 2\cos(\omega_c/2) - 1, & \omega_c \leq 2\pi/3 \\ z^2, & 2\pi/3 \leq \omega \leq \pi \end{cases}$$

The root locus of both optimal polynomials is plotted in Fig.2.

2.6. *Numerical solution for higher degrees*

For n > 1, no analytical solution can be obtained and a numerical algorithm has thus been chosen. The problem at hand amounts to solving the system of equations (1) and (2), which can be rewritten as

$$F_i(\alpha_1,\alpha_2,\ldots,\alpha_{n+1},\lambda_1,\lambda_2,\ldots,\lambda_\omega,\delta\lambda_0) = 0 \qquad i=1,2,\ldots,n+1+\ell+\delta$$

where $\alpha_i (i=1,2,\ldots,n+1)$ are the roots of $\varepsilon(z)$.

Starting from a set of non-optimal parameters $\{\bar{\alpha}_1, \bar{\alpha}_2,\ldots,\bar{\alpha}_{n+1},\bar{\lambda}_2,\ldots,\bar{\lambda}_\ell,\delta\bar{\lambda}_0\}$, one first computes the abscissae $\cos\omega_j$ at which $M(\cos\omega)$ is maximum. The perturbations for the initial approximations are then obtained by the generalized Newton-Raphson method.

$$\sum_{j=1}^{n+1} \frac{\partial F_i(\bar{\alpha}_1,\bar{\alpha}_2,\ldots)}{\partial \alpha_j} \Delta\alpha_j + \sum_{j=1}^{\ell} \frac{\partial F_i(\bar{\alpha}_1,\bar{\alpha}_2,\ldots)}{\partial \lambda_j} \Delta\lambda_j + \delta \frac{\partial F_i(\bar{\alpha}_1,\bar{\alpha}_2,\ldots)}{\partial \lambda_0} \Delta\lambda_0 = -F_i(\bar{\alpha}_1,\bar{\alpha}_2,\ldots)$$

$$i=1,2,\ldots,n+1+\ell+\delta$$

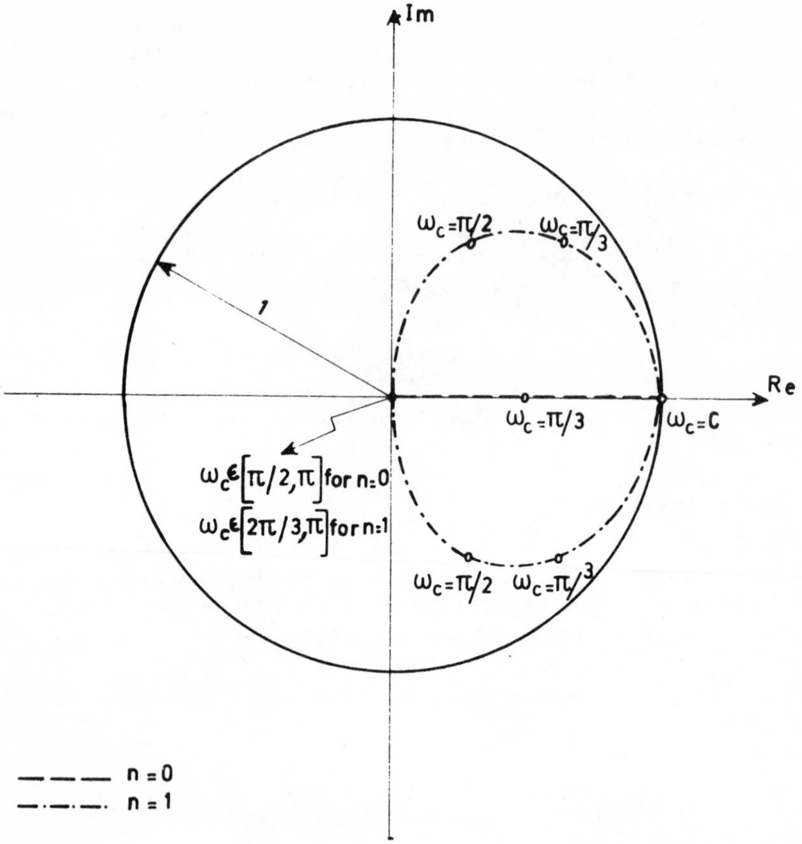

Fig. 2. Root locus of error $\varepsilon(z)$ for n=0 and 1.

With the new values $\{\bar{\alpha}_1 + \Delta\alpha_1, \bar{\alpha}_2 + \Delta\alpha_2, \ldots\}$ the process is then iterated until the corrections are smaller than a fixed tolerance.

By way of illustration, the optimal error is represented in Fig.3 for $\omega_c = \pi/2$ and various degrees.

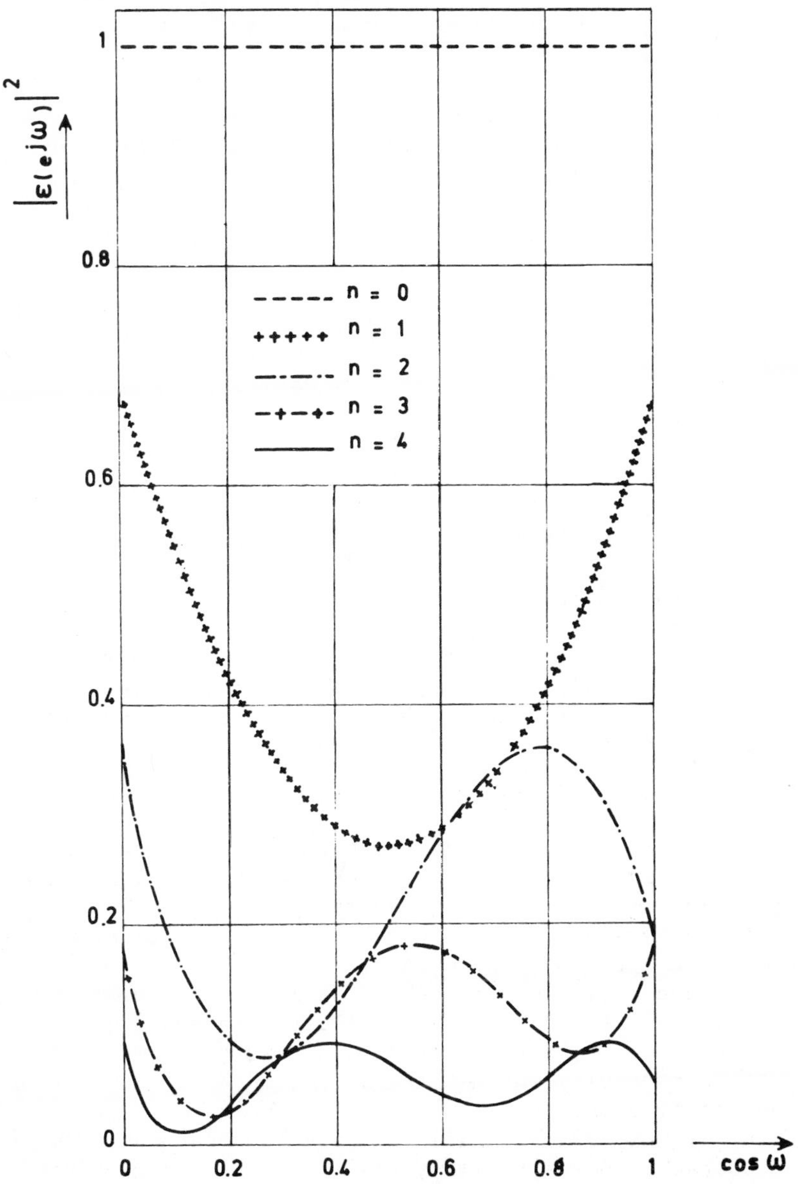

Fig. 3. Optimal error for $\omega_c = \pi/2$ and for degrees n=0 to 4.

3. STABILITY AND FREQUENCY CHARACTERISTICS OF OPTIMAL FILTERS

The foregoing approximation procedure does not ensure the stability of the optimal filters. One can check only a posteriori whether all roots of $P(z^{-1})$ are outside the unit circle or, conversely, whether all roots of $Q(z) = z^n P(z^{-1})$ are inside this circle.

3.1. Maximally flat solutions

By the expression(4) of the optimal error, the Q-roots are readily deduced from

$$(z-1)^{n+1} - z^{n+1} = 0$$

Thus

$$z_k = 0.5 + j\cot g \frac{k\pi}{n+1} \qquad k=1,2,\ldots,n$$

As their modulus is $\{2\sin[k\pi/(n+1)]\}^{-1}$, all roots lie inside the unit circle if $\sin[k\pi/(n+1)] > 2^{-1}$ or $k\pi/(n+1) > \pi/6$ for $k=1,2,\ldots,n$. Therefore, as for continuous filters [7], the upper bound of the degree is 4. Fig. 4 gives the frequency characteristics, attenuation and delay, of the four stable filters.

3.2. Chebyshev solutions

The number of solutions resulting from the approximation procedure is restricted, on the one hand, by the constraint of stability, and on the other hand by a realistic upper bound of the error norm. If this one is chosen equal to 10%, there exist no stable solutions for any ω_c when n is greater than 9, as indicated in Fig. 5. For lower degrees, some stable filters are found when ω_c is above a value which depends on n. As an example, the root locus of Q(z) for n=6 is plotted in Fig. 6: the roots move nearly along straight lines and the cutoff angular frequency $\omega_c = .91$ separates the stable solutions from the unstable ones. Finally, Fig. 7 and Fig. 8 give the attenuation and delay characteristics of all stable filters for $\omega_c = .5$ and 1. It can be noticed, in particular, that the closer the solution is to the stability bound-

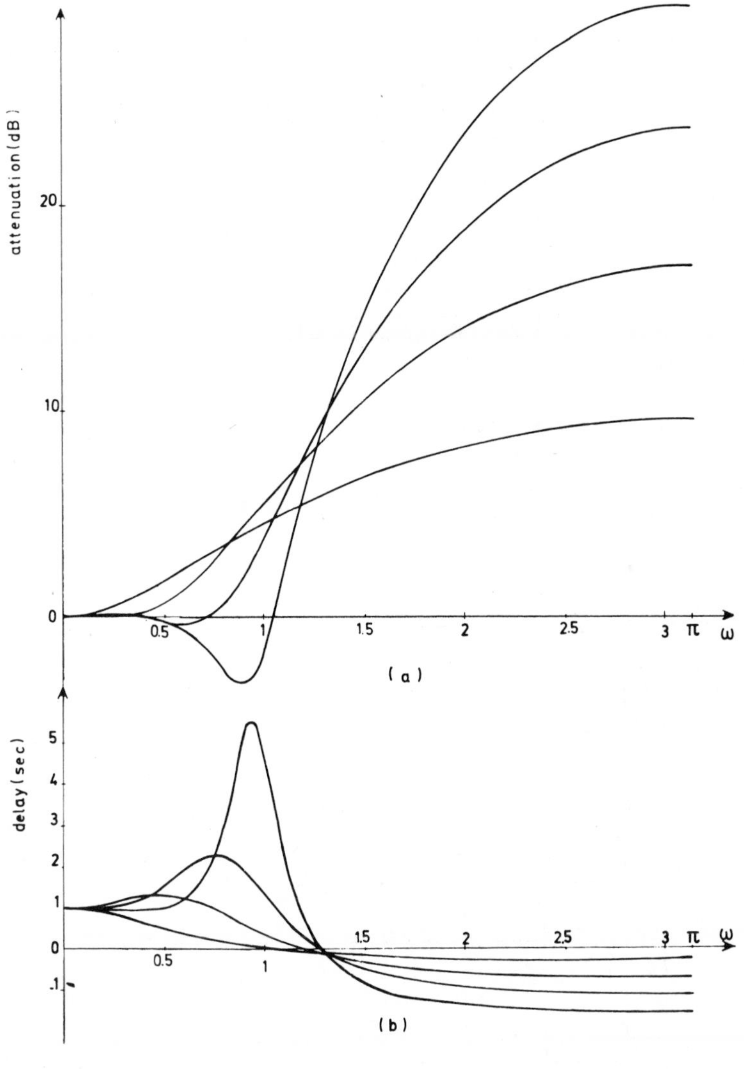

Fig. 4. Frequency characteristics of the maximally flat filters;
(a) attenuation, (b) delay.

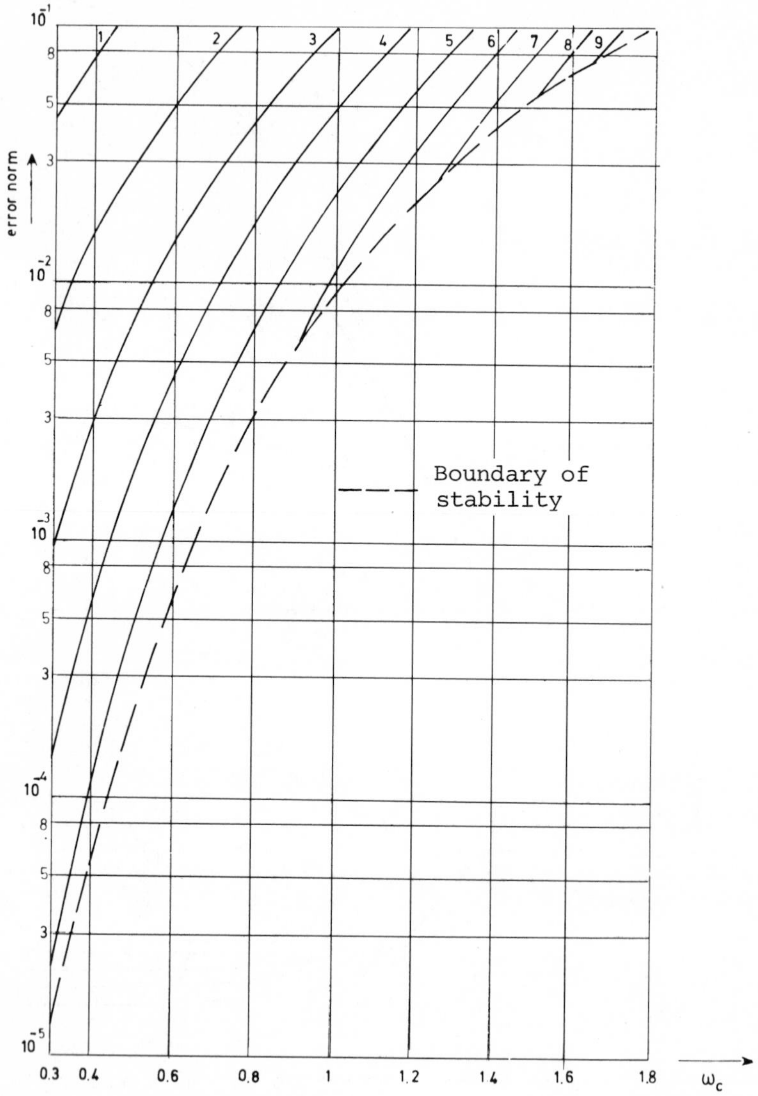

Fig. 5. Error norm versus cutoff angular frequency ω_c.

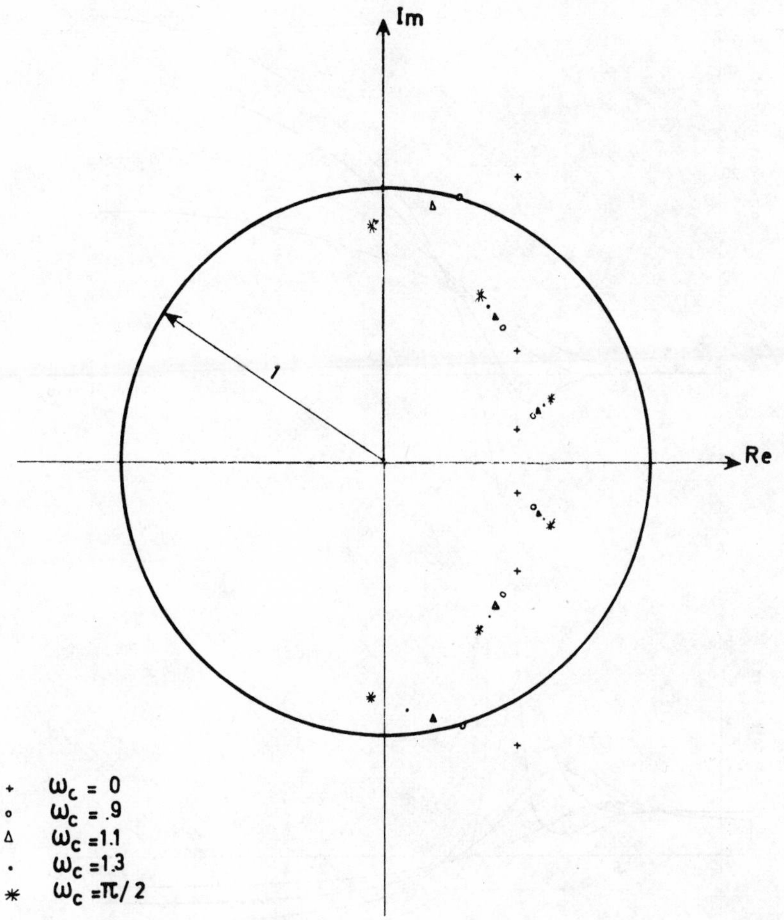

Fig. 6. Root locus of Q(z) for n=6 and ω_c ranging from 0 to $\pi/2$.

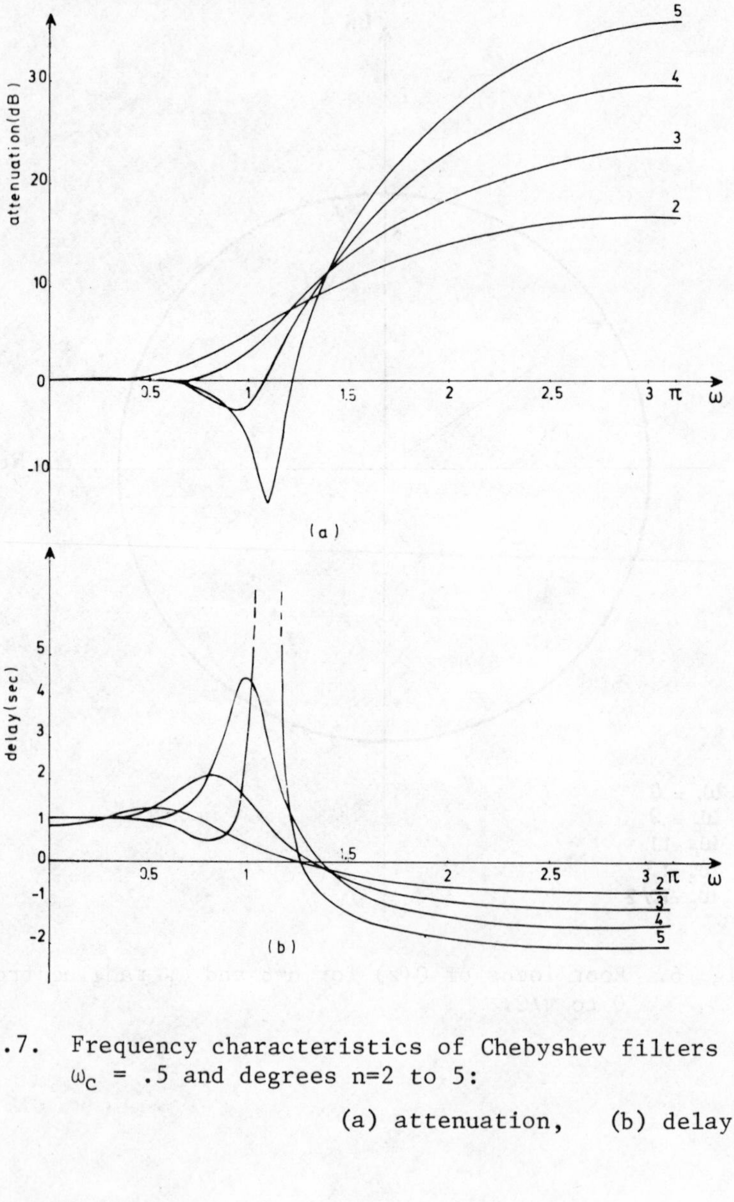

Fig.7. Frequency characteristics of Chebyshev filters for $\omega_c = .5$ and degrees n=2 to 5:

(a) attenuation, (b) delay.

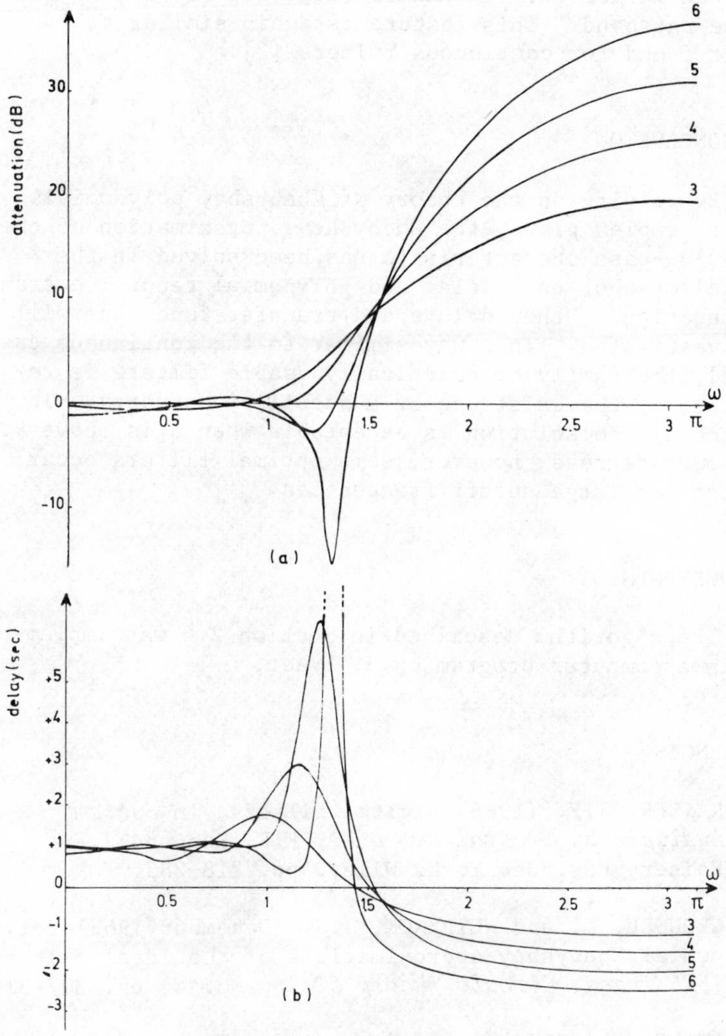

Fig. 8. Frequency characteristics of Chebyshev filters for $\omega_c = 1$ and degrees n=3 to 6: (a) attenuation, (b) delay.

ary, the larger the attenuation dip just beyond the end of the passband. This feature is again similar to the result found for continuous filters [2].

4. CONCLUSION

By relying on the theory of Chebyshev polynomials in the complex plane, the Chebyshev approximation of the ideal low-pass characteristic has been solved in the special case of unit delay and polynomial recursive transfer function. Other delays and transfer functions will be investigated. In a way similar to the continuous case [7, 2], the family of practically usable filters is restricted by the existence of a stability boundary. In particular, no solution is acceptable when n is above 9. For lower degrees, however, some optimal filters occur for not too large cutoff frequencies.

ACKNOWLEDGEMENT

The algorithm described in Section 2.6 was implemented in a computer program by J. Praet.

REFERENCES

1. KAISER, J.F. (1966) Digital filters, in *System Analysis by Digital Computer*, F.F. Kuo and J.F. Kaiser, Eds. New York, Wiley, pp. 218-285.

2. CROUSEL, L. and NEIRYNCK, J.J. (December 1968) Polynomial Chebyshev approximations of the ideal filter, *IEEE Trans. Circuit Theory CT-15*, No. 4, pp. 307-315.

3. HILLE, E. (1962) *Analytic Function Theory*, Vol. II, Boston, Ginn and Company.

4. DAVIS, P.J. (1963) *Interpolation and Approximation*, New York, Blaisdell.

5. HOLTZ, H. and LEONDES, C.T. (April 1966) The Synthesis of recursive filters, *J. ACM 13*, No. 2, pp. 262-280.

6. DWIGHT, H.B. (1957) *Tables of Integrals and other Mathematical Data*, New York, Macmillan.

7. GOLAY, M. (September 1960) Polynomials of transfer functions with poles only satisfying conditions at the orgin, *IEEE Trans. Circuit Theory CT-7*, pp. 224-229.

VARIATIONAL PRINCIPLES FOR LC NETWORKS. A REVIEW

V. AMOIA

Istituto di Elettrotecnica ed Elettronica
Politecnico di Milano, Milan, Italy

1. INTRODUCTION

The variational approach to physical systems theory is classical; the Lagrange and Hamilton formulations of physical theories have been known for a long time and form the foundation for much of modern theoretical physics.*

Many different reasons motivate variational formulations of physical theories.

First of all they may be useful in summarizing a subject and, consequently, they may, as Morse and Feshbach [3] remark, suggest fruitful analogies and generalizations.

In particular, in network theory, conservative and more generally Lagrangian [4] networks have been suggested as devices for generating extremals associated with variational problems [5].

Moreover a variational principle directly gives a scalar function which can play the role of a Lyapunov

---

*An introduction to this topic can be found in Guillemin [1] for linear electrical networks and in Stern [2] for nonlinear networks.

function in the study of stability criteria [6].

The direct methods of the calculus of variations may yield existence and uniqueness proofs for solutions and general information about their behavior [7].

Finally, a variational formulation leads to the so-called variational method of obtaining approximate numerical solutions [8].

Recently it has been pointed out that any variational formulation is strictly related, from the mathematical point of view, to the self-adjointness of the set of equations describing the physical system [9].

The approach adopted here is founded upon such and further results in differential operator theory (see Section 2), and upon the particular structure exhibited by the network basic equations (see Section 3).

A systematic derivation of variational principles, which in some respect seems to be new, is thus made possible. Some of them are well-known; others have received little notice so far or can be considered as the network counterparts of principles already known for a long time in other branches of physics [10].

In a few instances, special forms of such principles are possibly recorded here for the first time.

2. MATHEMATICAL BACKGROUND

Let H^n be the n-dimensional Hilbert space of vectors **v**, time dependent, continuously differentiable and with derivatives of suitable order*.

---

*According to the degree of the differential operators to be considered later.

We assume as the *scalar product* of two vectors **u**, **v** ε H^n the quantity*

$$(\mathbf{u},\mathbf{v}) \triangleq \int_{t_1}^{t_2} \mathbf{u}^T \cdot \mathbf{v}\, dt = \int_{t_1}^{t_2} \langle \mathbf{u},\mathbf{v} \rangle dt \tag{2.1}$$

t being the time and t_1, t_2 the time interval of interest for our considerations.

A set of differential equations may be written as

$$\mathcal{T}\mathbf{u} = \mathbf{f} \tag{2.2}$$

where **u** and **f** are both n-vectors (the n-tuples of unknown and forcing functions, respectively), and $\mathcal{T}$ is an $n \times n$ matrix which will be referred to as the *formal differential operator* associated with the set of equations [11]. When the range of variability, the functional class and the initial or boundary conditions are prescribed for **u**(t), the *domain* $D(T) \subset H^n$ of the operator is said to have been defined. In this case, the formal operator plus the domain

$$\begin{cases} \mathcal{T}\mathbf{u} = \mathbf{f} \\ \mathbf{u} \in D(T) \end{cases} \triangleq T\mathbf{u} = \mathbf{f} \tag{2.3}$$

is referred to as the *problem* or the *operator* T**.

The *Fréchet differential* $T'\phi$ of an operator in a given direction ϕ is defined as [7]

---

*The superscript T indicates the transpose of a vector or matrix. The angular brackets denote the ordinary scalar product of vectors, $\langle \mathbf{a},\mathbf{b} \rangle \triangleq \mathbf{a}^T \cdot \mathbf{b}$.

**We denote a differential operator by a block capital letter and the formal differential operator by the corresponding script capital letter.

$$T'\phi \overset{\Delta}{=} \lim_{h\to 0} \frac{T(\mathbf{u}+h\phi) - T\mathbf{u}}{h} = \frac{\partial}{\partial h} T(\mathbf{u} + h\phi)\Big|_{h=0} \tag{2.4}$$

T' is referred to as the *Fréchet derivative* of the operator T.

It is worth noting that, if T is linear,

$$T' = T \quad \forall\phi,\ \mathbf{u} \in D(T) \tag{2.5}$$

In a completely analogous fashion it is possible to define the Fréchet derivative of a formal differential operator:

$$\mathcal{T}'\phi = \lim_{h\to 0} \frac{\mathcal{T}(\mathbf{u} + h\phi) - \mathcal{T}\mathbf{u}}{h} = \frac{\partial}{\partial h} \mathcal{T}(\mathbf{u} + h\phi)\Big|_{h=0} \tag{2.6}$$

A linear formal operator $\tilde{L}$ is said to be the *formal adjoint* of a formal operator L when [11]

$$(\mathbf{u}, L\mathbf{v}) = (\tilde{L}\mathbf{u}, \mathbf{v}) + \{\text{boundary terms}\},\ \forall\ \mathbf{u},\mathbf{v} \in H^n \tag{2.7}$$

Two very common operators are

a) a constant square matrix A
b) the derivative of first order $\frac{d}{dt}(\cdot)$

It can be shown that:

$$\tilde{A} = A^T \quad ; \quad (\tilde{\frac{d}{dt}}(\cdot)) = -\frac{d}{dt}(\cdot) \tag{2.8}$$

A case of peculiar interest occurs when

$$L = \tilde{L} \tag{2.9}$$

For such operators, which are said to be *formally symmetric* or *formally self-adjoint*, some fundamental theorems can be proved.

Theorem 2.1

The set

$$\mathcal{L}\mathbf{u} = \mathbf{f} \tag{2.10}$$

of linear differential equations is derivable as the condition for a functional to be stationary, with respect to arbitrary variations of **u** *vanishing on the boundary with their derivatives, iff* $\mathcal{L}$ *is formally self-adjoint* [12].

Corollary 2.1

Under the hypothesis of Theorem 2.1, the set (2.10) *can be deduced as the Euler-Lagrange equations associated with the problem of making the functional*

$$J(\mathbf{u}) = \tfrac{1}{2}(\mathbf{u}, \mathcal{L}\mathbf{u}) - (\mathbf{u}, \mathbf{f}) \tag{2.11}$$

stationary [12].

Theorem 2.2

Two sets of linear differential equations

$$\begin{cases} \mathcal{L}\mathbf{u} = \dfrac{\partial}{\partial \mathbf{v}} f[\mathbf{u},\mathbf{v}] \\ \mathcal{M}\mathbf{v} = \dfrac{\partial}{\partial \mathbf{u}} f[\mathbf{u},\mathbf{v}] \end{cases} \tag{2.12}$$

are simultaneously derivable as the conditions for a functional to be stationary, with respect to arbitrary variations of **u** *and* **v**, *vanishing on the boundary with their derivatives, iff* [13]

$$\mathcal{L} = \tilde{\mathcal{M}} \tag{2.13}$$

Corollary 2.2

Under the hypothesis of Theorem 2.2, the set 2.12 *can be deduced as the Euler-Lagrange equations associated with the problem of making the functional*

$$J(\mathbf{u},\mathbf{v}) = (\mathbf{v},L\mathbf{u}) - \int_{t_0}^{t_1} f[\mathbf{u},\mathbf{v}]\, dt \tag{2.14}$$

stationary [13].

Theorem 2.3

The set

$$N\mathbf{u} = \mathbf{f} \tag{2.15}$$

of nonlinear differential equations is derivable as the condition for a functional to be stationary, with respect to variations of **u***, vanishing on the boundary with their derivatives, iff*

$$(N'\phi,\psi) = (\phi,N'\psi) \qquad \forall\ \phi,\psi,\mathbf{u} \in H^n \tag{2.16}$$

that is iff the Fréchet derivative of N *is formally symmetric.*

This theorem is a direct derivation from the fundamental one given by Tonti in [13].

Corollary 2.3

Under the hypothesis of Theorem 2.3, the set 2.15 can be deduced as the Euler-Lagrange equations associated with the problem of making the functional

$$J(\mathbf{u}) = (\mathbf{u}, \int_0^1 N\ (\lambda\mathbf{u})\ d\lambda) \tag{2.17}$$

stationary [13].

Theorem 2.4

The formal operator

$$L\ S\ \tilde{L}\ , \tag{2.18}$$

L *being a linear formal differential operator and* S *a constant square matrix, is formally self-adjoint if and only if* S *is symmetric.*

This theorem can be proved directly using simple properties of the scalar product.

Theorem 2.5

If a nonlinear operator $N(\mathbf{u})$ *is symmetric, i.e.,*

$$(N'\phi,\psi) = (\phi,N'\psi) \qquad \forall\ \phi,\psi \in D(N) \tag{2.19}$$

the following relationship holds [4]:

$$N(\mathbf{u}) = \frac{\partial}{\partial \mathbf{u}} < \mathbf{u} \ , \int_0^1 N(\lambda \mathbf{u})\ d\lambda > \tag{2.20}$$

This theorem can be easily derived from results contained in Tonti [13].

In the following, the inverse of a vector-valued function is sometimes required. Its existence and uniqueness is subordinated to the following theorem [14].

Theorem 2.6

Let the vector-valued function $f(\mathbf{x}) = \mathbf{y}$ *be continuously differentiable throughout the* $\mathbf{x}$*-space. Then there exists a continuously differentiable function* $f^{-1}(\mathbf{y})$, *the* inverse *of* f, *satisfying the conditions*

$$\begin{cases} \mathbf{x} = f^{-1}\ [f(\mathbf{x})] & \forall\ \mathbf{x} \\ \mathbf{y} = f\ [f^{-1}(\mathbf{y})] & \forall\ \mathbf{y} \end{cases} \tag{2.21}$$

if and only if

$$\begin{cases} \det \left(\frac{\partial f}{\partial \mathbf{x}}\right) \neq 0 & \forall\ \mathbf{x} \\ ||f(\mathbf{x})|| \to \infty \quad \text{as} \quad ||\mathbf{x}|| \to \infty \end{cases} \tag{2.22}$$

3.   NETWORK BASIC EQUATIONS

In this paper we restrict our considerations to networks containing linear and nonlinear capacitors (Fig. 1), inductors (Fig. 2), and fixed voltage and current sources; capacitive and inductive coupling is also allowed among some branches.

$$\left\{\begin{array}{l} \underset{\sim}{v} = S\underset{\sim}{q} \\ \underset{\sim}{i} = \dot{\underset{\sim}{q}} \end{array}\right.$$

a)   b)

$$\left\{\begin{array}{l} \underset{\sim}{q} = C\underset{\sim}{v} \\ \underset{\sim}{i} = \dot{\underset{\sim}{q}} \end{array}\right.$$

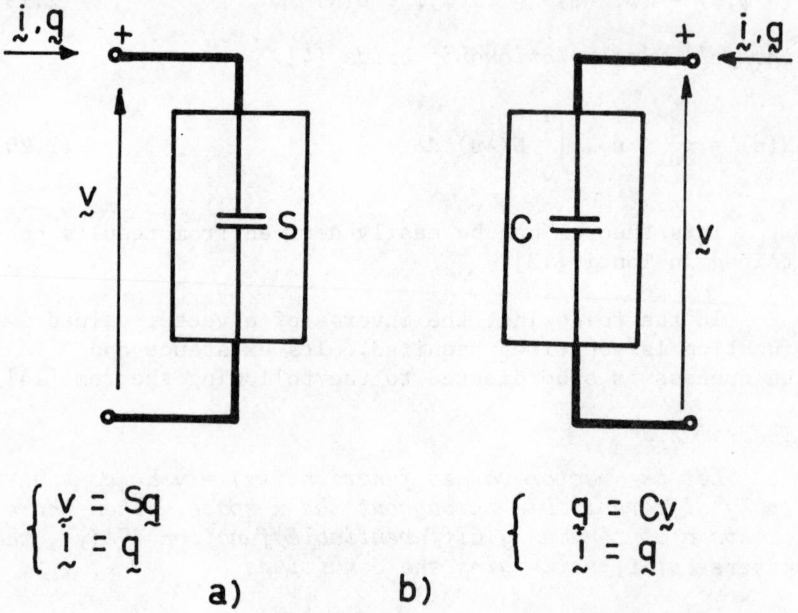

Fig. 1.   (a) current- and (b) charge-dependent condensers.
          Heavy lines are shorthand way of denoting n ter-
          minal pairs. q is the charge vector, S and C
          are the elastance and capacitance matrices.
          Note that these matrices are to be considered
          as operators acting on **q** and **v** respectively.
          When the network is linear, S and C are constant
          matrices; in the nonlinear case, S and C are
          vector valued functions of **q** and **v** respectively.

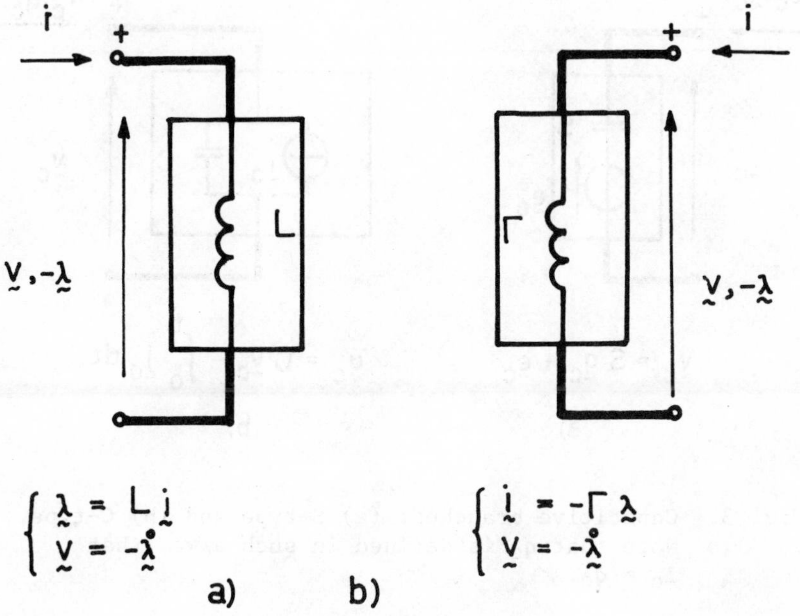

Fig. 2. (a) current- and (b) fluxlinkage-dependent inductors.
Heavy lines are a shorthand way of denoting n terminal pairs. λ is the fluxlinkage vector, L and Γ are the inductance and inverse inductance matrices. Note that these matrices are to be considered as operators acting on **q** and **v** respectively. When the network is linear, L and Γ are constant matrices; in the non-linear case, L and Γ are vector valued real functions of **i** and λ respectively.

The network branches will be classified as either *inductive* (L) or *capacitive* (C).

A capacitive branch can be

- S-type: it consists of a charge dependent condenser in series with a voltage source (Fig. 3a)

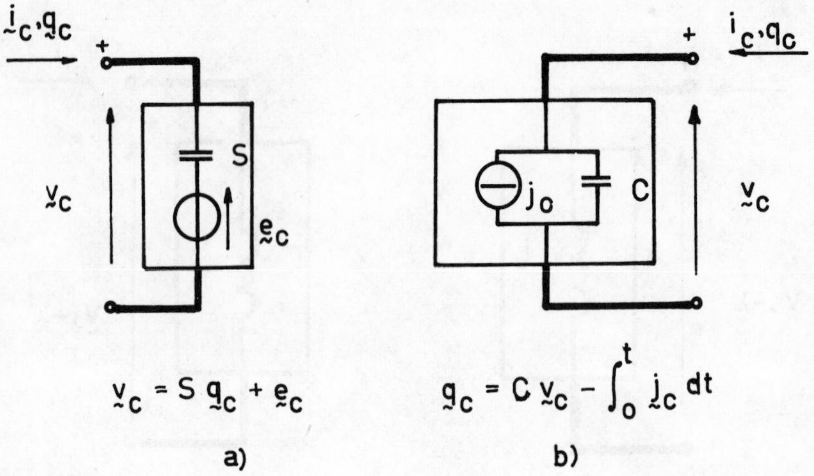

Fig. 3. Capacitive branches: (a) S-type and (b) C-type. Note that q_c is defined in such a way that $i_c = \dot{q}_c$.

or

- C-type: it consists of a voltage dependent condenser in parallel with a current source (Fig. 3b).

Similarly, an inductive branch can be

- L-type: it consists of a current dependent inductor in series with a voltage source (Fig. 4a)

- Γ-type: it consists of a flux linkage dependent inductor in parallel with a current source (Fig. 4b).

They are shown schematically in Figs. 3 and 4, where the *branch relations*, relating the branch variables, are also given.

The functions appearing in the matrices S, L, C, Γ will be assumed to be real, single valued, continuous, and differentiable unless otherwise stated.

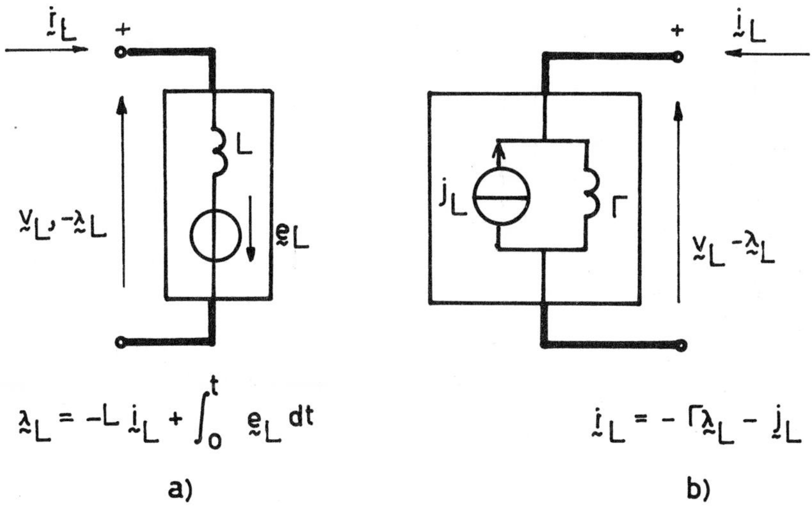

Fig. 4. Inductive branches: (a) L-type and (b) Γ-type. Note that λ_L is defined in such a way that $v_L = -\dot{\lambda}_L$.

Let us now assume that the rows and columns of the *fundamental loop matrix* B and of the *fundamental cut set matrix* D have been so arranged that we may partition them in the form

$$B = [B_C, B_L]; \qquad D = [D_C, D_L] \tag{3.1}$$

B_C, D_C and B_L, D_L being submatrices pertinent to capacitive and inductive branches respectively.

Then Kirchhoff's voltage law (KVL) and Kirchhoff's current law (KCL) may be written as:

$$\left[B_C, \ -\frac{d}{dt}B_L\right] \cdot \begin{bmatrix} v_C \\ \lambda_L \end{bmatrix} = 0 \tag{3.2}$$

and

$$\left[\frac{d}{dt} D_C, D_L\right] \cdot \begin{bmatrix} q_C \\ \\ i_L \end{bmatrix} = 0 \tag{3.3}$$

Equivalently, KVL and KCL may be expressed as

$$\begin{bmatrix} \mathbf{v}_C \\ \\ \lambda_L \end{bmatrix} = \begin{bmatrix} -D_C^T \frac{d}{dt} \\ \\ D_L^T \end{bmatrix} \phi + \begin{bmatrix} \mathbf{v}_C(0^+) \\ \\ \lambda(0^+) \end{bmatrix} \tag{3.4}$$

and

$$\begin{bmatrix} q_C \\ \\ i_L \end{bmatrix} = \begin{bmatrix} B_C^T \\ \\ B_L^T \frac{d}{dt} \end{bmatrix} \psi + \begin{bmatrix} q_C(0^+) \\ \\ i_L(0^+) \end{bmatrix} \tag{3.5}$$

where

$$\phi \triangleq - \int_0^t \mathbf{v}\, dt \tag{3.6}$$

is the *node-pair flux linkage vector* ($\mathbf{v}$ being *the node-pair voltage vector*), and

$$\psi \triangleq \int_0^t \mathbf{i}\, dt \tag{3.7}$$

is the *loop-charge vector* ($\mathbf{i}$ being *the loop-current vector*).

Observing that the operator appearing in eq. (3.5) is the formal adjoint of that appearing in eq. (3.2), and that the same applies to eqs. (3.3) and (3.4), network fundamental equations may be presented as shown in Table 1.

TABLE 1. Network basic equations.

$$u = \tilde{\mathfrak{n}} \underset{\sim}{\phi} + u$$

$$\underset{\sim}{c} = \tilde{\mathfrak{m}} \underset{\sim}{\psi} + \bar{\underset{\sim}{c}}$$

$$\underset{\sim}{u} = P \underset{\sim}{c} + \underset{\sim}{\varepsilon}$$

$$\underset{\sim}{c} = Q u + \underset{\sim}{\chi}$$

$$\mathfrak{m} \underset{\sim}{u} = 0$$

$$\mathfrak{n} \underset{\sim}{c} = 0$$

All the symbols used in Table 1 are explained in Table 2 in terms of branch quantities, as defined in Figs. 1 to 4.

The structure of network basic equations as appearing from Table 1 is well suited for a variational approach as will be given in the next sections.

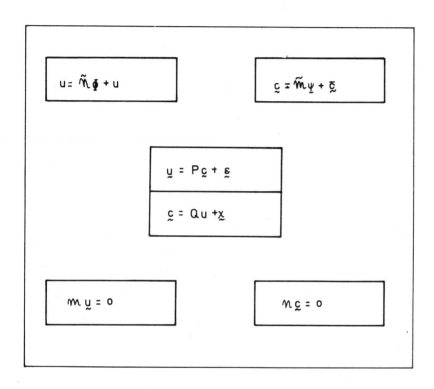

Table 2. Explaining Table 1 symbols.

$\underset{\sim}{v}$=node-pair voltage vector	$\underset{\sim}{i}$=loop-charge vector
$\underset{\sim}{\varrho} = -\int_0^t \underset{\sim}{v}\, dt$	$\underset{\sim}{\gamma} = \int_0^t \underset{\sim}{i}\, dt$
$\underset{\sim}{u} = \begin{bmatrix} \underset{\sim}{v}_C \\ \underset{\sim}{\lambda}_L \end{bmatrix}$	$\underset{\sim}{c} = \begin{bmatrix} \underset{\sim}{q}_C \\ \underset{\sim}{i}_L \end{bmatrix}$
$\bar{\underset{\sim}{u}} = \begin{bmatrix} \underset{\sim}{v}_C(o^+) \\ \underset{\sim}{\lambda}_L(o^+) \end{bmatrix}$	$\bar{\underset{\sim}{c}} = \begin{bmatrix} \underset{\sim}{q}_C(o^+) \\ \underset{\sim}{i}_L(o^+) \end{bmatrix}$
$\underset{\sim}{\varepsilon} = \begin{bmatrix} e_C \\ \int_0^t \underset{\sim}{e}_L\, dt \end{bmatrix}$	$\underset{\sim}{\chi} = \begin{bmatrix} -\int_0^t \underset{\sim}{j}_C\, dt \\ -\underset{\sim}{j}_L \end{bmatrix}$
$P = \begin{bmatrix} S & 0 \\ 0 & -L \end{bmatrix}$	$Q = \begin{bmatrix} c & o \\ o & -r \end{bmatrix}$
$\mathfrak{M} = [B_C, -\frac{d}{dt} B_L]$	$\mathfrak{n} = [\frac{d}{dt} D_C, D_L]$

4. VARIATIONAL THEOREMS FOR LINEAR LC NETWORKS

In this section we put together the previous results from operator theory (Section 2) and the network basic equations (Section 3) to get variational theorems.

A set of eight theorems is obtained. They are classified into two groups, denoted with the letters a and b. For each theorem of the first group, a dual can be found in the second group, and vice versa.

We give detailed proofs for the theorems of the first group only; theorems of the second group can be proved in a completely similar fashion.

For the sake of simplicity, we assume first that the network is linear and time-invariant. At the end of this Section the assumption of time-invariance is released.

I. *Classical Action Principle*

Referring to Table 1, let us perform the chain of substitutions schematically shown in Fig. 5a. The result is the set of *loop equations*:

$$\mathcal{M}P\tilde{\mathcal{M}}\psi + \mathcal{M}P\,\bar{c} + \mathcal{M}\varepsilon = 0 \qquad (4.1)$$

As a consequence of Theorem 2.4, eq. (4.1) satisfies Theorem 2.1 if and only if P is symmetric, i.e., if matrices S and L are symmetric. Then, using Corollary 2.1, we can state the following

Theorem 4.1a

Let the state of a linear, time-invariant, LC, *reciprocal network be described by the loop-charge vector* ψ. *If the network moves from state* ψ_1 *at time* t_1 *to state* ψ_2 *at time* t_2, *then among all possible paths between these end-states, the path the network follows must be one for which*

$$\delta[\tfrac{1}{2}(\psi,\mathcal{M}P\tilde{\mathcal{M}}\psi) + (\psi,\mathcal{M}P\bar{c}) + (\psi,\mathcal{M}\varepsilon)] = 0 \qquad (4.2)$$

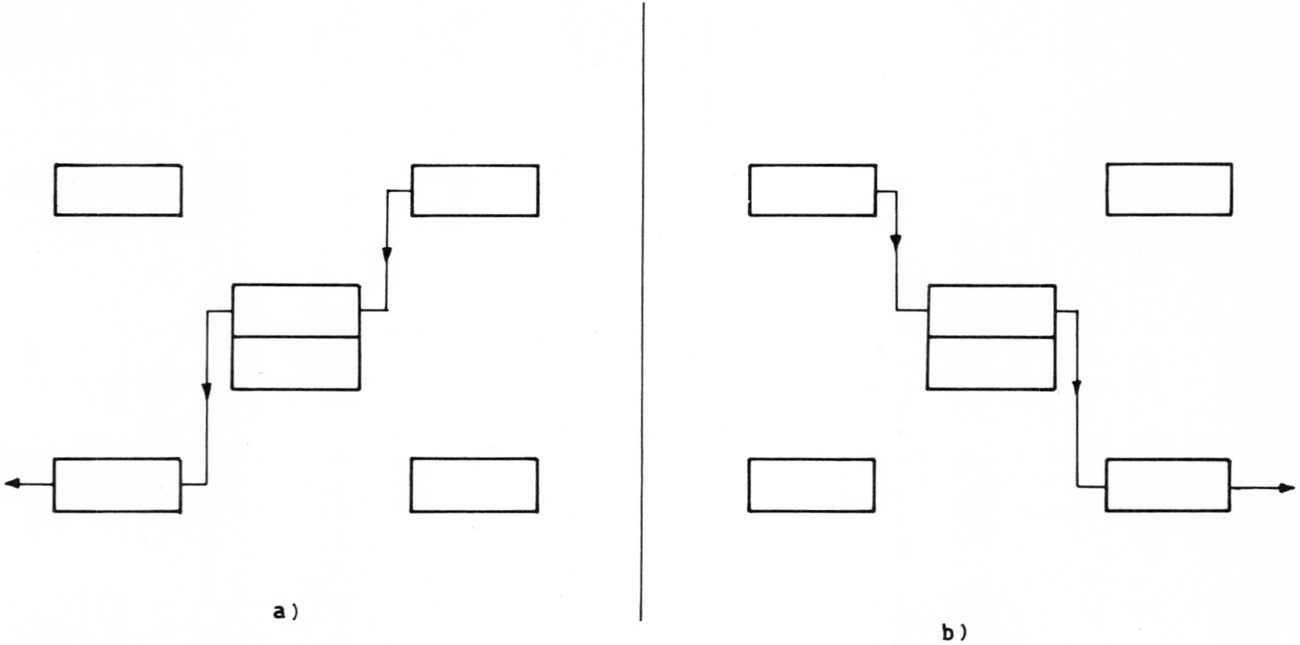

Fig. 5

Using simple algebra and Table 1, eq.(4.2) may be written as

$$\delta[\tfrac{1}{2}\tilde{M}\psi, P\tilde{M}\psi) + (\tilde{M}\psi, P\bar{c}) + (\tilde{M}\psi, \varepsilon)] = 0 \tag{4.3}$$

and subsequently* as

$$\delta_{\psi}[\tfrac{1}{2}(\mathbf{c},P\mathbf{c}) - \tfrac{1}{2}(\bar{c},P\bar{c}) + (c,\varepsilon) - (\bar{c},\varepsilon)] = 0 \tag{4.4}$$

Thus the *network Lagrangian* $\ell[\psi]$ can be expressed in terms of electric energy, magnetic co-energy and source potential function associated with the network as follows:

$$\ell[\psi] \overset{\Delta}{=} \tfrac{1}{2}\langle \mathbf{c},P\mathbf{c}\rangle - \tfrac{1}{2}\langle \bar{\mathbf{c}},P\bar{\mathbf{c}}\rangle + \langle \mathbf{c},\varepsilon\rangle - \langle \bar{\mathbf{c}},\varepsilon\rangle =$$

$$\omega_e[\psi] - \bar{\omega}_m[\psi] - \bar{\omega}_e[0^+] + \bar{\omega}_m[0^+] + z[\psi,\dot{\psi}] \tag{4.5}$$

where

$\omega_e[\psi] = \tfrac{1}{2}\langle \mathbf{q}_c, S\mathbf{q}_c\rangle$ is the electric energy associated with all the condensers and expressed as a function of ψ.

$\bar{\omega}_m[\psi] = \tfrac{1}{2}\langle \mathbf{i}_L, L\mathbf{i}_L\rangle$ is the magnetic co-energy associated with all the inductors and expressed as a function of ψ.

$\omega_e(0^+)$ is the electric energy initially stored in the condensers.

$\bar{\omega}_m(0^+)$ is the magnetic co-energy initially associated with the inductors.

$z[\psi,\dot{\psi}] = \langle \mathbf{q}_c, \mathbf{e}_c\rangle + \langle \mathbf{i}_L, \int_0^t \mathbf{e}_L\, dt\rangle$ is the potential function associated with the voltage sources and expressed in terms of ψ.

---

*The notation $\delta_x[\cdot]$ indicates that the variation of the functional between square brackets is to be taken with respect to x. This notation is used only when the functional is not explicitly expressed as a function of x.

Eq. (4.5) provides a useful physical interpretation of network Lagrangian $\ell(\psi)$ and suggests to consider eq. (4.2) as a particular condition for the energy distribution in LC reciprocal networks.

The result contained in Theorem 4.1a is the network counterpart of Hamilton's principle of dynamics. In the theory of electromagnetic fields it is known as classic action principle and attributed to Larmor [15]. In the theory of elastostatic fields it is referred to as the principle of potential energy [16].

Operating on basic network equations as indicated schematically in Fig. 5b, the set of *node equations*

$$NQ\tilde{N}\phi + NQ\,\bar{\mathbf{u}} + N\chi = 0 \qquad (4.6)$$

is found.

Starting from eq.(4.6) we can prove the following Theorem 4.1b which is the dual of Theorem 4.1a:

Theorem 4.1b

Let the state of a linear, time-invariant, LC, *reciprocal network be described by the flux linkage vector* ϕ. *If the network moves from state* ϕ_1 *at time* t_1 *to state* ϕ_2 *at time* t_2, *then among all possible paths between these end-states, the path the network follows must be one for which*

$$\delta[\tfrac{1}{2}(\phi,NQ\tilde{N}\phi) + (\phi,NQ\,\bar{\mathbf{u}}) + (\phi,N\chi)] = 0 \qquad (4.7)$$

The *network Lagrangian* is now:

$$\ell[\phi] \overset{\Delta}{=} \tfrac{1}{2}\langle\mathbf{u},Q\mathbf{u}\rangle - \tfrac{1}{2}\langle\bar{\mathbf{u}},Q\bar{\mathbf{u}}\rangle + \langle\mathbf{u},\chi\rangle - \langle\bar{\mathbf{u}},\mathbf{u}\rangle =$$

$$\bar{\omega}_e[\phi] - \omega_m[\phi] - \bar{\omega}_e[0^+] + \omega_m[0^+] + y[\phi,\dot{\phi}] \qquad (4.8)$$

where

$\bar{\omega}_e = \tfrac{1}{2}\langle\mathbf{v}_c, C\,\mathbf{v}_c\rangle$ is the electric co-energy associated with all the condensers.

$\omega_m = \frac{1}{2}\langle\lambda_L, \Gamma\ \lambda_L\rangle$ is the magnetic energy associated with all the inductors.

$\bar{\omega}_e(0^+)$ is the electric co-energy initially associated with the condensers.

$\omega_m(0^+)$ is the magnetic energy initially stored in the inductors.

$y[\phi,\dot{\phi}] = -\langle\lambda_L, \dot{\mathbf{j}}_L\rangle - \langle\mathbf{v}_C, \int_0^t \mathbf{j}_C\, dt\rangle$ is the potential function associated with the current sources.

Theorem 4.1b is the network counterpart of Castigliano's principle* for elastostatic fields [16] and of a variational theorem proved by Pratelli for electromagnetic fields [17].

II. *Canonical Form of Action Principle*

Using Table 2, network Lagrangian $\ell[\psi]$ can be made more explicit. Starting from eq. (4.2), it is simple to get

$$\ell[\psi,\dot{\psi}] \triangleq \tfrac{1}{2}\langle\psi, MP\tilde{M}\psi\rangle + \langle\psi, MP\bar{\mathbf{c}}\rangle + \langle\psi, M\varepsilon\rangle =$$

$$\tfrac{1}{2}\langle\psi, B_C S B_C^T\psi\rangle - \tfrac{1}{2}\langle\dot{\psi}, B_L L B_L^T\dot{\psi}\rangle + \langle\psi, B_C S\mathbf{q}_C(0^+)\rangle - \tag{4.9}$$

$$\langle\dot{\psi}, B_L L\mathbf{i}_L(0^+)\rangle + \langle\psi, B_C\mathbf{e}_C\rangle + \langle\dot{\psi}, B_L \int_0^t \mathbf{e}_L\, dt\rangle$$

Let us now introduce the *conjugate momenta*

$$\nu \triangleq \left(\frac{\partial\ell[\psi,\dot{\psi}]}{\partial\dot{\psi}}\right)^T = -B_L L B_L^T\dot{\psi} - B_L L\mathbf{i}_L(0^+) + B_L \int_0^t \mathbf{e}_L\, dt. \tag{4.10}$$

which are associated with the generalized velocities $\dot{\psi}$.

---

*It is also referred to as Menabrea's principle.

By solving* eq. (4.10) for $\dot{\psi}$ and substituting in eq. (4.9), we obtain:

$$\ell[\psi,\nu] = -\tfrac{1}{2}\langle\nu,(B_L L B_L^T)^{-1}\nu\rangle + \tfrac{1}{2}\langle\psi,B_C S B_C^T\psi\rangle + \langle\psi,B_C S\mathbf{q}_C(0^+)\rangle +$$

$$\tfrac{1}{2}\langle\mathbf{i}_L(0^+), L\mathbf{i}_L(0^+)\rangle + \langle\psi,B_C\mathbf{e}_C\rangle - \langle\mathbf{i}_L(0^+), \int_0^t \mathbf{e}_L\, dt\rangle +$$

$$\tfrac{1}{2}\langle \int_0^t \mathbf{e}_L\, dt,\ L^{-1}\int_0^t \mathbf{e}_L\, dt\rangle \qquad (4.11)$$

It is well known from classical dynamics [18] that a new variational theorem can be proved with reference to $\ell[\psi,\nu]$. It may be stated as follows:

Theorem 4.2a

Let the state of a linear, time-invariant, LC, *reciprocal network be described in the phase space by the position coordinates* ψ *and momenta* ν. *If the network moves from state* ψ_1,ν_1 *at time* t_1 *to state* ψ_2,ν_2 *at time* t_2, *then among all possible paths between these end-states, the path the network follows must be one for which*

$$\delta\int_{t_1}^{t_2} \ell[\psi,\nu]dt = 0 \qquad (4.12)$$

This theorem is often referred to as *the canonical form of Hamilton's principle* and the action integral in (4.12) is called the *canonical integral.*

In a completely similar fashion, starting from eq. (4.7), one can prove the following dual:

Theorem 4.2b

Let the state of a linear time-invariant, LC, *reciprocal network be described in the phase space by the position coordinates* ϕ *and momenta*

---

*We assume that Theorem 2.6 holds.

$$\mu \triangleq \left(\frac{\partial \ell[-\dot{\phi},\phi]}{\partial[-\dot{\phi}]}\right)^T = D_C C\, D_C^T \dot{\phi} + D_C C \mathbf{v}_C(0^+) + D_C \int_0^t \mathbf{j}_C\, dt \tag{4.13}$$

If the network moves from state ϕ_1, μ_1 *at time* t_1 *to state* ϕ_2, μ_2 *at time* t_2*, then among all possible paths between these end-states, the path the network follows must be one for which*

$$\delta \int_{t_1}^{t_2} \ell[\phi,\mu]\, dt = 0 \tag{4.14}$$

where*

$$\ell[\phi,\mu] = \tfrac{1}{2}\langle \mu, (D_C C D_C^T)^{-1}\mu\rangle - \tfrac{1}{2}\langle \phi, D_L \Gamma D_L^T \phi\rangle$$

$$- \langle \phi, D_L \Gamma \lambda(0^+)\rangle - \tfrac{1}{2}\langle \mathbf{v}_C(0^+), C\mathbf{v}_C(0^+)\rangle - \langle \phi, D_L \mathbf{J}_L\rangle \tag{4.15}$$

$$+ \langle \mathbf{v}_C(0^+), \int_0^t \mathbf{J}_C\, dt\rangle - \tfrac{1}{2}\langle \int_0^t \mathbf{J}_C\, dt, C^{-1} \int_0^t \mathbf{j}_C\, dt\rangle$$

It is worth mentioning that Theorem 4.1a (Theorem 4.1b) leads to a system of n simultaneous differential equations of the second order, the Lagrangian equations. Theorem 4.2a (Theorem 4.2b) leads to a system of 2n simultaneous differential equations of the first order, the Hamiltonian equations.

III. *Generalized Canonical Form of Action Principle*

Let us accomplish the chain of substitutions schematically shown in Fig. 6a, with reference to Table 1. We get two sets of equations:

$$\begin{cases} \tilde{M}\psi + \bar{\mathbf{c}} = Q\mathbf{u} + \chi \\ \\ M\mathbf{u} = 0 \end{cases} \tag{4.16}$$

---

*Again, assuming that Theorem 2.6 holds.

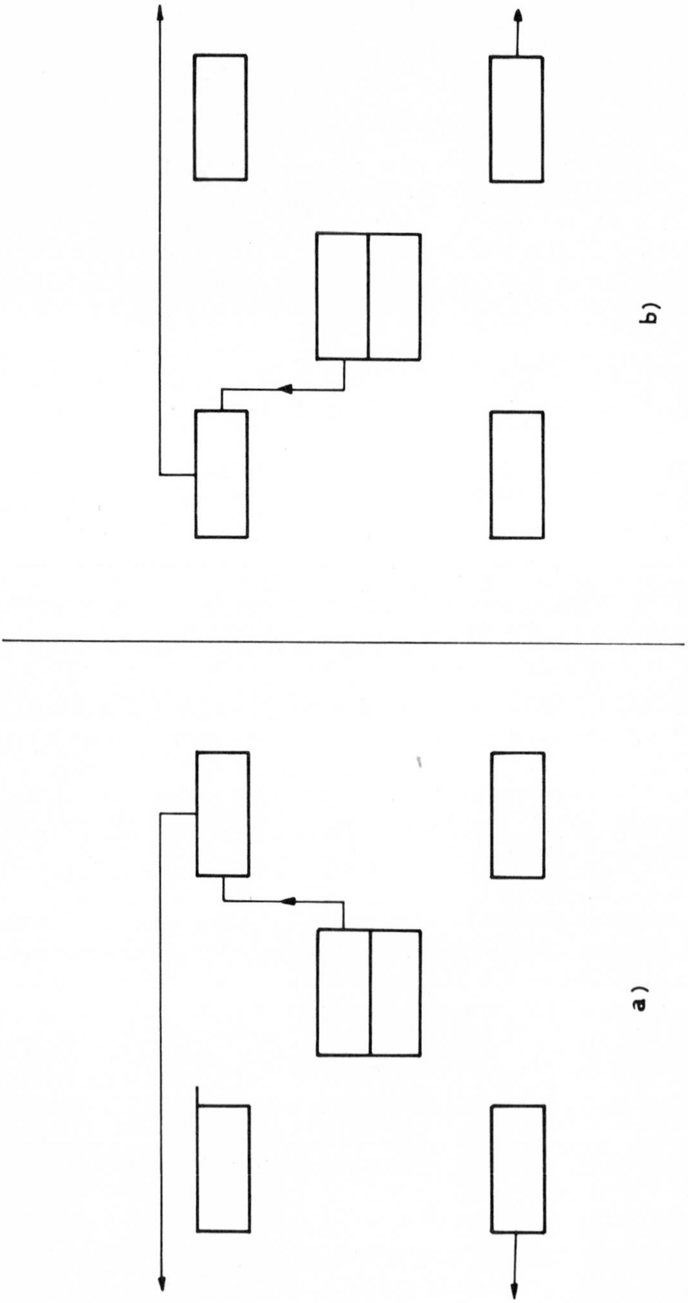

Fig. 6.

which may be written as:

$$\begin{cases} \tilde{M}\psi = \dfrac{\partial h[\psi,\mathbf{u}]}{\partial \mathbf{u}} \\ \\ M\mathbf{u} = \dfrac{\partial h[\psi,\mathbf{u}]}{\partial \psi} \end{cases} \tag{4.17}$$

by introducing the scalar-valued function:

$$h[\psi,\mathbf{u}] \stackrel{\Delta}{=} \tfrac{1}{2}\langle\mathbf{u},Q\mathbf{u}\rangle + \langle\mathbf{u},\chi\rangle - \langle\mathbf{u},\bar{\mathbf{c}}\rangle \tag{4.18}$$

and by assuming that Q is a symmetric matrix.

Eq. (4.17) satisfies Theorem 2.2; then, by using Corollary 2.2, we can state the following

Theorem 4.3a

Let the state of a linear, time-invariant, reciprocal, LC *network be described by the couple of variables* $\psi,\mathbf{u}$. *If the network moves from state* $\psi_1,\mathbf{u}_1$ *at time* t_1 *to state* $\psi_2,\mathbf{u}_2$ *at time* t_2*, then among all possible paths between these end-states, the path the network follows must be one for which*

$$\delta[(\psi,M\mathbf{u}) - \int_{t_1}^{t_2} h[\psi,\mathbf{u}]\,dt] = 0 \tag{4.19}$$

Using Table 2 and simple algebra, eq. (4.19) may be written as

$$\delta_{\mathbf{u},\psi}\left[((\tilde{M}\psi + \bar{\mathbf{c}}),\mathbf{u}) - \tfrac{1}{2}(\mathbf{u},Q\mathbf{u}) - (\mathbf{u},\chi)\right] = \delta_{\mathbf{u},\psi}\left[\tfrac{1}{2}(\mathbf{u},\mathbf{c})\right] = 0 \tag{4.20}$$

Then the network Lagrangian is the function

$$\ell[\mathbf{u},\psi] \stackrel{\Delta}{=} \tfrac{1}{2}\langle\mathbf{u},\mathbf{c}\rangle = \tfrac{1}{2}\langle\mathbf{c},P\mathbf{c}\rangle + \tfrac{1}{2}\langle\mathbf{c},\varepsilon\rangle = \omega_e[\mathbf{u},\psi] - \bar{\omega}_m[\mathbf{u},\psi] + \tfrac{1}{2}z[\mathbf{u},\psi] \tag{4.21}$$

where

$\omega_e[\mathbf{u},\psi]$ is the electric energy associated with all the condensers and expressed in terms of $\mathbf{u},\psi$.

$\bar{\omega}_m[\mathbf{u},\psi]$ is the magnetic co-energy associated with all the inductors and expressed in terms of $\mathbf{u},\psi$.

$z[\mathbf{u},\psi] = \langle \mathbf{q}_C, \mathbf{e}_C \rangle + \langle \mathbf{i}_L , \int_0^t \mathbf{e}_L \, dt \rangle$ is the potential function associated with the voltage sources and expressed in terms of $\mathbf{u},\psi$.

In a completely similar fashion, following the scheme shown in Fig. 6b, we get two sets of equations

$$\begin{cases} \tilde{N}\phi + \bar{u} = Pc + \varepsilon \\ \\ Nc = 0 \end{cases} \tag{4.22}$$

from which the following theorem can be obtained.

Theorem 4.3b

Let the state of a linear, time-invariant, reciprocal, LC *network be described by the couple of variables* $\phi,\mathbf{c}$. *If the network moves from state* $\phi_1,\mathbf{c}_1$ *at time* t_1 *to state* $\phi_2,\mathbf{c}_2$ *at time* t_2, *then among all possible paths between these end-states, the path the network follows must be one for which*

$$\delta[(\phi,N\mathbf{c}) - \int_{t_1}^{t_2} h[\phi,\mathbf{c}] \, dt] = 0 \tag{4.23}$$

where

$$h[\phi,\mathbf{c}] \overset{\Delta}{=} \tfrac{1}{2}\langle \mathbf{c},P\mathbf{c} \rangle + \langle \mathbf{c},\varepsilon \rangle - \langle \mathbf{c},\bar{u} \rangle \tag{4.24}$$

The network Lagrangian is now

$$\ell[\phi,\mathbf{c}] \overset{\Delta}{=} \tfrac{1}{2}\langle \mathbf{c},\mathbf{u} \rangle = \bar{\omega}_e[\phi,\mathbf{c}] - \omega_m[\phi,\mathbf{c}] + \tfrac{1}{2}y[\phi,\mathbf{c}] \tag{4.25}$$

where

$\omega_e[\phi,\mathbf{c}]$ is the electric co-energy associated with all the condensers and expressed in terms of $\phi,\mathbf{c}$.

$\omega_m[\phi,\mathbf{c}]$ is the magnetic energy associated with all the inductors and expressed in terms of $\phi,\mathbf{c}$.

$y[\phi,\mathbf{c}] = \langle\lambda_L,\mathbf{j}_L\rangle + \langle\mathbf{v}_C , \int_0^t \mathbf{j}_C\, dt\rangle$ is the potential function associated with the current sources and expressed in terms of $\phi,\mathbf{c}$.

Theorems 4.3a and 4.3b are known as the generalized canonical form of the action principle, because they can be deduced from Theorems 4.1a and 4.1b, respectively, by introducing the *generalized conjugate momenta**

$$\eta \triangleq \left(\frac{\partial \ell[\tilde{M}\psi]}{\partial(\tilde{M}\psi)}\right)^T , \quad \theta \triangleq \left(\frac{\partial \ell[\tilde{N}\phi]}{\partial(\tilde{N}\phi)}\right)^T \tag{4.26}$$

and using standard technique (follow the derivation of Theorems 4.2a and 4.2b).

Theorem 4.3a is the network counterpart of Belinfante's principle of electromagnetic fields [19]**, and of Reissner's principle of elastostatic fields [21].

Theorem 4.3b is the network counterpart of theorems proved by Marx for electromagnetic fields [22] and by Tonti for elastostatic fields [23].

IV. *A More General Principle*

A more general theorem and its dual can be obtained

---

*The Lagrangians $\ell[\tilde{M}\psi]$ and $\ell[\tilde{N}\phi]$ appear in (4.2) and (4.7) respectively. They are:

$$\ell[\tilde{M}\psi] \triangleq \tfrac{1}{2}\langle\tilde{M}\psi,P\tilde{M}\psi\rangle + \langle\tilde{M}\psi,P\bar{c}\rangle + \langle\tilde{M}\psi,\varepsilon\rangle \tag{4.3a}$$

$$\ell[\tilde{N}\phi] \triangleq \tfrac{1}{2}\langle\tilde{N}\phi,Q\tilde{N}\phi\rangle + \langle\tilde{N}\phi,Q\bar{u}\rangle + \langle\tilde{N}\phi,\chi\rangle \tag{4.8a}$$

**See also [20].

if three simultaneous sets of equations are considered:

$$\begin{cases} \tilde{M}\boldsymbol{\psi} - \mathbf{c} + \bar{\mathbf{c}} = 0 \\ P\mathbf{c} - \mathbf{u} + \boldsymbol{\varepsilon} = 0 \\ M\mathbf{u} = 0 \end{cases} \tag{4.27}$$

Their derivation from Table 1 is shown in Fig. 7a. By multiplying the first set by $\delta\mathbf{u}$, the second by $\delta\mathbf{c}$, the third by $\delta\boldsymbol{\psi}$ and adding we get

$$\begin{aligned} &<\tilde{M}\boldsymbol{\psi},\delta\mathbf{u}> - <\mathbf{c},\delta\mathbf{u}> + <\bar{\mathbf{c}},\delta\mathbf{u}> + <P\mathbf{c},\delta\mathbf{c}> \\ &\qquad - <\mathbf{u},\delta\mathbf{c}> + <\boldsymbol{\varepsilon},\delta\mathbf{c}> + <M\mathbf{u},\delta\boldsymbol{\psi}> = 0 \end{aligned} \tag{4.28}$$

And integrating over the time interval t_1, t_2 we obtain

$$\begin{aligned} &(\tilde{M}\boldsymbol{\psi},\delta\mathbf{u}) + (M\mathbf{u},\delta\boldsymbol{\psi}) - (\mathbf{c},\delta\mathbf{u}) - (\mathbf{u},\delta\mathbf{c}) + \\ &(P\mathbf{c},\delta\mathbf{c}) + (\boldsymbol{\varepsilon},\delta\mathbf{c}) + (\bar{\mathbf{c}},\delta\mathbf{u}) = 0 \end{aligned} \tag{4.29}$$

Assuming that $\delta\mathbf{u}$ and its derivatives vanish on the boundary, and that P is symmetric, eq. (4.29) may be written as:

$$\delta[(M\mathbf{u},\boldsymbol{\psi}) - (\mathbf{c},\mathbf{u}) + \tfrac{1}{2}(\mathbf{c},P\mathbf{c}) + (\boldsymbol{\varepsilon},\mathbf{c}) + (\bar{\mathbf{c}},\mathbf{u})] = 0 \tag{4.30}$$

This result can be embodied in the following

Theorem 4.4a

The motion of a linear, time-invariant, reciprocal, LC *network occurs in such a way that the definite integral in (4.3a) becomes stationary for arbitrary possible variations of the network configuration, provided that the initial and final configurations are prescribed.*

The network Lagrangian is now the scalar valued function

$$\begin{aligned} &\ell[\mathbf{u},\boldsymbol{\psi},\mathbf{c}] \triangleq <\mathbf{u},\tilde{M}\boldsymbol{\psi}> - <\mathbf{c},\mathbf{u}> + \tfrac{1}{2}<\mathbf{c},P\mathbf{c}> + <\boldsymbol{\varepsilon},\mathbf{c}> + <\bar{\mathbf{c}},\mathbf{u}> = \\ &\tfrac{1}{2}<\mathbf{c},P\mathbf{c}> + <\mathbf{c},\boldsymbol{\varepsilon}> = \omega_e[\mathbf{u},\boldsymbol{\psi},\mathbf{c}] - \bar{\omega}_m[\mathbf{u},\boldsymbol{\psi},\mathbf{c}] + z[\mathbf{u},\boldsymbol{\psi},\mathbf{c}] \end{aligned} \tag{4.31}$$

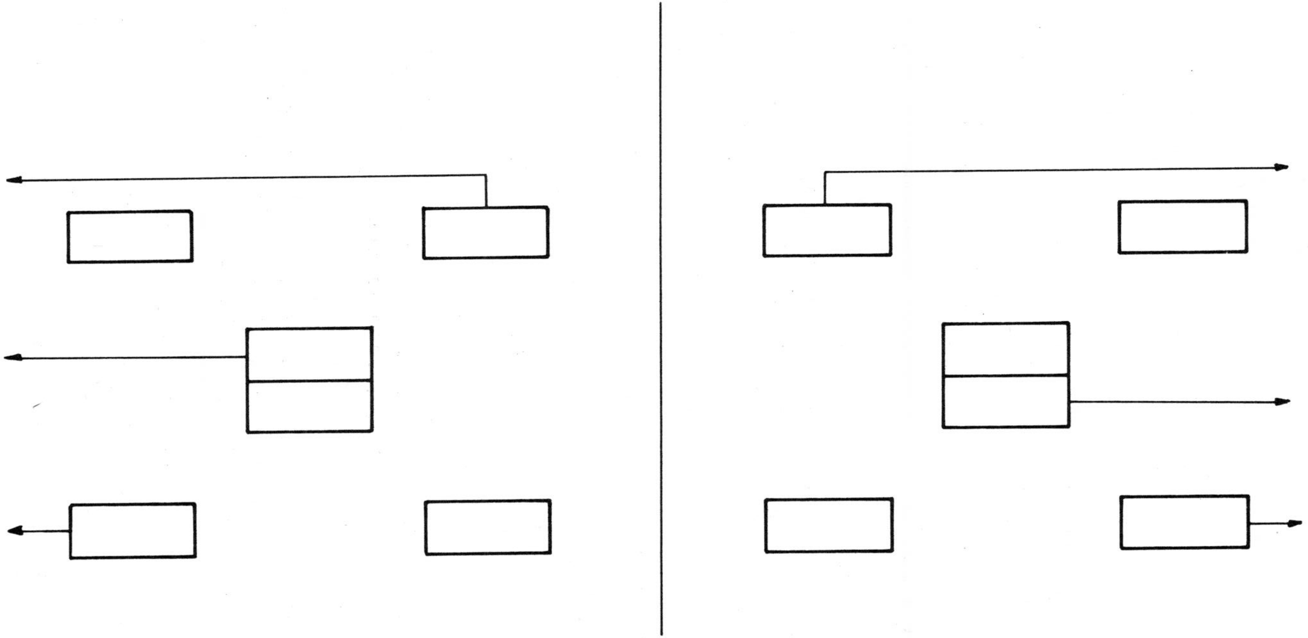

a) b)

Fig. 7

In a completely similar fashion, following the scheme shown in Fig. 7b, we get

$$\begin{cases} \tilde{N}\phi - \mathbf{u} + \bar{\mathbf{u}} = 0 \\ Q\mathbf{u} - \mathbf{c} + \chi = 0 \\ N\mathbf{c} = 0 \end{cases} \tag{4.32}$$

From eq. 4.32, if Q is symmetric, we obtain:

$$\delta[(N\mathbf{c},\phi) - (\mathbf{u},\mathbf{c}) + \tfrac{1}{2}(\mathbf{u},Q\mathbf{u}) + (\mathbf{u},\chi) + (\mathbf{c},\bar{\mathbf{u}})] = 0 \tag{4.33}$$

and consequently

Theorem 4.4b

The motion of a linear, time-invariant, reciprocal, LC *network occurs in such a way that the definite integral in 4.33 becomes stationary for arbitrary possible variations of the network configuration, provided that the initial and final configurations are prescribed.*

The network Lagrangian is the function

$$\ell[\mathbf{c},\phi,\mathbf{u}] \triangleq \langle \mathbf{c},\tilde{N}\phi\rangle - \langle \mathbf{u},\mathbf{c}\rangle + \tfrac{1}{2}\langle \mathbf{u},Q\mathbf{u}\rangle = \bar{\omega}_e[\mathbf{c},\phi,\mathbf{u}] - \omega_m[\mathbf{c},\phi,\mathbf{u}] + y[\mathbf{c},\phi,\mathbf{u}] \tag{4.34}$$

Theorem 4.4a is the network counterpart of the one proved by Hu and Washizu for elastostatic fields [24,25] and by Baar for elastodynamics [27]. It is also very similar to a variational theorem obtained by Tonti for electromagnetic fields*.

Theorem 4.4b is the network counterpart of the one proved by Tonti for elastostatic fields [23] and for electromagnetic fields also [10].

V. *An extension to time-varying networks*

For the sake of simplicity, we have maintained up

---

*Private communication.

to this point the assumption that the network is time-invariant. We now drop this hypothesis.

If we go through our derivation, starting from the beginning, it is not difficult to see that every theorem can be restated, provided that the network, although time-varying, remains reciprocal, i.e.,

$$C(t) = C^T(t) \;;\; \Gamma(t) = \Gamma^T(t)$$
$$S(t) = S^T(t) \;;\; L(t) = L^T(t) \qquad \forall\, t \in [t_1, t_2] \tag{4.35}$$

It is easy to see that many time-varying networks of engineering interest fall into this class.

5. VARIATIONAL THEOREMS FOR NONLINEAR LC NETWORKS

In this section an extension is given of the previous variational theorems (Section 4) to cover LC nonlinear networks.

For their derivation we follow the same scheme adopted previously, the only difference being that now the operators involved are often nonlinear.

Again we found eight theorems, which can be presented in two dual groups.

I. *Classical Action Principle*

Following the scheme shown in Fig. 5a we get the set of *loop equations*:

$$\mathcal{M}P(\tilde{\mathcal{M}}\psi + \bar{\mathbf{c}}) + \mathcal{M}\varepsilon = 0 \tag{5.1}$$

where now $P = P(\mathbf{c})$ is a vector-valued function of the vector

$$\mathbf{c}(\psi) = \tilde{\mathcal{M}}\psi + \bar{\mathbf{c}} \tag{5.2}$$

The Fréchet differential with respect to ψ in a

direction α of the nonlinear differential operator in eq. 5.1 is

$$\frac{\partial}{\partial h} MP(\tilde{M}(\psi + h\alpha) + \bar{c})\Big|_{h=0}$$

$$= M \frac{\partial}{\partial \mathbf{c}} P(\mathbf{c}) \cdot \frac{d}{dh} (\tilde{M}(\psi + h\boldsymbol{\alpha}) + \bar{\mathbf{c}})\Big|_{h=0} \tag{5.3}$$

$$= M \frac{\partial}{\partial \mathbf{c}} P(\mathbf{c})\tilde{M} \cdot \alpha$$

Let us now assume that the *Jacobian matrix*

$$\frac{\partial}{\partial \mathbf{c}} P(\mathbf{c}) = \begin{bmatrix} \frac{\partial S}{\partial \mathbf{q}_C} & 0 \\ 0 & -\frac{\partial L}{\partial \mathbf{i}_L} \end{bmatrix} \tag{5.4}$$

is symmetric, i.e., that

$$\begin{bmatrix} \frac{\partial S}{\partial \mathbf{q}_C} & 0 \\ 0 & -\frac{\partial L}{\partial \mathbf{i}_L} \end{bmatrix} = \begin{bmatrix} \left(\frac{\partial S}{\partial \mathbf{q}_C}\right)^T & 0 \\ 0 & -\left(\frac{\partial L}{\partial \mathbf{i}_L}\right)^T \end{bmatrix}$$

$$\forall\ \mathbf{q}_C,\ \mathbf{i}_L$$

From the network point of view, condition (5.5) means that the *incremental elastance* and the *incremental inductance matrices* are required to be symmetric, and in turn this implies that the network is reciprocal.

Thus, using Theorem 2.4, we found that the Fréchet derivative of the nonlinear operator appearing in eq. (5.1) is formally symmetric, if the network is reciprocal.

Theorem 2.3 thus holds, and, using Corollary 2.3, we can state the following

Theorem 5.1a

Let the state of a nonlinear, reciprocal, LC *network be described by the loop-charge vector* ψ. *If the network moves from state* ψ_1 *at time* t_1 *to state* ψ_2 *at time* t_2*, then among all possible paths between these end-states, the path the network follows must be one for which*

$$\delta[(\tilde{M}\psi, \int_0^1 P(\tilde{M}\psi\lambda + \bar{c})\, d\lambda) + (\tilde{M}\psi, \varepsilon)] = 0 \qquad (5.6)$$

In a completely similar fashion, we can prove the dual

Theorem 5.1b

Let the state of a nonlinear, reciprocal, LC *network be described by the flux linkage vector* ϕ. *If the network moves from state* ϕ_1 *at time* t_1 *to state* ϕ_2 *at time* t_2*, then among all possible paths between these end-states, the path the network follows must be one for which*

$$\delta[(\tilde{N}\phi, \int_0^1 Q(\tilde{N}\phi\lambda + \bar{u})\, d\lambda) + (\tilde{N}\phi, \chi)] = 0 \qquad (5.7)$$

Theorems 5.1a and 5.1b are the network counterparts of a theorem reported in [16] and another proved in [26] for elastostatic fields with nonlinear stress-strain relation.

II. *Canonical Form of Action Principle*

Following section 4.II, we have now

$$\ell[\tilde{M}\psi] \triangleq \langle\tilde{M}\psi, \int_0^1 P(\tilde{M}\psi\lambda + \bar{c})\, d\lambda\rangle + \langle\tilde{M}\psi, \varepsilon\rangle = \qquad (5.8)$$

$$<\psi, B_C \int_0^1 S(\lambda B_C^T \psi + \mathbf{q}_C(0^+))\, d\lambda>$$

$$- <\dot{\psi}, B_L \int_0^1 L(\lambda B_L^T \dot{\psi} + \mathbf{i}_L(0^+))\, d\lambda> \tag{5.8}$$

$$+ <\psi, B_C \mathbf{e}_C> + <\dot{\psi}, \int_0^t \mathbf{e}_L\, dt> = \ell[\psi\ \dot{\psi}]$$

We introduce then the conjugate momenta*

$$\nu \triangleq \left(\frac{\partial \ell[\psi\ \dot{\psi}]}{\partial \dot{\psi}}\right)^T = -B_L\ L(B_L^T \dot{\psi} + \mathbf{i}_L(0^+)) + B_L \int_0^t \mathbf{e}_L\, dt \tag{5.9}$$

By solving** eq. 5.8 for $\dot{\psi}$ and substituting in eq. 5.7 it is possible to obtain the network Lagrangian as a function of ψ and ν.

So we arrive at the following

Theorem 5.2a

Let the state of a non-linear, reciprocal, LC *network be described in the phase space by the position coordinates* ψ *and momenta* ν. *If the network moves from state* ψ_1, ν_1 *at time* t_1 *to state* ψ_2, ν_2 *at time* t_2, *then among all possible paths between these end-states, the path the network follows must be one for which*

$$\delta \int_{t_1}^{t_2} \ell[\psi, \nu] = 0 \tag{5.10}$$

In a similar fashion, introducing the conjugate momenta

---

*We use Theorem 2.5.

**Assuming that Theorem 2.6 holds.

$$\mu \triangleq \left\{\frac{\partial \ell[\phi,\dot{\phi}]}{\partial(-\dot{\phi})}\right\}^T = D_c C(D_c^T \dot{\phi} + \mathbf{v}_c(0^+)) - D_c \int_0^t \mathbf{j}_c \, dt \qquad (5.11)$$

we arrive at the dual

Theorem 5.2b

Let the state of a nonlinear, reciprocal, LC *network be described in the phase space by the position coordinates* ϕ *and momenta* μ. *If the network moves from state* ϕ_1, μ_1 *at time* t_1 *to state* ϕ_2, μ_2 *at time* t_2, *then among all possible paths between these end-states, the path the network follows must be one for which*

$$\delta \int_{t_1}^{t_2} \ell[\phi,\mu] \, dt = 0 \qquad (5.12)$$

III. *Generalized Canonical Form of Action Principle*

Operating on network basic equations as indicated schematically in Fig. 6a, we get the two sets of equations:

$$\begin{cases} \tilde{M}\psi + \bar{\mathbf{c}} = Q(\mathbf{u}) + \chi \\ M\mathbf{u} = 0 \end{cases} \qquad (5.13)$$

where $Q(\mathbf{u})$ is now a nonlinear vector-valued function.

Introducing the scalar-valued function:

$$h[\psi,\mathbf{u}] \triangleq \langle \mathbf{u}, \int_0^1 Q(\lambda\mathbf{u}) \, d\lambda \rangle + \langle \mathbf{u},\chi \rangle - \langle \mathbf{u},\bar{\mathbf{c}} \rangle \qquad (5.14)$$

and assuming that the Jacobian $\frac{\partial Q(\mathbf{u})}{\partial \mathbf{u}}$ is a symmetric matrix, eq.(5.13) may be written* as eq.(4.17).

In a very similar fashion, following Fig. 6b we get

---

*Using Theorem 2.5.

$$\begin{cases} \tilde{N}\phi + \bar{\mathbf{u}} = P(\mathbf{c}) + \varepsilon \\ \\ N\mathbf{c} = 0 \end{cases} \tag{5.15}$$

Introducing the scalar-valued function:

$$h[\phi,\mathbf{c}] \triangleq \langle \mathbf{c}, \int_0^1 P(\lambda \mathbf{c})\, d\lambda \rangle + \langle \mathbf{c},\varepsilon \rangle - \langle \bar{\mathbf{u}},\mathbf{c} \rangle \tag{5.16}$$

and assuming that the Jacobian $\frac{\partial P(\mathbf{c})}{\partial \mathbf{c}}$ is a symmetric matrix, eq. (5.15) may be written* as

$$\begin{cases} N\phi = \dfrac{\partial h[\phi,c]}{\partial c} \\ \\ N\mathbf{c} = \dfrac{\partial h[\phi,c]}{\partial \phi} \end{cases} \tag{5.17}$$

Thus,

Theorems 5.3a and 5.3b

Theorems 4.3a and 4.3b still apply to nonlinear, reciprocal, LC *networks, provided that* $h[\psi,\mathbf{u}]$ and $h[\phi,\mathbf{c}]$ *are defined according to eq.* (5.14) *and eq.* (5.16) *respectively.*

IV. *A More General Principle*

We consider now three simultaneous sets of equations:

$$\begin{cases} M\psi - \mathbf{c} + \bar{\mathbf{c}} = 0 \\ P(\mathbf{c}) - \mathbf{u} + \varepsilon = 0 \\ M\mathbf{u} = 0 \end{cases} \tag{5.18}$$

By multiplying the first set by $\delta\mathbf{u}$, the second by $\delta\mathbf{c}$, the third by $\delta\psi$, adding and integrating over the time

*Using Theorem 2.5.

interval t_1, t_2 we obtain

$$(\tilde{M}\psi,\delta\mathbf{u}) + (M\mathbf{u},\delta\psi) - (\mathbf{c},\delta\mathbf{u}) - (\mathbf{u},\delta\mathbf{c}) +$$
$$(P(\mathbf{c}),\delta\mathbf{c}) + (\varepsilon,\delta\mathbf{c}) + (\bar{\mathbf{c}},\delta\mathbf{u}) = 0 \tag{5.19}$$

Assuming that $\delta\mathbf{u}$ and its derivatives vanish on the boundary, and that the Jacobian matric $\partial P(\mathbf{c})/\partial\mathbf{c}$ is symmetric, eq. (5.19) may be written* as:

$$\delta[(\tilde{M}\psi,\mathbf{u}) - (\mathbf{c},\mathbf{u}) + (\mathbf{c}, \int_0^1 P(\lambda\mathbf{u})d\lambda) + (\varepsilon,\mathbf{c}) + (\bar{\mathbf{c}},\mathbf{u})] = 0 \tag{5.20}$$

This result can be embodied in the following

Theorem 5.4a

The motion of a nonlinear, reciprocal, LC *network occurs in such a way that the definite integral in eq.* (5.20) *becomes stationary for arbitrary possible variations of the network configuration, provided that the initial and final configurations are prescribed.*

In a very similar fashion, starting from the dual set

$$\begin{cases} \tilde{N}\phi - \mathbf{u} + \bar{\mathbf{u}} = 0 \\ Q(\mathbf{u}) - \mathbf{c} + \chi = 0 \\ N\mathbf{c} = 0 \end{cases} \tag{5.21}$$

we can prove the following

Theorem 5.4b

The motion of a nonlinear, reciprocal, LC *network occurs in such a way that the definite integral in*

---

*We use Theorem 2.5 and the simple rule: $\delta f(\mathbf{u}) = \langle \frac{\partial f(\mathbf{u})}{\partial \mathbf{u}}, \delta\mathbf{u} \rangle$, $f(\mathbf{u})$ being a scalar-valued function.

eq. (5.22)

$$(\tilde{N}\phi, c) - (u, c) + \left(u, \int_0^1 Q(\lambda c)\, d\lambda\right) + (\chi, u) + (c, \bar{u}) \qquad (5.22)$$

becomes stationary for arbitrary possible variations of the network configuration, provided that the initial and final configurations are prescribed.

Theorem 5.4a is the network counterpart of one proved in [28] for elastodynamic fields.

6. CONCLUSIONS

A systematic presentation has been made of a set of variational principles for linear and nonlinear LC reciprocal networks.

The vital role of self-adjoint operators in connection with variational formulations of physical theories has been emphasized.

A set of different network Lagrangians has been found, from which network differential equations can be deduced; all of them can be expressed in terms of magnetic and electric energy or co-energy (according to the case) and source potential functions. From this standpoint each variational principle can be viewed as stating a different, particular condition for the energy distribution in LC networks.

Although many of the variational principles reported are already known in network theory or in other branches of physics, the present unified derivation is in some respect new.

In a few instances and particularly for nonlinear networks, special forms of the principles are recorded here possibly for the first time.

ACKNOWLEDGMENT

The author gratefully acknowledges many helpful discussions with Dr. E. Tonti of the Department of Mathematics, Politecnico di Milano.

REFERENCES

1. GUILLEMIN, E.A. (1953) *Introductory Circuit Theory*, J. Wiley, New York.

2. STERN, T.E. (1965) *Theory of Nonlinear Networks and Systems*, Addison Wesley, Reading.

3. MORSE, P.M. and FESHBACH, H. (1953) *Methods of Theoretical Physics*, McGraw-Hill, New York.

4. ADAMS, K.M. (1968) Some Basic Concepts in Nonlinear Network Theory, *Network and Switching Theory*, G. Biorci (ed.), Academic Press, New York.

5. STERN, T.E. (1969) Some Network Interpretations of System Problems, *System Theory*, L.A. Zadeh and E. Polak (eds.), McGraw-Hill, New York, pp. 127-176.

6. LASALLE, J.P. and LEFSCHETZ, S. (1961) *Stability by Lyapunov's Direct Method with Applications*, Academic Press, New York.

7. VAINBERG, M.M. (1963) *Variational Methods for the Investigation of Nonlinear Operators*, Holden Day, S. Francisco.

8. MIKHLIN, S.G. (1964) *Variational Methods in Mathematical Physics*, Pergamon Press, Oxford.

9. LANCZOS, C. (1961) *Linear Differential Operators*, Van Nostrand, London.

10. DUINKER, S. (1968) The Relationship between Various Energy Distribution Theorems, *Network and Switching Theory*, G. Biorci (ed.), Academic Press, New York.

11. DUNFORD, N. and SCHWARTZ, J.T. (1964) *Linear Operators in Hilbert Space*, Interscience, New York.

12. TONTI, E. (1968) Variational Formulation for Linear Equations of Mathematical Physics, *Rend. Acc. Lincei, Cl. Sci. Fis. Mat., Serie VIII*, vol. XLIV, No. 1.

13. TONTI, E. (1969) Variational Formulation of Nonlinear Differential Equations, *Bulletin de la Classe des Sciences, Académie Royale de Belgique*, 5e Série, Tome LV, pp. 137-165.

14. PALAIS, R.S. (1959) Natural Operations on Differential Forms, *Trans. Am. Math. Soc. 92*, pp. 125-141.

15. CHERRY, E.C. *The Application of the Methods of Classical Dynamics to Lumped Electric Circuits*, Lecture Notes, Imperial College, London.

16. FUNG, Y.C. (1965) *Foundations of Solid Mechanics*, Prentice Hall, London.

17. PRATELLI, A. (1953) Principi Variazionali del Campo Elettromagnetico, *Ann. Scuola Normale Sup. di Pisa*, Serie III, vol. VII.

18. LANCZOS, C. (1949) *The Variational Principles of Mechanics*, University of Toronto Press, Toronto.

19. BELINFANTE, E.I. (1948) A First-Order Variational Principle for Classical Electrodynamics, *Phys. Rev. 7*, 779.

20. GEHENIAU, I. (1938) *Mécanique Ondulatoire de l'Electron et du Photon*, Gauthier-Villars, Paris.

21. REISSNER, E. (1950) On a Variational Theorem in Elasticity, *Journal of Mathematics and Physics 29*, pp. 90-95.

22. MARX, E. (1967) Classical Electrodynamics, *American Journal of Physics 1*, 35, p. 14.

23. TONTI, E. (1967) Principi Variazionali nell'Elastostatica, *Rend. Lincei*, vol. XLII, pp. 390.

24. HAY-CHANG HU (1955) On Some Variational Principles in the Theory of Elasticity and Plasticity, *Scientia Sinica*, Vol. IV, pp. 33-35.

25. WASHIZU, K. (1955) *On the Variational Principles of Elasticity and Plasticity*, MIT Tech. Rep. 25-18.

26. FINZI, B. (1940) Principio Variazionale della Meccanica dei Continui, *Accademia d'Italia 9*, vol. VII, p. 412.

27. BAAR, A.D.S. (1966) An extension of the Hu-Washizu Variational Principle in Linear Elasticity for Dynamic Problems, *Journ. Appl. Mech. 33*, Series E, No.2, p. 465.

28. YI-YUAN YU (1964) Generalized Hamilton's Principle and Variational Equation of Motion in Nonlinear Elasticity Theory with Application to Plate Theory, *Journ. Acoustical Society of America 36*, pp. 111-120.